Ultra Performance Liquid Chromatography Mass Spectrometry

Evaluation and Applications in Food Analysis

Ultra Performance Liquid Chromatography Mass Spectrometry

Evaluation and Applications in Food Analysis

Edited by

Mu. Naushad

Mohammad Rizwan Khan

CRC Press is an imprint of the
Taylor & Francis Group, an **informa** business

CRC Press
Taylor & Francis Group
6000 Broken Sound Parkway NW, Suite 300
Boca Raton, FL 33487-2742

First issued in paperback 2021
First issued in hardback 2019

ISBN 13: 978-1-03-223701-5 (pbk)
ISBN 13: 978-1-4665-9154-7 (hbk)

Library of Congress Cataloging-in-Publication Data

Ultra performance liquid chromatography mass spectrometry : evaluation and
applications in food analysis / edited by Mu Naushad, Mohammad Rizwan Khan.
pages cm
Includes bibliographical references and index.
ISBN 978-1-4665-9154-7 (hardback)
1. Food--Analysis. 2. Food additives--Analysis--Methodology. 3. High performance liquid chromatography. 4. Mass spectrometry. I. Naushad, Mu, editor of c ompilation. II. Khan, Mohammad Rizwan, editor of compilation.

TX548.2.L55U58 2014
664'.07--dc23 2013048950

Visit the Taylor & Francis Web site at
http://www.taylorandfrancis.com

and the CRC Press Web site at
http://www.crcpress.com

Contents

Preface

The development of many novel techniques that make our existence so comfortable has been intimately associated with the accessibility of suitable analytical methods. Liquid chromatography–mass spectrometry (LC–MS) is a powerful technique used for various applications based upon its very high sensitivity and selectivity. Generally, its applications are oriented toward the detection and identification of chemicals in a complex mixture. Preparative LC–MS systems can be used for fast and mass-directed purification of natural product extracts and new molecular entities important to food, pharmaceutical, agrochemical, and other industries.

This book presents a unique collection of up-to-date UPLC–MS/MS (ultra performance liquid chromatography–tandem mass spectrometric) methods for the separation and quantitative determination of pesticides, capsaicinoids, heterocyclic amines, aflatoxin, perfluorochemicals, acrylamide, procyanidins and alkaloids, lactose content, phenolic compounds, vitamins, and aroma and flavor compounds in a wide variety of foods and food products.

This book is the result of the remarkable contributions of 36 experts in interdisciplinary fields of science, with comprehensive, in-depth, and up-to-date research works and reviews. Composed of 16 chapters and compiled using more than 78 figures and 50 tables, this text delivers practical information to all people involved in the research and development, production, or routing analysis of foods and food products.

Mu. Naushad and Mohammad Rizwan Khan
Editors

Preface

Acknowledgment

We are most obligated to the elegance of the Almighty God, who inspires all of humanity to knowledge, and who blessed us with the needed favor to complete this work.

This book is the result of outstanding contributions from experts in the field of ultra-performance liquid chromatography and food science, with up-to-date and comprehensive reviews and research works. We are grateful to all of the contributing authors and their coauthors for their valued contribution to this book. We would also like to thank all of the publishers, authors, and others who have granted us permission to use their figures or tables. Sincere efforts have been made to reproduce them and to include citations with the reproduced materials. I would still like to offer my deep apologies to any copyright holder whose rights may have been mistakenly overlooked.

The authors would like to extend their sincere appreciation to King Saud University, Deanship of Scientific Research, College of Science Research Center for its supporting of this book.

Mu. Naushad and Mohammad Rizwan Khan
Editors

Editors

Dr. Mu. Naushad is assistant professor in the Department of Chemistry, King Saud University, Saudi Arabia. He earned his master's of science and PhD in analytical chemistry from Aligrah Muslim University, India in 2002 and 2007, respectively.

Dr. Naushad is the author of more than 60 research articles and several book chapters of international repute. He is editor/editorial member of more than 30 international journals, and is also the editor of a number of books, including, *Petroleum Production from Polystyrene Waste Plastic and Standard Polystyrene, Science & Technology Publishing, USA* and *A Book on Ion Exchange, Adsorption and Solvent Extraction, Nova, USA.*

Dr. Naushad has completed several major research projects, one of which is highly recognized by The National Plan for Science and Technology (NPST), Saudi Arabia. He has also been awarded State Merit Scholarships in California, Washington, and Michigan.

Dr. Naushad is a lifetime member of National Environmental Science Academy, India and Saudi Chemical Society, Saudi Arabia.

Mohammad Rizwan Khan is assistant professor in the Department of Chemistry, King Saud University, Saudi Arabia. He earned his bachelor's and master's degrees in chemistry from the Aligrah Muslim University, India in 2000 and 2002, respectively. For higher studies, he moved to the University of Barcelona, Spain, where in 2009 he received his PhD in analytical chemistry.

Dr. Khan has extensive research experience in multidisciplinary fields of analytical chemistry and environmental science. He has published several book chapters and research papers in international peer-reviewed journals.

Contributors

M. I. Alarcón-Flores
Department of Chemistry and Physics
University of Almería
Almería, Spain

Zeid Abdullah Alothman
King Abdullah Institute
for Nanotechnology
and
Department of Chemistry
King Saud University
Riyadh, Kingdom of Saudi Arabia

Asma'a Al-Rifai
Department of Chemistry
King Saud University
Riyadh, Kingdom of Saudi Arabia

Ibrahim Hotan Alsohaimi
Department of Chemistry
King Saud University
Riyadh, Kingdom of Saudi Arabia

Ahmad Aqel
King Abdullah Institute
for Nanotechnology
and
Department of Chemistry
King Saud University
Riyadh, Kingdom of Saudi Arabia

Jayashree Arcot
Food Science and Technology
School of Chemical Engineering
University of New South Wales
Sydney, Australia

Aadil Bajoub
Department of Analytical Chemistry
University of Granada
Granada, Spain

and

Agro-pôle Olivier
National School of Agriculture in Meknes
Meknès, Morocco

Martin P. Bucknall
Bioanalytical Mass Spectrometry
Facility
Mark Wainwright Analytical Centre
and
School of Chemistry
University of New South Wales
Sydney, Australia

Rosa Busquets
Nanoscience and Nanotechnology Group
University of Brighton
Brighton, United Kingdom

Alegría Carrasco-Pancorbo
Department of Analytical Chemistry
University of Granada
Granada, Spain

Maria V. Chandra-Hioe
Food Science and Technology
School of Chemical Engineering
University of New South Wales
Sydney, Australia

Alberto Fernández-Gutiérrez
Department of Analytical Chemistry
University of Granada
Granada, Spain

A. Garrido Frenich
Department of Chemistry and Physics
University of Almería
Almería, Spain

H. Gallart-Ayala
LUNAM
Ecole Nationale Vétérinaire
Agroalimentaire et de l'Alimentation Nantes Atlantique (Oniris)
Nantes, France

Maria Carla Gennaro
Department of Science and Technological Innovation—DISIT
University of Piemonte Orientale
Alessandria, Italy

Ayman Abdel Ghfar
Department of Chemistry
King Saud University
Riyadh, Kingdom of Saudi Arabia

María Gómez-Romero
Computational and Systems Medicine
Department of Surgery and Cancer
Imperial College London
London, United Kingdom

Fabio Gosetti
Department of Science and Technological Innovation—DISIT
University of Piemonte Orientale
Alessandria, Italy

Elena Hurtado-Fernández
Department of Analytical Chemistry
University of Granada
Granada, Spain

Mahamudur Islam
Department of Chemistry
Purushottam Institute of Engineering and Technology
Orissa, India

Cristina C. Jacob
LUNAM
Ecole Nationale Vétérinaire
Agroalimentaire et de l'Alimentation Nantes Atlantique (Oniris)
Nantes, France

Mohammad Rizwan Khan
Department of Chemistry
King Saud University
Riyadh, Kingdom of Saudi Arabia

Paolo Lucci
Department of Nutrition and Biochemistry
Pontificia Universidad Javeriana
Bogotà D.C., Colombia

Alba Macià
Department of Food Technology
Universitat de Lleida
Lleida, Spain

Emilio Marengo
Department of Science and Technological Innovation—DISIT
University of Piemonte Orientale
Alessandria, Italy

Oleg A. Mayboroda
Leiden Centre for Proteomics and Metabolomics
Leiden University Medical Center
Leiden, The Netherlands

Eleonora Mazzucco
Department of Science
and Technological
Innovation—DISIT
University of Piemonte Orientale
Alessandria, Italy

Lubinda Mbundi
Blond McIndoe Research Foundation
Queen Victoria Hospital NHS Trust
East Grinstead
West Sussex, United Kingdom

Maria José Motilva
Department of Food Technology
Universitat de Lleida
Lleida, Spain

Mu. Naushad
Department of Chemistry
King Saud University
Riyadh, Kingdom of Saudi Arabia

Oscar Núñez
Department of Analytical Chemistry
University of Barcelona
Barcelona, Spain

Nàdia Ortega
R+D+i Department
La Morella Nuts
Tarragona, Spain

Noureddine Ouazzani
Agro-pôle Olivier
National School of Agriculture
in Meknes
Meknès, Morocco

Tiziana Pacchiarotta
Leiden Centre for Proteomics and
Metabolomics
Leiden University Medical Center
Leiden, The Netherlands

R. Romero-González
Department of Chemistry
and Physics
University of Almería
Almería, Spain

J. L. Martínez Vidal
Department of Chemistry and Physics
University of Almería
Almería, Spain

Saikh Mohammad Wabaidur
Department of Chemistry
King Saud University
Riyadh, Saudi Arabia

Kareem Yusuf
Department of Chemistry
King Saud University
Riyadh, Kingdom of Saudi Arabia

1 History and Introduction of UPLC/MS

Mu. Naushad, Mohammad Rizwan Khan, and Zeid Abdullah Alothman

CONTENTS

1.1 HISTORY AND INTRODUCTION OF CHROMATOGRAPHY

Chromatography is the collective term for a set of laboratory techniques for the separation of mixtures. It involves the separation of mixtures due to differences in the distribution coefficient (equilibrium distribution) of sample components between two different phases that are immiscible. One of these phases is mobile while the other is stationary. Russian–Italian botanist Mikhail Tswett was the first to use the term "chromatography," derived from two Greek words *chroma*, meaning color, and *graphein*, meaning to write. He used finely divided $CaCO_3$-packed glass columns to separate yellow, orange, and green plant pigments (xanthophylls, carotenes, and chlorophylls, respectively) [1]. In 1906, he first used the term *chromatography* in his two papers about chlorophyll in the German botanical journal, *Berichte der Deutschen Botanischen Gesellschaft.* In 1907, Tswett demonstrated his chromatograph for the German Botanical Society. For several reasons, Tswett's work was long ignored. Willstater and Stoll later tried to repeat Tswett's experiments, but they used an overly aggressive adsorbent (destroying the chlorophyll) and they were unable to do so. Willstater and Stoll published their results, and Tswett's chromatography method soon fell into insignificance. After Tswett's work, chromatography methods changed little until the explosion of mid-twentieth century research in new techniques, particularly thanks to the work of German scientist Edgar Lederer [2] as well as the work of A.J. Martin and R.L. Synge [3]. Martin and Synge developed partition chromatography to separate chemicals on the basis of slight differences in partition coefficients between two liquid solvents. Martin was effectively the founder of modern chromatography, a method of separating different compounds in a mixture. Mikhail Tswett is usually credited with inventing chromatography at the beginning of the twentieth century, but the form of chromatography he developed was absorption chromatography, and it was rarely used for more than two decades after his early death, and is hardly ever used in the early twenty-first century.

In contrast, Martin invented three forms of chromatography: partition chromatography, paper chromatography, and gas–liquid chromatography, which were rapidly adopted and are still used extensively today. Even the other two forms of chromatography that resulted from Martin's work are in common use: thin-layer chromatography and high-performance liquid chromatography. Martin's development of chromatography was a remarkable achievement, and earned him a Nobel Prize in Chemistry in 1952. On June 7, 1941, Martin and Synge demonstrated partition chromatography at a meeting of the Biochemical Society, held at the National Institute for Medical Research, Hampstead. Their practical achievement was supported by their development of the theory of partition chromatography; within a short time, Martin was able to model the behavior of peptides and other compounds with a fair degree of precision. In the paper on this new technique (published in *Biochemical Journal* in November 1941), Martin and Synge suggested that a carrier gas could be used instead of a liquid for the mobile phase, and Martin went on to develop gas–liquid chromatography a decade later. In the mid-1970s, they also proposed the use of fine particles and high pressures to improve the separation, which remains the main feature of high-pressure liquid chromatography.

1.2 BASIC PRINCIPLE AND TYPES OF CHROMATOGRAPHY

Chromatography is a physical process and generally composed of two primary components: mobile phase and stationary phase. The mobile phase is the phase that moves in a definite direction. It consists of the sample being separated/analyzed and the solvent that moves the sample through the column. In high-performance liquid chromatography, the mobile phase consists of a nonpolar solvent(s) such as hexane in normal phase or polar solvents in reverse-phase chromatography. The mobile phase moves through the chromatography column (the stationary phase) where the sample interacts with the stationary phase and is separated. The stationary phase is the part of the chromatographic system through which the mobile phase flows, where distribution of the solutes between the phases occurs. It may be a solid or a liquid that is immobilized or adsorbed on a solid, and may consist of particles (porous or solid), the walls of a tube (e.g., capillary), or a fibrous material (e.g., paper). Chromatography works by allowing the molecules present in the mixture to distribute themselves between a stationary and a mobile phase.

$$\text{Distribution coefficient } (K_d) = \frac{\text{Concentration of component A in stationary phase}}{\text{Concentration of component A in mobile phase}}$$

Classification of chromatography according to a mobile phase:

- Liquid chromatography: mobile phase is a liquid (liquid–liquid chromatography (LLC), liquid–solid chromatography (LSC)).
- Gas chromatography: mobile phase is a gas (gas–solid chromatography (GSC), gas–liquid chromatography (GLC)).

Classification according to the packing of the stationary phase:

- Thin-layer chromatography (TLC): stationary phase is a thin layer supported on glass, plastic, or aluminum plates.
- Paper chromatography (PC): stationary phase is usually a piece of high-quality filter paper.
- Column chromatography (CC): stationary phase is a solid adsorbent placed in a vertical glass column.

Classification according to the force of separation:

- Adsorption chromatography
- Partition chromatography
- Ion-exchange chromatography
- Gel filtration chromatography
- Affinity chromatography

1.3 ADVANCEMENT IN LIQUID CHROMATOGRAPHY

Liquid chromatography is a technique used to separate a sample into its individual parts. This separation occurs on the basis of the interactions of the sample with the mobile and stationary phases. There are many stationary/mobile phase combinations that can be employed when separating a mixture. The components of a mixture are separated in a column based on each component's affinity for the mobile phase. So, if the components are of different polarities and a mobile phase of a distinct polarity is passed through the column, one component will migrate through the column faster than the other. It is apparent from Figure 1.1 that, in the first step, the mixture of two components (a and b) is at the top of the column. As the mobile phase passes through the column, the two components begin to separate into bands. In this example, component "b" has a stronger affinity for the mobile phase while component "a" remains relatively fixed in the stationary phase. The relative polarities of these two components are determined based on the polarities of the stationary and mobile phases. If this experiment were done as normal phase chromatography, component "b" would be less polar than component "a." On the other hand, this result yielded from reverse-phase chromatography would show that component "b" is more polar than component "a" [4].

Liquid chromatography is also called high-performance liquid chromatography (HPLC), which is a popular quantitative analytical technique. HPLC has several uses, including medical (e.g., detecting vitamin D levels in blood serum), legal (e.g., detecting performance-enhancement drugs in urine), research (e.g., separating the

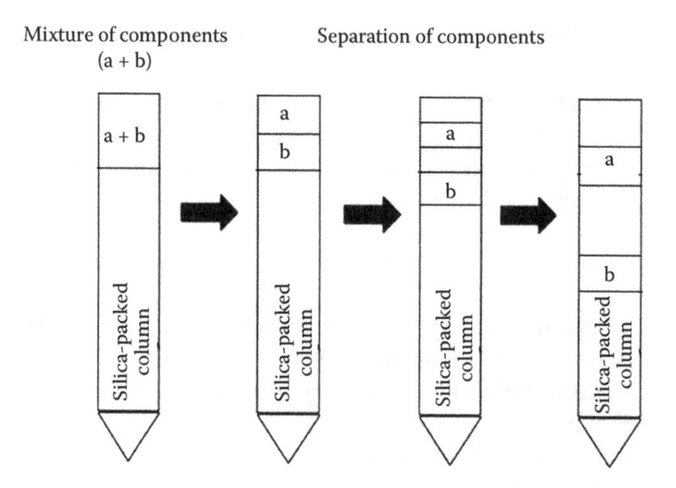

FIGURE 1.1 Migration of two components within a mixture.

components of a complex biological sample, or of similar synthetic chemicals from each other), and manufacturing (e.g., during the production process of pharmaceutical and biological products) [5]. HPLC is renowned from traditional ("low-pressure") liquid chromatography because operational pressures are significantly higher (50–350 bar), while ordinary liquid chromatography typically relies on the force of gravity to pass the mobile phase through the column. Due to the small sample amount separated in analytical HPLC, typical column dimensions are 2.1–4.6 mm diameter, and 30–250 mm length. Also, HPLC columns are made with smaller sorbent particles (2–5 μm in average particle size), which give HPLC a superior resolving power when separating mixtures. The schematic of an HPLC instrument (Figure 1.2) normally includes a sampler, pumps, and a detector. The sampler brings the sample mixture into the mobile-phase stream, which carries it into the column. The pumps

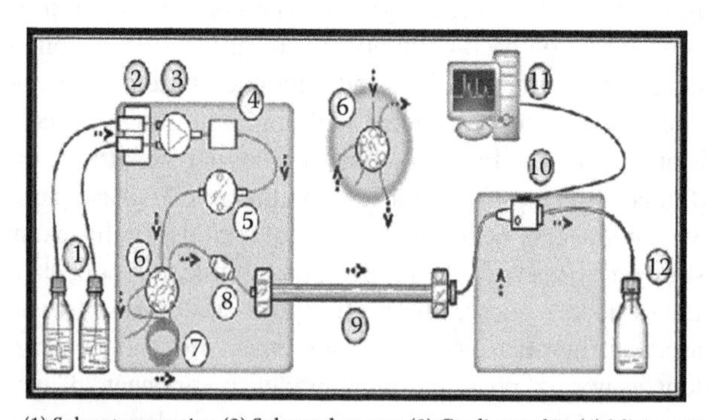

(1) Solvent reservoirs, (2) Solvent degasser, (3) Gradient valve, (4) Mixing vessel for delivery of the mobile phase, (5) High-pressure pump, (6) Switching valve in "inject position," (6) Switching valve in "load position," (7) Sample injection loop, (8) Pre-column (guard column), (9) Analytical column, (10) Detector (i.e., IR, UV), (11) Data acquisition, (12) Waste or fraction collector.

FIGURE 1.2 Schematic representation of an HPLC unit.

deliver the desired flow and composition of the mobile phase through the column. The detector generates a signal proportional to the amount of sample component emerging from the column, hence allowing for quantitative analysis of the sample components. A digital microprocessor and user software control the HPLC instrument and provide data analysis. Some models of mechanical pumps in an HPLC instrument can mix multiple solvents together in ratios changing in time, generating a composition gradient in the mobile phase. Various detectors are in common use, such as UV/vis, photodiode array (PDA), or mass spectrometry (MS) based.

1.3.1 Modes of Liquid Chromatography

Different modes of liquid chromatography have evolved for the analyses of a variety of compounds in various types of matrices. The modes of liquid chromatography include: normal-phase liquid chromatography (NPLC), reversed-phase liquid chromatography (RPLC), ion-exchange liquid chromatography (IELC), and size-exclusion chromatography (SELC). Selection of a liquid chromatographic mode for a particular analysis also requires selection of a column (stationary phase) and solvents for the mobile phase. Whichever mode is selected for development of a particular method, it is important that certain criteria for method validation are fulfilled.

1.3.1.1 Normal-Phase Chromatography

In normal-phase chromatography, molecules separate on the basis of differences in the strength of their interaction with polar stationary phase. The stationary phase is polar (e.g., silica gel, cyanopropyl-bonded, amino-bonded, etc.) and the mobile phase is nonpolar organic solvents (dehydrated), such as iso-octane, methylene chloride, methanol, ethanol, 2-propanol, acetonitrile, ethyl acetate, tetrahydrofuran, carbon tetrachloride, hexane, or a binary or tertiary solvent system based on two or three nonpolar organic solvents. Silica is the most common of the nonbonded stationary phases and can provide very high selectivity for many applications, but water adsorption by the silica can make reproducible retention times difficult. There are two major types of silica in use. Older, less pure silica have some trace metal content and have highly acidic sites on the surface. Newer-type silica columns are made up of high-purity silica and have fewer acidic sites, and are recommended for separating highly polar or basic compounds. Another nonbonded phase, alumina, has unique selectivity, but it is rarely used because it has problems such as low theoretical plate number (N), variable retention times, and low sample recovery. Of the bonded phases, cyano columns are the best for general analysis because they are the most stable and are more convenient to use than silica columns. Diol and amino columns can offer different selectivities but are less stable than cyano columns. The composition of a mobile phase can be optimized for a particular separation by selecting suitable solvents with "the right solvent strength." The solvent strength parameter ($\varepsilon°$) for n-pentane is 0; those for other organic solvents, increasing in order, are methanol > ethanol > 2-propanol > acetonitrile > ethyl acetate > tetrahydrofuran > carbon tetrachloride > hexane. In normal-phase chromatography, the stationary phase is hydrophilic and therefore has a strong affinity for hydrophilic molecules in the mobile phase. Thus, the hydrophilic molecules in the mobile phase tend to bind (or "adsorb") to the column, while the hydrophobic molecules pass through the column and are eluted first, and hydrophilic molecules can be eluted from the column by increasing the polarity of the solution in the mobile phase.

NPLC technique is mainly useful for

- Water-sensitive compounds
- Geometric isomers
- *cis–trans* isomers
- Chiral compounds

There are four main factors involved in the choice of solvents for normal-phase chromatography: solvent strength, localization, basicity, and UV cutoff. Solvents with high ε° values are strong solvents, and will more easily elute the more polar analytes. Localization is a measure of the interaction of the solvent with the stationary phase. Solvent molecules with polar functional groups will prefer a specific position relative to a nearby silanol group. Solvents that are not polar or weakly polar interact with the stationary phase very weakly and the coverage of the surface is random. Basicity is one of the axes of the solvent selectivity triangle. Selectivity can be changed by the use of basic solvents such as methyl tert-butyl ether or nonbasic solvents such as acetonitrile. UV cutoff is important when using a UV detector, as some normal-phase solvents have relatively high UV cutoffs.

If a molecule has several functional groups, then retention is based on the most polar one. Normal phase, using silica is also an excellent method of separation compounds with different functional groups compared to reversed-phase chromatography using C_{18}. Retention in normal-phase appears to occur by an adsorption process (analytes interact with the polar groups on the surface of the column packing). Because these surface sites are fixed, their location and spacing have an effect on separation. This feature allows for the separation of molecules that are chemically similar but physically different, so normal-phase chromatography is often used for the separation of isomers. The adsorption of an analyte is based on the type of functional group present as well as steric factors, which makes is similar to chiral or affinity chromatography, the difference being that the adsorption sites on silica are not very specific. Hydrophobic analytes are more soluble in organic solvents than they would be in the aqueous mobile phases used in reversed-phase chromatography. Very hydrophobic molecules are strongly retained in reversed-phase chromatography. This may result in long chromatographic runs and selectivity can sometimes be low, resulting in poor separations. Very hydrophobic molecules can be analyzed using normal-phase chromatography. Hydrophilic compounds are often not retained under reversed-phase conditions, but they are usually well retained under normal phase. One problem is that very hydrophilic compounds are not very soluble in the solvents used for normal phase. This problem can be solved by the use of special normal-phase columns with aqueous mobile phases.

1.3.1.2 Reversed-Phase Chromatography

In reversed-phase chromatography, the polarities of the mobile and stationary phases are opposite those of normal-phase chromatography. The most popular column-packing material is octadecylsilyl silica (ODS-C_{18}), in which silica is covalently modified by C_{18} functional group. Octadecyl (C_{18}), octyl (C_8), hexyl (C_6), propyl (C_3), ethyl (C_2), methyl (C_1), phenyl, and cyclohexyl functional groups bonded to a silica surface render silica (stationary surface) nonpolar and hydrophobic. The synthesis of reversed-phase packing material takes place like this:

Si—OH
Silica support particle
① Cl_3-Si-C_{18}
② H_2O
OH
Si—O—Si—C_{18} + 3 HCl
OH
Covalent bond

In RPLC, the mobile phase is more polar than in the stationary phase; water and water-miscible organic solvents such as methanol, acetonitrile, and tetrahydrofuran are normally used. Since most organic molecules have some nonpolar regions, retention in RPLC arises because water in the mobile phase repels the nonpolar regions of solute molecules and facilitates their interaction with the nonpolar functional groups of the silica (stationary phase). Solute molecules are eluted in order of increasing hydrophobicity or decreasing polarity. Decreasing the mobile phase polarity by adding more organic solvent reduces the hydrophobic interaction between the solute and the solid support, resulting in desorption. The more hydrophobic the molecule the more time it will spend on the solid support and the higher the concentration of organic solvent that is required to promote desorption. A simple reversed-phase mechanism can be shown as:

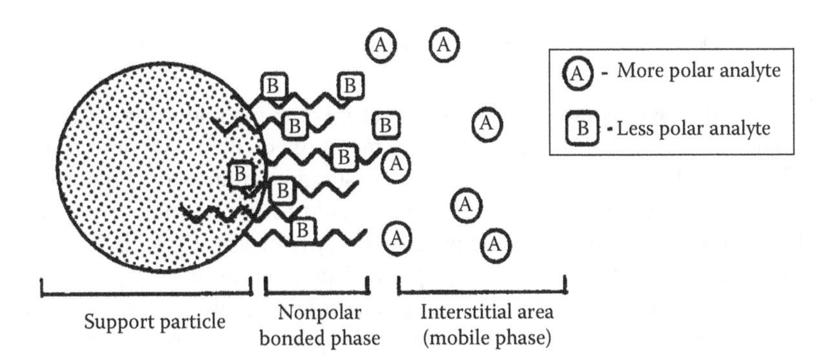

Reversed-phase chromatography is the most popular separation technique at analytical scale, because reversed-phase chromatography

- Is a robust technique and the columns are efficient and stable
- Applies to a very wide range of molecules, including charged and polar molecules
- Allows precise control of variables such as organic solvent type and concentration, pH, and temperature

At process scale, RPLC is not typically used for protein purification due to the presence of the organic solvent, which can cause denaturation of proteins and destroys their biological activity. Reversed-phase chromatography is the most dominant analytical HPLC technique and there are many different stationary phases available for method optimization.

1.3.2 Theories of Chromatography

There are two theories to explain chromatography.

- *Plate theory:* Older theory, developed by Martin and Synge in 1941.
- *Rate theory:* Proposed by Van Deemter in 1956 and recently applied in use.

Plate theory supposes that the chromatographic column contains a large number of imaginary theoretical plates, and within each theoretical plate analyte completely equilibrate between stationary phase and mobile phase. Chromatography columns with high numbers of theoretical plates produce very narrow peaks, resulting in better separation. The above equation shows that efficiency (N) can be increased by using longer columns, or using columns with small plate height. If the length of the column is L, then the HETP will be given as

$$\text{HETP} = L/N \qquad (1.1)$$

where N = number of theoretical plates.

The number of theoretical plates that a real column possesses can be found by examining a chromatographic peak after elution as

$$N = 5.54\,(t_r/w_{1/2})^2 \qquad (1.2)$$

where $w_{1/2}$ is the peak width at half-height and t_r is the retention time of the compound. The values of t_r and $w_{1/2}$ are often both measured on the x-axis of a chromatogram. They could both typically have units measured in minutes or actual distance measured with a ruler, in which case the units would be in centimeters or a similar unit. The width of a chromatographic peak is typically measured at a point halfway between the baseline and the top of the peak. This is because the peak broadens rapidly near the baseline and it is difficult to measure accurately. (The fact that the peak width is measured at 1/2 height instead of at the baseline, so 5.54 is taken in the equation.) If the width is measured at the baseline, then the equation becomes

$$N = 16(t_r/w)^2 \qquad (1.3)$$

where w is the peak width at the baseline and t_r is the retention time of the compound.

The rate theory, on the other hand, describes the migration of molecules in a column. This includes band shape, broadening, and the diffusion of a solute. Rate theory follows the van Deemter equation, which is the most appropriate for prediction of dispersion in liquid chromatography columns. It does this by taking into account the various pathways that a sample must travel through a column. Using the van Deemter equation, it is possible to find the optimum velocity and a minimum plate height;

$$H = A + B/u = Cu \tag{1.4}$$

where A is the eddy diffusion, B is the longitudinal diffusion, C is the mass transfer, and u is the linear velocity.

The A term is independent of velocity and represents "eddy" mixing. It is smallest when the packed column particles are small and uniform. The B term represents axial diffusion or the natural diffusion tendency of molecules. This effect is diminished at high flow rates and so this term is divided by v. The C term is due to kinetic resistance to equilibrium in the separation process. The kinetic resistance is the time lag involved in moving from the gas phase to the packing stationary phase and back again. The greater the flow of gas, the more a molecule on the packing tends to lag behind molecules in the mobile phase. Thus the term is proportional to v. Modern liquid chromatography is also known as high-performance or high-pressure liquid chromatography, or simply liquid chromatography. The potential advantages of modern liquid chromatography first came to the attention of a wide audience in early 1969, when a special session on liquid chromatography was organized as part of the Fifth International Symposium on Advances in Chromatography [6]. Nevertheless, modern liquid chromatography had its beginnings in the late 1950s, with the introduction of automated amino acid analysis by Spackman et al. [7]. This was followed by the pioneering work of Hamilton and Giddings [8] on the fundamental theory of high-performance liquid chromatography and Waters Associates in the mid-1960s. The abbreviation HPLC was coined by Professor Csaba Horvath for his 1970 Pittcon paper, which indicated that high pressure was used to generate the flow required for liquid chromatography in packed columns. In the beginning, pumps only had a pressure capability of 500 psi (35 bar). These new HPLC instruments could develop up to 6000 psi (400 bar) of pressure, and incorporated improved injectors, detectors, and columns. HPLC really began to take hold in the mid-to-late 1970s. In principle, liquid chromatography and HPLC work the same way; however, the speed, efficiency, sensitivity, and ease of operation of HPLC is vastly superior.

There are five major components of HPLC:

1. *Pump*: The role of the pump is to force a liquid (called the mobile phase) through the liquid chromatograph at a specific flow rate (1–2 mL/min). Typical pumps can reach pressures in the range of 6000–9000 psi (400–600 bar). During the chromatographic experiment, a pump can deliver a constant mobile phase composition (isocratic) or an increasing mobile phase composition (gradient).
2. *Injector*: The injector serves to introduce the liquid sample into the flow stream of the mobile phase. Typical sample volumes are 5–20 μL. An autosampler is the automatic version, which is used when the user has many samples to analyze or when manual injection is not practical.
3. *Column*: The column is considered the "heart of the chromatograph." The column's stationary phase separates the sample components of interest

using various physical and chemical parameters. The small particles inside the column cause high backpressure at normal flow rates. There are mainly four types of columns used in HPLC:
Analytical column: 1.0–4.6 mm; length 15–250 mm
Preparative column: >4.6 mm; length 50–250 mm
Capillary column: 0.1–1.0 mm; various length
Nano column: <0.1 mm, or sometimes stated as <100 μm

4. *Detector*: The detector identifies the individual molecules that come out from the column. It serves to measure the amount of those molecules so that the chemist can quantitatively analyze the sample components.
5. *Computer*: The computer not only controls all the modules of the HPLC instrument, but it takes the signal from the detector and uses it to determine the time of elution (retention time) of the sample components (qualitative analysis) and the amount of sample (quantitative analysis).

HPLC is now one of the most powerful tools in analytical chemistry. It has the ability to separate, identify, and quantitate the compounds that are present in any sample that can be dissolved in a liquid. Today, compounds in trace concentrations as low as parts per trillion (ppt) may easily be identified. HPLC can be, and has been, applied to just about any sample, such as pharmaceuticals, food, cosmetics, environmental matrices, forensic samples, and industrial chemicals.

In 2004, further advances in instrumentation and column technology were made to achieve very significant increases in resolution, speed, and sensitivity in liquid chromatography. Columns with smaller particles (1.7 μm) and instrumentation with specialized capabilities designed to deliver mobile phase at 15,000 psi (1000 bar) were needed to achieve a new level of performance. A new system had to be holistically created to perform ultra-performance liquid chromatography, now known as ultra-performance liquid chromatography (UPLC) technology. The original UPLC system debuted at the 2004 Pittcon exhibition. Acquity UPLC was the first liquid chromatograph designed at the system level to fully exploit the performance advantages of columns packed with a sub-2-μm stationary phase. Since backpressure is inversely proportional to the square of the particle diameter, a consequence of the smaller particle size was very high backpressures, in the range of 7000–10,000 psi. Since then, other vendors have introduced versions of high-pressure systems employing sub-2-μm particle columns. Waters was honored in 2004 with the prestigious Pittcon Editors Gold Award for the best new product at the Pittsburgh Conference on Analytical Chemistry and Applied Spectroscopy (Pittcon) for Acquity UPLC.

The UPLC is based on the same principle as HPLC, namely adsorption chromatography using an adsorbent of very fine size, to increase the surface area and thus adsorption. The stationary phase in UPLC consists of particles less than 2.0 μm, while HPLC columns are typically filled with particles of 3–5 μm. The underlying principles of this evolution are also governed by the van Deemter equation, which is an empirical formula that describes the relationship between linear velocity (flow rate) and plate height (HETP or column efficiency). Naturally, the column length has to be less and much higher pressure is needed to maintain percolation of the developing solvent. Efficiency is three times greater with 1.7 μm particles compared to 5 μm particles, and two times greater compared to 3.5 μm particles. Resolution is 70% higher than with 5 μm particles and 40% higher than with 3.5 μm particles. High speed is obtained because column length with 1.7 μm particles can be reduced by a factor of 3 compared to 5 μm particles for the same efficiency, and flow rate can be three times higher. This means separations can be nine times faster with equal resolution. Sensitivity increases because less band spreading occurs during migration through a column with smaller particles. Pumps in conventional HPLC systems reach a pressure of maximum 400 bar, while in UPLC systems can reach pressures of 1000 bar and more. The use of smaller particles allows for

- Better resolution (separation efficiency)
- Faster chromatography or a combination of both
- Increased sensitivity, due to sharper (narrower) and higher peaks

At high pressures, frictional heating of the mobile phase can be quite significant and must be considered. With column diameters typically used in HPLC (3.0–4.6 mm), a consequence of frictional heating is the loss of performance due to temperature-induced nonuniform flow. To minimize the effects of frictional heating, smaller diameter columns (1–2.1 mm) are typically used for UPLC [9,10]. Finally, we can say that UPLC was developed to get faster results with more resolution, more information, and more robust method, and more samples can be analyzed per system per scientist.

1.3.2.1 Theory of Separations Using Small Particles

The potentials of the van Deemter equation cannot be fulfilled without smaller particles than those traditionally used in HPLC. The design and development of sub-2 mm particles is a significant challenge, and researchers have been active in this area for some time to capitalize on their advantages [11–13]. According to van Deemter equation, smaller particles provide not only increased efficiency but also the ability to do work at increased linear velocity without a loss of efficiency, providing both resolution and speed. Efficiency is the primary separation parameter behind UPLC since it relies on the same selectivity and retentivity as HPLC. In the fundamental resolution equation [14], resolution is proportional to the square root of N.

$$Rs = \frac{\sqrt{N}}{4}\left(\frac{\alpha - 1}{\alpha}\right)\left(\frac{k}{k+1}\right) \tag{1.5}$$

where N is the number of theoretical plates, α is the selectivity factor, and k is the mean retention factor. However, N is inversely proportional to particle size (dp):

$$N \propto \frac{1}{dp} \tag{1.6}$$

As the particle size is lowered by thrice, that is, from 5 to 1.7 mm, N is increased by 3 and the resolution by square root of 3, that is, 1.7 N is also inversely proportional to the square of the peak width [14].

$$N \propto \frac{1}{w} \tag{1.7}$$

This elucidates that the narrower the peaks, the easier they are to separate from each other. Also, peak height is inversely proportional to the peak width (w):

$$H \propto \frac{1}{w} \tag{1.8}$$

So, as the particle size decreases, N increases. Subsequently, an increase in sensitivity is obtained, since narrower peaks are taller peaks. Narrower peaks also mean more peak capacity per unit time in gradient separations, desirable for many applications such as natural extracts, peptide maps, and so on [15]. Still, another equation comes into force from the van Deemter plot when moving toward smaller particles:

$$F_{opt} \propto \frac{1}{dp} \tag{1.9}$$

As particle size decreases, the optimum flow rate (F_{opt}) to reach maximum N increases. However, since back pressure is proportional to flow rate, smaller particle sizes require much higher operating pressure and a system properly designed for the same. Higher resolution and efficiency can be taken a level further when analysis speed is the primary objective. Efficiency is proportional to column length and inversely proportional to the particle size:

$$N \propto \frac{L}{dp} \tag{1.10}$$

Therefore, the column can be shortened by the same factor as the particle size, without the loss of resolution. Using a flow rate three times higher due to smaller particles and shortening the column by one-third, the separation is completed in 1/9th the time while maintaining resolution [15].

Although columns with high-efficiency, nonporous 1.5 mm particles are commercially available, they suffer from poor loading capacity and retention due to low surface area. Silica-based particles have good mechanical strength but can suffer from a number of disadvantages, which include a limited pH range and tailing of basic analytes. Polymeric columns can overcome pH limitation, but they have their own issues, including low efficiency, limited loading capacity, and poor mechanical strength. In 2000, XTerra® columns were introduced, a first-generation hybrid chemistry that took advantage of the best of both the silica and polymeric column.

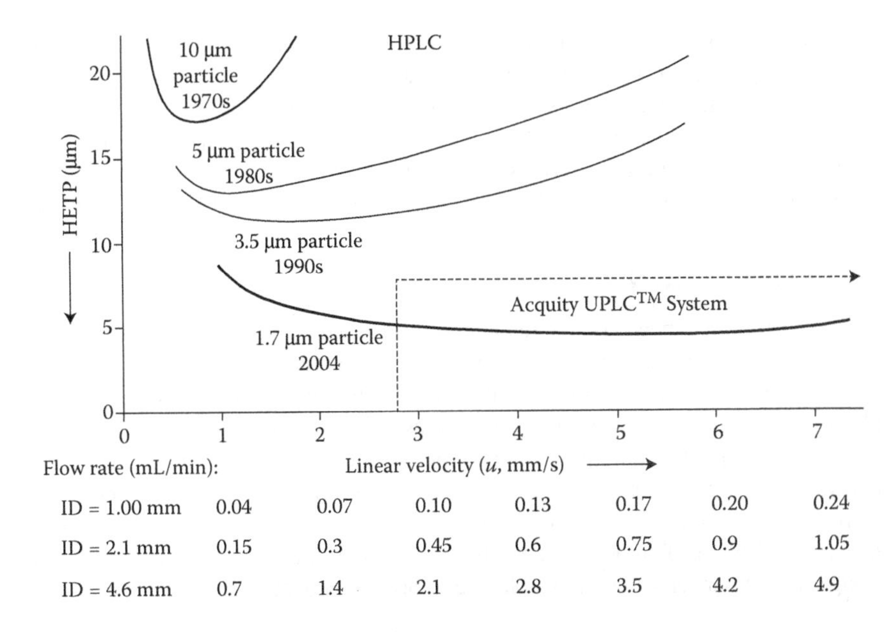

FIGURE 1.3 Van Deemter plot, illustrating the evolution of particle sizes over the last three decades. (From Waters Inc. *An Introduction to UPLC Technology: Improve Productivity and Data Quality*, presented at the American Association of Pharmaceutical Scientists conference, 2007. With permission.)

XTerra columns are mechanically strong, with high efficiency, and operate over an extended pH range, but in order to provide the necessary mechanical stability for UPLC, a second-generation bridged ethyl hybrid (BEH) technology was developed [16, 17]. It is apparent from Figure 1.3 that as the particle size decreases to less than 2.5 mm, there is not only a significant gain in efficiency, but the efficiency does not diminish at increased flow rates or linear velocities. By using smaller particles, speed and peak capacity (number of peaks resolved per unit time) can be extended to new limits, termed ultra-performance liquid chromatography (UPLC).

1.4 UPLC COUPLED WITH MS

The basis of MS is the production of ions that are subsequently separated or filtered according to their mass-to-charge (*m/z*) ratio and detected. The resulting mass spectrum is a plot of the (relative) abundance of the produced ions as a function of the *m/z* ratio. Coupling of MS to chromatographic techniques has always been desirable due to the sensitive and highly specific nature of MS compared to other chromatographic detectors. The coupling of LC with MS (LC–MS) was an obvious extension, but progress in this area was limited for many years due to the relative incompatibility of existing MS ion sources with a continuous liquid stream. Several interfaces were developed, but they were cumbersome to use and unreliable, so uptake by clinical laboratories was very limited. This situation changed when Fenn developed the electrospray ion source in the 1980s [18]. Manufacturers rapidly developed instruments equipped with electrospray sources, which had a great impact on protein and peptide biochemistry. Fenn, along with Koichi Tanaka, was awarded the Nobel Prize in 2002 for developing the matrix-assisted laser desorption ionization, another extremely useful MS ionization technique for the analysis of biological molecules. UPLC coupled with MS technology provides parent and fragment mass information in one chromatographic run; thus, provide an attractive alternative to current LC methods.

1.5 CONCLUSION

Owing to the fast speed, better resolution, and sensitivity, UPLC system presents the possibility to expand the scope of chromatographic separations beyond those achieved by HPLC. This system is not only useful because of these properties, but also because it reduces the noise and improves signal-to-noise ratio. Due to very narrow and sharp peaks, more peaks may appear in less time, which may facilitate the analysis of complex mixtures and may give more information regarding the sample to be analyzed. UPLC removed the barrier of traditional chromatographic packing material by development of new, highly efficient, mechanically strong, 1.7 mm bridge hybrid particles that are stable over a broad pH operating range.

ACKNOWLEDGMENT

The authors would like to extend their sincere appreciation to King Saud University, Deanship of Scientific Research, College of Science Research Center for its supporting of this book chapter.

REFERENCES

1. Tswett, M. 1906. Physikalisch–Chemische Studien über das Chlorophyll. Die Adsorption. (Physical–chemical studies of chlorophyll. Adsorption.) *Berichte der Deutschen botanischen Gesellschaft*, 24:316–326.
2. Kuhn, R., Winterstein, A., Lederer, E. 1931. Zur Kenntnis der Xanthophylle. (On our knowledge of xanthophyll.) *Zeitschrift für physiologische Chemie*, 197:141–160.
3. Martin, A.J., Synge, R.L. 1941. A new form of chromatography employing two liquid phases. *Biochem. J.*, 35:1358–1368.
4. http://chemwiki.ucdavis.edu/Analytical_Chemistry/Instrumental_Analysis /Liquid_Chromatogr.
5. Gerber, F., Krummen, M., Potgeter, H., Roth, A., Siffrin, C., Spoendlin, C. 2004. Practical aspects of fast reversed-phase high-performance liquid chromatography using 3 μm particle packed columns and monolithic columns in pharmaceutical development and production working under current good manufacturing practice. *J. Chromatogr. A*, 1036:127–133.
6. Zlatkis, A. ed., *Advances in Chromatography*, 1969, Preston Technical Abstracts Co.
7. Spackman, D.H., Stein, W.N., Moore, S. 1958. Automatic recording apparatus for use in the chromatography of amino acids. *Anal. Chem.*, 30:1190–1205.
8. Giddings, J.C. 1965. *Dynamics of Chromatography*. Dekker, New York.
9. MacNair, J.E., Lewis, K.C., Jorgenson, J.W. 1997. Ultrahigh-pressure reversed-phase liquid chromatography in packed capillary columns. *Anal. Chem.*, 69:983–989.

10. Colon, L.A., Citron, J.M., Anspach, J.A., Fermier, A.M., Swinney, K.A. 2004. Very high pressure HPLC with 1 m mid columns. *Analyst*, 129:503–504.
11. Jerkovitch, A.D., Mellors, J.S., Jorgenson, J.W. 2003. The use of micron sized particles in ultrahigh-pressure liquid chromatography. *LC-GC North America*, 16:20–23.
12. Wu, N., Lippert, J.A., Lee, M.L. 2001. Practical aspects of ultrahigh pressure capillary liquid chromatography. *J. Chromotogr. A.*, 911:1–12.
13. Unger, K.K., Kumar, D., Grun, M., Buchel, G., Ludtke, S., Adam, T., Scumacher, K., Renker, S. 2000. *J. Chromatogr. A.*, 892:47–55.
14. Skoog, D.A., Holler, F.J., Nieman, T.A. 1998. *An Introduction to Chromatographic Separations. Principals of Instrumental Analysis*. 5th ed. Florida: Saunders College Publishing, 674–697.
15. Swartz, M.E. 2004. Ultra performance liquid chromatography: Tomorrow's HPLC technology today. *LabPlus Int.,* 18:6–9.
16. MacNair, J.E., Patel, K.D., Jorgenson, J.W. 1997. Ultrahigh-pressure reversed-phase liquid chromatography in packed capillary columns. *Anal Chem.*, 69:983–989.
17. MacNair, J.E., Patel, K.D., Jorgenson, J.W. 1999. Ultrahigh-pressure reversed-phase capillary liquid chromatography: Isocratic and gradient elution using columns packed with 1.0 mm particles. *Anal Chem.*, 71:700–708.
18. Fenn, J.B., Mann, M., Meng, C.K., Wong, S.F., Whitehouse, C.M. 1989. Electrospray ionization for mass spectrometry of large biomolecules. *Science*, 246:64–71.

2 UHPLC–MS(/MS) Analysis of Pesticides in Food

Paolo Lucci, Rosa Busquets, and Oscar Núñez

CONTENTS

2.1 INTRODUCTION

Pesticide residues in food are a growing concern, both for consumers and for legislation. They comprise the deposits of pesticide active ingredients and their metabolites or breakdown products present in some component of the environment after their application, spillage, or dumping. Pesticides must undergo extensive efficacy, environmental, and toxicological testing to be registered by governments for legal use in specified applications. The applied chemicals and/or their degradation products may remain as residues in agricultural products, which become a concern for human exposure. Among the public, but also among experts, the perception of risks related to pesticide presence in food is quite high. As reported by Grob et al. [1], if you ask educated consumers about the principal source of food contamination they will list pesticides as the first item. This is understandable, since pesticides are used worldwide on a broad variety of crops to control pests and prevent diseases in order to increase agricultural production, improve the quality, and extend the storage life of food crops [2]. Thus, the monitoring of pesticide residues in food is nowadays a priority objective in order to get extensive evaluation of food quality and to avoid possible risks to human health. For instance, in Europe, the European Union (EU) operates a country-by-country monitoring program that oversees the quality of imported food products [3]. In addition, the setting of low EU-harmonized maximum residue levels (MRLs) for unregistered pesticide/sample combinations and the introduction of very low residue limits (10 μg/kg) in fruits and vegetables intended for baby food production [4–7] has increased the necessity of developing analytical methodologies to monitor pesticides at low concentration levels [2]. Unfortunately, not all farmers follow legal practices, and due to the tremendous number of pesticides and crops in production worldwide, analytical methodologies designed to determine known and unknown multiple pesticide residues are necessary.

Traditionally, the analysis of pesticides in food has been accomplished by gas chromatography–mass spectrometry (GC–MS) techniques [8–12], where the use of conventional library searching routines is well established. Typically, single quadrupole analysis and selected ion monitoring (SIM) is used, although gas chromatography–tandem mass spectrometry (GC–MS/MS) methods are becoming more prominent. Nevertheless, today, liquid chromatography coupled with tandem mass spectrometry (LC–MS/MS) has become a powerful tool for pesticide residue analysis in a variety of complex matrices [11,13–19] due to its selectivity and sensitivity, a substantial reduction of sample-treatment steps compared with other methodologies such as GC–MS(/MS), and its reliable quantification and confirmation at the low concentration levels required. Moreover, the demands of high sample throughput in short time frames have given rise to high efficiency and fast or even ultra-fast LC methods, which are becoming very popular for the analysis of pesticides in food [19]. Among the several modern approaches in HPLC methods that enable the reduction of the analysis time without compromising resolution and separation efficiency, UHPLC methods either using sub-2 μm particle-packed columns [20,21] or porous shell columns (with sub-3 μm superficially porous particles) [21–23] are the most popular. Fast chromatographic separations in UHPLC can be achieved either by increasing the mobile phase flow rate, by decreasing the column length, or by reducing the column particle diameter. In conventional 3 or 5 μm particle size columns the efficiency decreases with the increase in mobile phase flow rate, as can be expected by the van Deemter theory [24]. On the other hand, a reduction of column length also improves the analysis time, because the retention of the analytes decreases, but a reduction in column efficiency is observed. This can be solved by decreasing the particle diameter of the column packing material—the smaller the particle diameter, the higher the column efficiency [24]. However, the use of small particles induces a high pressure drop. Thus, new ultra-high-pressure-resistant systems are necessary in order to profit fully from the advantages of the use of sub-2 μm particles. Alternatively, fast chromatographic and high-efficiency separations can also be achieved using columns packed with superficially porous particles, also known as fused-core (or porous shell) columns [21,25]. These silica particles consists of a 1.7–1.9 μm (depending on the brand company) fused core and a 0.35–0.5 μm layer of porous silica coating, obtaining 2.6–2.7 μm particles, which exhibit efficiencies that are comparable to sub-2 μm porous particles, but with modest backpressures.

Nevertheless, due to the increased number of pesticides used worldwide and the variety and complexity of food matrices, the use of ultra-fast separations is not enough to develop fast analytical methods for the analysis of pesticide residues in food. Besides, multi-residue screening methods able to analyze not only target but nontarget or even unknown compounds, minimizing the sample manipulation, are demanded. Thus, sample extraction and clean-up treatments must also be optimized when considering reducing the total analysis time. In 2003, Anastassiades et al. [26] developed an ideal sample extraction and clean-up procedure named QuEChERS, acronym of "*Quick, Easy, Cheap, Effective, Rugged, and Safe*," which has become particularly popular to determine moderately polar pesticide residues in various food matrices [27,28]. Other modern trends in sample preparation for the analysis of pesticide residues in food include the use of online solid-phase extraction (SPE) methods, or the use of more SPE-based selective approaches such as molecularly imprinted polymers (MIPs).

In addition, when dealing with the multi-residue analysis of pesticides in such complex matrices as foodstuff, the reduction of the total analysis time because of both fast UHPLC separations and simple sample treatments may create new analytical challenges. More matrix-related compounds may be introduced into the chromatographic system due to the simplification on sample treatment and, although high resolution and separation efficiency is achieved by UHPLC, matrix effects (ion suppression or ion enhancement) may occur. MS or MS/MS is then mandatory, but, in some cases, alternative confirmation and identification strategies [17] and HRMS will be required [29].

Finally, one of the most difficult and challenging aspects in this field is to address the analysis of nontarget pesticide residues [30]. For this kind of analysis HRMS combined with databases and accurate mass measurements to achieve confirmation and identification are necessary. This approach works quite well for nontarget and known pesticide residues, thus, a standard is available for identification. However, for unknown substances, confirmation and identification are more difficult and will usually require HRMS and accurate mass measurements of both the protonated molecule and its fragment ions.

The aim of this chapter is to present the state of the art on UHPLC–MS(/MS) analysis of pesticides in food. It includes a selection of the most relevant papers recently published regarding instrumental and column technology focusing on UHPLC analysis with sub-2 μm and novel porous shell particle-packed columns. Sample treatment procedures such as QuEChERS, MIPs, and online SPE will also be addressed. MS strategies for the analysis of pesticide residues as well as to guarantee confirmation and identification such as the use of HRMS or alternative confirmation strategies will be discussed with relevant application examples.

2.2 SAMPLE TREATMENT

Today, pesticide residues in food are among the leading food safety issues. For this reason, monitoring these agrochemicals in food is extremely important to ensure human dietary intakes at acceptable levels and to check that the MRLs are not exceeded. To fulfill these aims, rapid, sensitive, and accurate analytical methods are strictly required. Because pesticides are a very heterogeneous group of chemicals, the tendency nowadays is to develop and use multi-residual methods for simultaneous analysis of a large number of pesticides in food samples of different origins. Usually, the sequence steps in pesticide analysis are matrix modification, extraction, clean-up, and determination [31]. Within this context, sample treatment is probably the most tedious, laborious, expensive, and time-consuming step in many analytical procedures and, despite the significant advances in chromatographic separations and MS techniques, it is still one of the most important parts of the analytical process for achieving good analytical results [21]. As a consequence, over the last few years, considerable efforts have been made to simplify this stage and to develop modern methods that allow the determination of pesticide residues and their degradation products without compromising the integrity of the extraction process [27].

In pesticide residue analysis, the choice of the proper sample treatment methodology depends on several aspects, such as the composition of the food matrix and the physical–chemical properties of pesticides. Taking into consideration that these compounds are usually detected at very low levels (ng/g), sample treatment undoubtedly represents a necessary and challenging task for obtaining adequate preconcentration and clean-up efficiencies. According to the literature, a multiplicity of sample preparation techniques has been used over the years for the extraction of pesticides in food matrices, with liquid extraction (LE) being among the most popular techniques. In LE, food samples are usually blended with organic solvent alone, solvent mixtures, water, or pH-adjusted water [32]. Acetone, methanol, acetonitrile, ethyl acetate, and diethyl ether are among the most-used organic solvents. For instance, acetonitrile is a good choice for low fatty matrices because it allows good recoveries of a wide polarity range of pesticides, avoiding the coextraction of highly nonpolar fats and highly polar proteins, sugars, and salt [32]. On the other hand, the extraction of pesticides from fatty foods (>2%) such as animal products often requires a complete extraction of the fats in the sample (i.e., using petroleum ether) and a subsequent separation of the target chemicals from lipids using various approaches. However, other alternatives to conventional LE and Soxhlet extraction, such as microwave-assisted extraction (MAE) [33,34], pressurized liquid extraction (PLE) [35,36], and supercritical fluid extraction (SFE) [37,38], have recently sparked considerable interest due to their potential to provide considerable reduction in time and solvent consumption and high throughput of samples, thus overcoming the drawbacks of some of the traditional approaches. For instance, Fuentes et al. [33] proposed a method using MAE to assist the liquid–liquid extraction of pesticides in avocado oil and olive oil, whereas PLE, also called accelerated solvent extraction (ASE), have been successfully employed for pesticides extraction in several matrices such as fish, vegetable, fruit, grain, and cereal samples. PLE represents an extremely effective extraction technique based on an extraction under elevated temperature (50–200°C) and pressure (5–200 atm) conditions for short time periods (5–10 min) [39]. Advantages of this procedure lie in the smaller solvent volume required, the reduced handling of the sample, and the possibility of full automation. However, a weakness of PLE is the decrease of analyte extraction efficiency when using hydrophobic organic solvents in samples with relatively high water content [40]. Hence, to solve this problem, a desiccation step to increase the penetration of the solvent in the matrix is usually required, thus inevitably reducing the benefits associated with this technique [41].

Whichever technique is used for extraction, an additional clean-up procedure is usually included prior to chromatographic analysis in order to minimize matrix effects caused by the presence of coextracted compounds. Sorbent-based (extraction and clean-up) techniques such as SPE [42,43], solid-phase microextraction (SPME) [44–46], matrix solid-phase dispersion (MSPD) [47,48], and MIPs [49,50] are among the most commonly used. SPE can be used as a direct extraction technique for liquid samples such as juice, soft drinks, wine, and honey, or as a clean-up method for organic solvent extracts after analyte isolation by LE. SPE methods using octyl- (C_8) and octadecyl- (C_{18}) bonded silicas, hydrophilic–lipophilic balanced polymeric sorbents (Oasis-HLB®), porous graphitic carbon, as well as cation exchangers have been extensively reported in literature. The choice of the SPE sorbent obviously depends on the physical–chemical properties of the pesticide and on its possible interaction with the sorbent. For instance, moderately polar and nonpolar pesticides can be extracted from polar solutions by using C_8, C_{18}, CH, CN, and nonpolar silica sorbents. On the other hand, higher specificity and selectivity together with satisfactory extraction efficiency can be obtained using sorbents based on MIPs or immunoaffinity columns (IAC) [51]. MIPs have been successfully employed for the determination of trace dichlorvos residues in vegetables [52], metolcarb in cabbage, cucumber, and pear [53], as well as for the extraction of thiabendazole from citrus fruits and orange juice samples [54]. Another interesting future of SPE is the possibility of complete automation of the process. SPE technique can be easily coupled online to LC systems, thus overcoming many of the limitations associated with the classical offline SPE and making possible the development of faster methods together with high preconcentration factors and recoveries [21]. In several published applications, online SPE systems have been employed for the determination of pesticides in drinking water samples [55]. For instance, Garcia-Ac et al. [56] have recently developed and validated an online SPE of large-volume injections coupled to LC–MS/MS method for the analysis (quantitation and confirmation) of 14 selected trace organic contaminants, including herbicides (atrazine, cyanazine, and simazine) and two of their transformation products (deethylatrazine and deisopropylatrazine), in drinking and surface water. On the other hand, Riediker et al. [57] reported an online SPE–LC–ESI–MS–MS method for the determination of chlormequat and mepiquat in pear, tomato, and wheat flour.

Finally, considering that the development of multi-residual methods is becoming more and more popular in many laboratories, QuEChERS procedure is probably today's sample treatment of choice for pesticide analysis, and several works can be encountered in the literature dealing with the simultaneous extraction and clean-up of more than 100 pesticides with satisfactory recoveries [25,28,58,59]. QuEChERS is a fascinating alternative sample preparation technique entailing solvent extraction with acetonitrile, ethyl acetate, or other organic solvents, and partitioning with magnesium sulfate alone or in combination with other salts, followed by a dispersive solid-phase extraction (d-SPE) clean-up step, typically using primary secondary amine (PSA) [27]. The detailed QuEChERS method was first published in 2003 by Anastassiades et al. [26], demonstrating its great potential in the extraction of pesticide residues in fruits and vegetables. However, up until now numerous modifications of the original method have been reported in the literature in order to successfully apply this methodology not only on several nonfatty foods (<2%) but also on fatty food matrices (2–20%), such as milk, meat-based baby foods, eggs, olive oil, flaxseeds, and avocado [60–62]. For instance, Jeong et al. [63] have optimized the QuEChERS sample preparation approach for the extraction and clean-up of 14 pesticides found in milk samples. The developed method, which was optimized in terms of amounts of sodium acetate (Na acetate), PSA, and octadecylsilane (C_{18})

showed satisfactory recoveries (82.01–98.84%) for 10 of the pesticides (vinclozolin, penconazole, dieldrin, myclobutanil, endosulfan sulfate, bifenthrin, fenpropathrin, cyhalothrin, permethrin, and fenbuconazole) analyzed and lower, but still acceptable recovery rates (<80%) for the four lipophilic pesticides (2,4′-DDE, 4,4′-DDE, 2,4′-DDT, and 4,4′-DDT), thus proving the suitability of the QuEChERS method for detecting pesticides in fatty food matrices.

2.3 UHPLC–MS(/MS) METHODS

2.3.1 Chromatographic Separations

Chromatography has experienced a boost in its capabilities over the last 10 years, after a steady period with limited innovation. Fully porous particles of 5 μm, C_{18} stationary phases and column lengths of >150 mm dominated the field until the first decade of this millennium. Advances in material science, the need of higher analytical throughput, and competition by nongranular chromatographic phases, which threatened to take part of the market, have driven liquid separation toward the higher chromatographic efficiency that laboratories enjoy today [64–67]. The improvement in efficiency in LC has been achieved mostly on the basis of a reduction of the particle size of the stationary phase, which had made possible by working at high linear velocities in the mobile phase with minor sacrifice in plate height or chromatographic efficiency. As an example, Figure 2.1 shows the benefit of decreasing particle size on the peak efficiency for a fast chromatography of acetophenone [68].

This “revolution” in chromatography, which was not a reality until the late 1990s [69,70], from where UHPLC is taking off, has its basis in the van Deemter equation

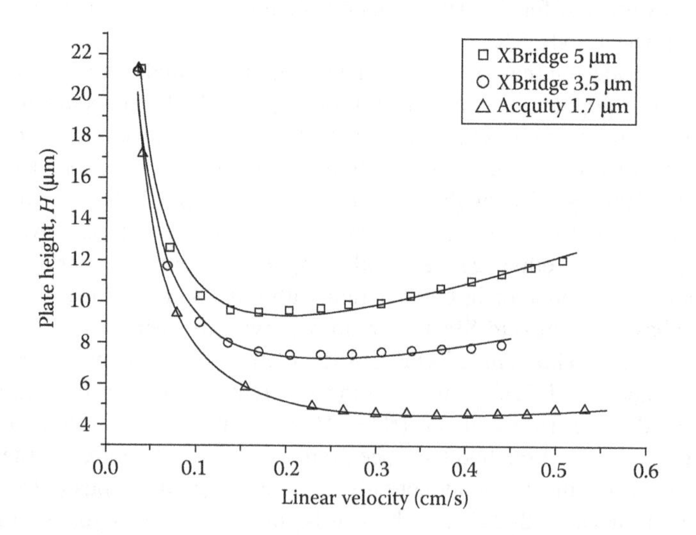

FIGURE 2.1 *H–u* plots obtained for acetophenone on Acquity and XBridge columns. Columns: Acquity BEH C18,1.7 μm, 10 cm × 2.1 mm ID; XBridge C18,3.5 μm, 150 × 4.6 mm ID; XBridge C18,5 μm, 25 cm × 4.6 mm ID. (Adapted from De Villiers, A. et al. 2006. *J. Chromatogr. A* 1127:60–69. With permission.)

[24], which describes the mechanisms involved in band broadening and provides the keys to improve chromatographic quality. The decrease in column permeability caused by the use of sub-2 μm particles and the increase in the flow rate has required working with elevated inlet pressures and temperatures to keep the pressure in the system within the limits of the chromatographs, which was redesigned to cope with ultra-high-pressure conditions. An alternative to the use of totally porous sub-2 μm particles for achieving high-efficiency separations with conventional chromatographs and pressures below 250 bar is the use of partially porous particles, which keep the improvements in efficiency and provide higher permeability. The latest developments in fully and partially sub-2 μm porous materials have been recently reviewed [71–73].

The application of chromatography in the analysis of pesticides in food has pulled chromatography to the stage where it is today; the need of fast analysis, low limits of detection required to reach statutory levels, and the limited sensitivity given by detection systems have left no other option but to improve the chromatographic part. Today the UHPLC market is dominated by chromatographic columns developed by Waters®, which produces over 60 types of UHPLC columns. Table 2.1 shows a selection of recent works and conditions optimized for the analysis of pesticides in food [74–81]. It can be observed that columns packed with 1.7 μm particles with a C_{18} reverse-phase chemically bound to an ethylene-bridged hybrid (BEH) structure, with different column lengths (50,100,150 mm), have been the prevalent option. In this type of column, the packing is constituted by a combination of inorganic (silica) and organic polymer phases and provides high pH stability (pH 1–12) and good peak symmetry for basic compounds, features that are important to achieving a multiresidue analysis where over 20 compounds are typically separated. Columns with different selectivity to Acquity BEH C_{18}, Acquity HSS T3 [81], and Acquity BEH Shield RP C_{18} [77] have also been used, but in fewer applications. Compared to Acquity BEH C_{18}, the stationary phase from the column HSS T3 has lower pore size, higher surface area, half-ligand density, and narrower pH tolerance range (pH 2–8). The Acquity Shield RP C_{18} column incorporates a polar group (carbamate) bound to the silica particles, which gives a degree of hydrophilicity to the reversed phase and allows working at a broad pH range (pH 2–11). Analysis times have been below 15 min despite the high number of compounds being monitored (Table 2.1). In a work where four diacylhydrazine insecticides were analyzed, the analysis time was lower than 3 min [82].

The hyphenation of UHPLC to MS, carried out generally with electrospray (Table 2.1), requires volatile and low-viscosity solvents to favor the ionic evaporation, as was the case in HPLC–MS, and to keep the pressures low, which is particularly important in UHPLC. These requirements restrict the composition of the mobile phase; however, the separation of high number of compounds is still achieved. The conditions compiled in Table 2.1 show that the aqueous phase is generally constituted by water, 0.02–0.2% formic in water, or 2–10 mM ammonium acetate, and methanol or acetonitrile are the solvents chosen for the organic phase, like in conventional LC. The separation has been carried out at a higher temperature than room temperature, up to 50°C [75], to keep viscosity of the solvent low, which reduces the pressure in the chromatograph.

TABLE 2.1
Summary of Analysis Conditions for the Determination of Pesticides in Food from Selected Studies from the Period 2010 to 2013

Target Compounds	Application Field	Sample Treatment	Column	Mobile Phase/Flow Rate	Detection and Acquisition Mode	Analysis Time	Reference
			Multiresidue Analysis				
150 pollutants including herbicides and fungicides	Water	SPE (HLB)	UPLC BEH C_{18} (100 × 2.1 mm,1.7 µm particle size)	0.01% formic acid in H_20 and 0.01% formic acid in methanol/0.3 mL/min	UHPLC-QTOF (MS^{E} [a])	12 min	[74]
71 pesticides and metabolites	Fruits and vegetables	Pressurized liquid extraction (1,1,1,2-tetrafluoroethane and toluene)	Acquity UPLC BEH C_{18} (100 × 2.1 mm, 1.7 µm particle size)	2 mM ammonium acetate solution and methanol, 50°C/0.45 mL/min	UHPLC-QqQ (MRM)	15 min	[75]
13 pesticides	Milk	QuEChERS[b]	Acquity UPLC BEH C_{18} column	0.1% formic acid in H_2O and acetonitrile	UHPLC-Q-TOF-(MS^E)	–	[76]
65 pesticides	Tea	QuEChERS, SPE	Acquity UPLC BEH Shield C_{18} (150 × 2.1 mm, 1.7 µm particle size)	0.02% formic acid in H_2O and 0.02% fromic acid in acetonitrile, 30°C/0.3 mL/min	UHPLC-QqQ (MRM)	12 min	[77]

29 pesticides	Fruits and vegetables	QuEChERS	Acquity UPLC BEH C_{18} (100 × 2.1 mm, 1.7 µm particle size)	0.1% formic acid in H_2O and methanol, 40°C/0.3 mL/min	UHPLC-QqQ (MRM)	8 min	[78]
			Pesticide Type Targeted Analysis				
Difenoconazole	Tomato products	QuEChERS	Acquity UPLC BEH C_{18} column (50 × 2.1 mm, 1.7 µm particle size)	0.2% formic acid in H_2O and methanol, 45°C/0.3 mL/min.	UHPLC-QqQ (MRM)	4.5 min	[79]
Carbamate pesticides (15)	*Radix Glycyrrhizae*	Pressurized liquid extraction and QuEChERS	Acquity UPLC C_{18} column (50 × 2.1 mm, 1.7 µm particle size)	0.1% formic acid and ammonium acetate in water (5 mmol/L) and acetonitrile/0.3 mL/min.	UHPLC-QqQ (MRM)	6.2 min	[80]
Carbamate pesticides (18)	Nuts	Gel permeation column	Acquity UPLC C_{18} HSS T3 (50 × 2.1 mm, 1.8 µm particle size)	10 mM ammonium acetate/acetonitrile	UHPLC-QqQ (MRM)	8 min	[81]

[a] MS^E (E represents collision energy) acquisition performed to obtain accurate relative molecular masses and fragment ions.
[b] QuEChERS (Quick, Easy, Cheap, Effective, Rugged, Safe).
– Not defined.
ESI was the ionization source used in all the works

2.3.2 Mass Spectrometry

The use of a wide range of chemicals to control pests in crops requires, at the end of the food production chain, high-throughput monitoring tools with great sensitivity and characterization power. MS uses the mass-to-charge ratio of ions directly related to the pesticides and/or their fragmentation products as characterization tools. The analysis is carried out from two different approaches: targeted quantitative analysis, for which triple quadrupoles are the main type of analyzers used [75,77–81,83], and identification/semi-quantitative analysis, where determinations with high resolution have prevailed [74,76,84]. In any case, mass spectrometers with high scan velocity are indispensable when hyphenated to UHPLC, and early generations of mass spectrometers have lost their applicability in this field.

Ways to approach a multiclass-wide scope screening of pesticides have been discussed in several works [74,75,84]. The objective was to develop semi-quantitative methods for the analysis of a number of pesticides in a single run, which should be able to capture enough information not only from the target compounds but also from others that may have relevant toxicity, such as transformation products. Hence, single-stage analysis would contain data that could be examined retrospectively and in a relatively short analysis time [74,75,84]. The solution proposed when using a UHPLC–(Q)TOF was acquired in full scan with high resolution, complemented by the fragmentation of all the ions generated in the ESI interface at a low and high energy (collision energy), acquired with high resolution (using a narrow mass window of 0.02 Da), at the expected retention time (2.5% retention time deviation tolerance), and a minimum peak width of 5s. The authors carried out the identification at two concentration levels (0.1 and 1 μg/L) to provide a semi-quantitative screening [74]. Similarly, in a lineal ion trap, the optimal solution found was to perform a full-scan, high-resolution acquisition with alternating scan events with/without fragmentation. The detection of analytes was based on the extraction of the exact mass (±5 ppm) of the major adduct ion at a retention time previously studied (±30 s) and the presence of a second ion related with the target compound, which could be an adduct, isotope or fragment, option that led to 0.3% false negatives [84]. High-resolution systems, an ion trap hybrid mass system interfaced with an UHPLC (UHPLC-qTrap), UHPLC–TOF–MS, and UHPLC/OrbitrapHR-MS have been compared for untargeted and targeted searches for characteristic accurate mass ions for over 170 pesticides in ginseng and spinach. At a level of 25 ng/g, the systems identified successfully over 89% of the compounds, and the UHPLC-qTrap produced the lowest false-negative rates. According to the authors, technical improvements were needed for nontargeted screening [85]. Identification based on the ration of two ions (20% accuracy) [86] has been followed by some authors [75], whereas others using a Q-TOF system stated that that ratio could easily be altered at low concentration levels and the presence of two accurate mass measured ions in any of the acquisition functions (low collision energy function and/or high collision energy function) using a narrow mass window of 0.02 Da with mass errors below 2 mDa, at the expected retention time, were used as identification criteria. Other examples of high-resolution MS will be discussed in depth in the next section.

The MS analyzers used in a selected range of applications are given in Table 2.1. Alternative detection systems capable to cope with fast chromatography techniques are those based on spectroscopy methods; however, due to sensitivity issues, they are not being applied in the detection of pesticides in food. Strategies based on two stages, a first injection for screening of positive samples and a second to quantify and confirm with two multiple-reaction monitoring (MRM) transitions have not been the usual approach [77]. When working with triple quadrupoles, the selectivity of the analysis given by such analyzers has prevailed over possibility to give a general overview of the compounds in the sample due to the limited sensitivity when carrying out a full scan. In these systems, two MRM assays were used for each pesticide. Collision energies were optimized for two selective ion transitions for every pesticide and both MRM transitions were used for confirmation analysis to meet the EU Decision (2002/657/EC, 2002) [86]. The most sensitive transitions were selected for quantification analysis MRM [75,77–80,83,85,87,88]. An example of this type of analysis is shown in Figure 2.2, where a UHPLC chromatogram of 15 carbamate pesticides analyzed by UHPLC-QqQ (MRM) is presented.

A matrix has been shown to affect the signal for most of the pesticides analyzed in multiple matrices, despite the high resolution offered by UHPLC and sample treatment based on QuEChERS, pressurized solvent extraction, SPE, or a combination of both [75,77,78,80,82,83,89,90]. The most common way to overcome this issue has been quantifying with a matrix matched in agreement with the SANCO guidelines [91], which has been the quantification strategy followed by many authors. The high number of pesticides analyzed in multi-residue analysis does not make feasible the use of standard addition, which would increase the analysis time and materials needed or the use of labeled internal standards, which would make the analysis very expensive as well as increase the number of transitions to follow when triple

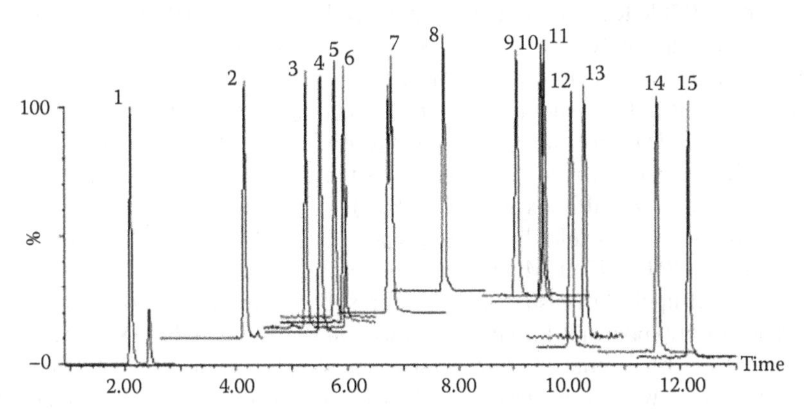

FIGURE 2.2 UHPLC–MS of pesticide residues spiked in tea (0.1 mg/kg) obtained with an Aquity UPLC BEH Shield C18 (2.1 × 150 mm, 1.7 μm) (Waters) and a QqQ mass analyser (Quattro Micro API) working in MRM mode. The compounds separated are (1) Propamocarb; (2) Primicarb; (3) Carbofuran-3-hydroxy; (4) Mevinphos; (5) Acetamiprid; (6) Thiofanox-sulphon; (7) Spiroxamine; (8) Triasulfuron; (9) Bromoxynil; (10) Promecarb; (11) Triadimefon; (12) Fenhexamid; (13) Fenoxycarb; (14) Clethodim; 15 Flufenoxuron. (Adapted from Chen, G., Cao, P., and Liu, R. 2011. *Food Chem.* 125:1406–1411. With permission.)

quadrupole (QqQ) instruments are used. Besides, some authors have observed that it is possible to obtain a similar matrix effect among some of the matrices, such as zucchini and cucumber, for the analysis of some pesticides [75], and that could be used to reduce the number of analyses and resources committed to the analysis.

2.3.3 High-Resolution MS

As previously mentioned, UHPLC–MS/MS using triple quadrupole instruments is frequently proposed for the multiresidue analysis of pesticides by monitoring two SRM transitions. Such a performance helps the analyst comply with the EU directive 2002/657/EC and confidently report a positive or negative finding [86]. The criteria to report a positive or a negative result are based mainly on the monitoring of these two transitions: the deviation of the relative intensity of the recorded product ions (which must not exceed a certain percentage of the one achieved with a reference standard) and the retention time of the precursor ion (variation from 1% to 5% when comparing with the retention time of a standard) [30,92]. However, the application of these criteria does not completely eliminate the possibility to report a false-positive or a false-negative result [93]. Because the resolving power of triple quadrupole mass spectrometers is often insufficient to baseline-resolve isobaric ions [29,94,95], the presence of an interfering compound coeluting with the target pesticide residue may induce a false-positive assignment. The possibility of reporting a false negative will be even more problematic because the presence of a possible harmful compound would be ignored.

High-resolution MS and accurate mass measurements are emerging as the best options for the analysis of pesticide residues in food in order to guarantee confirmation and identification. Basically, there are four types of high-resolution MS instrumentation: magnetic sector, time-of-flight (TOF), Orbitrap, and Fourier transform ion cyclotron (FTICR). The most frequently used with LC or UHPLC methods are TOF and Orbitrap analyzers. Magnetic sector instruments were developed, and are mainly used, for GC systems and FTICR instrumentation. Although a very powerful research tool [96], it is seldom used in the screening and monitoring of pesticides [97]. In general, TOF instruments present a resolving power (instrument's ability to measure the mass of two closely related ions precisely) of approximately 10,000–40,000 full-width at half-maximum (FWHM) with accuracies in the mass determination of 1–5 ppm, while the resolving power of Orbitrap instruments is in the range 10,000–140,000 FWHM with 1–2 ppm mass accuracy (for comparison, a conventional quadrupole MS instrument shows resolving powers of 1000 FWHM and accuracies of 500 ppm) [30]. Recent advances in both LC–TOF–MS and Orbitrap mass analyzers have reduced instrument costs, simplified analysis, and improved accuracy, and today these advances offer bench-top instrumentation that is amenable to screening and identification of pesticides in food, not only of target pesticides, but also of nontarget pesticide residues [98–100].

Accurate mass measurements improve the probability of a correct analysis when monitoring *m/z* ions in an MS spectrum, as the number of possible compounds with a specific *m/z* value will decrease as the accuracy is improved (lower ppm errors). Moreover, high-resolution MS will allow removing interferences from other compounds and their fragment ions, adducts, and stable isotopic peaks. So, as the mass resolution increases, the more selective the method will be, at least to remove interferences. Nevertheless, it should be pointed out that accurate mass measurements allow us to only determine the molecular formula for a target or nontarget pesticide residue, and a unique molecular formula separated from every possible interference is not a unique compound [30]. Several compounds may have the same formula, even those that are not structural isomers. Consequently, alternative information such as fragmentation spectra and chromatographic separation are needed to guarantee unambiguous identification. Once the molecular formula for a compound is reached, additional resolving power and mass accuracy will be irrelevant. The studies performed by Thurman's research group [98,99,101] have shown that in the analysis of pesticide residues and for many vegetable matrices a resolving power between 6000 and 10,000 FWHM is enough to remove interferences based on the complexity of the food samples. This mass resolving power is easily attainable by the TOF and Orbitrap MS instruments available.

Some recent applications of UHPLC–TOF–MS [102–106] and UHPLC–Orbitrap–MS [107–110] methods for the analysis of pesticide residues in food are summarized in Table 2.2. As can be seen, the current trends in the analysis of pesticide residues by HRMS focus as much as possible on the multiresidue analysis of target pesticides. For instance, Wang et al. proposed a UHPLC–HRMS method for the determination of more than 140 pesticide residues in several fruit and vegetable matrices, either using quadrupole-TOF (QTOF) instruments [102,103,111] or Orbitrap MS [110]. Quantification, within an analytical range from 5 to 500 μg/kg, was achieved using matrix-matched standard calibration curves with isotopically labeled standards or a chemical analog as internal standards. In general, recoveries after extraction by QuEChERS were between 80% and 110%. The authors also compared the proposed UHPLC–TOF–MS method with a conventional LC–ESI–MS(/MS) method in a triple quadrupole instrument [102,103]. Compared to the LC–ESI–MS(/MS) method, the UHPLC–TOF–MS showed relatively poor repeatability and large measurement uncertainty, while 86% of the analyzed pesticides provided an intermediate precision lower than 20%, and 83% of the pesticides showed measurement uncertainty of ≤40%. Thus, the authors proposed LC–ESI–MS(/MS) as the first choice for quantification or for pretarget analysis because of its superior sensitivity and good repeatability, but UHPLC–TOF–MS was necessary to provide accurate mass measurements and, thus, was an ideal tool for post-target screening and confirmation analysis (not always achieved by the triple quadrupole instrument).

Grimalt et al. [106] also investigated and compared the potential of three MS analyzers (triple quadrupole, TOF, and QTOF) for quantification, confirmation, and screening purposes in pesticide residue analysis of fruit and vegetable samples. For this purpose, 11 target pesticides were taken as model and the proposed UHPLC method was validated in nine food matrices for the three mass analyzers. In all cases, limits of quantification around 10 μg/kg were achieved, fulfilling the most restrictive case of baby food analysis [4–7]. As expected, the lower detection limits were obtained with the triple quadrupole instrument, but QTOF showed the highest confirmatory capacity. TOF MS was also investigated by the authors for screening purposes. For this purpose, they analyzed around 50 commercial fruits and vegetables samples, searching for more than 400 pesticides. As an example, Figure 2.3

TABLE 2.2
UHPLC–TOF–MS and UHPLC–Orbitrap–MS Methods for the Analysis of Pesticide Residues in Food

Compounds	Sample	Sample Treatment	Chromatographic Separation	MS Detection	Analysis Time[a]	Reference
148 pesticide residues	Fruits and vegetables	QuEChERS	Acquity UPLC BEH C_{18} column (100 × 2.1 mm i.d., 1.7 µm particle size) Gradient elution A: acetonitrile B: 10 mM ammonium acetate aqueous solution Flow-rate: 0.4 mL/min Column temperature: 45°C	Q-TOF instrument Full-scan MS acquisition mode (*m/z* 50–950) Resolving power: 15,000 FWHM ESI(+) Lock mass reference: leucine enkephalin infused with a lock-spray of 5 µL/min.	14 min	[102,103]
212 pesticide residues	Food plants	QuEChERS	Acquity UPLC HSS T3 column (100 × 2.1 mm i.d., 1.8 µm particle size) Gradient elution A: methanol B: 5 mM ammonium formate aqueous solution Flow-rate: 0.3 mL/min Column temperature: 40°C	TOF instrument Full-scan MS acquisition mode (*m/z* 50–950) Resolving power: >11,000 FWHM ESI(+) and ESI(−) Lock mass reference: leucine enkephalin infused with a lock-spray	24 min[b]	[104]
Organic pollutants including pesticides	Food (orange and banana)	Offline SPE HLB cartridges	Acquity UPLC BEH C_{18} column (150 × 2.1 mm i.d., 1.8 µm particle size) Gradient elution A: 0.01% formic acid in methanol B: 0.01% formic acid in water Flow-rate: 0.3 mL/min Column temperature: 60°C	Q-TOF instrument Full-scan MS acquisition mode (*m/z* 50–1000) Resolving power: 10,000 FWHM ESI(+) and ESI(−) Lock mass reference: leucine enkephalin infused with a lock-spray of 30 µL/min.	18 min	[105]

11 target pesticide residues Screening of 423 pesticide residues	Fruits and vegetables	Extraction with methanol:water (80:20 v/v)	Acquity UPLC BEH C_{18} column (50 × 2.1 mm i.d., 1.7 µm particle size) Gradient elution A: 0.5 mM ammonium acetate in methanol B: 0.5 mM ammonium acetate in water Flow-rate: 0.3 mL/min	QTOF instrument Full-scan MS and full product ion scan acquisition modes (*m/z* 50–1000) Resolving power: 10,000 FWHM ESI(+) Lock mass reference: leucine enkephalin infused with a lock-spray of 30 µL/min.	8 min	[106]
Carbamate and organophosphate pesticides	Food-related matrices (whole milk, muscle tissue, liver tissue, corn silage)	QuEChERS	Hypersil Gold AQ C_{18} column (50 × 2.1 mm i.d., 1.9 µm particle size) Gradient elution A: 0.1% formic acid in acetonitrile B: 0.1% formic acid in water Flow-rate: 0.38 mL/min Column temperature: 35°C	Exactive Orbitrap Full-scan MS acquisition mode (*m/z* 80–1000) Resolving power: 50,000 FWHM H-ESI(+)	12 min	[107]
72 organic pollutants including acidic pesticides	Aqueous samples (drinking water)	Online SPE with a C_{18} Hypersil Gold column (20 × 2.1 mm i.d., 12 µm) Flow-rate: 1 mL/min	C_{18} Core-shell Accucore RP-MS column (50 × 2.1 mm i.d., 2.6 µm particle size) Gradient elution A: 0.1% formic acid in methanol B: 1% methanol and 0.1% formic acid in water Flow-rate: 0.6 mL/min Column temperature: 30°C	Exactive Orbitrap Full-scan MS acquisition mode (*m/z* 103–500) Resolving power: 25,000 FWHM H-ESI(+) and H-ESI(−) in scan-to-scan polarity switching	15 min[c]	[108]
350 Pesticides and veterinary drugs	Honey	Extraction with water and 1% formic acid in acetonitrile	Hypersil Gold AQ C_{18} column (100 × 2.1 mm i.d., 1.7 µm particle size) Gradient elution A: 0.1% formic acid and 4 mM ammonium formate in methanol	Exactive Orbitrap Full-scan MS acquisition mode Resolving power: 25,000 FWHM	14 min	[109]

continued

TABLE 2.2 (continued)
UHPLC–TOF–MS and UHPLC–Orbitrap–MS Methods for the Analysis of Pesticide Residues in Food

Compounds	Sample	Sample Treatment	Chromatographic Separation	MS Detection	Analysis Time[a]	Reference
			B: 0.1% formic acid and 4 mM ammonium formate in water Flow rate: 0.3 mL/min Column temperature: 30°C	MS/MS acquisition mode (higher collisional dissociation (HCD) cell with collision energy of 30 eV) Resolving power: 10,000 FWHM H-ESI(+) and H-ESI(−)		
166 pesticide residues	Fruits and vegetables	QuEChERS	Acquity UPLC BEH C_{18} column (100 × 2.1 mm i.d., 1.7 µm particle size) Gradient elution A: acetonitrile B: 10 mM ammonium acetate aqueous solution Flow-rate: 0.4 mL/min Column temperature: 45°C	Q-Exactive Orbitrap Full sacn MS acquisition mode (*m/z* 65–950) Resolving power: 70,000 FWHM Full MS/dd-MS^2 (TopN) acquisition mode (m/z 65–950) (HCD at 35% normalized collision energy) Resolving power: 70,000 FWHM H-ESI(+)	14 min	[110]

[a] Not including sample treatment.
[b] Considering two analyses (in ESI(+) and in ESI(−) ionization modes).
[c] Including online SPE.

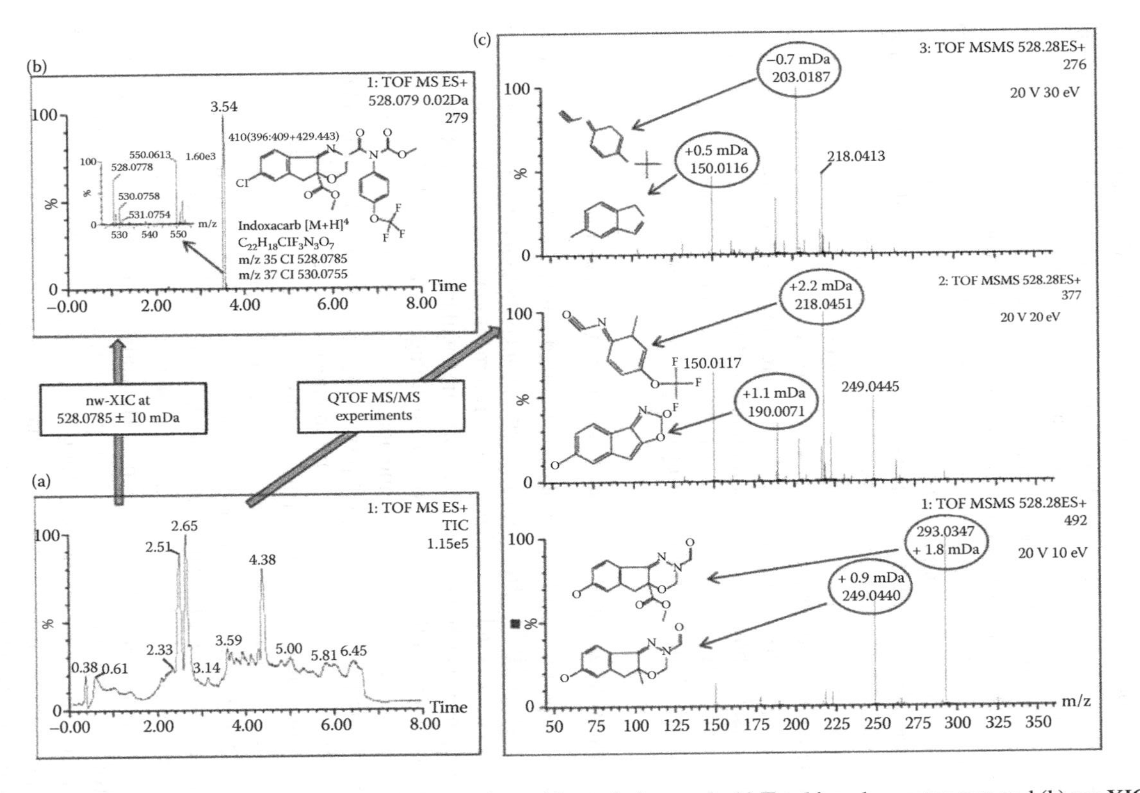

FIGURE 2.3 UHPLC–ESI–TOF chromatograms of a grape sample positive to indoxacarb. (a) Total ion chromatogram and (b) nw-XIC (±10 mDa) and combined mass spectrum for the peak at 3.54 min. (c) Combined product ion spectra obtained by QTOF MS/MS experiments performed at different collision energies. Mass errors and proposed chemical structures are included for the six product ions evaluated. (Adapted from Grimalt, S. et al. 2010. *J. Mass Spectrom.* 45:421–436. With permission.)

shows the UHPLC–ESI–TOF chromatograms of a grape sample positive to indoxacarb, as well as the MS and product ion MS spectra obtained by QTOF–MS(/MS) at three different energies. The confirmation of a potential positive of indoxacarb in this sample was approached first by checking the isotopic distribution, as this pesticide presents a chlorine atom. Data fit satisfactorily with the isotope model, as well as for the isotopic distribution corresponding to the sodium adduct. However, the QTOF instrument allowed the authors to guarantee confirmation by performing MS/MS experiments at different collision energies. They were able to confirm the presence of indoxacarb in this grape sample by studying its fragmentation pathway, obtaining the accurate masses of up to six abundant product ions.

Several authors are also analyzing pesticides when developing UHPLC–HRMS methods for the screening of organic contaminants in general. For instance, Filigenzi et al. [107] proposed a UHPLC–Orbitrap–MS method for the analysis of 118 target compounds, including plant alkaloids, carbamate and organophosphate pesticides, and several types of veterinary drugs in several food-related matrices such as whole milk, muscle and liver tissue, and corn silage. Most of the compounds were detectable at low μg/kg levels with a mass accuracy error lower than 2 parts per million (ppm). Their results showed that the combination of a generic extraction procedure (QuEChERS) and UHPLC with full-scan HRMS in an Orbitrap instrument can be proposed as a useful method for screening complex matrices.

2.3.4 Alternative Acquisition and Confirmation Systems

The necessity of performing fast sample screening of hundreds of pesticide residues in food samples by UHPLC–HRMS with accurate mass measurements requires databases of exact masses (and even of exact mass fragment ions) in order to rapidly look for these substances via appropriate data software available with the HRMS instruments. Moreover, one of the most important advantages of using HRMS in full-scan MS acquisition mode (but also in combination with several fragmentation acquisition modes) is the possibility of a post-processing analysis of previously analyzed samples if new target pesticide residues are intended to be determined. Thus, today the creation of the database is fundamental in this type of analysis [30]. For instance, Gómez-Pérez et al. [109] created a database for the simultaneous analysis of more than 350 pesticides and veterinary drugs (including antibiotics) in honey samples using UHPLC–Orbitrap–MS. The developed database included exact masses of the target ions and retention time data, and the probable adduct ions, and allowed the automatic search of the included compounds. The authors used the database for qualitative analysis, but it was also evaluated for quantitative purposes in routine analysis after a generic extraction method (see Table 2.2). Figure 2.4 shows an example of the report provided by ToxID™ software (automatic compound screening software, Thermo Scientific) used for screening purposes. For pesticides, LODs were in general lower than the MRLs established by the EU in honey. The method was applied to the analysis of 26 real honey samples and some pesticides (azoxystrobin, coumaphos, dimethoate, and thiacloprid) were detected in four samples, while veterinary drugs were not detected in any sample. Axosystrobin and coumaphos were quantified in two different organic honey samples at 1.5 and 5.1 μg/kg, respectively.

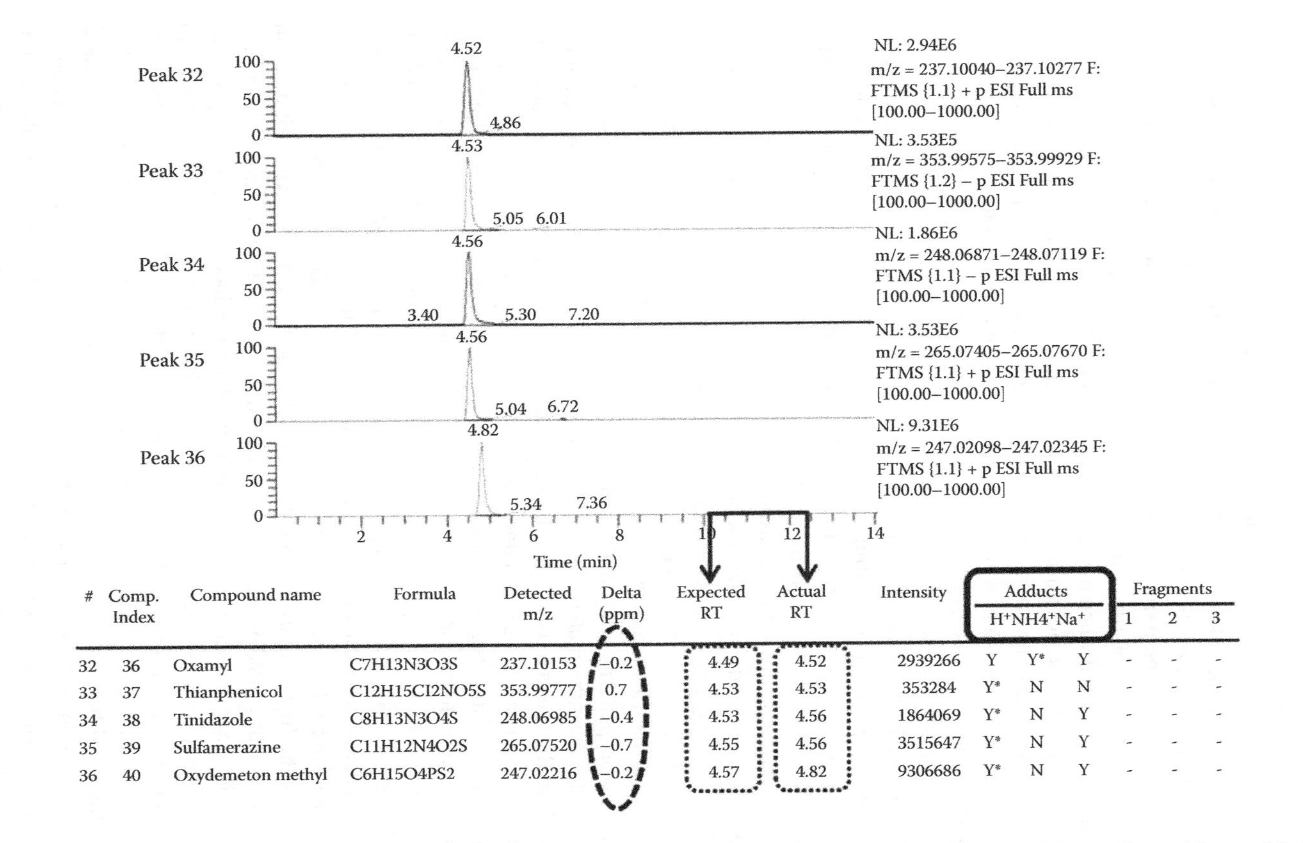

#	Comp. Index	Compound name	Formula	Detected m/z	Delta (ppm)	Expected RT	Actual RT	Intensity	Adducts			Fragments		
									H⁺	NH4⁺	Na⁺	1	2	3
32	36	Oxamyl	C7H13N3O3S	237.10153	−0.2	4.49	4.52	2939266	Y	Y*	Y	-	-	-
33	37	Thianphenicol	C12H15CI2NO5S	353.99777	0.7	4.53	4.53	353284	Y*	N	N	-	-	-
34	38	Tinidazole	C8H13N3O4S	248.06985	−0.4	4.53	4.56	1864069	Y*	N	Y	-	-	-
35	39	Sulfamerazine	C11H12N4O2S	265.07520	−0.7	4.55	4.56	3515647	Y*	N	Y	-	-	-
36	40	Oxydemeton methyl	C6H15O4PS2	247.02216	−0.2	4.57	4.82	9306686	Y*	N	Y	-	-	-

FIGURE 2.4 Example of the report provided by ToxIDTM software used for screening purposes. (Adapted from Gomez-Perez, M. L. 2012. *J. Chromatogr. A* 1248:130–138. With permission.)

The match probabilities of these databases are based on the exact mass being measured correctly and compared against the one in the database. This will depend on the stability and accuracy of the HRMS instrument mass calibration. For more accurate mass measurements, TOF and QTOF instruments frequently use a lock mass correction (see Table 2.2), which consists of the infusion of a reference compound such as leucine enkephalin ($[M + H]^+ = 556.2771$) at a low flow rate (5–30 μL/min) by means of a LockSpray probe. Database software then correct the *m/z* values for any target or nontarget pesticide, taking into account the variation obtained between leucine enkephalin exact mass and the experimental one. Moreover, some skill in MS is required when using databases in order to use the $M + 1$ and $M + 2$ isotopic information to achieve a correct formula, and to lower the probability of error as much as possible to low ppm values [30]. To help in this process, several database programs also give the isotope table and the isotope abundances for each match, and many instruments provide software to help with isotope identification and matching, which is quite useful for correct formula identification.

The use of library search engines is also an alternative identification and confirmation system that is becoming popular in the analysis of pesticide residues. As mentioned in the introduction section, working with conventional library searching routines is well established when GC–MS is used for pesticide analysis, and the use of HRMS with accurate mass measurements is helping in improving library search engines for target and nontarget compounds. The strength of the library search is the speed of screening hundreds of pesticides in minutes, including both their protonated molecules and fragment ions, with accurate mass measurements of all these ions. At the moment, the use of classical fragmentation libraries based on the comparison to fragmentation patterns is probably not needed when accurate mass measurements are used, because the matches are based on the presence of the ion rather than the intensity of the ion and its fragmentation pattern [30]. This helps in eliminating the problem of instrument variation and matrix effects on the fragmentation.

Library searching engines can also be used with conventional triple quadrupole instruments able to perform data-acquisition experiments. For instance, in a recent work, Núñez et al. [17] proposed an alternative strategy for the multiresidue analysis of 100 pesticides in fruits and vegetables using LC–triple quadrupole–MS by performing a data-dependent analysis. For this purpose, highly selective selected reaction monitoring (H-SRM) acquisition mode, which provides high sensitivity, was selected as the first scan event. When the H-SRM ions were detected with a signal higher than a threshold value of 10^3, a second scan, enabling the acquisition of a product ion MS/MS spectrum, was activated. This product ion MS/MS spectrum was not obtained with a defined collision energy, but by using reversed energy ramp (RER) mode and applying collision energies from 90 to 25 eV. The results were then a product ion scan spectrum average of spectra at the different collision energies, which will better help in confirmation and identification of pesticide residues. For example, Figure 2.5 shows the analysis of an orange sample by data-dependent scan analysis where three pesticides, azoxystrobin, imazalil, and methidathion were detected, and the RER product ion scan spectrum of imazalil (second scan event). The advantage of this RER product ion scan spectrum is that it provides higher structural information than a simple product ion scan spectrum obtained at specific

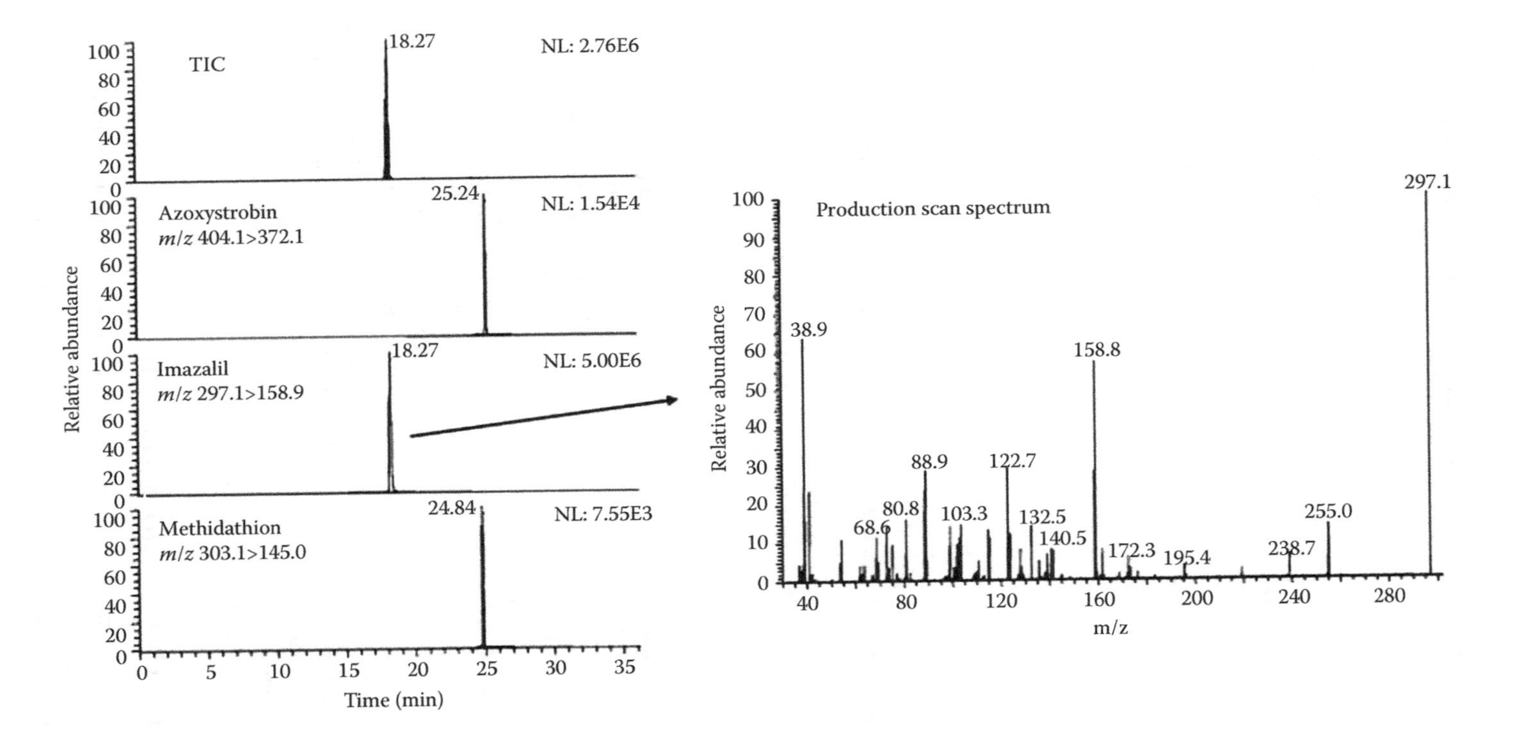

FIGURE 2.5 Analysis of an orange sample by data-dependent analysis. Pesticides detected in H-SRM mode (first scan event) and, as an example, the RER product ion scan spectrum of imazalil (second scan event). (Adapted from Núñez, O. et al. 2012. *J. Chromatogr. A* 1249:164–180. With permission.)

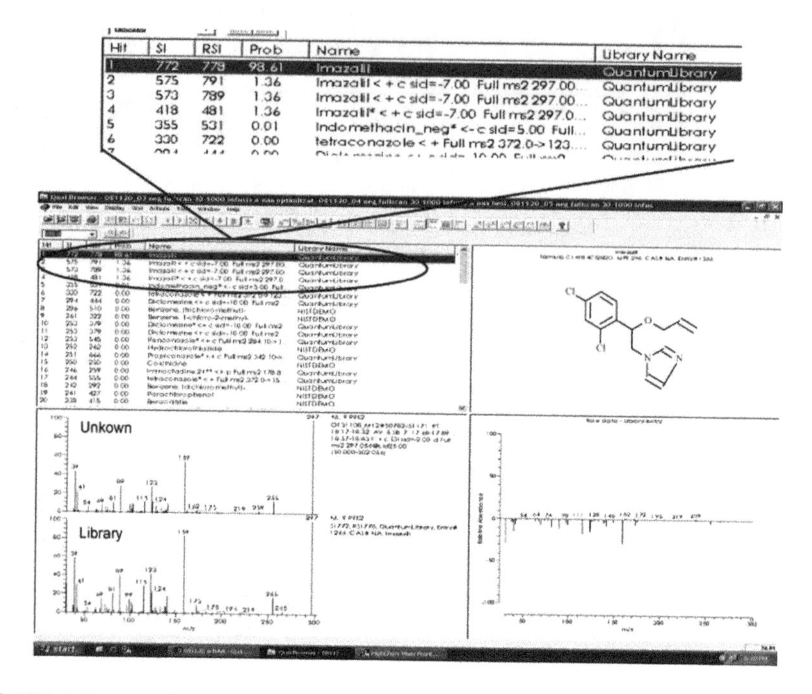

FIGURE 2.6 Results generated by the library search engine with the RER product ion scan spectrum of imazalil in an orange sample. (Adapted from Núñez, O. et al. 2012. *J. Chromatogr. A* 1249:164–180. With permission.)

collision energy. Since it is well known that low collision energies are too weak to adequately fragment the precursor ions, while at high collision energies few product ions are generated, the use of RER is more likely to generate fragment-rich MS/MS spectra that will be optimal for library entries and searching purposes.

The authors then created a product ion scan spectra library and loaded the RER product ion scan spectra for each pesticide into the Quantum Library of the Xcalibur software (Thermo Scientific), and a user library for routine library searching was generated. Figure 2.6 shows the results generated by the software library search engine for the analysis of pesticides in an orange sample. The library compares RER product ion scan spectra of the target sample (in the example the pesticide imazalil) with those of the user library previously generated, and, as can be seen, a match (with a probability of 98.61) was found, allowing the confirmation of the presence of imazalil pesticide in the sample.

This data-dependent analysis combined with library search engines can then be proposed as a further confirmatory strategy in the analysis of pesticide residues.

2.4 LEVELS OF PESTICIDES IN FOOD

Because many commonly used pesticides are potentially harmful to the environment and consequently to human beings through the consumption of pesticide contaminated food, legal action levels (e.g., MRLs or tolerances) for many of these agrochemicals in food have so far been established globally by different regulatory organizations. However, despite the efforts made over the past years to harmonize their use between countries, with great concern about international trade in agricultural commodities and food products, MRL regulations worldwide have not yet been fully harmonized and the current status of different normatives should be considered. For instance, for the European community, the MRLs for all crops and pesticides can be found in the MRL database on the Commission website [112], whereas the U.S. Environmental Protection Agency (EPA) has developed a website [113] that provides information about organic certification application and regulations, as well as pesticide labeling and the tolerance limits (maximum permission limits) of pesticide products. In general, MRLs are in the range of 0.01–10 mg/kg, depending on the combination commodity and pesticides, whereas specific rules on the presence of pesticide residues in infant, cereal-based baby food, and baby food have been established [32]. For instance, the use of certain very toxic pesticides has been prohibited in the production of baby food and levels lower than the general maximum level of 0.01 mg/kg for a few other very toxic pesticides have been set [6,7,32].

Several studies about the occurrence of pesticides in food and food products have been reported in the literature, although most monitoring reports have been published over the last 15 years. Moreover, since the general population is mainly exposed to pesticide residues through the ingestion of contaminated vegetables, fruits, and cereals, most of the papers published focus on these types of foods, and scarce reports have dealt with the monitoring of pesticide residues in food and food products of animal origins. Organochlorines (OCs) such as DDT (**d**ichloro**d**iphenyl**t**richloroethane), endrin, aldrin, dieldrin, HCH (hexachlorocyclohexane) isomers, and hexachlorobenzene are chemically stable compounds that have been broadly used as insecticides as effective weapons in crop protection since the introduction of pesticides. However, due to their environmental persistence, most OCs are currently banned worldwide and synthetic pyrethroid and organophosphate (OP) compounds have been introduced as replacements, most of them being commercially available in several countries [114]. The results obtained from the 2009 EU-coordinated program have shown that 1.2% of the 10,553 samples analyzed exceeded the MRL [115]. Residues of 338 different pesticides were found in vegetables and 319 in fruit and nuts, while in cereals the presence of 93 different pesticide residues has been detected. The most common pesticide/crop combinations were imazalil/bananas (49.5%), chlormequat/wheat (42.3%), and fenhexamid/table grapes (23.8%), while the highest percentage of samples exceeding the MRL was observed for table grapes (2.8%), followed by peppers (1.8%), aubergines (1.7%), peas (1.0%), wheat (0.8%), butter (0.6%), cauliflower (0.5%), and bananas (0.4%). On the other hand, the analysis of the results of the 2010 EU-coordinated program showed that the highest percentage of samples exceeding the MRL was identified for oats (5.3%), followed by lettuce (3.4%), strawberries (2.8%), and peaches (1.8%) [116]. In another recent study in which 10 fruits and vegetables purchased from a local market were analyzed, it was shown that the following pesticides were present: quinalphos was the major component, detected in 40% of samples (tomato, chili, guava, strawberry) at mean 29.6 μg/L; profenofos and chlorpyrifos were present in 30% of monitored samples at mean 17.3 and 13.1 μg/L, respectively; malathion was found in 20% of the samples at mean 55.1 μg/L, and diazinon was determined to be in only tomato and guava samples at mean 8.2 μg/L [117]. In all samples the residue levels of the pesticides determined were found to be lower than the MRLs specified by the EU. Malathion was also found in orange samples (~50 μg/kg) while residues of iprodione have been detected in lettuce, strawberry, and plum samples in the range of 40.93–241.06 μg/kg [118]. Residues of β-endosulfan have been found in avocado (25 μg/kg) [41], cucumber (19 μg/kg), and pepper (18–32 μg/kg) [119], whereas α-endosulfan has been detected in Korean cabbage (45 μg/kg) [120]. Procymidone has been reported in pepper (76–86 μg/kg) [119], grape (2.41 mg/kg), [121], and cabbage (10–20 μg/kg) [120]. Other pesticides that are commonly found in fruits and vegetables are metalaxyl (grape) [121,122], pyriproxyfen (orange, tangerine, grapefruit, lemon, and red pepper) [123,124], buprofezin (citrus fruit, green pepper, and tomato) [119,123,124], dimethoate (cucumber, potato, and grape) [122,124], and carbendazim (orange) [125]. The pesticide residues that are frequently detected in tea include fenvalerate (50–250 μg/kg), cypermethrin (10–50 μg/kg), fenpropathrin (30–300 μg/kg), buprofezin (60–250 μg/kg), and triazophos (20–200 μg/kg) [126]. However, pesticides can not only be detected in conventional crops, but also in organic ones. In fact, even if the use of synthetic pesticides is not permitted in organic agriculture, there can never be a guarantee that organically grown plants are totally pesticide free. Therefore, organic production also needs to be monitored. For instance, Walorczyk et al. [127] determined the pesticide residue levels in Polish organic crops from 2007 to 2010. A total of 528 samples of differing matrices (fruits, vegetables, cereals, plant leaves, and other green parts) were analyzed, of which 4.4% samples contained pesticide residues higher than 0.01 mg/kg. A total of 20 different pesticide residues were determined in the samples with compounds, such as pirimiphos-methyl, dichlofluanid, flusilazole, tolyfluanid, cyprodinil, pyrimethanil, chlorothalonil, cypermethrin, DDT, and fenitrothion, which was detected more than once.

Finally, even if most of the pesticides worldwide are used in fruit and vegetable crops, data on pesticide residues in animal products are also essential, taking into account that livestock can be easily exposed to pesticides directly or through residues in their feed. For instance, the 2009 European Union Report revealed the presence of 34 different pesticides in animal products [115]. García de Llasera et al. [128] revealed the presence of chlofenvinphos and chlorpyrifos in liver samples, whereas α-endosulfan, endosulfan sulfate, and dichloran have been reported in pork and lamb samples (<10 μg/kg) [129]. Bolaños et al. [130] studied the occurrence of organochlorine pesticides and polychlorinated biphenyls in chicken eggs. Benzene hexachloride and 28 polychlorinated biphenyl were detected in only one of the 30 samples at concentrations of 15 and 10 ng/g, respectively. However, five samples also contained traces of OC and PCB residues, even if at concentration levels below the limit of quantification. Finally, α-endosulfan and β-endosulfan were found in commercial milk-based infant formulas at concentration levels from 1.18 to 5.03 μg/kg [35]. The same study also showed the presence of fenitrothion, chlorpyrifos ethyl, and bifenthrin at maximum concentrations of 0.23,1.30, and 0.68 μg/kg, respectively.

2.5 SUMMARY AND CONCLUSIONS

An overview of several recent applications of UHPLC–MS(/MS) methods for the multi-residue analysis of pesticides in food products has been presented. In order to cope with the necessity of high throughput and fast analysis of pesticide residues in food, all aspects of analytical method development, that is, sample extraction and clean-up, chromatographic analysis, and quantitation and confirmation aspects, must be taken into account.

The most recent approach in fast LC methodology for the analysis of pesticide residues in food has been presented. UHPLC, using either sub-2 μm particle size column or fused-core column technologies, is becoming a reliable alternative to GC–MS methodologies for the analysis of pesticides, with sub-2 μm columns being the most popular ones.

Despite the important advances in fast LC, food matrices are very complex, and although in general multi-residue methods with minimal sample manipulation are demanded, sample extraction and clean-up treatments must be carefully developed to also reduce the total analysis time. The most recently introduced sample treatment methodologies for pesticide residue analysis have also been addressed, with QuEChERS being the most popular one for its easy application and good results. However, other alternatives, such as online SPE or the use of more selective methods such as MIP, are also being applied for the analysis of pesticides.

Quantitation and confirmation aspects regarding MS have also been addressed. When triple quadrupole instruments are employed, two SRM transitions are frequently used for quantitation and confirmation aspects. However, sometimes this strategy is not enough. Today, high-resolution MS, either using TOF or Orbitrap instruments, is becoming a powerful technique for the analysis of pesticide residues in food. Among the advantage of high-resolution and accurate mass measurements, the possibility of a post-processing data analysis is one of the most powerful advantages of this kind of methodology. Additionally, examples of alternative quantitation and confirmation methods, such as the use of pesticide spectra libraries, have also been discussed.

Finally, recent levels of pesticides in several food matrices were presented and discussed.

ACKNOWLEDGMENTS

Rosa Busquets acknowledges The Intra European Fellowship (Polarclean project, N°. 274985) for financial support.

REFERENCES

1. Grob, K., Biedermann, M., Scherbaum, E., Roth, M., and Rieger, K. 2006. Food contamination with organic materials in perspective: Packaging materials as the largest and least controlled source? A view focusing on the European Situation. *Crit. Rev. Food Sci. Nutr.* 46:529–535.
2. Fernandez-Alba, A. R. and Garcia-Reyes, J. F. 2008. Large-scale multi-residue methods for pesticides and their degradation products in food by advanced LC-MS. *TrAC, Trends Anal. Chem.* 27:973–990.

3. European Union. 1999. Council Directive 91/414/EEC of 15 July 1991 concerning the placing of plant protection products on the market, European Commission, Brussels. *Official Journal L* 230:1.
4. European Commission. 1999. Commission Directive 1999/39/EC of 6 May 1999 amending Directive 96/5/EC on processed cereal-based foods and baby foods for infants and young children. European Commission, Brussels.
5. European Commission. 1999. Commission Directive 1999/50/EC of 25 May 1999 amending Directive 96/5/EC on infant formulae and follow-on formulae. European Commission, Brussels.
6. European Commission 2003. Commission Directive 2003/13/EC of 6 May 1999 amending Directive 96/5/EC on processed cereal based foods and baby foods for infants and young children. European Commission, Brussels.
7. European Commission 2003. Commission Directive 2003/14/EC of 25 May 1999 amending Directive 91/321/EEC on infant formulae and follow-on formulae. European Commission, Brussels.
8. Toledano, R. M., Cortes, J. M., Andini, J. C., Villen, J., and Vazquez, A. 2010. Large volume injection of water in gas chromatography–mass spectrometry using the through oven transfer adsorption desorption interface: Application to multiresidue analysis of pesticides. *J. Chromatogr. A* 1217:4738–4742.
9. Koesukwiwat, U., Lehotay, S. J., and Leepipatpiboon, N. 2011. Fast, low-pressure gas chromatography triple quadrupole tandem mass spectrometry for analysis of 150 pesticide residues in fruits and vegetables. *J. Chromatogr. A* 1218:7039–7050.
10. Matsuoka, T., Akiyama, Y., and Mitsuhashi, T. 2011. Application of multi-residue analytical method for determination of 496 pesticides in frozen gyoza dumplings by GC–MS and LC–MS. *J. Pestic. Sci.* 36:486–491.
11. Vukovic, G., Shtereva, D., Bursic, V., Mladenova, R., and Lazic, S. 2012. Application of GC-MSD and LC-MS/MS for the determination of priority pesticides in baby foods in Serbian market. *LWT— Food Sci. Technol.* 49:312–319.
12. Kwon, H., Lehotay, S. J., and Geis-Asteggiante, L. 2012. Variability of matrix effects in liquid and gas chromatography–mass spectrometry analysis of pesticide residues after QuEChERS sample preparation of different food crops. *J. Chromatogr. A* 1270:235–245.
13. Alder, L. 2011. Targeted pesticide residue analysis using triple quad LC-MS/MS. *Methods Mol. Biol.* 747:173–191.
14. Kujawski, M. W. and Namiesnik, J. 2011. Levels of 13 multi-class pesticide residues in Polish honeys determined by LC-ESI-MS/MS. *Food Control* 22:914–919.
15. Kmellar, B., Pareja, L., Ferrer, C., Fodor, P., and Fernandez-Alba, A. R. 2011. Study of the effects of operational parameters on multiresidue pesticide analysis by LC-MS/MS. *Talanta* 84:262–273.
16. Sack, C., Smoker, M., Chamkasem, N., Thompson, R., Satterfield, G., Masse, C., Mercer, G. et al. 2011. Collaborative validation of the QuEChERS procedure for the determination of pesticides in food by LC-MS/MS. *J. Agric. Food Chem.* 59:6383–6411.
17. Núñez, O., Gallart-Ayala, H., Ferrer, I., Moyano, E., and Galceran, M. T. 2012. Strategies for the multi-residue analysis of 100 pesticides by liquid chromatography–triple quadrupole mass spectrometry. *J. Chromatogr. A* 1249:164–180.
18. Camino-Sanchez, F. J., Zafra-Gomez, A., Oliver-Rodriguez, B., Ballesteros, O., Navalon, A., Crovetto, G., and Vilchez, J. L. 2010. UNE-EN ISO/IEC 17025:2005-accredited method for the determination of pesticide residues in fruit and vegetable samples by LC-MS/MS. *Food Addit. Contam., Part A* 27:1532–1544.
19. Di Stefano, V., Avellone, G., Bongiorno, D., Cunsolo, V., Muccilli, V., Sforza, S., Dossena, A., Drahos, L., and Vekey, K. 2012. Applications of liquid chromatography–mass spectrometry for food analysis. *J. Chromatogr. A* 1259:74–85.

20. D'Orazio, G., Rocco, A., and Fanali, S. 2012. Fast-liquid chromatography using columns of different internal diameters packed with sub-2 Î1/4m silica particles. *J. Chromatogr. A* 1228:213–220.
21. Núñez, O., Gallart-Ayala, H., Martins, C. P. B., and Lucci, P. 2012. New trends in fast liquid chromatography for food and environmental analysis. *J. Chromatogr. A* 1228:298–323.
22. Fekete, S. and Fekete, J. 2011. Fast gradient screening of pharmaceuticals with 5 cm long, narrow bore reversed-phase columns packed with sub-3 micro m core-shell and sub-2 micro m totally porous particles. *Talanta* 84:416–423.
23. Fekete, S., Ganzler, K., and Fekete, J. 2011. Efficiency of the new sub-2 micro m core-shell (Kinetex?) column in practice, applied for small and large molecule separation. *J. Pharm. Biomed. Anal.* 54:482–490.
24. Van Deemter, J. J., Zuiderweg, F. J., and Klingengerg, A. 1956. Longitudinal diffusion and resistance to mass transfer as causes of nonideality in chromatography. *J. Chem. Eng. Sci.* 5:271–289.
25. Wang, J., Chow, W., and Cheung, W. 2011. Application of a tandem mass spectrometer and core-shell particle column for the determination of 151 pesticides in grains. *J. Agric. Food Chem.* 59:8589–8608.
26. Anastassiades, M., Lehotay, S. J., Stajnbaher, D., and Schenck, F. J. 2003. Fast and easy multiresidue method employing acetonitrile extraction/partitioning and "dispersive solid-phase extraction" for the determination of pesticide residues in produce. *J. AOAC Int.* 86:412–431.
27. Lucci, P., Pacetti, D., Núñez, O., and Frega, N.G. 2012. Current trends in sample treatment techniques for environmental and food analysis. In *Chromatography: The Most Versatile Method of Chemical Analysis*, ed. L.A. Calderon, InTech Publisher, Rijeka (CR), pp. 127–164, ISBN: 978-953-51-0813-9.
28. Lehotay, S. J. 2011. QuEChERS sample preparation approach for mass spectrometric analysis of pesticide residues in foods. *Methods Mol. Biol.* 747:65–91.
29. Schurmann, A., Dvorak, V., Cruzer, C., Butcher, P., and Kaufmann, A. 2009. False-positive liquid chromatography/tandem mass spectrometric confirmation of sebuthylazine residues using the identification points system according to EU directive 2002/657/EC due to a biogenic insecticide in tarragon. *Rapid Commun. Mass Spectrom.* 23:1196–1200.
30. Ferrer, I., Thurman, E. M., and Zweigenbaum, J. 2011. LC/TOF-MS analysis of pesticides in fruits and vegetables: The emerging role of accurate mass in the unambiguous identification of pesticides in food. *Methods Mol. Biol.* 747:193–218.
31. Ahmed, F. E. 2001. Analyses of pesticides and their metabolites in foods and drinks. *TrAC, Trends Anal. Chem.* 20:649–661.
32. Picó, Y., Font, G., Ruiz, M. J., and Fernandez, M. 2006. Control of pesticide residues by liquid chromatography–mass spectrometry to ensure food safety. *Mass Spectrom. Rev.* 25:917–960.
33. Fuentes, E., Baez, M. E., and Quinones, A. 2008. Suitability of microwave-assisted extraction coupled with solid-phase extraction for organophosphorus pesticide determination in olive oil. *J. Chromatogr. A* 1207:38–45.
34. Hernandez-Borges, J., Ravelo-Perez, L. M., Hernandez-Suarez, E. M., Carnero, A., and Rodriguez-Delgado, M. A. 2008. Determination of abamectin residues in avocados by microwave-assisted extraction and HPLC with fluorescence detection. *Chromatographia* 67:69–75.
35. Mezcua, M., Repetti, M. R., Aguera, A., Ferrer, C., Garcia-Reyes, J. F., and Fernandez-Alba, A. R. 2007. Determination of pesticides in milk-based infant formulas by pressurized liquid extraction followed by gas chromatography tandem mass spectrometry. *Anal. Bioanal. Chem.* 389:1833–1840.

36. Blasco, C., Font, G., and Pico, Y. 2005. Analysis of pesticides in fruits by pressurized liquid extraction and liquid chromatography–ion trap–triple stage mass spectrometry. *J. Chromatogr. A* 1098:37–43.
37. Garces-Garcia, M., Brun, E. M., Puchades, R., and Maquieira, A. 2006. Immunochemical determination of four organophosphorus insecticide residues in olive oil using a rapid extraction process. *Anal. Chim. Acta* 556:347–354.
38. Rissato, S. R., Galhiane, M. S., Knoll, F. R. N., and Apon, B. M. 2004. Supercritical fluid extraction for pesticide multiresidue analysis in honey: Determination by gas chromatography with electron-capture and mass spectrometry detection. *J. Chromatogr. A* 1048:153–159.
39. Richter, B. E., Jones, B. A., Ezzell, J. L., Porter, N. L., Avdalovic, N., and Pohl, C. 1996. Accelerated solvent extraction: A technique for sample preparation. *Anal. Chem.* 68:1033–1039.
40. Gilbert-Lopez, B., Garcia-Reyes, J. F., and Molina-Diaz, A. 2009. Sample treatment and determination of pesticide residues in fatty vegetable matrices: A review. *Talanta* 79:109–128.
41. Fernandez Moreno, J. L., Arrebola Liebanas, F. J., Garrido Frenich, A., and Martinez Vidal, J. L. 2006. Evaluation of different sample treatments for determining pesticide residues in fat vegetable matrices like avocado by low-pressure gas chromatography-tandem mass spectrometry. *J. Chromatogr. A* 1111:97–105.
42. Economou, A., Botitsi, H., Antoniou, S., and Tsipi, D. 2009. Determination of multi-class pesticides in wines by solid-phase extraction and liquid chromatography–tandem mass spectrometry. *J. Chromatogr. A* 1216:5856–5867.
43. Yang, X., Zhang, H., Liu, Y., Wang, J., Zhang, Y. C., Dong, A. J., Zhao, H. T., Sun, C. H., and Cui, J. 2011. Multiresidue method for determination of 88 pesticides in berry fruits using solid-phase extraction and gas chromatography–mass spectrometry: Determination of 88 pesticides in berries using SPE and GC-MS. *Food Chem.* 127:855–865.
44. Durovic, R., Milinovic, J., Markovic, M., and Markovic, D. 2007. Headspace solid phase microextraction in pesticide residues analysis: 2. Apple samples. *Pestic. Fitomed.* 22:173–176.
45. Blasco, C., Fernandez, M., Pico, Y., and Font, G. 2004. Comparison of solid-phase microextraction and stir bar sorptive extraction for determining six organophosphorus insecticides in honey by liquid chromatography–mass spectrometry. *J. Chromatogr. A* 1030:77–85.
46. Blasco, C., Font, G., Manes, J., and Pico, Y. 2003. Solid-phase microextraction liquid chromatography/tandem mass spectrometry to determine postharvest fungicides in fruits. *Anal. Chem.* 75:3606–3615.
47. Gilbert-Lopez, B., Garcia-Reyes, J. F., Lozano, A., Fernandez-Alba, A. R., and Molina-Diaz, A. 2010. Large-scale pesticide testing in olives by liquid chromatography–electrospray tandem mass spectrometry using two sample preparation methods based on matrix solid-phase dispersion and QuEChERS. *J. Chromatogr. A* 1217:6022–6035.
48. Silva, M. G. D., Aquino, A., Dorea, H. S., and Navickiene, S. 2008. Simultaneous determination of eight pesticide residues in coconut using MSPD and GC/MS. *Talanta* 76:680–684.
49. Pereira, L. A. and Rath, S. 2009. Molecularly imprinted solid-phase extraction for the determination of fenitrothion in tomatoes. *Anal. Bioanal. Chem.* 393:1063–1072.
50. Xin, J., Qiao, X., Ma, Y., and Xu, Z. 2012. Simultaneous separation and determination of eight organophosphorous pesticide residues in vegetables through molecularly imprinted solid-phase extraction coupled to gas chromatography. *J. Sep. Sci.* 35:3501–3508.

51. Wang, Y., Zhang, Q., Li, P., Zhang, W., Li, Y., and Ding, X. 2011. Selective sample cleanup by immunoaffinity chromatography for determination of fenvalerate in vegetables. *J. Chromatogr. B: Anal. Technol. Biomed. Life Sci.* 879:3531–3537.
52. Xu, Z., Fang, G., and Wang, S. 2009. Molecularly imprinted solid phase extraction coupled to high-performance liquid chromatography for determination of trace dichlorvos residues in vegetables. *Food Chem.* 119:845–850.
53. Qian, K., Fang, G., He, J., Pan, M., and Wang, S. 2010. Preparation and application of a molecularly imprinted polymer for the determination of trace metolcarb in food matrices by high performance liquid chromatography. *J. Sep. Sci.* 33:2079–2085.
54. Barahona, F., Turiel, E., Cormack, P. A. G., and Martin-Esteban, A. 2011. Synthesis of core-shell molecularly imprinted polymer microspheres by precipitation polymerization for the inline molecularly imprinted solid-phase extraction of thiabendazole from citrus fruits and orange juice samples. *J. Sep. Sci.* 34:217–224.
55. Viglino, L., Aboulfadl, K., Mahvelat, A. D., Prevost, M., and Sauve, S. 2008. On-line solid phase extraction and liquid chromatography/tandem mass spectrometry to quantify pharmaceuticals, pesticides and some metabolites in wastewaters, drinking, and surface waters. *J. Environ. Monit.* 10:482–489.
56. Garcia-Ac, A., Segura, P. A., Viglino, L., Fuertoes, A., Gagnon, C., Prevost, M., and Sauve, S. 2009. On-line solid-phase extraction of large-volume injections coupled to liquid chromatography–tandem mass spectrometry for the quantitation and confirmation of 14 selected trace organic contaminants in drinking and surface water. *J. Chromatogr. A* 1216:8518–8527.
57. Riediker, S., Obrist, H., Varga, N., and Stadler, R. H. 2002. Determination of chlormequat and mepiquat in pear, tomato, and wheat flour using on-line solid-phase extraction (Prospekt) coupled with liquid chromatography–electrospray ionization tandem mass spectrometry. *J. Chromatogr. A* 966:15–23.
58. Koesukwiwat, U., Lehotay, S. J., Miao, S., and Leepipatpiboon, N. 2010. High throughput analysis of 150 pesticides in fruits and vegetables using QuEChERS and low-pressure gas chromatography-time-of-flight mass spectrometry. *J. Chromatogr. A* 1217:6692–6703.
59. Kmellar, B., Abranko, L., Fodor, P., and Lehotay, S. J. 2010. Routine approach to qualitatively screening 300 pesticides and quantification of those frequently detected in fruit and vegetables using liquid chromatography tandem mass spectrometry (LC-MS/MS). *Food Addit. Contam.* 27:1415–1430.
60. Lehotay, S. J., Mastovska, K., and Yun, S. J. 2005. Evaluation of two fast and easy methods for pesticide residue analysis in fatty food matrixes. *J. AOAC Int.* 88:630–638.
61. Koesukwiwat, U., Lehotay, S. J., Mastovska, K., Dorweiler, K. J., and Leepipatpiboon, N. 2010. Extension of the QuEChERS method for pesticide residues in cereals to flaxseeds, peanuts, and doughs. *J. Agric. Food Chem.* 58:5950–5958.
62. Garcia-Reyes, J. F., Ferrer, C., Gomez-Ramos, M. J., Fernandez-Alba, A. R., Garcia-Reyes, J. F., and Molina-Diaz, A. 2007. Determination of pesticide residues in olive oil and olives. *TrAC, Trends Anal. Chem.* 26:239–251.
63. Jeong, I. S., Kwak, B. M., Ahn, J. H., and Jeong, S. H. 2012. Determination of pesticide residues in milk using a QuEChERS-based method developed by response surface methodology. *Food Chem.* 133:473–481.
64. Fields, S. M. 1996. Silica xerogel as a continuous column support for high-performance liquid chromatography. *Anal. Chem.* 68:2709–2712.
65. Minakuchi, H., Nakanishi, K., Soga, N., Ishizuka, N., and Tanaka, N. 1996. Octadecylsilylated porous silica rods as separation media for reversed-phase liquid chromatography. *Anal. Chem.* 68:3498–3501.
66. Laemmerhofer, M., Svec, F., Frechet, J. M. J., and Lindner, W. 2000. Chiral monolithic columns for enantioselective capillary electrochromatography prepared by

copolymerization of a monomer with quinidine functionality. 2. Effect of chromatographic conditions on the chiral separations. *Anal. Chem.* 72:4623–4628.

67. Premstaller, A., Oberacher, H., Walcher, W., Timperio, A. M., Zolla, L., Chervet, J. P., Cavusoglu, N., van Dorsselaer, A., and Huber, C. G. 2001. High-performance liquid chromatography-electrospray ionization mass spectrometry using monolithic capillary columns for proteomic studies. *Anal. Chem.* 73:2390–2396.
68. De Villiers, A., Lestremau, F., Szucs, R., Gelebart, S., David, F., and Sandra, P. 2006. Evaluation of ultra performance liquid chromatography. Par I. Possibilities and limitations. *J. Chromatogr. A* 1127:60–69.
69. Plumb, R. S., Rainville, P., Smith, B. W., Johnson, K. A., Castro-Perez, J., Wilson, I. D., and Nicholson, J. K. 2006. Generation of ultrahigh peak capacity LC separations via elevated temperatures and high linear mobile-phase velocities. *Anal. Chem.* 78:7278–7283.
70. MacNair, J. E., Lewis, K. C., and Jorgenson, J. W. 1997. Ultrahigh-pressure reversed-phase liquid chromatography in packed capillary columns. *Anal. Chem.* 69:983–989.
71. Wang, Y., Ai, F., Ng, S. C., and Tan, T. T. Y. 2012. Sub-2 μm porous silica materials for enhanced separation performance in liquid chromatography. *J. Chromatogr. A* 1228:99–109.
72. Wang, X., Barber, W. E., and Long, W. J. 2012. Applications of superficially porous particles: High speed, high efficiency or both? *J. Chromatogr. A* 1228:72–88.
73. Fekete, S., Olah, E., and Fekete, J. 2012. Fast liquid chromatography: The domination of core-shell and very fine particles. *J. Chromatogr. A* 1228:57–71.
74. Diaz, R., Ibanez, M., Sancho, J. V., and Hernandez, F. 2013. Qualitative validation of a liquid chromatography–quadrupole–time of flight mass spectrometry screening method for organic pollutants in waters. *J. Chromatogr. A* 1276:47–57.
75. Bakirci, G. T. and Hisil, Y. 2012. Fast and simple extraction of pesticide residues in selected fruits and vegetables using tetrafluoroethane and toluene followed by ultrahigh-performance liquid chromatography/tandem mass spectrometry. *Food Chem.* 135:1901–1913.
76. Gao, F., Zhao, Y., Shao, B., and Zhang, J. 2012. Determination of residues of pesticides and veterinary drugs in milk by ultra performance liquid chromatography coupled with quadrupole-time of flight mass spectrometry. *Se Pu* 30:560–567.
77. Chen, G., Cao, P., and Liu, R. 2011. A multi-residue method for fast determination of pesticides in tea by ultra performance liquid chromatography–electrospray tandem mass spectrometry combined with modified QuEChERS sample preparation procedure. *Food Chem.* 125:1406–1411.
78. Queiroz, S. C. N., Ferracini, V. L., and Rosa, M. A. 2012. Multiresidue method validation for determination of pesticides in food using QuEChERS and UPLC-MS/MS. *Quim. Nova* 35:185–192.
79. Kong, Z., Dong, F., Xu, J., Liu, X., Zhang, C., Li, J., Li, Y., Chen, X., Shan, W., and Zheng, Y. 2011. Determination of difenoconazole residue in tomato during home canning by UPLC-MS/MS. *Food Control* 23:542–546.
80. Yang, R. Z, Wang, J. H., Wang, M. l., Zhang, R., Lu, X. Y., and Liu, W. H. 2011. Dispersive solid-phase extraction cleanup combined with accelerated solvent extraction for the determination of carbamate pesticide residues in radix glycyrrhizae samples by UPLC-MS-MS. *J. Chromatogr. Sci.* 49:702–708.
81. Lin, Q. B., Xue, Y. Y., and Song, H. 2010. Determination of the residues of 18 carbamate pesticides in chestnut and pine nut by GPC cleanup and UPLC-MS-MS. *J. Chromatogr. Sci.* 48:7–11.
82. Liu, X. G., Xu, J., Dong, F. S., Li, Y. B., Song, W. C., and Zheng, Y. Q. 2011. Residue analysis of four diacylhydrazine insecticides in fruits and vegetables by Quick, Easy, Cheap, Effective, Rugged, and Safe (QuEChERS) method using ultra-performance liquid chromatography coupled to tandem mass spectrometry. *Anal. Bioanal. Chem.* 401:1051–1058.

83. Xu, J., Dong, F., Liu, X., Li, J., Li, Y., Shan, W., and Zheng, Y. 2012. Determination of sulfoxaflor residues in vegetables, fruits and soil using ultra-performance liquid chromatography/tandem mass spectrometry. *Anal. Methods* 4:4019–4024.
84. Mol, H. G. J., Zomer, P., and de Koning, M. 2012. Qualitative aspects and validation of a screening method for pesticides in vegetables and fruits based on liquid chromatography coupled to full scan high resolution (Orbitrap) mass spectrometry. *Anal. Bioanal. Chem.* 403:2891–2908.
85. Hayward, D. G., Wong, J. W., Zhang, K., Chang, J., Shi, F., Banerjee, K., and Yang, P. 2011. Multiresidue pesticide analysis in ginseng and spinach by nontargeted and targeted screening procedures. *J. AOAC Int.* 94:1741–1751.
86. European Commission 2002. Commission Decision of 12 August 2002 implementing Council Directive 96/23/EC concerning the performance of analytical methods and the interpretation of results. European Commission, Brussels.
87. Chen, L., Song, F., Zheng, Z., Xing, J., Liu, Z., and Liu, S. 2012. Studies on the determination method of pesticide multi-residues in ginseng by ultra performance liquid chromatography tandem mass spectrometry. *Huaxue Xuebao* 70:843–851.
88. Chung, S. W. C. and Lam, C. H. 2012. Development and validation of a method for determination of residues of 15 pyrethroids and two metabolites of dithiocarbamates in foods by ultra-performance liquid chromatography–tandem mass spectrometry. *Anal. Bioanal. Chem.* 403:885–896.
89. Botitsi, H. V., Garbis, S. D., Economou, A., and Tsipi, D. F. 2011. Current mass spectrometry strategies for the analysis of pesticides and their metabolites in food and water matrices. *Mass Spectrom. Rev.* 30:907–939.
90. Zhang, X. Z., Luo, F. J., Liu, G. M., Lou, Z. Y., and Chen, Z. M. 2011. Determination of diafenthiuron and its metabolites in tea and soil by ultra performance liquid chromatography–electrospray ionization tandem mass spectrometry. *Fenxi Huaxue* 39:1329–1335.
91. SANCO/2009/10684. 2010. European Council No. Guidance document on method validation and quality control procedures for pesticide residues analysis in food and feed.
92. Van Eeenoo, P. and Delbeke, F. T. 2004. Criteria in chromatography and mass spectrometry—A comparison between regulations in the field of residue and doping analysis. *Chromatogr. Suppl.* 59:S39–S44.
93. Pozo, O. J., Sancho, J. V., Ibanez, M., Hernandez, F., and Niessen, W. M. A. 2006. Confirmation of organic micropollutants detected in environmental samples by liquid chromatography tandem mass spectrometry: Achievements and pitfalls. *TrAC, Trends Anal. Chem.* 25:1030–1042.
94. Kaufmann, A. and Butcher, P. 2006. Strategies to avoid false negative findings in residue analysis using liquid chromatography coupled to time-of-flight mass spectrometry. *Rapid Commun. Mass Spectrom.* 20:3566–3572.
95. Jiwan, J. L. H., Wallemacq, P., and Herent, M. F. 2011. HPLC-high resolution mass spectrometry in clinical laboratory? *Clin. Biochem.* 44:136–147.
96. Rogers, R. P., Schaub, T. M., and Marshall, A. G. 2005. MS returns to its roots. *Anal. Chem.* 77:A21-A27.
97. Thurman, E.M. and Ferrer, I. 2003. Comparison of quadrupole, time-of-flight, triple quadrupole, and ion-trap mass spectrometry for the analysis of emerging contaminants. In *Liquid Chromatography/Mass Spectrometry/Mass Spectrometry and Time-of-Flight Mass Spectrometry for the Analysis of Emerging Contaminants.* American Chemical Society Symposium Vol. 850, pp. 14–31.
98. Thurman, E. M., Ferrer, I., and Fernandez-Alba, A. R. 2005. Matching unknown empirical formulas to chemical structure using LC/MS TOF accurate mass and database searching: Example of unknown pesticides on tomato skin. *J. Chromatogr. A* 1067:127–134.
99. Thurman, E. M., Ferrer, I., Garcia-Reyes, J. F., Zweigenbaum, J., Woodman, M., and Fernandez-Alba, A. R. 2005. Discovering metabolites of post harvest fungicides in

citrus with liquid chromatography/time-of-flight mass spectrometry and ion trap tandem mass spectrometry. *J. Chromatogr. A* 1082:71–80.

100. Kellmann, M., Muenster, H., Zomer, P., and Mol, H. 2009. Full scan MS in comprehensive qualitative and quantitative residue analysis in food and feed matrices: How much resolving power is required? *J. Am. Soc. Mass Spectrom.* 20:1464–1476.
101. Ferrer, I., Thurman, E. M., and Fernandez-Alba, A. R. 2005. Quantitation and accurate mass analysis of pesticides in vegetables by LC/TOF-MS. *Anal. Chem.* 77:2818–2825.
102. Wang, J., Leung, D., and Chow, W. 2010. Applications of LC/ESI-MS/MS and UHPLC QqTOF MS for the determination of 148 pesticides in berries. *J. Agric. Food Chem.* 58:5904–5925.
103. Wang, J., Chow, W., and Leung, D. 2010. Applications of LC/ESI-MS/MS and UHPLC QqTOF MS for the determination of 148 pesticides in fruits and vegetables. *Anal. Bioanal. Chem.* 396:1513–1538.
104. Lacina, O., Urbanova, J., Poustka, J., and Hajslova, J. 2010. Identification/quantification of multiple pesticide residues in food plants by ultra-high-performance liquid chromatography–time-of-flight mass spectrometry. *J. Chromatogr. A* 1217:648–659.
105. Diaz, R., Ibanez, M., Sancho, J. V., and Hernandez, F. 2012. Target and non-target screening strategies for organic contaminants, residues and illicit substances in food, environmental and human biological samples by UHPLC-QTOF-MS. *Anal. Methods* 4:196–209.
106. Grimalt, S., Sancho, J. V., Pozo, O. J., and Hernandez, F. 2010. Quantification, confirmation and screening capability of UHPLC coupled to triple quadrupole and hybrid quadrupole time-of-flight mass spectrometry in pesticide residue analysis. *J. Mass Spectrom.* 45:421–436.
107. Filigenzi, M. S., Ehrke, N., Aston, L. S., and Poppenga, R. H. 2011. Evaluation of a rapid screening method for chemical contaminants of concern in four food-related matrices using QuEChERS extraction, UHPLC and high resolution mass spectrometry. *Food Addit. Contam., Part A* 28:1324–1339.
108. Wode, F., Reilich, C., van Baar, P., Duennbier, U., Jekel, M., and Reemtsma, T. 2012. Multiresidue analytical method for the simultaneous determination of 72 micropollutants in aqueous samples with ultra high performance liquid chromatography–high resolution mass spectrometry. *J. Chromatogr. A* 1270:118–126.
109. Gomez-Perez, M. L., Plaza-Bolanos, P., Romero-Gonzalez, R., Martinez-Vidal, J. L., and Garrido-Frenich, A. 2012. Comprehensive qualitative and quantitative determination of pesticides and veterinary drugs in honey using liquid chromatography–Orbitrap high resolution mass spectrometry. *J. Chromatogr. A* 1248:130–138.
110. Wang, J., Chow, W., Leung, D., and Chang, J. 2012. Application of ultrahigh-performance liquid chromatography and electrospray ionization quadrupole orbitrap high-resolution mass spectrometry for determination of 166 pesticides in fruits and vegetables. *J. Agric. Food Chem.* 60:12088–12104.
111. Wang, J., Chow, W., and Leung, D. 2011. Applications of LC/ESI-MS/MS and UHPLC/Qq-TOF-MS for the determination of 141 pesticides in tea. *J. AOAC Int.* 94:1685–1714.
112. Pesticide EU-MRLs Database. 2005. Regulation (EC) n. 396/2005. http://ec.europa.eu/sanco_pesticides/public/index.cfm
113. U.S. EPA Office of Pesticide Programs. What the Pesticide Residue Limits Are on Food. Available at http://www.epa.gov/pesticides/food/viewtools.html
114. Pérez-Parada, A., Colazzo, M., Besil, N., Dellacassa, E., Cesio, V., Heinzen, H., and Fernández-Alba, A.R. 2011. Pesticide residues in natural products with pharmaceutical use: occurrence, analytical advances and perspectives. In *Pesticides in the Modern World–Trends in Pesticide Analysis,* ed. L.M. Stoytcheva, InTech Publisher, Rijeka (CR), pp. 357–390. ISBN: 978-953-307-437-5.
115. European Food Safety Authority. 2011. The 2009 European Union Report of Pesticide Residues in Food. *EFSA Journal* 9(11):2430.

116. European Food Safety Authority. 2013. The 2010 European Union Report of Pesticide Residues in Food. *EFSA Journal* 11(3):3130.
117. Chai, M. K. and Tan, G. H. 2009. Validation of a headspace solid-phase microextraction procedure with gas chromatography-electron capture detection of pesticide residues in fruits and vegetables. *Food Chem.* 117:561–567.
118. Huskova, R., Matisova, E., Hrouzkova, S., and Svorc, L. 2009. Analysis of pesticide residues by fast gas chromatography in combination with negative chemical ionization mass spectrometry. *J. Chromatogr. A* 1216:6326–6334.
119. Arrebola, F. J., Martinez Vidal, J. L., Gonzalez-Rodriguez, M. J., Garrido-Frenich, A., and Sanchez Morito, N. 2003. Reduction of analysis time in gas chromatography. Application of low-pressure gas chromatography-tandem mass spectrometry to the determination of pesticide residues in vegetables. *J. Chromatogr. A* 1005:131–141.
120. Nguyen, T. D., Yu, J. E., Lee, D. M., and Lee, G. H. 2008. A multiresidue method for the determination of 107 pesticides in cabbage and radish using QuEChERS sample preparation method and gas chromatography mass spectrometry. *Food Chem.* 110:207–213.
121. Rial Otero, R., Cancho Grande, B., and Simal Gandara, J. 2003. Multiresidue method for fourteen fungicides in white grapes by liquid–liquid and solid-phase extraction followed by liquid chromatography–diode array detection. *J. Chromatogr. A* 992:121–131.
122. Venkateswarlu, P., Mohan, K. R., Kumar, C., and Seshaiah, K. 2007. Monitoring of multi-class pesticide residues in fresh grape samples using liquid chromatography with electrospray tandem mass spectrometry. *Food Chem.* 105:1760–1766.
123. Soler, C., Manes, J., and Pico, Y. 2005. Routine application using single quadrupole liquid chromatography–mass spectrometry to pesticides analysis in citrus fruits. *J. Chromatogr. A* 1088:224–233.
124. Fenoll, J., Hellin, P., Martinez, C. M., Miguel, M., and Flores, P. 2007. Multiresidue method for analysis of pesticides in pepper and tomato by gas chromatography with nitrogen-phosphorus detection. *Food Chem.* 105:711–719.
125. Zamora, T., Pozo, O. J., Lopez, F. J., and Hernandez, F. 2004. Determination of tridemorph and other fungicide residues in fruit samples by liquid chromatography–electrospray tandem mass spectrometry. *J. Chromatogr. A* 1045:137–143.
126. Huang, Z., Li, Y., Chen, B., and Yao, S. 2007. Simultaneous determination of 102 pesticide residues in Chinese teas by gas chromatography–mass spectrometry. *J. Chromatogr. B: Anal. Technol. Biomed. Life Sci.* 853:154–162.
127. Walorczyk, S., Drozdzynski, D., Kowalska, J., Remlein-Starosta, D., Ziolkowski, A., Przewozniak, M., and Gnusowski, B. 2013. Pesticide residues determination in Polish organic crops in 2007–2010 applying gas chromatography–tandem quadrupole mass spectrometry. *Food Chem.* 139:482–487.
128. Garcia de Llasera, M. P. and Reyes-Reyes, M. L. 2009. A validated matrix solid-phase dispersion method for the extraction of organophosphorus pesticides from bovine samples. *Food Chem.* 114:1510–1516.
129. Garrido, F. A., Martinez, V. J. L., Cruz, S. A. D., Gonzalez, R. M. J., and Plaza, B. P. 2006. Multiresidue analysis of organochlorine and organophosphorus pesticides in muscle of chicken, pork and lamb by gas chromatography-triple quadrupole mass spectrometry. *Anal. Chim. Acta* 558:42–52.
130. Bolaños, P. P., Frenich, A. G., and Vidal, J. L. M. 2007. Application of gas chromatography-triple quadrupole mass spectrometry in the quantification-confirmation of pesticides and polychlorinated biphenyls in eggs at trace levels. *J. Chromatogr. A* 1167:9–17.

3 Ultra-High-Performance Liquid Chromatography in Food Metabolomics
Food Quality and Authenticity

Cristina C. Jacob and H. Gallart-Ayala

CONTENTS

3.1 INTRODUCTION

Metabolomics is an emerging field in analytical chemistry and can be regarded as the end point of the "omics" cascade. In its most ambitious global form, metabolomics tries to comprehensively analyze known and unknown metabolites in a given biological sample—metabolome—dealing with the analysis of low molecular mass compounds (<1500 Da) such as amino acids, peptides, organic acids, vitamins, polyphenols alkaloids, nucleic acids, carbohydrates, and other small molecules [1].

Until recently, this approach has focused mainly on the study of human diseases, pharmaceuticals, plant metabolism analysis, and toxicology. However, nowadays it is being extended to food science for food component analysis, food quality and authenticity, and food consumption monitoring in order to improve consumer health and confidence. Because metabolomics allows the simultaneous characterization of large numbers of molecules, this approach has been recently introduced by food and nutrition scientists to obtain a comprehensive fingerprint of food and nutrition. In this context, Cifuentes [2] introduced the term "foodomics" as a discipline that studies food and nutrition domains through the application of advanced "omics" technologies.

Usually, three different analytical approaches can be defined in metabolomics studies: untargeted, class-specific nontargeted and targeted metabolomics. Identification and quantification of the compounds from the whole metabolome is defined as untargeted metabolomic profiling. The sample cleanup for untargeted approaches is not generally rigorous, since it may result in the loss of some low-molecular weight compounds. The analytical methodology employed should provide maximum coverage over classes of compounds while offering relatively high throughput. By this approach, large datasets are often obtained, making it impossible to treat the information manually. Therefore, several informatics tools and software have been developed to help with data processing. Generally, untargeted approaches involve the use of multivariate statistical data analysis—principal component analysis (PCA), partial least square (PLS), and orthogonal partial least square (OPLS)—to identify the relevant spectral features that distinguish sample classes (usually disease or treatment). In food science, this approach has been applied for comparisons between different samples of a specific food type, for example, to detect new or unexpected food contamination or to reveal time trends [3].

In contrast to untargeted metabolomics, which aims to separate and identify as many compounds as possible, class-specific nontargeted metabolomics is focused on a group of chemically related compounds, aiming to find previously unknown metabolites belonging to that class. By this approach, the sample preparation step may be optimized for the class of investigated compounds, resulting in both a preconcentration and reduced matrix effect.

Targeted metabolomics has the most modest goal of quantifying selected metabolites, most typically dozens to hundreds of known compounds. This requires the ability to differentiate the target analytes from other interfering compounds. Typically, targeted metabolomics is performed when target compounds have been previously established as putative biomarkers in a nontargeted or untargeted approach.

Metabolomics studies only became possible as a result of recent technology breakthroughs in small molecule separation and identification. Metabolite signatures in metabolomics research could be achieved using a range of analytical platforms, including nuclear magnetic resonance (NMR), gas chromatography (GC), high-performance liquid chromatography (HPLC), and ultra-high-performance liquid chromatography (UHPLC) hyphenated with mass spectrometry (MS).

NMR-based metabolomics has been successfully applied in food science [4]. However, one of the main limitations of this technique is its sensitivity in the study of mixtures, especially when it is compared with the higher MS sensitivity and dynamic range. GC coupled with MS provides efficient and reproducible analysis and has also been used for food metabolomics studies [5]. The major issue, however, is that GC is by its nature limited to the analysis of small volatile molecules and molecules that can be made volatile.

The use of LC with conventional LC columns (5 μm) is a common and well-established technique in food analysis. However, the limitations of the use of this type of column do not solve all analytical problems. This fact is especially critical in food metabolomics because of the complexity of food matrices and the high number of compounds of interest. Therefore, the use of ultra-high-pressure liquid chromatography (UHPLC) can overcome the limitations of conventional LC, providing high-efficiency and high-resolution chromatographic separations in combination with an analysis time reduction. In order to improve analysis confidence, UHPLC is generally coupled to MS because of its sensitivity, selectivity, and high dynamic range. Different mass analyzers, depending on the goal, such as triple quadrupoles (QqQ), ion trap (IT), time of flight (TOF), and Orbitrap, are generally used.

This chapter will focus on the use of UHPLC techniques in food metabolomics. It includes a selection of the most relevant works recently published. A discussion regarding the use of different stationary phases, advantages, and drawbacks of the technique will be addressed. Moreover, some applications of UHPLC in food metabolomics will also be discussed.

3.2 UHPLC IN FOOD METABOLOMICS

The recent introduction of UHPLC using sub-2 μm porous particles size results in higher peak capacity, improved resolution, and increased sensitivity compared with conventional HPLC columns, therefore making it suitable in several applications such as food and environmentals, pharmaceuticals, bio-analysis, and metabolomics (Figure 3.1) [6]. When analyzing complex mixtures with LC–MS, as is often the case in metabolomics investigations, UHPLC can be of great advantage over conventional HPLC columns in that more components can be detected. Indeed, in an untargeted approach, it has been demonstrated that UHPLC offers higher retention time reproducibility, similar response reproducibility, more detected features, better separation as a result of the narrow peaks produced, and better signal-to-noise when compared to HPLC [7]. In addition, shorter run analysis without loss of resolution is another advantage of UHPLC over HPLC, permitting for an increase in the number of samples to analyze in one batch.

UHPLC has been used recently in food metabolomics. Table 3.1 summarizes recent publications of UHPLC separation mode applied in this field. Examples of different applications such as food quality and authenticity and food component

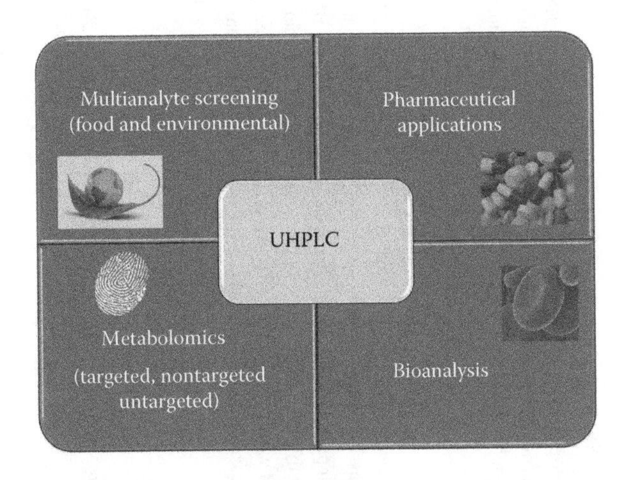

FIGURE 3.1 Main fields of applications of UHPLC.

TABLE 3.1
UHPLC in Food Metabolomics

Case Study	Matrix	Liquid Chromatography		Detection	Reference
		Column	Mobile Phase and Flow Rate		
Multiple class phenolic compounds	Fruits and beverages	Acquity HSS T3 (100 mm × 2.1 mm ID, 1.8 µm)	ACN:Water both with 0.1% HCOOH Flow rate : 400 µL/min	QqQ mass spectrometer SRM scan mode	[8]
Phytohormonal	Tomato plant	Nucleodur Gravity C18 (50 mm × 2.1 mm ID, 1.8 µm)	MeOH:Water + 0.1% HCOOH Flow rate: 300 µL/min T: 30°C	Orbitrap HRMS Full scan at a mass resolving power of 100,000 FWHM	[9]
Red wine classification	Red wine	Kinetex C18 (150 mm × 4.6 mm ID, 2.6 µm)	ACN:Water + 0.1% HCOOH Flow rate: 1.3 mL/min	UV, fluorescence, and QqQ Full scan MS	[10]
Comprehensive analysis	Green tea	Acquity BEH C18 (150 mm × 2.1 mm ID, 1.7 µm)	ACN :Water both with 0.1% HCOOH Flow rate : 300 µL/min	TOF HRMS Full scan	[11]
Red wine classification	Red wine	Zorbax eclipse Plus C18 (100 mm × 2.1 mm ID, 1.8 µm)	MeOH :5 mM Ammonium formate pH 3.8 for ESI + MeOH :5 mM Ammonium acetate pH 4.5 for ESI- Flow rate: 300 µL/min T: 40°C	Q-TOF HRMS Full scan mode at mass resolving power 12,000 FWHM	[12]
Sterol profile	Virgin olive oil	Acquity BEH C18 (50 mm × 2.1 mm ID, 1.7 µm)	ACN:Water both with 0.1% CH_3COOH Flow rate: 800 µL/min T: 10°C	Single quadrupole (Q) SIM mode	[13]
Metabolomic	Ginseng	Acquity BEH C18 (100 mm × 2.1 mm ID, 1.7 µm)	ACN:Water both with 0.1% HCCOH Flow rate: 500 µL/min T: 35°C	Q-TOF HRMS Full-scan mode	[14]
Phenolic profile	Buckwheat	Acquity BEH C18 (50 mm × 2.1 mm ID, 1.7 µm)	ACN:Water both with 0.1% TFA Flow rate: 300 µL/min	Q-TOF HRMS Full scan and MS/MS	[15]

Metabolomics	Wine	HSS T3 (100 mm × 2.1 mm ID, 1.8 μm)	ACN :Water both with 0.1% HCOOH Flow rate : 400 μL/min T: 40°C	LTQ FT-ICR-ultra full-scan HRMS	[16]
Metabolomics	Avocado fruit	Acclaim RSLC 120 C18 (100 mm × 2.1 mm ID, 1.8 μm)	ACN :Water both with 0.1% HCOOH Flow rate : 600 μL/min T: 50°C	TOF Full-scan HRMS	[17]
Veterinary drugs	Honey-based products	TFC : Cyclone P (50 × 0.5 mm, 60 μm) Analytical column: Hypersil Gold aQ C18 (100 mm × 2.1 mm ID, 1.7 μm)	MeOH:water both with 4 mM ammonium formate buffer + 0.1% HCOOH Flow rate: 300 μL/min	Exactive Orbitrap Full-scan HRMS and all ions fragmentation HCD	[18]
Metabolite profile	Brazilian *Lippia* species	Acquity BEH C18 (150 mm × 2.1 mm ID, 1.7 μm)	ACN:Water both with 0.1% HCOOH Flow rate: 600 μL/min T: 60°C	TOF HRMS Full scan	[19]
Metabolic profile	Tea	RRHD SB-C18 (150 mm × 2.1 mm ID, 1.8 μm)	ACN:Water both with 0.1% HCCOH Flow rate: 400 μL/min T: 25°C	Exactive Orbitrap Full-scan HRMS	[20]
Phenolic profile and other polar compounds	Watermelon	RP C18 (150 mm × 4.6 mm ID, 1.8 μm)	ACN:Water with 0.5% CH_3COOH Flow rate: 800 μL/min T: 25°C	QTOF HRMS Full scan and MS/MS	[21]
Pigment profile	Wine	Acquity BEH C18 (150 mm × 2.1 mm ID, 1.7 μm)	MeOH:Water both with 5% HCOOH Flow rate: 400 μL/mi T: 40°C	QqQ SRM acquisition mode	[22]
Phenolic profile	Olive oil	RP C18 (150 mm × 4.6 mm ID, 1.8 μm)	MeOH:Water with 0.25 CH_3COOH Flow rate: 800 μL/min	TOF HRMS Full scan	[23]
Phenolic profile	Fruit skin of *Prunus domestica* plums	Kinetex PFP (150 mm × 4.6 mm ID, 2.6 μm)	MeOH:Water with 5% HCOOH Flow rate: 500 μL/min	UV detection	[24]
Contaminants	Orange juice	Acclaim RSLC C18 (100 mm × 2.1 mm ID, 2 μm)	MeOH:5 mM Ammonium Formate + 0.02% HCOOH Flow rate: Gradient from 200 to 450 μL/min	QTOF HRMS Full scan	[3]

analysis are included [3,8–24]. Information regarding the metabolomic approach, matrix analyzed, UHPLC column, mobile phase composition, and detection system are included and discussed.

3.2.1 LC System: Modern Approaches

The use of monoliths columns, LC at high temperatures, and LC at high pressures using sub-2 μm particle-packed columns have enabled the reduction of analytical time without compromising the resolution and the separation efficiency [25,26]. Although monolith columns enable the use of high flow rates without generating back pressure, and, in addition, efficiencies are generally equivalent to those measured with traditional HPLC columns, this technology still suffers from a restricted number of column chemistries and geometries [27]. At temperatures higher than 80°C (high-temperature LC) the mobile phase viscosity decreases, allowing the application of high flow rates with limited back pressure. However, the number of stationary phases stable at elevated temperatures is limited. The use of small particles in ultra-high-pressure systems presents another option in performing rapid analyses. Indeed, as already demonstrated, the use of sub-2 μm particles induces an increase in efficiency, optimal velocity, and improvements in mass transfer. Therefore, efficient separations can be performed with shorter analysis times when sub-2 μm particles are used. However, small particles induce a high-pressure drop, and according to Darcy's law, the pressure drop is inversely proportional to particle size at the optimum linear velocity [25]. However, improvements of the chromatographic system have been made to address this inconvenience, and instrumentation that can withstand pressure beyond 400 bar has been commercialized, allowing for full profit from the advantages in using sub-2 μm particle-packed columns.

Quality and stability of the packed stationary phases are essential for performing efficient separation at ultra-high pressures. Developments in stationary phases have led to high chemical (a wide range of pH) and mechanical stability (high back pressures) and also to the introduction of new selectivity chemistries [28]. Significant developments in UHPLC technology were recently observed, and there is currently a wide variety of stationary phases packed with sub-2 μm particles (Table 3.2). Both hybrid and silica materials were used as stationary phase supports and they were in addition modified by various chemistries, including C8, C18, Phenyl, Cyano, Shield, and others polarities. These different stationary phases generally provide complementary chromatographic separations.

Although there are a huge number of stationary phases available for UHPLC, the choice of the UHPLC column I.D. is more restricted compared to conventional LC because both frictional heating effects under very high pressures and solvent consumption should be considered. Indeed, frictional heating phenomenon can be observed when using columns packed with sub-2 μm particles at elevated mobile phase linear velocity, thus generating very high-pressure drop. This high pressure is induced by friction of the mobile phase percolating through the stationary phase, generating heat that can negatively affect the chromatographic separation. In order to improve this inconvenient, 2.1 mm ID columns are the reference dimension for UHPLC operations, and experiments are generally performed at flow rates between

TABLE 3.2
Currently Commercial Available sub-2 μm Analytical Columns and Their Properties

Column Name	Stationary Phase Support	Particle Size (μm)	Limitations			Manufacturer
			pH	Column Chemistry	Temperature	
Acquity BEH	Hybrid	1.7	1–12	C8, C18, Phenyl	20–90	Waters
			2–11	Shield		
			1–8	Silica for HILIC		
			2–11	Amide, Glycan		
Acquity HSS	Silica	1.8	2–8	T3, C18 SB	20–45	Waters
			1–8	C18		
Altima HP	Hybrid	1.5	1–10	HILIC	20–60	Alltech
Platinum	Silica	1.5	2–8	C8, C18	20–60	Alltech
GP series	Silica	1.8	2–8.5	C8, C18, C4	20–60	Sepax
HP series	Silica	1.8	2–8.5	PHE, CN, NH2, SCX, SAX, silica, HILIC		Sepax
Poly-RP	Hybrid (PS/DVB)	1.0/1.7	1–14	PHE	20–60	Sepax
HypersilGold	Silica	1.9	1–11	C18	25–60	Thermo Electron
			2–9	C8, Q		
			2–8	PFP		
Nucleodur	Silica	1.8	1–11	C8, C18	Up to 85	Machery Nagel
			1–10	C18 isis, Sphinx RP		
			1–9	C18 pyramid		
Pathfinder	Hybrid	1.5	1–12	AS, AP, PS, MR	Up to 250	Shimadzu

continued

TABLE 3.2 (continued)
Currently Commercial Available sub-2 μm Analytical Columns and Their Properties

Column Name	Stationary Phase Support	Particle Size (μm)	Limitations			Manufacturer
			pH	Column Chemistry	Temperature	
Pinnacle DB	Silica	1.9	2.5–7.5	C18, PFP-propyl, silica, acqueous C18, CN, C8, PAH, X3-C18	Up to 80	Restek
Pronto		1.8 total porous	2–8	C18, C8, PHE	20–60	Bischoff
Pearl		1.5 nonporous		C18		
TSKgel Super ODS	Silica	2.0	2–7.5	C18, C8, PHE	20–60	Tosoh
YMC ultra-fast	Silica	2.0	2–8	C18, Hydro C 18	20–60	YMC
Zorbax	Silica	1.8	2–9	Eclipse plus C8, C18, Eclipse XDB-C18,	Up to 60	Agilent
			2–8	C8, PHE	80–100	
			1–6	Eclipse phenyl-hexyl, PAH, Elipse	StableBond	
			2–11.6	XDB-CN		
				StableBond		
				Extend-C18		

Source: Adapted from Novakova, L. and Vlckova, H. 2009. *Anal. Chim. Acta* 656:8–35. With permission.

500 and 1000 μL/min, depending on the acceptable efficiency loss and the size of the investigated compounds [28,29]. Although better peak shapes are generally obtained using sub-2 μm particle columns, the use of high flow rates (>600 μL/min) resulted in less detected compounds due to matrix effects and chromatographic coelutions being not appropriated in untargeted metabolomics approaches. For this reason, in some publications the use of lower flow rates (300–400 μL/min) using sub-2 μm columns are preferred (Table 3.1).

The narrow peaks produced by UHPLC require a small detection volume and fast acquisition rate to ensure high efficiency. Most commercial UHPLC instruments are equipped with a modified UV detector, which ensures the optimal peak capture. In order to minimize the extra-column volume, the flow cell volume is usually much lower than for traditional HPLC. However, smaller flow cells would reduce the path-length upon which the signal depends. Moreover, a reduction in cross-section means the light path is reduced conducting to a reduced transmission and increased noise. Therefore, light-guided flow cells may be used in order to maintain the path length. Fast detector time constant (<0.1 s) and a high data-acquisition rate must be achieved by the detector software in order to ensure enough data points for a narrow peak [28]. The same concept is applied for MS detectors, where low dwell times and low inter-channel and inter-scan delays are required. Other detectors, including fluorescence detection (FD) and evaporative light scattering detector (ELSD), are also currently available for coupling with UHPLC.

In food metabolomics, UHPLC technology has been applied in food authentication and food quality according to recent publications reported in the literature (Table 3.1). For instance, Hurtado-Fernandez et al. [17] have developed an untargeted metabolomics UHPLC method using a sub-2 μm particle column for the analysis of different varieties of avocado at two different ripening degrees. In this study, different gradient elution and the influence of flow rate was carefully studied. The best results were a compromise between the chromatographic separation and analysis time. Therefore, a mobile phase composition of ACN:water, both solvents with 0.1% of formic acid at a flow rate of 600 μL/min was used. However, in order to decrease the column back-pressure, column temperature was set at 50°C. Under these conditions a good LC profile for the analysis of avocado was obtained (Figure 3.2). Moreover, the use of unsupervised multivariate analysis method—PCA—showed that ripening degree was the most important source of data variation (Figure 3.3a), while a supervised multivariate PLS model (Figure 3.3b) was created in order to obtain the tentative biomarkers.

The use of fast UHPLC chromatography using sub-2 μm particle size column at high flow rates is only applied in a targeted food metabolomics approach when a limited number of compounds are monitored. This is the case of the method proposed by Lerma-Garcia et al. [13] for the sterol profile of olive oil. In this work the authors proposed a fast UHPLC separation (analysis time <5 min) for the analysis of 14 sterols in olive oil. This method was applied to classify extra virgin olive oils according to the genetic variety. Although the limited number of compounds analyzed and the low selectivity of the detection method applied, good discrimination between different genetic varieties of extra virgin olive oil was demonstrated.

Recently, another type of stationary phase known as "fused-core" or "core-shell" technology came onto the market in 2007 and it provides similar performance to

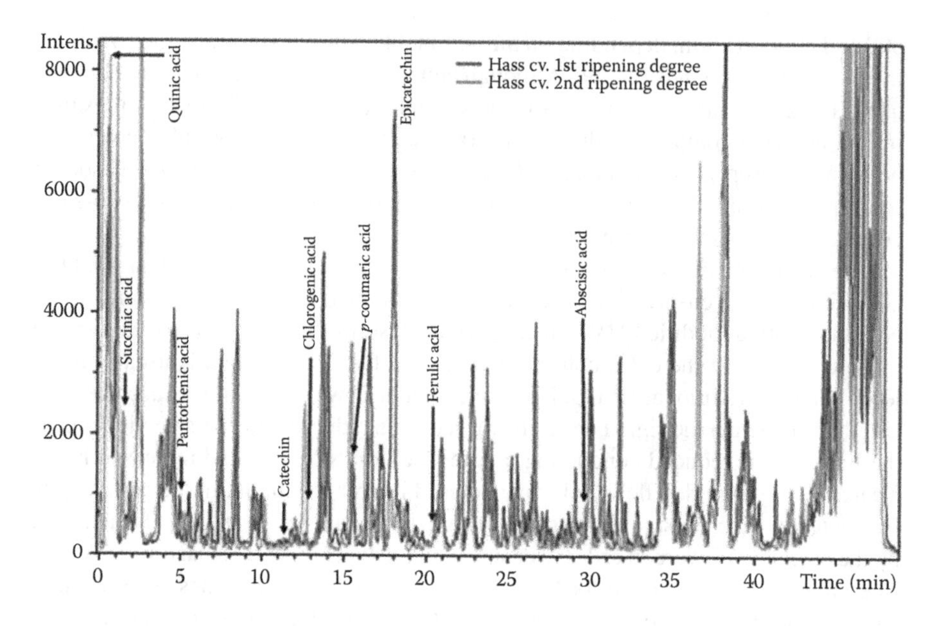

FIGURE 3.2 UHPLC–MS profile of avocado methanolic extract achieved by a sub-2 μm particle size column. (Adapted from Hurtado-Fernandez, E. et al. 2011. *J. Chromatogr. A* 1218:7723–7738. With permission.)

UHPLC with sub-2 μm particles but with a 2–3 times lower pressure [27], allowing the use of conventional HPLC systems, although to obtain good chromatographic efficiencies it is important to consider the fluidic energy of the LC system. This technology consists of columns packed with 2.6–2.7 μm superficially porous particles composed of a 1.7–1.9 μm solid inner core and a 0.35–0.5 μm porous outer core. Nowadays, this type of column technology is the main competitor of sub-2 μm columns [30].

Moreover, the use of 4.6 mm internal diameter columns instead of 2.1 mm in combination with the use of "core shell" particles permits to increase the flow rate without an increase in the back pressure. This type of column was selected by Serrano-Lourido et al. [10] for the analysis and classification of red wines from different Spanish appellations of origin. The use of LC–UV–vis and LC-fluorescence signals obtained in combination with chemometric analysis by PCA and PLS-DA was successfully applied for Spanish wine classification.

3.2.2 UHPLC Coupled to Mass Spectrometry

The LC–MS system can be improved on two levels. Improvements in the LC part are beneficial for throughput and/or resolution, as previously mentioned, while MS detection improvements have a positive impact on sensitivity and selectivity. Significant progress has been made in MS during the last decade, which has provided more sensitive, robust, user-friendly, and faster (fast duty cycle) systems. However, the coupling UHPLC–MS leads to another issue, which is related to the

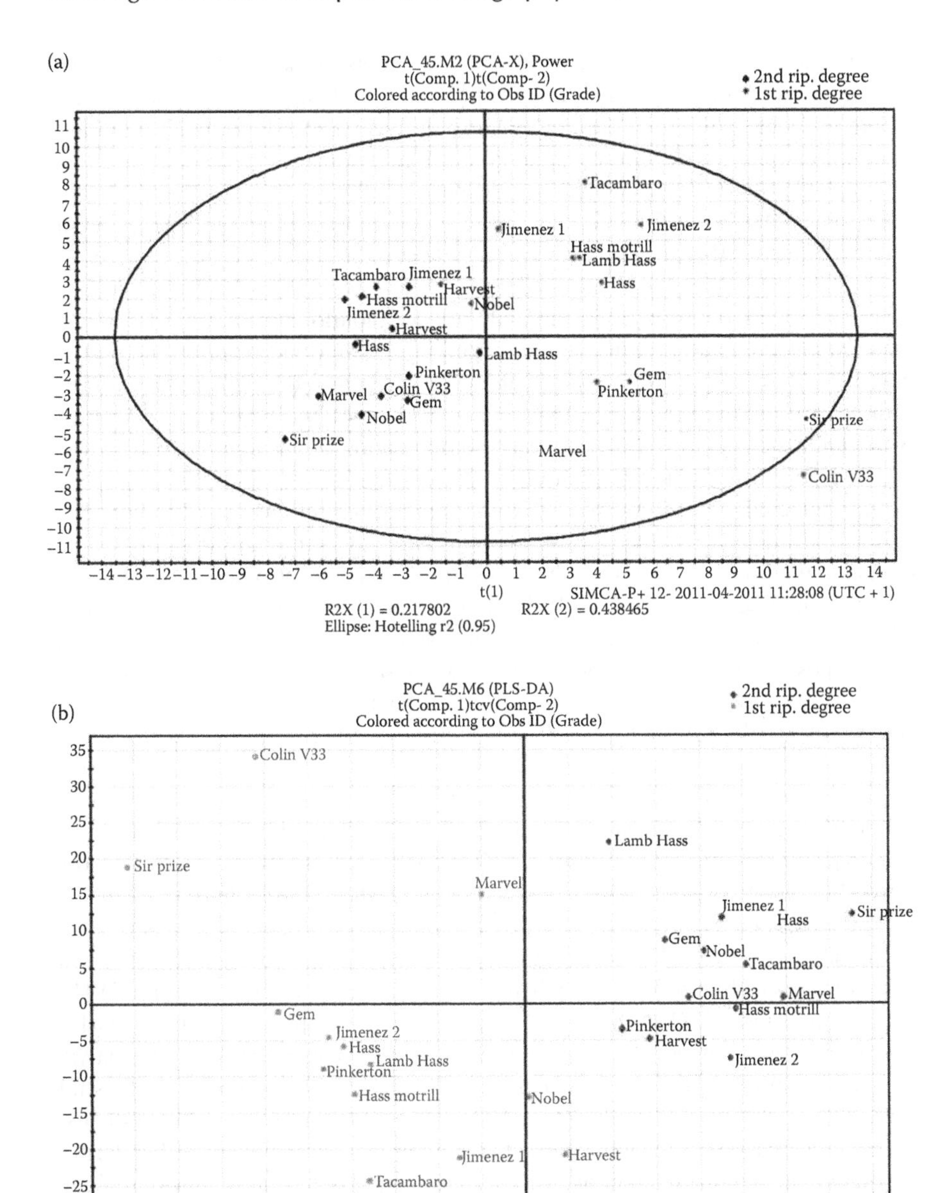

FIGURE 3.3 (a) PCA and (b) PLS modeling of UHPLC–TOF–MS for the analysis of avocado ripening. (Adapted from Hurtado-Fernandez, E. et al. *J. Chromatogr. A* 1218:7723–7738. With permission.)

narrow peaks that are generated when using columns, packed with sub-2 μm particles at elevated pressures. In HPLC, the peak widths at baseline are approximately 10 s on average, while they are reduced to only 2–4 s in UHPLC under a gradient elution. For quantitative purposes, at least 10–15 acquisition points per peak are recommended to correctly define the chromatographic peak and achieve suitable performance. However, this criterion may be difficult to meet and it depends on the duty cycle (acquisition speed) of the employed device.

In regular food analysis, the analytes of interest are selected before making measurements. Commonly, the methods aim to monitor ppb levels of the chosen compounds, and QqQ instruments operating in selected reaction monitoring (SRM) mode are the most common analyzers used in food analysis. Despite its high sensitivity and selectivity, the use of SRM mode in QqQ is limited by the cycle time when dealing with hundreds of compounds, and a significant drawback to this type of analyzer is that those molecules that are not initially anticipated are not detected, regardless of how high their concentrations may be. For these reasons, nowadays to solve the problems related to both cycle time and target screening method, LC coupled to high-resolution mass spectrometry (LC–HRMS) is being implemented in food analysis. In untargeted metabolomics studies, where full-scan MS spectra and accurate mass measurements are needed for identification purposes, the use of high-resolution mass spectrometers such as times-of-flight (TOF), Orbitrap, and ion cyclotron Fourier transform resonance (ICR-FT/MS) are especially useful.

ICR-FT/MS is the most powerful commercially available instrument with regard to mass accuracy and resolution; however, TOF- and Orbitrap-based technologies are currently the most common analyzers used in LC–HRMS. Indeed, the long scanning time required for ultra-high-resolution ICR-FT-MS (mass resolving power >750,000 FWHM at $m/z = 400$) contradicts coupling to LC, when at least 10 data points per peak are required to minimally define a chromatographic peak [31]. Nevertheless, the possibility to operate different mass-resolving powers with this kind of instrument allows coupling to UHPLC when a mass-resolving power of 50,000 FWHM was selected. Indeed, Cuadros-Inostroza et al. [16] developed a UHPLC-LTQ-FT-ICR method operating at a mass resolving power of 50,000 FWHM for the analysis and classification of Chilean red wines. Figure 3.4 shows the flowchart of steps followed in this work as well as the LC–HRMS chromatograms obtained in both positive and negative ESI. The analysis of wine samples was performed in both polarities in order to increase the number of analyzed substances. In fact, as is shown in Figure 3.4d the same number of *m/z* signals were obtained in both polarities, demonstrating their complementarities.

In the untargeted food metabolomics approach, the coupling of UHPLC–HRMS using a TOF analyzer has been preferentially used, mainly because TOF–MS is currently one of the best-suited approaches in terms of speed scan to be coupled with UHPLC systems (Table 3.1). However, TOF analyzers fail when accurate mass measurements and elemental composition assignments are essential for the characterization of small molecules. On the other hand, UHPLC coupled to the Orbitrap-based technology has also recently seen a growing number of applications. The

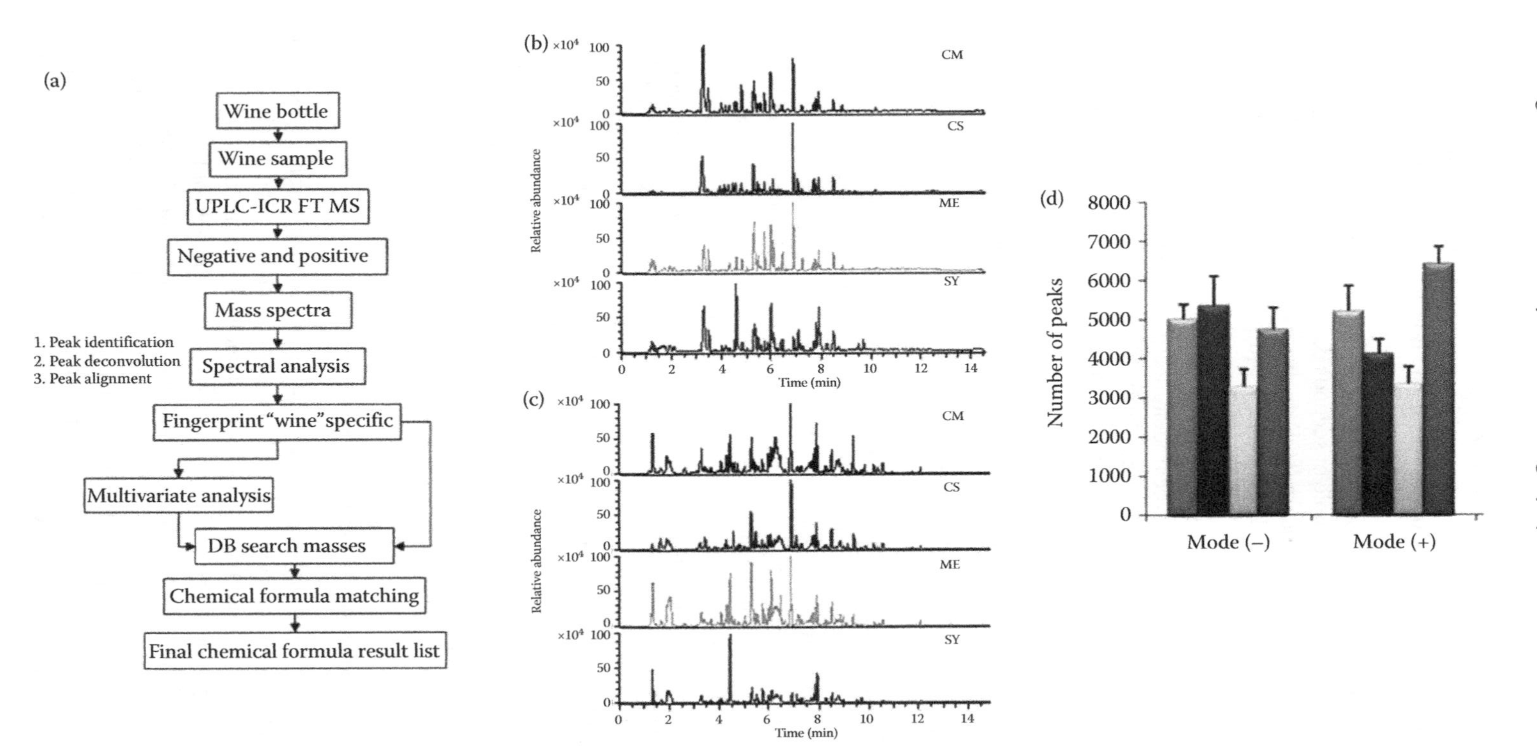

FIGURE 3.4 Metabolomic profiling of wine samples. (a) Flowchart of the steps involved in metabolomic profiling of wine samples. (b) Negative and (c) positive base peak chromatograms representing the wine cultivars carmenère (CM), cabernet sauvignon (CS), merlot (ME), and syrah (SY). (d) Number of ion traces between 100 and 1000 *m/z* in both negative [mode (–)] and positive [(mode +)] mode. (Adapted from Cuadros-Inostroza, A. et al. 2010. *Anal. Chem.* 82:3573–3580. With permission.)

most recent generation of Orbitrap analyzer (Exactive Plus and Q Exactive) is remarkably fast and able to operate at acquisition rates of 12 Hz. Nevertheless, this speed is to the detriment of performance (resolving power of 17,500 FWHM at *m/z* 200 at this speed) compared to the maximal resolution that can be theoretically achieved (>140,000 FWHM at 1 Hz), therefore the full potential of both platforms is not exploited (peaks that last a couple of seconds do not offer the finest conditions for high MS). For this reason, modern TOF analyzers offer a comparable performance at fast acquisition speeds, lower space requirements, and more favorable prices [29,32].

Nevertheless, sometimes the use of full-scan MS operating at HRMS in both polarities is not sufficient for unequivocal identification purposes. For instance, for a relative high molecular weight compounds (400–1000 Da), even operating at mass resolving power of 50,000 FWHM (*m/z* 200), using a mass error window of 5 ppm, and considering an elemental composition composed by $C \leq 200$, $H \leq 400$, $O \leq 60$, $N \leq 5$, $P \leq 2$, and $Na \leq 1$ more than 50 different elemental compositions may be possible to attribute to the masses measured [33]. Therefore, other acquisition modes such as the use of MS/MS are recommended for identification purposes. This fact makes attractive the use of hybrid mass spectrometers such as Q-TOF, ion-trap–time-of-flight (IT–TOF), linear ion-trap quadrupole, Orbitrap (LTQ-Orbitrap), or quadrupole-Orbitrap (Q-Exactive) to be use in food metabolomics. Generally, as it is observed in Table 3.1, Q-TOF instruments are the preferred ones. In this case, two different experiments are generally conducted, a first one in full-scan mode and a second one in MS/MS mode selecting as precursor ions the ones identified as tentative biomarkers for identification purposes. This methodology was successfully employed by Abu-Reidah et al. [21] for the phenolic profile and other polar compounds in watermelon. In this study, a total of 71 phenolic and other polar compounds have been tentatively characterized by using HRMS and MS/MS data provided by the Q-TOF-MS. Recently, a new concept in MS/MS, "all-ion fragmentation" (AIF), performed in Orbitrap instruments or MS^E in Q-TOF have been introduced in order to obtain both precursor ions and product ions in the same spectra. Moreover, these new acquisition modes allowed performing a retrospective analysis of both precursors and product ions. For instance, Aguilera-Luiz et al. [18] used a UHPLC–HRMS method using an Orbitrap mass analyzer operating in full-scan and AIF mode for the quantitative analysis of veterinary drugs in honey.

3.3 CONCLUSIONS AND FUTURE TRENDS

Metabolomics approaches have been successfully applied in many areas of food science and nutrition research, including food component analysis and food quality/authenticity assessment. Additionally, it has been showed that the use of UHPLC coupled with MS represents a powerful instrumental tool in the field of food metabolomics. In a targeted metabolomics way and when using QqQ as analyzers, UHPLC has proven its higher resolution and higher efficiency over the conventional HPLC technique. Although commonly used, this technology is limited when dealing with hundreds of compounds, especially due to the limited cycle time of QqQ analyzers. Another inconvenience associated with targeted analysis is that only the target molecules can be detected, therefore HRMS coupled to LC in full-scan mode is being used in food metabolomics to improve these issues.

To date, the required chromatographic and spectrometric resolutions move in opposite directions. Orbitrap technology provides a high and reproducible mass accuracy along the all-mass range, making it well suited for untargeted metabolomics. However, when coupled with UHPLC and in order to fully exploit the chromatographic potential, Orbitrap analyzers are not used at its maximum resolution. On the other hand, TOF analyzers are much faster but in practice very unstable, and their stability becomes worse over time with slight temperature fluctuations. TOF analyzers seem to be currently sufficient when targeted food metabolomics studies are conducted and an ample database support and reference materials are available. For untargeted metabolomics, however, a high stable mass accuracy over time is required in order to reach more reliable results. Therefore, for untargeted food metabolomics, sensitivity and dynamic range of all instruments are finally limited by the efficiency of the LC–MS interfaces and the ion capacity of mass analyzers.

Generally, the applications in food metabolomics reported in the literature are conducted using reverse-phase LC. However, the combination of hydrophilic interaction chromatography (HILIC) gives the possibility of a better separation of polar compounds, extending food metabolic information.

REFERENCES

1. Nicholson, J. K., Lindon, J. C., and Holmes, E. 1999. "Metabonomics": Understanding the metabolic responses of living systems to pathophysiological stimuli via multivariate statistical analysis of biological NMR spectroscopic data. *Xenobiotica* 29:1181–1189.
2. Cifuentes, A. 2009. Food analysis and foodomics foreword. *J. Chromatogr. A* 1216:7109–7109.
3. Tengstrand, E., Rosen, J., Hellenas, K. E., and Aberg, K. M. 2013. A concept study on non-targeted screening for chemical contaminants in food using liquid chromatography-mass spectrometry in combination with a metabolomics approach. *Anal. Bioanal. Chem.* 405:1237–1243.
4. Mannina, L., Sobolev, A. P., and Capitani, D. 2012. Applications of NMR metabolomics to the study of foodstuffs: Truffle, kiwifruit, lettuce, and sea bass. *Electrophoresis* 33:2290–2313.
5. Fang, G. H., Goh, J. Y., Tay, M., Lau, H.F., and Li, S. F. Y. 2013. Characterization of oils and fats by H-1 NMR and GC/MS fingerprinting: Classification, prediction and detection of adulteration. *Food Chem.* 138:1461–1469.
6. Theodoridis, G. A., Gika, H. G., Want, E. J., and Wilson, I. D. 2012. Liquid chromatography-mass spectrometry based global metabolite profiling: A review. *Anal. Chim. Acta* 711:7–16.
7. Nordström, A., O'Maille, G., Qin, C., and Siuzdak, G. 2006. Nonlinear data alignment for UPLC-MS and HPLC-MS based metabolomics: Quantitative analysis of endogenous and exogenous metabolites in human serum. *Anal. Chem.* 78:3289–3295.
8. Vrhovsek, U., Masuero, D., Gasperotti, M., Franceschi, P., Caputi, L., Viola, R., and Mattivi, F. 2012. A versatile targeted metabolomics method for the rapid quantification of multiple classes of phenolics in fruits and beverages. *J. Agr. Food Chem.* 60:8831–8840.
9. Van Meulebroek, L., Vanden Bussche, J., Steppe, K., and Vanhaecke, L. 2012. Ultra-high performance liquid chromatography coupled to high resolution Orbitrap mass

spectrometry for metabolomic profiling of the endogenous phytohormonal status of the tomato plant. *J. Chromatogr. A* 1260:67–80.

10. Serrano-Lourido, D., Saurina, J., Hernandez-Cassou, S., and Checa, A., 2012. Classification and characterisation of Spanish red wines according to their appellation of origin based on chromatographic profiles and chemometric data analysis. *Food Chem.* 135:1425–1431.
11. Pongsuwan, W., Bamba, T., Harada, K., Yonetani, T., Kobayashi, A., and Fukusaki, E. 2008. High-throughput technique for comprehensive analysis of Japanese green tea quality assessment using ultra-performance liquid chromatography with time-of-flight mass spectrometry (UPLC/TOF MS). *J. Agr. Food Chem.* 56:10705–10708.
12. Vaclavik, L., Lacina, O., Hajslova, J., and Zweigenbaum, J. 2011. The use of high performance liquid chromatography-quadrupole time-of-flight mass spectrometry coupled to advanced data mining and chemometric tools for discrimination and classification of red wines according to their variety. *Anal. Chim. Acta* 685:45–51.
13. Lerma-García, M. J., Simó-Alfonso, E. F., Méndez, A., Lliberia, J. L., and Herrero-Martínez, J. M. 2011. Classification of extra virgin olive oils according to their genetic variety using linear discriminant analysis of sterol profiles established by ultra-performance liquid chromatography with mass spectrometry detection. *Food Res. Int.* 44:103–108.
14. Kim, N., Kim, K., Choi, B. Y., Lee, D., Shin, Y.-S., Bang, K.-H., Cha, S.-W. et al. 2011. Metabolomic approach for age discrimination of *Panax ginseng* using UPLC-Q-Tof MS. *J. Agr. Food Chem.* 59:10435–10441.
15. Kim, H.-J., Park, K.-J., and Lim, J.-H. 2011. Metabolomic analysis of phenolic compounds in buckwheat (*Fagopyrum esculentum* M.) sprouts treated with methyl jasmonate. *J. Agr. Food Chem.* 59:5707–5713.
16. Cuadros-Inostroza, A., Giavalisco, P., Hummel, J., Eckardt, A., Willmitzer, L., and Pena-Cortes, H. 2010. Discrimination of wine attributes by metabolome analysis. *Anal. Chem.* 82:3573–3580.
17. Hurtado-Fernandez, E., Pacchiarotta, T., Gomez-Romero, M., Schoenmaker, B., Derks, R., Deelder, A.M., Mayboroda, O.A., Carrasco-Pancorbo, A., and Fernandez-Gutierrez, A. 2011. Ultra high performance liquid chromatography-time of flight mass spectrometry for analysis of avocado fruit metabolites: Method evaluation and applicability to the analysis of ripening degrees. *J. Chromatogr. A* 1218:7723–7738.
18. Aguilera-Luiz, M. M., Romero-Gonzalez, R., Plaza-Bolanos, P., Vidal, J. L. M., and Frenich, A. G. 2013. Rapid and semiautomated method for the analysis of veterinary drug residues in honey based on turbulent-flow liquid chromatography coupled to ultra high-performance liquid chromatography-orbitrap mass spectrometry (TFC-UHPLC-Orbitrap-MS). *J. Agr. Food Chem.* 61:829–839.
19. Funari, C. S., Eugster, P. J., Martel, S., Carrupt, P.-A., Wolfender, J.-L., and Silva, D. H. S. 2012. High resolution ultra high pressure liquid chromatography-time-of-flight mass spectrometry dereplication strategy for the metabolite profiling of Brazilian *Lippia* species. *J. Chromatogr. A* 1259:167–178.
20. Fraser, K., Lane, G. A., Otter, D. E., Hemar, Y., Quek, S.-Y., Harrison, S. J., and Rasmussen, S. 2013. Analysis of metabolic markers of tea origin by UHPLC and high resolution mass spectrometry. *Food Res. Int.* 53:827–835.
21. Abu-Reidah, I. M., Arraez-Roman, D., Segura-Carretero, A., and Fernandez-Gutierrez, A. 2013. Profiling of phenolic and other polar constituents from hydro-methanolic extract of watermelon (*Citrullus lanatus*) by means of accurate-mass spectrometry (HPLC-ESI-QTOF-MS). *Food Res. Int.* 51:354–362.
22. Arapitsas, P., Perenzoni, D., Nicolini, G., and Mattivi, F. 2012. Study of sangiovese wines pigment profile by UHPLC-MS/MS. *J. Agr. Food Chem.* 60:10461–10471.

23. Bakhouche, A., Lozano-Sanchez, J., Beltran-Debon, R., Joven, J., Segura-Carretero, A., and Fernandez-Gutierrez, A. 2013. Phenolic characterization and geographical classification of commercial Arbequina extra-virgin olive oils produced in southern Catalonia. *Food Res. Int.* 50:401–408.
24. Treutter, D., Wang, D., Farag, M. A., Baires, G. D. A., Ruehmann, S., and Neumueller, M. 2012. Diversity of phenolic profiles in the fruit skin of *Prunus domestica* plums and related species. *J. Agr. Food Chem.* 60:12011–12019.
25. Nguyen, D. T. T., Guillarme, D., Rudaz, S., and Veuthey, J. L. 2006. Chromatographic behaviour and comparison of column packed with sub-2 μm stationary phases in liquid chromatography. *J. Chromatogr. A* 1128:105–113.
26. Wu, N. and Clausen, A. M. 2007. Fundamental and practical aspects of ultrahigh pressure liquid chromatography for fast separations. *J. Sep. Sci.* 30:1167–1182.
27. Brice, R. W., Zhang, X., and Colon, L. A. 2009. Fused-core, sub-2 μm packings, and monolithic HPLC columns: A comparative evaluation. *J. Sep. Sci.* 32:2723–2731.
28. Novakova, L. and Vlckova, H. 2009. A review of current trends and advances in modern bio-analytical methods: Chromatography and sample preparation. *Anal. Chim. Acta* 656:8–35.
29. Rodriguez-Aller, M., Gurny, R., Veuthey, J.-L., and Guillarme, D. 2013. Coupling ultra high-pressure liquid chromatography with mass spectrometry: Constraints and possible applications. *J. Chromatogr. A* 1292:2–18
30. Guillarme, D., Ruta, J., Rudaz, S., and Veuthey, J. L. 2010. New trends in fast and high-resolution liquid chromatography: A critical comparison of existing approaches. *Anal. Bioanal. Chem.* 397:1069–1082.
31. Forcisi, S., Moritz, F., Kanawati, B., Tziotis, D., Lehmann, R., and Schmitt-Kopplin, P. 2013. Liquid chromatography-mass spectrometry in metabolomics research: Mass analyzers in ultra high pressure liquid chromatography coupling. *J. Chromatogr. A* 1292:51–65.
32. Núñez, O., Gallart-Ayala, H., Martins, C. P. B., Lucci, P., and Busquets, R. 2013. State-of-the-art in fast liquid chromatography–mass spectrometry for bio-analytical applications. *J. Chromatogr. B*, 927:3–21.
33. Kind, T. and Fiehn, O. 2007. Seven Golden Rules for heuristic filtering of molecular formulas obtained by accurate mass spectrometry. *Bmc Bioinformatics*, 8:105.

4 UHPLC–MS-Based Methods for the Study of Foodborne Carcinogens

Lubinda Mbundi and Rosa Busquets

CONTENTS

4.1 INTRODUCTION

In a healthy individual, the integrity of the tissue and organs is maintained by complex inter- and intra-cellular activities. When tissue is damaged due to trauma or disease, these cellular mechanisms repair and regenerate the affected tissue. However, in some injury or diseases, these cellular mechanisms are affected and/or corrupted and often result in chronic illness and death. In this regard, carcinogenesis and mutagenesis phenomena have been advocated as major challenges in health sciences as they involve alterations in cellular activities that tightly control and regulate cell proliferation, survival, and programmed cell death.

Depending on an individual's genetic predisposition, some carcinogens may lead to cancerous changes after short-term or prolonged exposure to lower or higher levels of carcinogens. One of the most common ways organisms are exposed to carcinogens is through food intake. It is well established that a wide variety of food items can lead to the formation of some constituents that can undergo metabolic activation and cause genetic alterations (mutations) that have the potential to lead to oncogenesis [1–3]. These compounds that are present in food or formed during the preparation and storage thereof are referred to as foodborne carcinogens. Since there is a chronic exposure to some of these food products and only zero level of exposure will result in no risk, foodborne contaminants attract special interest from the public, scientific community, and food industry.

The health effects of foodborne carcinogens in humans are indirectly obtained from *in vitro* and animal studies. As such, the only link to establish a cause (intake of food contaminated with foodborne carcinogens) and effect (different types of cancer) relationship is through epidemiological studies and/or the quantification of validated biomarkers associated with exposure. To date, epidemiological studies have yet to establish an unequivocal link between the intake of most of the aforementioned toxicants and the onset of cancer processes, the variability of chronic exposure over a lifetime, and the exposure to factors affecting toxicant metabolism. Furthermore, the repair of the modified DNA and the fate of the foodborne pro-mutagens are also not well defined. Moreover, variations in the genetic makeup of the person exposed to foodborne carcinogens and the potential exposure and influence of other mutagens from elsewhere poses serious and complex challenges in endeavors to establish the cause–effect relationships and clinical validity [1,4–7]. However, the capacity of the foodborne carcinogens to react with the DNA, in spite of possible repair and detoxification mechanism, indicates a connection between the intake of these carcinogens and the development or aggravation of cancer [1,8].

The International Agency for Research on Cancer (IARC) [9], the National Toxicology Program [10], and the Environmental Protection Agency (EPA) [11] evaluate cancer-causing candidates, make reports on their toxicity every few years, and keep a database describing the health effects from the exposure to certain substances. The nomenclature used by these independent agencies in the classification of the toxicity of substances they evaluate is summarized in Table 4.1.

Despite the availability and knowledge of the dangers that foodborne carcinogens pose to the public, there are only a few companies in the food industry that are involved in foodborne carcinogen research, among them is Nestle [12,13]. Nonetheless, it is expected that the food industry will acknowledge and address the problem of the generation of pro-mutagens in food and adopt practices that control and limit the amount of foodborne toxicants in their products. For instance, the food industry could monitor and control levels of carcinogenic agents such as acrylamide (Group 2A) [5,14–16] in carbohydrate-rich cooked products such as roasted potatoes [17], potato chips [18],

TABLE 4.1
Denominations Given to Chemical Compounds after Being Assessed for Their Genotoxic Potential

IARC [9]	EPA [11]	National Toxicology Program [10]
Group 1: human carcinogen	Group A: carcinogenic to humans	Known to be human carcinogen
Group 2A: probably carcinogenic to humans	Group B: likely to be carcinogenic to humans	Reasonably anticipated to be human carcinogen
Group 2B: possibly carcinogenic to humans	Group C: suggestive evidence of carcinogenic potential	
Group 3: unclassifiable as to cause carcinogenicity in humans	Group D: inadequate information to assess carcinogenic potential	
Group 4: probably not carcinogenic to humans		

biscuits, breakfast cereals, olives [9–21], coffee [22], and chocolate [23]. This could be encouraged by the setting and enforcement of statutory levels of contaminants by governmental agencies, as is already the case with mycotoxins [24,25] such as aflatoxins (Group 1), ochratoxins (Group 2B), and fumonisins (Group 2B) [26], which are secondary metabolites produced by molds. The production of food items low in foodborne toxicants as a niche market could be further encouraged by public demand once the knowledge of these toxicants and how formation thereof can be controlled and reduced is readily available. Indeed, this knowledge could further inform the public on the best methods of food preparation and storage. Certainly, recent years have seen increasing interest for more information on foodborne toxicants and carcinogens, and where and how they occur. As a result, it is essential that the information produced from research accurately informs on the levels of foodborne carcinogens in food and variability in contamination levels of such toxicants due to processing, storage conditions, and composition of toxicant precursors in raw food [1,27] so as to reduce inaccuracies in epidemiological studies [4,6]. Furthermore, accurate information on length of exposure and/or dosage levels of toxicants that can lead to or aggravate cancer would be useful in informing preventative and or treatment measures.

As such, it is hereby envisaged that the food industry will need fast, robust, easy-to-use, and sensitive methods to quantify foodborne mutagens in their products as part of quality control. Currently, the optimal techniques used in the analysis of mutagenic compounds from food involve the separation of target compounds with liquid chromatography (LC) and detection by mass spectrometry (MS). In this regard, ultra-high-pressure liquid chromatography (UHPLC) has received increased attention due to increased efficiency and higher throughput and sensitivity. This chapter compiles the current approaches where UHPLC–MS has been employed for the analysis of foodborne carcinogens that are present or form in food as a consequence of physicochemical (Maillard reaction or nonenzymatic browning) [28] or biochemical reactions [2,29]. Genotoxic compounds from external sources, such as leached compounds from food packaging, are not included.

4.2 SAMPLE PRETREATMENT AND QUANTIFICATION FOR UHPLC–MS ANALYSIS

Owing to the complex nature of biological matrices (i.e., presence of a plethora of different biomolecules with varying physico-chemical properties), analysis of food matrices poses several challenges. One of which is the need to extract the target contaminants from the food items with enough extraction recovery and purity to guarantee sensitivity, accuracy, and precision in the determination by UHPLC–MS. However, several sample treatment steps are usually necessary to achieve that. In this regard, solid-phase extraction (SPE) has been the predominant sample treatment method for foodborne contaminants [30–38], even in matrices as simple as drinking water [39]. The injection of extracts purified with liquid–liquid extraction (LLE) and/or precipitation of macromolecules [40–42] and immunoaffinity columns [32,43] have also shown to be suitable for the analysis of the purified food extracts with UHPLC–MS. These sample treatment procedures, and others, applied to purify some foodborne carcinogens from a range of food matrices for their determination with UHPLC are compiled in Table 4.2.

TABLE 4.2
Sample Treatment Methodology Used in the Purification and Preconcentration of Some Foodborne Carcinogens

Compound	Matrix	Sample Treatment	Reference
Acetaldehyde	Apple (tannins)	Freeze-dried and extraction with methanol and water/acetone. Extracts eluted on fractogel chromatography (methyl acrylate copolymer in solution in aqueous ethanol 20%, HW-50 F). Elution with methanol/formic acid (9:1) and H_2O/acetone/acetic acid (59.5/39.5/1)	[41]
Acrylamide	Potato chips	Defatting with petroleum ether/extraction with 2 M NaCl(aq)/LLE ethyl acetate/SPE (OASIS HLB)	[35]
	Model: asparagine and sugars	LLE with ethyl acetate/dry with N_2/reconstitute with H_2O/SPE (HLB), elution with H_2O	[32]
Aflatoxins and ochratoxin A	Beer	SPE (Oasis HLB), elution with ACN	[37]
	Wines	pH adjusted to 7.2 with NaOH/immunoaffinity column (OCHRAPREP) –elution with methanol: acetic acid (98:2)/drying and reconstitution in methanol: H_2O (50:50). Analysis with UHPLC-fluorescence	[43]
Aflatoxins	Beer	Extraction with ethyl acetate/evaporation and reconstitution with ACN	[42]
	Peanuts	Extraction with ACN in H_2O (84:16)/homogenization, centrifugation/column purification (silica gel, alumina, kieselguhr) elution with methanol	[36]
	Corn and peanuts	Extraction with ACN/H_2O) (84:16)/filtration with paper/rotaevaporation and dilution with ACN and SPE (Mycosep#226 AflaZon+)	[44]
Multiresidue mycotoxins (32)	Corn, wheat, barley, dried grains	Extraction with a ACN/H_2O/formic acid (84.0:15.9:0.1)/centrifugation and dry with N_2 and reconstitute with 10 mM ammonia acetate and water/acetonitrile/formic acid (95.0:4.9:0.1)	[45]

TABLE 4.2 (continued)
Sample Treatment Methodology Used in the Purification and Preconcentration of Some Foodborne Carcinogens

Compound	Matrix	Sample Treatment	Reference
Multiresidue mycotoxins (11 toxins)	Cereals (rice, wheat, oat, and maize meal)	Extraction with ACN: water: acetic acid (79:20:1)/centrifugation and extraction using 0.5 mL of the final extract diluted with acetonitrile: water: acetic acid (20:79:1).	[40]
Glycidol	Edible palm oils	LLE with hexane/ACN (1:1)/SPE with Sep-Pak, elution with $CHCl_3$/SPE Sep-Pak Plus C_{18}, elution with ACN and ethyl acetate. Drying and reconstitution in ACN	[33]
Heterocyclic amines	Meat extract (spiked)	Homogenization in NaOH (1 M)/LLE with NaOH and dichloromethane on diatomaceous earth/SPE (PRS ion exchange), elution with 0.5 ammonium acetate (0.5 M)/SPE (C_{18}), elution with ammonia in methanol (1:9)	[31]
	Mutton Shashlik	Extraction with methanol/NaOH (1 N) (3:7)/SPE LiChrolut EN and elution with ethanol/dichloromethane (1:9)	[34]
	Beef, mutton, chicken, and fish	Homogenization in HCl (0.1 M)/centrifugation and neutralization of the supernatant with NaOH (pH 7)/filtration through paper and Amberlite XAD-2 resin, elution with acetone and methanol	[46]
Nitrosamine (*N*-nitrosodimethylamine)	Drinking water	Addition of sodium thiosulfate/filtration with glass fiber/adjust pH to 8.0 with sodium bicarbonate/SPE with carbon cartridges, elution with dichloromethane/diethylether solution (50:50)/fluorisil	[39]
Nitrosamines (9)	Drinking water	SPE (Resprep EPA 521), elution with dichloromethane	[47]

Note: The purified extracts generated were analyzed with UHPLC–MS(/MS), otherwise specified.

Although the sample pretreatment steps to extract contaminants usually produce extracts where the analytes are with improved purity, matrix components that are also present in the purified extract can coelute with the analyte after the chromatographic separation and alter the signal in the detection. This is much more of a problem when analyte detection is carried out with electrospray MS (ESI–MS). The extent of the effect of the matrix is greatly influenced by the nature of the ionization source design [48]. For instance, recent work by Jackson et al. (2012), where they assessed the analysis of 29 mycotoxins in corn, wheat, barley, and distillers dried grains demonstrated high variability in the recoveries between matrices and inaccuracy in the quantification when external calibration and ESI were used. Indeed, the analytes presented in the purified sample matrices were found to produce a different signal with respect to the signal obtained in pure solvent constituting the standards used in the quantification [45]. However, the quantification with isotopically labeled internal standards was found to improve the accuracy and precision of the analysis, making the quantification almost independent from the signal suppression/enhancement phenomena caused by each matrix [45] as well as increasing the throughput of the determination, factors that are particularly important when analyzing high number of samples with different matrices. This is one of the reasons that isotopic dilution approach is commonly used for the analysis of complex samples, such as food when the determination is carried out with MS. Indeed, several studies have employed this approach to analyze foodborne mutagens that have commercially available and affordable labeled standards such as acrylamide [35,49] in carbohydrate-rich food, *N*-nitrosodimethylamine (Group 2A) [50] in water [47], and some mycotoxins in leaves [38]. Other methods of choice that can correct for the effect of the matrices in MS detection include standard addition and matrix-matched approaches. Indeed, these approaches have been used in the analysis of compounds such as heterocyclic amines (HCAs; Groups 2A and 2B) [26] in thermally treated proteinaceous food [31]_ENREF_31, aflatoxins (Group 1 for AFB1 and Group 2B) [26] in beer [37,42] and peanuts [36], and T-2 toxin in rice [26,30]. However, standard addition and matrix-matched approaches require larger amounts of consumables and the analysis takes longer than the isotopic dilution approach. Moreover, matrix-matched quantification requires matrices that are free from analytes, but this is not possible in some cases, such as when toxicants are generated by thermal treatment (i.e., HCAs and acrylamide in fried meat and potato chips, respectively).

Another challenge in the analytical studies of foodborne carcinogens is that some of these toxicants are present in very low concentrations in food samples. For instance, the HCA IQ (2-amin-3-methylimidazo[4,5-f] quinolone, Group 2A) [26] at levels below 2 μg IQ/kg of meat in different types of cooked meat and fish [27] and aflatoxins at levels below 16 ng/L in beer have been reported [42,51]. In standard LC–MS, the concentration levels commonly analyzed range from 0.01 to 1 mg/L. Therefore, the analytes need to be preconcentrated in order to be detected with LC–MS. Although the sensitivity can be improved with UHPLC–MS, which will be explained later in this chapter, sample purification and preconcentration are still needed (see Table 4.1). An illustration of the capacity of UHPLC–MS to detect extremely low levels of aflatoxins in food samples (parts per trillion), after a preconcentration with LLE, is given in Figure 4.1. The determination was carried out with a triple quadrupole acquiring multiple reaction monitoring (MRM) in the most sensitive mode.

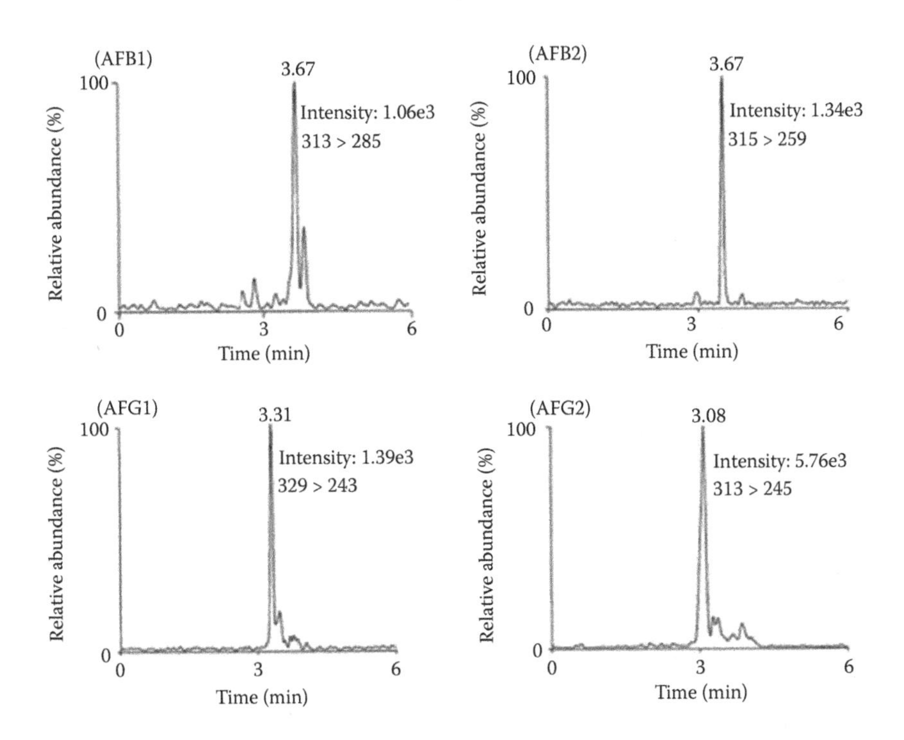

FIGURE 4.1 UPLC–MS/MS chromatograms of aflatoxins in Moussi beer sample. (Adapted from Khan, M. R. et al. 2013. *J. Sep. Sci.* 36: 572–577. With permission.)

In situations where low analyte concentration is not a problem, such as in the analysis of food samples where carcinogen agents produced by the cooking process (i.e., acrylamide in potato chips) are in relative abundance, other problems such as sample purity and poor resolution still persist. Acrylamide forms in carbohydrate-rich food and has been found at concentrations as high as 13 mg/kg in chips cooked under industrial conditions [52], and 0.1–9.2 mg/kg in a wider range of commercial products [20]. Despite its relatively higher concentration, acrylamide is not a trouble-free substance to analyze since the presence of fat from the cooking process in some food items and its removal in the sample treatment can compromise its recovery. Moreover, the low molecular weight of acrylamide limits its retention in reversed-phase chromatography and affects its separation from other compounds in the matrix in conventional chromatography columns. An example of the high importance of chromatographic resolution in food analysis is illustrated in Figure 4.2, which shows the chromatogram obtained from an extract from potato chips analyzed with HPLC using a graphitic carbon column[21]. The chromatographic conditions chosen in this work allowed the separation of two substances with the same mass-to-charge (*m/z*) ratio and molecular weight (peaks 1 and 2 in Figure 4.2a). Furthermore, the molecules had similar product ion scan spectra (Figure 4.2c) when the detection was performed with an ion trap mass spectrometer. As such, it was necessary to work with high-resolution conditions using a time-of-flight (TOF) mass spectrometer (Figure 4.2b) to find that the two peaks had slightly different *m/z* (Figure 4.2d), where peaks

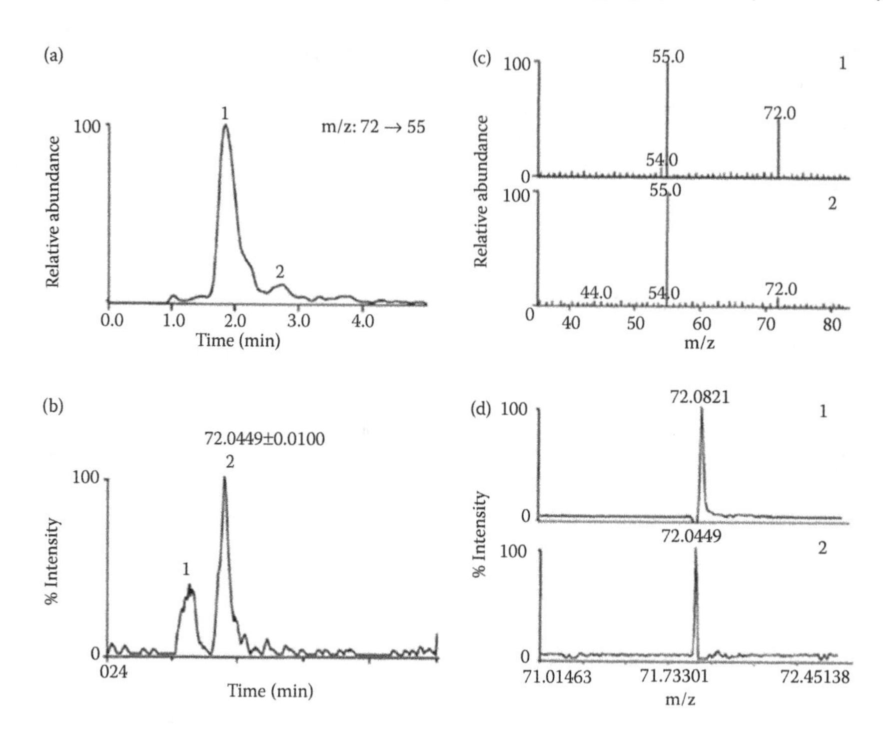

FIGURE 4.2 LC–APCI chromatograms and MS spectra of a potato chip sample obtained using (a) linear trap product ion scan; (b) time-of-flight MS acquisition mode; (c) linear trap product ion scan spectra; (d) time-of-flight spectra (1) interfering peak and (2) acrylamide. (Adapted from Bermudo, E. et al. 2008. *Talanta* 76: 389–394. With permission.)

1 and 2 corresponded to valine and acrylamide, respectively. If the chromatographic resolution between both peaks had been lower, a global peak resulting from the coelution of both valine and acrylamide could probably have been obtained, which would have led to an overestimation of the concentration of acrylamide or valine.

However, although high chromatographic resolution is one of the features of utmost importance in food analysis and other complex matrices by LC–MS, the identification and confirmation of an analyte can also be carried out by complementary identification points described in the EU directive 2002/657/EC [53], by using other analytical approaches.

4.3 CURRENT USES OF UHPLC–MS IN FOODBORNE CARCINOGEN ANALYSIS

The use UHPLC–MS for the analysis of food benefits from chromatographic efficiency and fast analysis. With regards to chromatographic efficiency, the characteristic narrow chromatographic peaks of UHPLC–MS can translate into higher signal and better detection limits. Besides, narrow peaks can potentially present higher resolution with matrix components that could otherwise affect the signal in the ionization source and reduce the accuracy and precision of the results. The reduction on the band broadening can be achieved based on the van Deemter theory [54], which was further developed by other authors [55,56] and is given by the van Deemter equation (Equation 4.1).

$$H = A + \frac{B}{u} + Cu = 2\lambda dp + \frac{2\gamma D_M}{u} + \frac{f(k)dp^2u}{D_M} \tag{4.1}$$

where

- H is the plate height (or chromatographic efficiency)
- u is the linear velocity of the mobile phase
- A is a constant related to the different lengths of time it takes solute molecules of the same type to follow different pathways within the stationary phase (Eddy diffusion)
- B is a constant for the longitudinal diffusion in the chromatographic column
- C is a constant for the mass transfer between the mobile and stationary phases
- dp is the particle diameter
- D_M is the analyte diffusion coefficient of the mobile phase
- λ is a structural factor of the packing material
- γ is a constant accounting for tortuosity
- k is the retention factor for an analyte

Particle size (dp) plays a major role in chromatographic efficiency, as it can be seen in the van Deemter equation, and it is on this that the improvements associated with UHPLC are fundamentally based. However, this approach has not been a reality for a long time until recently because chromatographic systems were yet to be redesigned to cope with the high pressure associated with the low permeability of beds packed with the fine particles of 1.7–1.9 μm that are currently used in UHPLC [57,58]. UHPLC also benefits from high temperature and low viscosity in the mobile phase [59]. However, as an alternative to the use of fully porous sub-2 μm particles in the stationary phase, high efficiency can also be achieved by using superficially porous particles that are bigger than the ones used in UHPLC and give higher permeability at lower inlet pressure [60]. The use of sub-2 μm and core–shell particles allows for working with high flow rates or linear velocities, which result in a high-throughput analysis without sacrificing peak efficiency.

To illustrate this evolution in chromatography we can quote the improvement achieved in the separation of HCAs. In the late 1990s, the separation of these mutagens was carried out with columns of dimensions 4.6 mm × 150 mm, particle size of 5 μm, and the analysis time was in the range of 35 min [61]. The development of UHPLC has allowed for chromatographic separations about 20 times faster [31]. The range of experimental conditions used in the separation and determination of foodborne carcinogens in food items by UHPLC–MS is given in Table 4.3. As can be seen in the table, reversed-phase columns have dominated the separation of the applications and the use of normal phases has not been attempted. Particularly the column Waters Acquity C_{18}, with 1.7 μm dp and an ethylene-bridge hybrid (BEH) binding the C_{18} chains to the silica particles is the option chosen in most of the works. In the second place, the column Waters Acquity HSS T3, which has different

TABLE 4.3
Separation and Detection Conditions in the Analysis of Some Foodborne Carcinogens

Compound	Matrix	Chromatographic Column/ Analysis Time	Mobile Phase	Source-Mass Spectrometer/ LOD	Reference
Acetaldehyde	Apple (tannins)	Waters Acquity HSS T3 (1.8 µm, 100 mm × 2.1 mm i.d.)/5.7 min	a. 80% ACN, 1% formic acid, 19% water b. 1% formic acid in water	ESI-Ion trap: amaZon X ESI Trap (Bruker Daltonics) LOD:	[41]
Acrylamide	Potato chips	Waters Acquity BEH C_{18} (1.7 µm, 50 mm × 2.1 mm i.d.)/7 min	a. formic acid in water	ESI-QqQ: Quattro Ultima (Micromass) LOD: 1 µg/kg (in matrix)	[35]
	Model: asparagine and sugars	Waters Acquity BEH C_{18} (1.7 µm, 50 mm × 2.1 mm i.d.)/4 min	a. 10% methanol, 0.1% formic acid in water	ESI-QqQ: Quattro Ultima (Micromass) LOD:	[32]
Aflatoxins and ochtratoxin A	Beer	Waters Acquity C_{18} (1.7 µm, 50 mm, 2.1 mm i.d.)/3.2 min	a. 0.1% formic acid in water b. methanol	ESI-QqQ: Quattro premier (Micromass) LOD: 0.1 µg/L (in matrix)	[37]
	Wines	Kinetex PFP column (2.6 µm, 100 mm, 3.0 i.d.)[a]/3 min	a. methanol: 10 mM ammonium acetate in water (70:30, v/v)	ESI-ion trap LCQ Advantage (Thermo-Fisher) LOD: 0.3 ng/L (in matrix)	[43]
Aflatoxins	Beer	Waters Acquity UPLC BEH C_{18} (1.7 µm, 100 mm × 2.1 mm i.d.)/6 min	a. methanol b. 1% formic acid in water	ESI-QqQ: Quattro premier (Micromass) LOD: 1 and 3 ng/L (in matrix)	[42]
	Peanuts	Waters Acquity HSS T3 (1.8 µm, 100 mm × 2.1 mm i.d.)/6.4 min	a. methanol/ACN (50:50, v/v) b. 0.1% formic acid in water	ESI-QqQ Quattro Ultima Pt (Micromass) 0.01–0.2 µg/kg (with standard)	[36]
	Corn and peanuts	Waters Acquity UPLC BEH C_{18} (1.7 µm, 50 mm, 2.1 mm i.d.)/6 min	a. ACN/methanol (50/50, v/v) b. 0.1% formic acid in water	UV[b] LOD 0.2–0.3 µg/kg (in matrix)	[44]

Multiresidue mycotoxins (32)	Corn, wheat, barley, and dried grains	Waters Acquity UPLC BEH C18 (1.7 µm, 100 mm, 2.1 mm i.d.)/16 min	a. 0.1% formic acid in methanol b. 0.1% formic acid in water	ESI-QqQ TDQ (Waters) LOD 0.2–640 µg/kg (in matrix)	[45]
Multiresidue mycotoxins (11 toxins)	Cereals (rice, wheat, oat, and maize meal)	Thermo Scientific C_{18} (1.9 µm, 50 mm × 2.1 mm i.d.)/8 min	a. methanol b. 0.1% acetic acid	ESI-QqQ: TSQ quantum ultra Mass (ThermoScientific) LOD 0.1–40 µg/kg	[40]
Glycidol fatty acid esters and 3-chloropropane-1,2-diol fatty acid esters	Edible palm oils	Waters Acquity UPLC BEH C_{18} (1.7 µm, 50 mm × 2.1 mm i.d.)	a. 0.2 mM sodium formate 97.5% methanol in water b. 0.2 mM sodium formate 85% water in methanol	ESI-TOF (MicroTOF, Bruker) LOD: 0.1–0.9 µg/L	[33]
Heterocyclic amines (HCAs)	Meat extract (spiked with 15 HCAs)	Waters Acquity BEH C_{18} (1.7 µm, 50 mm × 2.1 mm i.d.)/2 min	a. ACN b. 30 mM formic acid/ammonium formate in water (pH 4.75)	ESI-QqQ: Quattro Premier (Micromass) LOD: 5–51 ng/kg	[31]
	Mutton Shashlik (15 HCAs	Waters Acquity BEH C_8 (1.7 µm, 100 mm × 2.1 mm i.d.)	a. ACN b. 30 mM acetic acid/ammonium acetate (pH 4.5)	ESI-QqQ Quattro Premier XE triple quadrupole (Micromass)	[34]
	Beef, mutton, chicken, and fish (2 HCAs)	Waters Acquity UPLC UPLC HSS T3 (1.7 µm, 100 mm × 2.1 mm i.d.)/3.5 min	a. ACN b. 0.1% trifluoroacetic in water	MALDI-TOF (Bruker Daltonics Autoflex II TOF Instrument) LOD 0.2–40 µg/kg (in standards)	[46]
Nitrosamine (*N*-nitrosodimethylamine)	Drinking water	Waters Acquity UPLC BEH C_{18} (1.7 µm, 150 mm × 2.1 mm i.d.)/3.5 min	a. ACN b. 10 mM ammonium bicarbonate	ESI-QqQ TDQ (Waters) LOD: 1.2 ng/L (in standards)	[39]
Nitrosamines (9)	Drinking water	Waters Acquity BEH C_{18} (1.7 µm, 150 mm, 2.1 mm i.d.)/7.2 min	a. methanol b. 10 mM ammonium bicarbonate	ESI-QqQ Quattro Premier XE triple quadrupole (Micromass)	[47]

[a] Liquid chromatography column.
[b] UV detection system.

selectivity than the former with lower pore size, higher surface area, half ligand density, and narrower pH tolerance has also been used [36,41]. In particular, similar separation times have been achieved when analyzing compounds from the same family when using either BEH or HSS T3 sorbents [36,44]. Likewise, stationary phases with 1.9 μm particle size were just used in one of the works included in Table 4.3 and have shown to be capable of separating 11 mycotoxins within 8 min [40]. Ventura et al. separated five mycotoxins, including ochratoxin, with a column with sub-2 μm fully porous particles in about 3 min, with a peak width for ochratoxin was 0.1 min [37]. However, when the separation was carried out with superficially porous particles of 2.6 μm, a peak width 0.2 min was obtained for that mycotoxin and the analysis took place in less than 3 min [43]. Even though both types of stationary phases can provide similar results, UHPLC columns need special instrumentation, whereas columns packed with superficially porous particles can be used in conventional HPLC.

Although high throughput has been achieved with UHPLC–MS, there is still room for improvement. For instance, while the separation times for generally a few foodborne carcinogens (<10) have been kept in the range of 3–8 min, the analysis of a mixture of 15 HCAs was recently achieved in just 2 min, [31] suggesting that the analysis of a lower number of analytes could potentially be carried out with less time. However, the separation of a higher number of food toxicants (multi-residue) in a single chromatogram is widely carried out in multi-residue analysis of pharmaceuticals or pesticides in food. Nonetheless, multi-residue analysis is still not common for the analysis of most foodborne carcinogens with the exception of toxicant families such as mycotoxins, where 35 mycotoxins were recently separated with a total of 16 min [45]. This is so partly because most foodborne carcinogen families are made up of fewer substances.

With regard to the mobile phase composition, it has generally included methanol or acetonitrile and aqueous solution with a volatile acid, such as formic acid or acetic acid, or a base, such as ammonium acetate. However, a few works have included nonvolatile components such as sodium formate [33] or ammonium bicarbonate [39], which could lead to precipitates in the MS. In any case, the content of organic solvent is normally kept high to decrease the viscosity of the solvent and the pressure associated when working with these conditions.

With respect to mass analyzers, triple quadrupoles (QqQs) are the most popular in the analysis of foodborne carcinogens by UHPLC–MS (see Table 4.3) partly because they provide some of the highest sensitivities in targeted quantitative analysis compared to other types of analyzers. This "target-based" approach has been carried out across foodborne contaminants and matrices and has been the common strategy used in the determination of foodborne contaminants. Comparatively, analyzers that can provide higher resolution in the measurement of mass-to-charge ratio (*m*/*z*) such as TOF, quadrupole coupled to time-of-flight (QTOF), Orbitrap, and Fourier transform ion cyclotron (FTICR) are seldom coupled to UHPLC for the analysis of foodborne carcinogens. However, given the complexity of food samples, these analyzers, including ion trap, which has a highly sensitive full scan among other positive features, would be highly appropriate for screening for new foodborne carcinogens.

Despite being a separation technique, UHPLC has also contributed to obtain better sensitivity by minimizing band broadening. For instance, the limits of detection achieved for the analysis of aflatoxins in corn and peanuts were 0.2–0.3 μg/kg by UHPLC-UV [44], and similar detection limits (0.01–0.2 μg/kg) were achieved in the analysis of the same type of compounds in these matrices using MS/MS with a QqQ [36].

Fast analysis, superior chromatographic efficiency, and sensitive determination can be achieved with UHPLC–MS, where for instance a determination of 15 compounds in a complex matrix in <2 min is possible. The evolution of HPLC to UHPLC has been possible due to the development of chromatographic columns with smaller particle size, chromatographs able to work at higher pressure, and mass spectrometers with high-speed scan velocity. Furthermore, conventional sample treatment methods used to purify the analyses for HPLC-UV or MS have shown to be suitable when the determination technique is UHPLC. Currently, the analysis of foodborne contaminants follows a target-based approach by acquiring a highly selective acquisition mode in a QqQ analyzer. However, a multiclass wide-scope screening is still necessary in order to provide a broader picture of what is going on in the food matrix. UHPLC–MS is the best technique for the analysis of foodborne contaminants, and these hyphenated techniques will help to uncover secrets still hidden in the world of foodborne contaminants.

4.4 SUMMARY AND CONCLUSIONS

Given the amount of food types and preparations thereof that are associated with foodborne carcinogens, the risk presented in this mode of exposure is a real one, and research needs to be done to provide epidemiologically valid data. Indeed, accurate data would clearly inform the medical and socio-economic burden associated with foodborne toxicants, as it would inform the studies aimed at establishing the link between cancer and the chronic exposure to such toxicants. It is also important to encourage the food industry and public to become more involved in the efforts to understand and control the problem of foodborne carcinogens, possibly through improved and well-defined research findings that can inform official policies and health standards.

This may be accomplished by fast and sensitive analytical methods that can provide accurate data. In this regard, progress has been made with respect to HPLC, wherein new stationary phases with low particle size, optimized porosity, and surface chemistry are allowing reducing analysis times by about one order of magnitude. As a result, UHPLC has been born. In addition, lower peak widths are obtainable with UHPLC, owing to the improvements in the stationary phase, which leads to better resolution within matrix components and increased sensitivity. However, despite the advances in material science and chromatography, laborious sample treatment methods are still needed to achieve accurate determinations with high sensitivity and selectivity. Today, targeted analysis of known components is the approach generally followed when monitoring foodborne carcinogens by MS. However, semi-quantitative, nontargeted analyses by high-resolution MS could lead to the discovery of new foodborne contaminants. Thus far, UHPLC–MS is the most sensitive and reliable technique for the determination of carcinogens in food and food analysis in general.

ACKNOWLEDGMENT

Rosa Busquets acknowledges the IEF Marie Curie fellowship (n°274985) from the FP7 People program.

REFERENCES

1. Jägerstad, M. and Skog, K. 2005. Genotoxicity of heat-processed foods. *Mutat. Res.* 574: 156–172.
2. Hussein, H. S. and Brasel, J. M. 2001. Toxicity, metabolism, and impact of mycotoxins on humans and animals. *Toxicology.* 167: 101–34.
3. Sugimura, T., Wakabayashi, K., Nakagama, H., and Nagao, M. 2004. Heterocyclic amines: Mutagens/carcinogens produced during cooking of meat and fish. *Cancer Sci.* 95: 290–299.
4. Knize, M. G. 2006. Assessing human exposure to heterocyclic aromatic amines. In *Acrylamide and Other Hazardous Compounds in Heat-Treated Foods*, ed. J. Alexander, K. Skog, Cambridge and Boca Raton: Woodhead Publishing Limited and CRC Press LLC, pp. 231–46. ISBN 978-1-84569011-3.
5. Arribas-Lorenzo, G. and Morales, F. J. 2012. Recent insights in acrylamide as carcinogen in foodstuffs. In *Advances in Molecular Toxicology*, ed. C. Fishbein James, Elsevier, pp. 163–93, ISBN 978-0-44459389-4.
6. Turesky, R. J. and Marchand, L. L. 2011. Metabolism and biomarkers of heterocyclic aromatic amines in molecular epidemiology studies: Lessons learned from aromatic amines. *Chem. Res. Toxicol.* 24: 1169–214.
7. Busquets, R., Mitjans, D., Puignou, L., and Galceran, M. T. 2008. Quantification of heterocyclic amines from thermally processed meats selected from a small-scale population-based study. *Mol. Nutr. Food Res.* 52: 1408–1420.
8. Turesky, R. J. 2006. Genotoxocity metabolism, and biomarkers of heterocyclic amines. In *Acrylamide and Other Hazardous Compounds in Heat-Treated Foods*, ed. J. Alexander, K. Skog, Cambridge and Boca Raton: Woodhead Publishing Limited and CRC Press LLC, pp. 247–274. ISBN 978-1-84569011-3.
9. International Agency for Research on Cancer. Agents classified by the IARC Monographs IARC, http://monographs.iarc.fr/ENG/Classification/index.php.
10. National Toxicology Program. Department of Health and Human Services. Report on Carcinogenshttp://Ntp.Niehs.Nih.Gov/?Objectid = 03c9f0a4-B1c2-31de-Aba8508ae 9949c57.
11. US Environmental Protection Agency. Risk assessment for carcinogens. http://www.epa.gov/ttnatw01/toxsource/carcinogens.html.
12. Gross, G. A. and Grüter, A. 1992. Quantitation of mutagenic/carcinogenic heterocyclic aromatic amines in food products. *J. Chromatogr. A.* 592: 271–78.
13. O'Brien, J., Renwick, A. G., Constable, A., Dybing, E., Müller, D. J. G., Schlatter, J., Slob, W. et al. 2006. Approaches to the risk assessment of genotoxic carcinogens in food: A critical appraisal. *Food Chem. Toxicol.* 44: 1613–1635.
14. International Agency for Research on Cancer. 1994. Acrylamide. *IARC Monographs on the Evaluation of Carcinogen Risk to Humans*, 389. Lyon, France: International Agency for Research on Cancer.
15. US Environmental Protection Agency. Acrylamide. http://www.epa.gov/iris/subst/0286.htm.
16. US Department of Health and Human Services. 2011. Acrylamide. In *Report on carcinogens*: National Toxicology Program, Department of Health and Human Services.
17. Skog, K., Viklund, G., Olsson, K., and Sjöholm, I. 2008. Acrylamide in home-prepared roasted potatoes. *Mol. Nutr. Food Res.* 52: 307–312.

18. Viklund, G., Sjöholm, I., Skog, K., and Olsson, K. M. 2008. Variety and storage conditions affect the precursor content and amount of acrylamide in potato crisps. *J. Sci. Food Agr.* 88: 305–312.
19. Bermudo, E., Moyano, E., Puignou, L., and Galceran, M. T. 2006. Determination of acrylamide in foodstuffs by liquid chromatography ion-trap tandem mass-spectrometry using an improved clean-up procedure. *Anal. Chim. Acta* 559: 207–214.
20. Bermudo, E., Núñez, O., Puignou, L., and Galceran, M. T. 2006. Analysis of acrylamide in food samples by capillary zone electrophoresis. *J. Chromatogr. A.* 1120: 199–204.
21. Bermudo, E., Moyano, E., Puignou, L., and Galceran, M. T. 2008. Liquid chromatography coupled to tandem mass spectrometry for the analysis of acrylamide in typical Spanish products. *Talanta* 76: 389–394.
22. Bagdonaite, K., Derler, K., and Murkovic, M. 2008. Determination of acrylamide during roasting of coffee. *J. Agric. Food Chem.* 56: 6081–6086.
23. Ren, Y. P., Zhang, Y., Jiao, J. J., and Cai, Z. X. 2006. Sensitive isotope dilution liquid chromatography/electrospray ionization tandem mass spectrometry method for the determination of acrylamide in chocolate. *Food Addit. Contam.* 23: 228–36.
24. European Commision. 2010. Commision Regulation (Ec) No. 165/210 of 26 February 2010 Amending Regulation Ec No 1881/2006 Setting maximum levels for certain contaminants in foodstuffs as regards aflatoxins. *Official Journal of the European Union*, L50, 8–11, 2010.
25. European Commission. 2007. Commission Regulation 2007/1126/EC Amending Regulation (Ec) No. 1881/2006 Setting Maximum Levels for Certain Contaminants in Foodstuffs as Regards *Fusarium* Toxins in Maize and Maize Products. Brussels, 2007.
26. International Agency for Research on Cancer. 1993. Monographs on the Evaluation of Carcinogenic Risks to Humans. Some Natural Occurring Substances: Food Items and Constituents. Heterocyclic Amines and Mycotoxins. Published by World Health Organization and International Agency for Research on Cancer, Lyon, France. pp, 163–242.
27. Busquets, R. 2012. Food borne carcinogens: A dead end? de In *Carcinogen*, ed. M. Pesheva, Rijeka, Croatia: Intech Publisher, pp. 163–84, ISBN: 978-953-51-0658-6.
28. Hodge, J. E. 1953. Dehydrated foods, chemistry of browning reactions in model systems. *J. Agric. Food Chem.* 1: 928–943.
29. Ferrante, M., Sciacca, S., and Conti Oliveri, G. 2012. Carcinogen role of food by mycotoxins and knowledge gap. In *Carcinogen*, ed. M. Pesheva, Rijeka, Croatia: Intech Publisher pp. 133–162, ISBN: 978-953-51-0658-6.
30. Wang, Y., Cao, X., Li, Y., Yang, S., Shen, J., and Zhang, S. 2012. Simultaneous determination of type-a and type-B trichothecenes in rice by UPLC-Ms/Ms. *Anal. Methods* 4: 4077–4082.
31. Barcelo-Barrachina, E., Moyano, E., Galceran, M. T., Lliberia, J. L., Bago, B., and Cortes, M. A. 2006. Ultra-performance liquid chromatography-tandem mass spectrometry for the analysis of heterocyclic amines in food. *J. Chromatogr. A.* 1125: 195–203.
32. Zhang, Yu. and Zhang, Y. 2008. Effect of natural antioxidants on kinetic behavior of acrylamide formation and elimination in low-moisture asparagine-glucose model system. *J. Food Eng.* 85: 105–115.
33. Katsuhito, H., Koriyama, N., Omori, H., Kuriyama, M., Arishima, T., and Kazunobu Tsumura, K. 2012. Simultaneous determination of 3-MCPD fatty acid esters and glycidol fatty acid esters in edible oils using liquid chromatography time-of-flight mass spectrometry. *Lwt-Food Science and Technology* 48: 204–08.
34. Sun, Li., Zhang, F., Yong, W., Chen, S., Yang, M.-L., Ling, Y., Chu, X., and Lin, J.-M. 2010. Potential sources of carcinogenic heterocyclic amines in Chinese mutton Shashlik. *Food Chem.* 123: 647–652.
35. Zhang, Y., Jiao, J., Cai, Z., Zhang, Y., and Ren, Y. 2007. An improved method validation for rapid determination of acrylamide in foods by ultra-performance liquid

chromatography combined with tandem mass spectrometry. *J. Chromatogr. A.* 1142: 194–198.

36. Huang, B., Han, Z., Cai, Z., Wu, Y., and Ren, Y. 2010. Simultaneous determination of aflatoxins B1, B2, G1, G2, M1 and M2 in peanuts and their derivative products by ultra-high-performance liquid chromatography–tandem mass spectrometry. *Anal. Chim. Acta* 662: 62–68.
37. Ventura, M., Guillen, D., Anaya, I., Broto-Puig, F., Lliberia, J. L., Agut, M., and Comellas, L. 2006. Ultra-performance liquid chromatography/tandem mass spectrometry for the simultaneous analysis of aflatoxins B1, G1, B2, G2 and ochratoxin a in beer. *Rapid Commun Mass Spectrom.* 20: 3199–3204.
38. Han, Z., Ren, Y., Liu, X., Luan, L., and Wu, Y. 2010. A reliable isotope dilution method for simultaneous determination of fumonisins B1, B2 and B3 in traditional Chinese medicines by ultra-high-performance liquid chromatography-tandem mass spectrometry. *J. Sep. Sci.* 33 17–18: 2723–33.
39. Wang, W., Hu, J., Yu, J., and Yang, M. 2010. Determination of N-nitrosodimethylamine in drinking water by UPLC-Ms/Ms. *J. Environ. Sci.* 22: 1508–1512.
40. Soleimany, F., Jinap, S., Faridah, A., and Khatib. 2012. A UPLC–Ms/Ms for simultaneous determination of aflatoxins, ochratoxin A, zearalenone, DON, fumonisins, T-2 Toxin and HT-2 toxin, in cereals. *Food Control* 25: 647–653.
41. Mouls, L. and Fulcrand, H. UPLC-ESI-Ms. 2012. Study of the oxidation markers released from tannin depolymerization: Toward a better characterization of the Tannin evolution over food and beverage processing. *Journal Mass Spectrom.* 47: 1450–1457.
42. Khan, M. R., Alothman, Z. A., Ghfar, A. A., and Wabaidur, S., 2013. Analysis of aflatoxins in nonalcoholic beer using liquid–liquid extraction and ultraperformance LC-MS/MS. *J. Sep. Sci.* 36: 572–577.
43. Mikulíková, R., Běláková, S., Benešová, K., and Svoboda, Z. 2012. Study of ochratoxin a content in South Moravian and foreign wines by the UPLC method with fluorescence detection. *Food Chem.* 133: 55–59.
44. Fu, Z., Huang, X., and Min., S. 2008. Rapid determination of aflatoxins in corn and peanuts. *J. Chromatogr. A.* 1209: 271–274.
45. Jackson, L. C., Kudupoje, M. B., and Yiannikouris, A. 2012. Simultaneous multiple mycotoxin quantification in feed samples using three isotopically labeled internal standards applied for isotopic dilution and data normalization through ultra-performance liquid chromatography/electrospray ionization tandem mass spectrometry. *Rapid Commun. in Mass Spectrom* 26: 2697–2713.
46. Zaidi, R., Kumar, S., and Rawat, P. R. 2012. Rapid detection and quantification of dietary mutagens in food using mass spectrometry and ultra performance liquid chromatography. *Food Chem.* 135: 2897–2903.
47. Wang, W., Ren, S., Zhang, H., Yu, J., An, W., Hu, J., and Yang, M. 2011. Occurrence of nine nitrosamines and secondary amines in source water and drinking water: Potential of secondary amines as nitrosamine precursors. *Water Res.* 45: 4930–4938.
48. Ghosh, C., Shinde, C. P., and Chakraborty, B. S. 2012. Influence of ionization source design on matrix effects during LC–ESI-MS/MS analysis. *J. Chromatogr. B* 893–894: 193–200.
49. Zhang, Y., Fang, H., and Zhang, Y. 2008. Study on formation of acrylamide in asparagine-sugar microwave heating systems using UPLC-MS/MS analytical method. *Food Chemistry* 108: 542–550.
50. International Agency for Research on Cancer. 1998. Some N-nitroso compounds. *IARC Monographs on the Evaluation of Carcinogen Risk to Humans*, 19. Lyon, France: International Agency for Research on Cancer.

51. Benešová, K., Běláková, S., Mikulíková, R., and Svoboda, Z. Monitoring of selected aflatoxins in brewing materials and beer by liquid chromatography/mass spectrometry. *Food Control* 25: 626–630.
52. Viklund, G., Mendoza, F., Sjöholm, I., and Skog, K. 2007. An experimental set-up for studying acrylamide formation in potato crisps. *LWT—Food Science and Technology* 40: 1066–1071.
53. European Commission. 2002. Commission Decision of 12 August 2002 Implementing Council Directive 96/23/EC concerning the performance of analytical methods and the interpretation of results. European Commission, Brussels.
54. van Deemter, J. J., Zuiderweg, F. J., and Klinkenberg, A. 1956. Longitudinal diffusion and resistance to mass transfer as causes of nonideality in chromatography. *Chem. Eng. Sci.* 5: 271–289.
55. Knox, J. H. 1977. Practical aspects of LC theory. *J. Chromatogr Sci.* 15: 352–364.
56. Giddings, J. C. 1965. Comparison of theoretical limit of separating speed in gas and liquid chromatography. *Anal. Chem.* 37: 60–63.
57. de Villiers, A., Lestremau, F., Szucs, R., Gélébart, S., David, F., and Pat, S. 2006. Evaluation of ultra performance liquid chromatography: Part I. possibilities and limitations. *J. Chromatogr A* 1127: 60–69.
58. Wang, Y., Ai, F., Ng, S.-C., and Tan, T. T. Y. 2012. Sub-2 μm porous silica materials for enhanced separation performance in liquid chromatography. *J. Chromatogr. A.* 1228: 99–109.
59. Gritti, F. and Guiochon, G. 2012. The current revolution in column technology: How it began, where is it going? J. *Chromatogr.* A 1228: 2–19.
60. Wang, X., Barber, W. E., and Long, W. J. 2012. Applications of superficially porous particles: High speed, high efficiency or both? *J. Chromatogr A* 1228: 72–88.
61. Toribio, F., Puignou, L., and Galceran, M.T. 1999. Evaluation of different clean-up procedures for the analysis of heterocyclic aromatic amines in a lyophilized meat extract. *J.Chromatogr.* A 836: 223–233.

5 Ultra-Performance Liquid Chromatography–Mass Spectrometry for the Determination of Capsaicinoids in Capsicum Species

Saikh Mohammad Wabaidur

CONTENTS

5.1 INTRODUCTION

5.1.1 Capsicums

Hot or spicy peppers are among the most popular food additives around the world because of their sensory attributes of pungency, aroma, and color. They are found to be very commercially important, as vast quantities of the diverse varieties of peppers are consumed around the world. The food industry usually uses enormous amounts of capsicum species as coloring and flavoring agents in different types of soups, sauces, snacks, candies, soft drinks, processed meats, and alcoholic beverages. Pepper fruits (*Capsicum annuum* L.) are also considered important vegetables and have been vastly used as vegetable foods in various parts of the world, including as a source of food supplements in the form of vitamins C and E [1,2] as well as provitamin A and carotenoids compounds [3–5]. There are more than 200 varieties of *Capsicum* species that can be found throughout the world [6]. The capsicum fruits vary in their size, shape, flavor, and sensory heat level. Among the various species of capsicum, the main species are, (i) *Capsicum annuum* (comprising the NuMex, Jalapeño, and Bell varieties), (ii) *Capsicum frutescens* (Tabasco variety), (iii) *Capsicum chinense* (Habanero and Scotch Bonnet varieties), (iv) *Capsicum baccatum* (Aji varieties), and (v) *Capsicum pubescens* (Rocoto and Manzano varieties) [6].

5.1.2 Capsaicinoids and Their Importance

The pungent metabolites found in the fruits of *Capsicum* species are called capsaicinoids, which were first discovered and isolated by Christian Friedrich Bucholz in 1816. The capsaicinoid compounds are a family of natural products isolated from the fruits of hot peppers. These substances produce the characteristic sensations associated with the ingestion of spicy food and are particularly irritating to the eyes, skin, nose, tongue, and respiratory tract. The capsaicinoids are a group of alkaloids with a structure of vanillylamide of branched fatty acids with 9–11 carbons [7,8]. Capsaicinoids are present in different quantities in various types of genus *Capsicum* [9]. In recent years, these alkaloid compounds have attracted the interest of scientists and researchers because they show promise of being powerful antioxidants [5,10], which protect the human body by suppressing the formation of free radicals associated with the normal natural metabolism of aerobic cells [11,12]. They are also notable for antimutagenic and antitumoral properties [13,14], and are commonly utilized as topical analgesics for many painful clinical conditions such as postherpetic neuralgia [15,16]. These different physical and chemical properties and applications of capsaicinoids make them very interesting compounds. More than 12 different capsaicinoids have been found in nature, among the most abundant are capsaicin (C; *trans*-8 methyl-*N*-vanillyl-6-nonenamide) and dihydrocapsaicin (DHC; 8 methyl-*N*-vanillylnonanamide), which are responsible for about 80–90% of the spiciness [17–20], and the less-abundant capsaicinoids are nordihydrocapsaicin (n-DHC), norcapsaicin, homocapsaicin (h-C), homodihydrocapsaicin (h-DHC), nornorcapsaicin, nornordihydrocapsaicin, nonivamide, and others (Figure 5.1) [20]. An accurate determination of the levels of various capsaicinoids has become important because

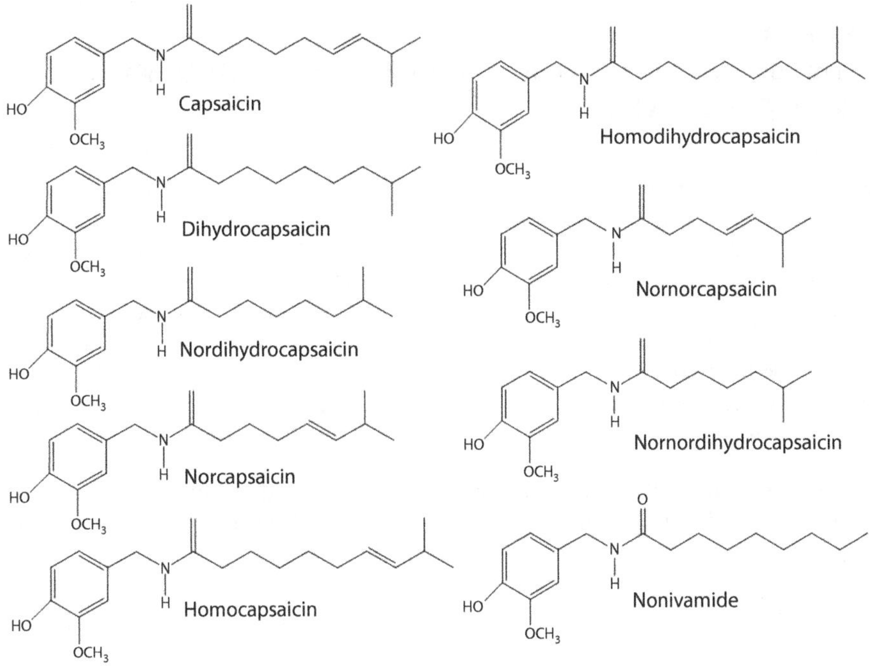

FIGURE 5.1 Chemical structures of capsaicinoids.

of the increasing demand by consumers for spicy foods and the increasing use in pharmaceutical sectors [21].

5.1.3 Pungency of Capsaicinoids

Pungency, which is also known as the organoleptic sensation of heat, is the most important determining factor of *Capsicum* species, and the pungency levels of peppers are generally expressed as Scoville heat units (SHU) [22]. The pepper contents in capsicum are calculated in microgram per gram and converted into SHUs [23] in order to classify them with respect to their various pungency levels. The conversion to SHU is done by multiplying the capsaicin content in pepper dry weight (g/g) by the coefficient corresponding to the heat value for pure capsaicin, which is a fixed value of 1.6×10^7. There are five levels of pungency have been classified in the literature using SHU, which are namely nonpungent, mildly pungent, moderately pungent, highly pungent, and very highly pungent [24].

The SHU officially measures the pungency level of a given capsicum sample. Among all other methods, the Scoville Scale remains the most widely used and well known across the globe. The greater the SHU value, the hotter the pepper. However, the Scoville heat varies from pepper to pepper depending on the nature of the pepper. A list of pungency levels of a few selected peppers according to SHU have been tabulated for convenience and better understanding (Table 5.1). Roughly, one part per million of chili heat is expressed as 1.5 Scoville units. There are also many

TABLE 5.1
List of Pungency Levels According to Scoville Heat Unit (SHU) of Few Selected Peppers Cultivars

Pepper	SHU	Pungency Level
Bell, Sweet Italian, Pepperoncini	0–500	Nonpungent
New Mexico, Ancho, Passila, Poblano, Sandia, Rocotillo	500–2500	Mildly pungent
Jalapeño, Chipotle	2500–10,000	Moderately pungent
Serrano	5000–23,000	Moderately pungent
De Arbol	15,000–30,000	Moderately pungent
Piquin, Aji, Cayenne	30,000–50,000	Highly pungent
Habenero, Scotch Bonnet	80,000–300,000 +	Very highly pungent
Hottest pepper recorded, Habenero	577,000	Very highly pungent

volatile compounds other than capsaicinoids that are found in capsicums; a detailed list of them in Habanero chili pepper cultivars (mg kg^{-1}dry fruit) is included in Table 5.2 [25].

5.1.4 Activities

In the past few decades, the various activities such as antioxidant, cell inflammation, antimutagenic, and toxicity of capsaicinoid compounds have been investigated and reported by many researchers and scientists. Materska and Perucka reported the determination of the antioxidant activity of the capsaicinoid compounds isolated from hot and semi-hot pepper fruits at two different maturity stages during their cultivation [26]. Reilly et al. have found that capsaicinoids are able to activate the vanilloid receptors, which are responsible for inflammation and epithelial cell death in animal and human bodies [27,28]. The characteristic pharmacological responses of capsaicinoids include severe irritation, inflammation, erythema, and transient hyper- and hypoalgesia at exposed sites such as, eyes, skin, nose, tongue, and respiratory tract. The vanilloid receptors are the calcium channel; when activated by capsaicinoids they produce the characteristic sensations and causes toxicity in many mammalian cell types [27–29].

It has also been observed that capsaicinoids causes cough reflex and neurogenic inflammation in respiratory tissues by promoting the amount of release of neuropeptides from neurons [30,31]. Recently, the U.S. Environmental Protection Agency (EPA) has confirmed that capsaicin works synergistically with particulate matter and neuropeptides to promote the production of inflammatory mediators by large number in the normal human body [30–32]. It was also reported that capsaicin, among all other capsaicinoids, inhibits the growth of immortalized and malignant cells [33,34] and induces apoptosis in various types of cancers [35–38]. A recent study demonstrated that capsaicinoid, especially capsaicin, induced apoptosis in human glioblastoma cells [39]. A few experimental analyses have demonstrated that capsaicin also inhibits NADH oxidase in melanoma cells [34]. The capsaicinoids, mainly capsaicin, has shown antimicrobial properties that indicate its potential as a natural inhibitor of

TABLE 5.2
List of Volatile Compounds Found in Habanero Chili Pepper Cultivars (mg/kg Dry Fruit)

Compound	KI	Red 1	Red 2	Red 3	Red 4	Orange1	Orange 2	Orange 3	Orange 4	Orange 5	Brown
Hexanal	800	t	t	t	t	t	0.01	0.01	t	t	0.01
(*E*)-2-Hexenal	854	–	0.01	0.01	t	0.02	0.02	0.03	0.01	0.02	0.03
(*Z*)-3-Hexenol	857	t	t	t	t	t	t	0.01	t	t	t
Hexanol	867	t	t	t	–	t	t	t	–	t	t
α-Pinene	939	–	t	–	–	t	0.01	t	–	–	t
Hexyl acetate	1009	–	–	–	t	t	t	–	–	–	0.02
Isobutyl isopentanoate	1018	–	–	–	–	–	t	t	–	–	t
Limonene	1031	t	t	t	t	t	t	t	–	–	t
Linalool	1098	t	t	t	–	t	0.01	0.01	t	0.01	t
Isopentyl 2-methylbutanoate	1100	–	–	–	–	–	t	t	–	t	0.01
Isopentyl isopentanoate	1103	0.01	t	–	–	0.02	0.06	0.06	0.02	0.04	0.06
2-Methylbutyl isopentanoate	1105	–	–	–	–	–	0.02	t	–	–	0.01
Hexyl isobutanoate	1149	0.02	0.04	–	0.01	0.05	0.07	0.06	0.02	0.02	0.09
Pentyl isopentanoate	1151	0.02	t	–	–	0.01	0.06	0.23	0.02	0.04	0.10
Isoprenyl pentanoate	1152	t	–	–	–	0.01	0.10	0.29	0.03	T	0.07
2-Isobutyl-3-methoxypyrazine	1186	–	–	–	–	0.01	0.01	0.01	0.01	0.01	0.01
α-Terpineol	1189	t	t	t	t	0.01	t	t	0.01	0.01	0.02
Methyl salicylate	1192	–	–	–	–	t	t	t	t	–	–
Hexyl 2-methylbutanoate	1236	0.08	0.25	0.09	0.07	0.24	0.37	t	0.18	0.40	0.39
Hexyl isopentanoate	1244	0.80	1.95	1.62	0.67	2.05	2.50	2.50	2.42	3.04	1.69
Heptyl butanoate	1291	0.10	0.11	0.10	0.11	0.12	0.10	0.10	0.11	0.11	0.13
(*Z*)-3-Hexenyl 2-methylbutanoate	1293	0.06	–	–	–	0.05	0.12	0.10	0.09	0.17	0.37
(*Z*)-3-Hexenyl isopentanoate	1295	0.31	0.77	0.24	0.16	0.18	0.63	0.72	0.54	1.34	1.40

continued

TABLE 5.2 (continued)
List of Volatile Compounds Found in Habanero Chili Pepper Cultivars (mg/kg Dry Fruit)

Compound	KI	Red 1	Red 2	Red 3	Red 4	Orange1	Orange 2	Orange 3	Orange 4	Orange 5	Brown
Hexyl pentanoate	1298	0.21	0.66	0.14	0.12	0.09	0.55	0.84	0.52	1.49	0.91
(*E*)-2-Hexenyl pentanoate	1299	0.06	0.07	–	–	0.01	0.13	0.21	0.10	0.21	0.14
Heptyl isobutanoate	1300	0.05	0.06	0.08	0.07	0.06	0.08	0.06	0.05	0.09	0.07
Pentyl isohexanoate	1303	–	–	–	–	0.01	t	t	0.01	t	t
9-Decanolide	1308	–	–	–	0.01	–	0.01	t	–	–	–
Hexyl tiglate	1328	–	–	–	–	t	–	–	–	–	t
Heptyl 2-methylbutanoate	1333	–	t	–	–	–	0.03	0.06	0.01	0.01	0.04
Heptyl isopentanoate	1338	0.03	0.04	–	–	0.02	0.13	0.21	0.02	0.07	0.04
Hexyl isohexanoate	1342	0.02	0.05	–	–	0.05	0.09	0.16	0.10	0.29	0.04
Heptyl pentanoate	1376	0.02	0.16	0.11	t	0.42	0.49	0.48	0.15	0.69	0.45
(*Z*)-3-Hexenyl hexanoate	1382	t	0.01	–	–	t	–	0.01	0.01	0.05	t
Hexyl hexanoate	1384	t	t	–	–	0.02	0.02	0.05	0.04	0.11	0.03
Decanoic acid	1385	t	t	–	–	t	t	0.01	0.01	0.02	0.01
β-Cubebene	1388	t	0.01	–	–	0.01	0.03	0.01	0.01	0.03	0.02
3,3-Dimethylcyclohexanol	1392	0.07	0.60	1.98	0.07	2.05	1.84	1.59	1.36	1.52	1.90
Benzyl pentanoate	1396	0.02	0.04	–	–	0.01	0.04	0.07	0.05	0.09	0.13
Octyl 2-methylbutanoate	1418	t	0.05	0.07	–	t	0.18	0.16	0.05	0.15	t
β-Caryophylene	1419	–	–	–	–	t	0.01	–	t	0.02	t
(*E*)-a-Ionone	1430	t	t	–	–	0.01	0.01	0.02	0.02	0.08	0.02
Octyl isopentanoate	1440	t	0.06	–	–	t	0.11	0.15	0.06	0.19	0.19
2-Methyl-1-tetradecene	1445	0.01	0.07	t	–	0.05	0.42	0.13	0.05	0.39	0.11

α-Himachalene	1447	–	0.01	–	–	t	0.02	0.01	0.05	t	t
Heptyl hexanoate	1449	–	t	t	–	t	0.01	0.01	0.02	0.01	0.01
α-Humulene	1455	–	–	–	–	t	–	t	t	t	–
(*E*)-b-Farnesene	1457	–	–	–	–	t	0.08	t	0.01	0.04	t
2-Methyltetradecane	1462	0.02	0.13	–	–	0.03	0.47	–	0.06	0.26	0.06
γ-Himachalene	1483	0.03	0.13	0.09	0.03	0.11	0.38	0.22	0.28	0.27	–
Germacrene D	1485	0.01	0.02	t	–	0.04	0.14	0.09	0.02	0.09	t
(*E*)-b-Ionone	1489	0.01	0.02	0.01	t	0.05	0.11	0.06	0.05	0.18	0.05
γ-Cadinene	1514	–	–	–	–	t	–	–	–	–	–
δ-Cadinene	1523	t	t	–	–	0.01	0.05	0.02	0.02	0.10	0.01
trans-Cadina-1(2),4-diene	1535	–	–	–	–	–	t	–	0.01	0.02	–
Hexyl benzoate	1575	–	–	t	t	0.01	t	t	t	t	t
Oxacyclotetradecan-2-one	1577	0.01	0.02	–	–	–	0.19	0.04	0.01	0.03	0.01
Tetradecanal	1610	0.01	–	–	–	0.01	0.09	0.12	–	0.11	0.01
Oxacyclopentadecan-2-one	1650	0.01	T	–	–	T	0.01	t	t	0.01	0.01
Pentadecanal	1707	T	T	T	–	0.01	0.02	t	0.01	t	t
Benzyl benzoate	1762	0.01	0.01	0.02	0.01	0.02	0.01	0.01	0.02	t	0.01
Tetradecanoic acid	1782	0.01	0.01	0.01	0.02	0.01	t	t	0.02	t	t
Hexadecanal	1811	0.01	0.02	0.02	0.02	0.02	0.01	t	0.02	0.01	0.01
Total volatiles		2.02	5.38	4.59	1.37	5.90	9.85	8.93	6.68	11.84	8.69

Source: Adapted from Pino, J. et al. 2007. *Food Chem.* 104:1682–1686. With permission.

Note: – Not detected; t means lower than 0.01 mg/kg dry fruit.

pathogenic microorganisms in various foods [40–43]. Additionally, it has also been used as a bird, animal, and insect repellent and as biochemical pesticide [44].

5.1.5 Human Exposure and Toxicological Effects of Capsaicinoids

People are exposed by capsaicinoids if they breathe in or come in contact with capsaicinoid products. They could also be exposed by eating or smoking a product that contains capsaicinoids. Another possibility for exposure is touching freshly sprayed plants of the capsaicinoid containing sprays. There is an abundance of literature on the studies of capsaicinoid toxicity, which demonstrated extreme differences in their toxicological effects depending upon the route of exposure [45]. Glinsukon et al. reported that oral and topical capsaicinoid exposures yielded lethal dosage (LD50) values in mice at 190 and 500 mg/kg, respectively, while the intravenous and intratracheal instillation routes produced LD50 values of 0.56 and 1.6 mg/kg, respectively [45]. LD50 values are a standard measurement of acute toxicity and are expressed in mg of pesticide per kg of body weight of tested animals. It represents the individual dose required to kill 50% of a population of test animals (e.g., rats, fish, mice, cockroaches). With the help of LD50 values it is possible to compare relative toxicities among pesticides. The lower the LD50 dose indicates that the pesticide is more toxic. Most importantly, the cause of death for animals due to exposure to capsaicinoids is very rapid onset of convulsions (within 0.05 [i.v. (intravenous)] to 3.38 [p.o. (oral therapy)] min) due to dysfunction and failure of cardiovascular and pulmonary systems [45].

Human beings frequently use capsaicinoid-containing products in various forms on a daily basis, and according to the majority of diverse people, uptake of capsaicinoids via both topical and oral routes is perceived as safe under normal conditions. However, extreme exposure to capsaicinoids may result in acute toxicity, severe injury, and fatality in the human body [46]. The extreme effect can be found in the case of a child undergoing homeopathic treatment due to a digestive disorder. During homeopathic treatment the excessive intakes of oleoresin capsicum resulted in death of the child [47]. On the other hand, harsh cardiovascular and pulmonary toxicities have also been found in people exposed to pepper sprays, particularly in those individuals who already had preexisting respiratory or cardiovascular diseases and individuals under the influence of illicit drugs. Although capsaicinoid-containing products are often used in cooking or eating, excessive consumption can lead to irritation of the mouth, stomach, and intestines, and people may experience vomiting and diarrhea. Coughing, breathing difficulty, teary eyes, nausea, nasal irritation, and temporary blindness can occur when people inhale capsaicinoid containing sprays. It can also cause severe irritation in the eye and skin of an animal or human body.

The main objectives of this chapter are to gain widespread knowledge on *Capsicum* and its extracted compounds, which are responsible for its pungent taste. This chapter includes a brief description on the extraction and various biological activities of the capsaicinoids isolated from pepper fruits. The chapter has also focused on the clinical applications and recent advances in the study of these capsaicinoids compounds. A detailed discussion of UPLC–MS technique used for the separation and determination of capsaicinoids in various capsicum species is also presented.

5.2 EXTRACTION METHODS

Hot peppers (e.g., *C. annum* L.) are usually grown in the agricultural field and collected from the local markets. The collected pepper samples are then rinsed with distilled water and patted dry in sunlight. Pepper fruits will normally be analyzed and picked at the stage of full ripeness, since during that period they can be easily divided into pericarps, placenta, and seeds. The sample size of the fresh peppers can be reduced by removing the stem and a portion of the pericarp since the stem and pericarp have been shown to contain only negligible concentrations of the capsaicinoids [48–50]. Then the capsaicinoids are extracted from the pericarps, placenta, and seeds and used for analysis. The extraction of capsaicinoids from the *Capsicum* samples is achieved by various methods [51–53].

A variety of methods have been used for the extraction and analysis of these capsaicinoids compounds. In particular, extraction of capsaicinoids from peppers has been reported, such as, maceration [54], magnetic stirring [55], Soxhlet [56,57], extraction by pressurized liquids [58], and enzymatic extraction [59], among others. These techniques are employed with an increasing demand since they allow the extraction process to be automated, extraction times to be shortened, the consumption of solvents to be reduced, and the exposure of operative personnel to solvents to be minimized. However, conventional extraction techniques need longer extraction times and consumption of a high volume of solvent [60]. To shorten extraction times and reduce the consumption of organic solvents, more efficient extraction techniques are employed, such as, ultrasound-assisted extraction (UAE) [61], microwave-assisted extraction (MAE) [62–64], and supercritical fluid extraction [65]. MAE is a process of using microwave energy to heat solvents in contact with a sample in order to separate target analytes from the sample matrix into the solvent. The ability to rapidly heat the sample solvent mixture is the main advantage of MAE techniques. The application of close vessels allows the extraction to be performed at elevated temperatures, which accelerates the mass transfer of target compounds from the sample matrix. Usually a typical MAE extraction procedure takes 15–30 min and employs small solvent volumes in the range of 10–30 mL. These volumes are much smaller than the volumes used by conventional extraction procedures, and several samples can be extracted simultaneously by MAE techniques. In MAE techniques, the dried plant material is used for extraction in most cases, but the minute microscopic traces of moisture among the plant cells serves as the target for microwave heating. When the moisture is being heated inside the plant cell due to application of microwaves, it evaporates and generates tremendous pressure on the cell wall, which facilitates leaching out of the active constituents from the ruptures cells to the surrounding solvent [66].

The UAE technique is based on the employment of the energy derived from ultrasounds that facilitate the extraction of analytes from the solid sample by the organic solvent [66]. The enhancement of extraction efficiency of organic compounds by UAE is attributed to the phenomenon of cavitation produced in the solvent by the passage of ultrasonic waves [67]. The uses of higher temperatures in UAE lead to an increase in the efficiency of the extraction process [67–69]. The ultrasound-assisted extraction of capsaicinoids from pepper has also been carried out by Barbero et al. in an ultrasonic bath employing different extraction solvents, such as methanol, ethanol, and acetone [70]. Alothman et al. extracted capsaicinoids from capsicum samples employing liquid–liquid extraction techniques using ethanol as an extracting solvent [71]. The extraction method can be explained briefly as follows. In the beginning, the peppers were peeled, and the peduncle and seeds were separated. The pericarp and the placenta of capsicum fruits were triturated with a conventional beater to prepare a homogeneous sample for the UPLC–MS analysis of capsaicinoids. Then the dried pepper was placed in 120 mL glass bottles with Teflon-lined lids in an identical ratio of sample/ethanol (1:1; g/mL). After that the bottles were capped and placed in an 80°C water bath for 4 h and manually swirled after each hour. Finally, the samples were removed from the water bath and cooled at room temperature. The supernatant of the samples (5 mL) was filtered through 0.45 μm filter paper using a 5 mL disposable syringe (Millipore Corporation) in a UPLC sample vial, capped and stored at 4°C in a refrigerator until analysis is done.

5.3 DETERMINATION METHODS

A large variety of analytical methods has been utilized for the analysis of capsaicinoids in capsicum species, including colorimetry [72–74], gas chromatography (GC) [75–78], complexation chromatography [79], gas chromatography/mass spectrometry (GC/MS) [25] liquid chromatography (LC) [76,80–84], gas chromatography/liquid chromatography–mass spectrometry (GC/LC–MS) [78,85], nuclear magnetic resonance (NMR)-flow probe analysis [86], spectrophotometry [7,87–90], amperometric titration [91], micellar electrokinetic capillary chromatography [21,92], and sensory methods such as an electronic nose [56]. In recent times, reversed-phase-high-performance liquid chromatography (RP-HPLC), which offers sufficient accuracy and precision, has become the most frequently used method for analysis of capsaicinoids because of its rapidity and reliability during analysis [93]. Most recently, HPLC methods combined with ultra violet (UV) [94], fluorescence [64], electrochemical [95], and MS [96–100] detection techniques and GC with flame ionization detection [25,101] have also been developed for the determination of capsaicinoids in peppers. Liu et al. also reported the capillary electrophoresis (CE) method, a complimentary separation technique to LC for determination of capsaicin and dihydrocapsaicin in *C. anuum* species [102].

More recently, ultra-high-performance liquid chromatography (u-HPLC) method coupled with MS has been adopted in many areas of food and biological analysis due to its rapid analysis and remarkabl separation capacity [103,104]. u-HPLC method is economical and environmentally friendly due to its extremely rapid analysis and the consumption of solvent for the mobile phase can be reduced 5- to 10-fold compared to the conventional HPLC method [9,105].

However, there are drawbacks. Longer analysis times and high solvent consumption of various separation methods are of main concern for the analytical chemist. Therefore, high speed and low sample consumption of analysis are being increasingly demanded in many areas where HPLC is applied, including pharmaceutical and food analysis, in order to increase throughput and reduce loss of time and extra sample volume. At the same time, the rapid separation of samples is an analytical stage that requires high efficiency as well as speed, due to the complexity of the sample matrix, and hence it is particularly challenging to achieve the goals. Therefore, the development of a rapid, sensitive, and reproducible method has been required for separation and determination of capsaicinoid compounds. The addition of ultra-performance liquid chromatography–mass spectrometry (UPLC–MS) method fulfilled these aforementioned demands and showed some complementary advantages to the conventional HPLC–MS, u-HPLC methods in terms of shorter analysis times, low sample volume, and much improved sensitivity [71]. Therefore, nowadays this UPLC–MS technique is routinely performed in pharmaceutical industries and related contract research institutes, laboratories concerned with biochemistry, biotechnology, environmental analysis, natural product research, and several other research fields. The UPLC–MS method has successfully been applied for the determination of n-DHC, C, DHC, h-C, and h-DHC present in the varieties of hot peppers [71].

5.3.1 Chromatographic Separations

LC is an important technique in modern separation chemistry. The most common forms of LC include reversed-phase, normal-phase, size-exclusion, and ion-exchange chromatography. In liquid chromatographic separations the components of a sample are distributed between stationary and mobile phases. Usually the stationary phase is a solid packed in a column, while the mobile phase is a liquid or a mixture of liquid. The LC separations are achieved based on differences in the physico-chemical properties of the molecules, such as size, mass, and volume. In recent times, the columns are packed with very small size particles (2–5 mm in average particle size), which allow a relatively high pressure (50–350 bar) and are referred to as high-performance LC. The smaller size particles provide larger surface area and produce better separations for the target analytes, but the pressure increases by the inverse of the particle diameter squared [106,107]. This instrumental technique has been used to separate a mixture of compounds in analytical chemistry and biochemistry with the purpose of identifying, quantifying, or purifying the individual components of the mixture. In HPLC the sample is forced by a liquid which is a mobile phase at high pressure through a column which is packed with a stationary phase composed of irregularly or spherically shaped particles, a porous monolithic layer, or a porous membrane. In recent times, HPLC has been considered among the most powerful tools in analytical chemistry, since compounds in trace concentration levels as low as parts per trillion (ppt) can be easily determined by this technique. HPLC has been applied to many sample types, such as pharmaceuticals, food, nutraceuticals, cosmetics, environmental matrices, forensic samples, and industrial chemicals.

5.3.2 Ultra-Performance Liquid Chromatography

Further advances in HPLC instrumentation and column technology were made in 2004, with significant increases to the resolution, speed, and sensitivity in liquid chromatographic separations. Columns packed with smaller particles (1.7 mm) and instrumentation with specialized capabilities designed to deliver mobile phase at 15,000 psi (1000 bars) were needed to achieve a new level of performance. UPLC is now considered a new direction for LC, designed with specialized capabilities to deliver mobile phases at up to 15,000 psi. It improves chromatography in three areas with respect to conventional LC, including speed, resolution, and sensitivity [108–112]. As the UPLC system uses fine particles, there is a significant increase in efficiency and it is not affected at increased linear velocities or flow rates responsible for peak broadening [108,113,114]. Also, the smaller particles extend the speed and peak capacity of the system to new limits while maintaining an acceptable loss of load, known as ultra performance. Usually very low volumes with minimal carryover are injected in the UPLC system, which is responsible for increased sensitivity [115]. The data in Table 5.3 shows the comparison between a UPLC and HPLC system.

In recent times, UPLC technique has routinely been performed for analysis of pharmaceutical formulations; natural products; foods and herbal medicines in various research institutes; laboratories concerned with biochemistry, biotechnology, environmental analysis; natural product research; and several other research fields. For the analysis of foods, natural products, or medicines, analytical scientist need to extend their understanding to provide evidence-based validation and effectiveness to establish safety parameters for their production. Also, the rapid separation of samples is an analytical stage that requires high efficiency as well as speed due to the complexity of the sample matrix, and hence it is particularly challenging to achieve. UPLC provides high-quality separations and detection capabilities to identify active compounds in highly complex samples that result from natural products and analyzed foods.

UPLC is widely used for analysis of foods, natural products, and herbal medicines. Purification and qualitative and quantitative chromatography and mass spectrometry are being applied to determine and characterize active food ingredients. UPLC provides high-quality separations and detection capabilities to identify active compounds in highly complex samples that results from various foods, food products, and traditional medicines. UPLC increases productivity in both chemistry and instrumentation by providing increased resolution, speed, and sensitivity for LC. When many scientists experience separation barriers with conventional HPLC, UPLC extends and expands the utility of chromatography. The main advantage is the reduction of analysis time, which also means reduced solvent consumption, which is very important in many analytical laboratories. The time spent for optimizing new methods can also be greatly reduced. The time needed for column equilibration while using gradient elution and during method validation is much shorter. Sensitivity can

TABLE 5.3
Comparison between HPLC and UPLC Systems

Parameters	HPLC	UPLC
Particle size	3–5 μm	Less than 2 μm
Maximum backpressure	35–40 MPa	103.5 MPa
Analytical column	Alltima C_{18}	Acquity UPLC BEH C_{18}
Column dimensions	150 × 3.2 mm	150 × 2.1 mm
Column temperature	30°C	65°C
Injection volume	5 μL (std. in100% MeOH)	2 μL (std. in100% MeOH)

TABLE 5.4
Advantages and Disadvantages of UPLC Techniques

Advantages of UPLC	Disadvantages of UPLC
i. Mainly UPLC techniques; decreases run time and increases sensitivity of the developed method	i. In UPLC methods, normally very high backpressure is required, consequently it requires more frequent maintenance of the columns and instruments that reduce the lifetime of the UPLC columns
ii. Provides much more selectivity, sensitivity, and dynamic range to the LC analysis	ii. Few papers have reported that even higher performance can be demonstrated by using stationary phases of particle size around 2 μm without the adverse effects of high backpressure, and less than 2 μm packing materials are generally nonregenerable, which are also the limitations of this technique and thus result limited use [115,117]
iii. Able to maintain resolution performance and the faster-resolving power quickly quantifies targeted and nontargeted compounds during the LC assay	
iv. Faster analyses are achieved through the use of novel material of very fine particle size (1.7 μm) as stationary phase	
v. Less solvent consumption, and by reducing process cycle times decreases the operation cost and minimizes carryover	
vi. Increases sample throughput and enables manufacturers to produce materials that consistently meet or exceeds the product specifications [108]	
vii. Delivers real-time analysis in step with manufacturing processes and can ensure end-product quality, including final release testing	

Source: Adapted from Srivastava, B. et al. 2010. *Int. J. Pharm. Qual. Assur.* 2:19–25. With permission.

be compared by studying the peak width at half height. The advantages and disadvantages of UPLC techniques are provided in Table 5.4 [108,115–117].

5.3.2.1 Acquisitions and Confirmation Techniques of Capsaicinoids

The capsaicinoids are extracted using different solvents and more recently ultrasound-assisted extraction [61], extraction by means of supercritical fluids [65], extraction by pressurized liquids [58] and enzymatic extraction [59], and analyzed by HPLC [64,93–95], GC [72–76], hyphenated systems as HPLC–MS [96–100], and GC–MS [77]. Normally the GC methods require derivatization of the compounds to make them sufficiently volatile for determination. There are many other reported papers have been found in the literature for the analytical separation, quantitation, and identification of naturally occurring capsaicinoids in different matrices. Select matrices are discussed here.

Choi et al. [118] analyzed eight capsaicinoids and two piperines in fresh Korean red peppers and pepper containing foods by HPLC, where analyses were carried out after 60 min of solvent extraction and run time was approximately 152.3 min. Ha et al. analyzed capsaicin in Gochujang (a fermented food from Korea) using liquid extraction and shaking over 3 h and after a large purification procedure finally injected in a GC combined with flame ionization detector (GC-FID) [119]. Spiecer and Almirall compared solid-phase micro extraction (SPME) with liquid–liquid extraction for the determination of capsaicin and dihydrocapsaicin in aerosol defense sprays from fabrics [120]. They concluded that SPME resulted in better recovery and identification of lower quantities of the compounds of interest when compared with solvent extraction, while the capsaicinoids are extracted with different solvents that take about 4 h and are analyzed by HPLC utilizing at least another 1 h.

Barbero et al. have achieved the separation and quantification of the capsaicinoids in different varieties of peppers by developing an HPLC method with fluorescence detection utilizing acidified water (0.1% acetic acid as solvent A) and acidified methanol (0.1% acetic acid as solvent B) as a mobile phase [64]. Figure 5.2 shows a typical chromatogram obtained by Barbero et al. using their proposed method. They performed the quantification of nordihydrocapsaicin, capsaicin, dihydrocapsaicin, homocapsaicin, and homodihydrocapsaicin present in three varieties of peppers: cayenne pepper (*C. frutescens*), long marble pepper (*C. annuum*), and round marble pepper (*C. annuum*).

Perucka and Oleszek have reported extraction and determination of capsaicinoids in fresh fruit of hot pepper using spectrophotometry and HPLC techniques [79]. Cooper et al. developed a reversed-phase HPLC method utilizing a conventional C_{18} column to separate capsaicin, dihydrocapsaicin, and nordihydrocapsaicin present in hot peppers [121]. The isocratic mobile phase (60:40 [v/v] methanol/water at a flow rate of 1.5 mL/min) was employed and achieved the separation of these three capsaicinoids in 28 min [121]. Krajewska and Power developed a reversed-phase HPLC

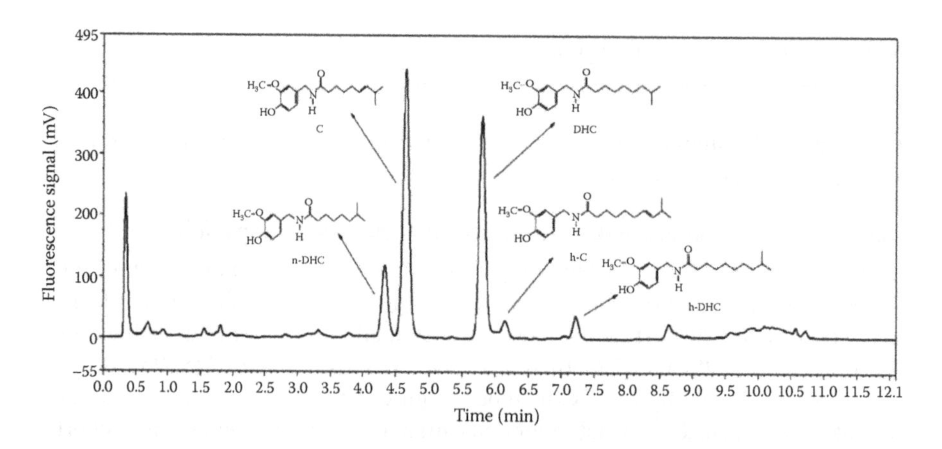

FIGURE 5.2 Chromatogram of pepper extract and chemical structures of nordihydrocapsaicin (n-DHC), capsaicin (C), dihydrocapsaicin (DHC), homocapsaicin (h-C), and homodihydrocapsaicin (h-DHC). Fluorescence detection: excitation 280 nm, emission 310 nm. (Adapted from Barbero, G. F., Palma, M. and Barroso, C. G. 2006. *Anal. Chim. Acta* 578:227–233. With permission.)

method utilizing a conventional C_{18} column to separate capsaicin and dihydrocapsaicin in *Capsicum* fruit extracts [122]. The mobile phase was a gradient regime of methanol and water at a flow rate changing from 0.9 to 1.8 mL/min and the separation of these two capsaicinoids was obtained in a time of 16 min [122]. Kozukue et al. have recently developed a separating method for homocapsaicin-I, homocapsaicin-II, homodihydrocapsaicin-I, homodihydrocapsaicin-II, and nonivamide with three minor capsaicinoids present in peppers by employing LC–MS technique [99]. They achieved the separation of these eight capsaicinoids in a time of 66 min, which is a really long time for analysis [99].

Liu et al. determined capsaicin and dihydrocapsaicin in *C. annuum* and related products by CE with a mixed surfactant system [102]. The mixed surfactant systems composed of two or more surfactants have shown unique selectivity because of synergism effect due to mixing of anionic and nonionic surfactants [123].

Laskaridou-Monnerville developed micellar electrokinetic capillary chromatography, which is a modified capillary chromatography technique and was applied to determine capsaicin and dihydrocapsaicin in different varieties of *C. frutescens* [19]. Using their method, they had separated the samples by differential partition between micelles (pseudo stationary stage) and the mobile phase (aqueous buffer at pH 9). Both of the capsaicinoid compounds, capsaicin and dihydrocapsaicin, were detected within 11 min with an excellent resolution [19].

5.3.2.2 Hyphenation of UPLC with MS Techniques

In the past decade, a number of analytical techniques have been developed for the determination and identification of capsaicinoids by using ionization techniques that permit HPLC–MS coupling, such as atmospheric pressure chemical ionization (APCI) [100,124] and electrospray interfaces (ESI) [71], using ion-trap or quadrupole mass spectrometers. UPLC coupled with MS has attracted much attention in the separation, quantification, and identification of various compounds. Since high speed and low sample consumption are being achieved, as well as the rapid separations of samples with high efficiency were acquired using ultra performance liquid chromatography-mass spectrometry technique. Different analytical methods have been developed for the quantification of capsaicinoids in different pepper samples. Ana et al. have studied the quantitative inheritance of capsaicin and dihydrocapsaicin contents in fruits in an intra-specific cross of *C. annuum* L. across two different environments, spring and summer [125]. They have used high-performance liquid chromatography–electrospray ionization/time-of-flight mass spectrometry (HPLC–ESI/MS[TOF]) as a method to identify and quantify capsaicin and dihydrocapsaicin in extracts of pepper fruits.

Most recently, Alothman et al. have developed an ultra-performance liquid chromatography–mass spectrometry method for the analysis of major capsaicinoids (capsaicin, dihydrocapsaicin) and minor (nordihydrocapsaicin, homocapsaicin, and homodihydrocapsaicin) compounds present in different *Capsicum* samples [71]. Extraction of capsaicinoids was achieved by employing liquid–liquid extraction techniques using ethanol as an extracting solvent. The chromatographic separations of capsaicinoids were achieved by reversed phase C_{18} column with gradient mobile phase (solvent A: acetonitrile and solvent B: water with 0.1% formic acid). Their developed method showed some complementary advantages over the conventional

HPLC–MS method in terms of shorter analysis times and improved sensitivity. First, an optimization process (Quan-optimization) for each analyte was performed to optimize cone voltage for all the individual capsaicinoids. This process was performed by infusing each individual capsaicinoid standard solution (50 μg/mL) to the ion source of the mass detector. The tuning of the method was performed to obtain the precursor ions with maximum intensity. Optimal MS conditions for each capsaicinoid were obtained using intellistart software program. The electrospray ionization source (Z-spray) was operated in positive ionization mode and the data acquisition was performed using selected ion reaction (SIR) monitoring. They achieved the chromatographic separation of capsaicinoids under ambient temperature conditions with a Waters Acquity UPLC system equipped with a quaternary pump (Waters, Mildford, MA, USA) and Acquity BEH C_{18} column (100 mm × 2.1 mm i.d., 1.7 μm particle size; Waters). The gradient elution program 0–8 min, 60–50% B with equilibration time 2 min was employed for better separation, while solvent A was acetonitrile and solvent B was water with 0.1% formic acid. The capsaicinoids were eluted according to their molecular size (Figures 5.3 and 5.4), wherein the relatively small molecules with polar groups are less retained on C_{18} stationary phase and were eluted first. The total run time to obtain the separation of these five capsaicinoids using the developed

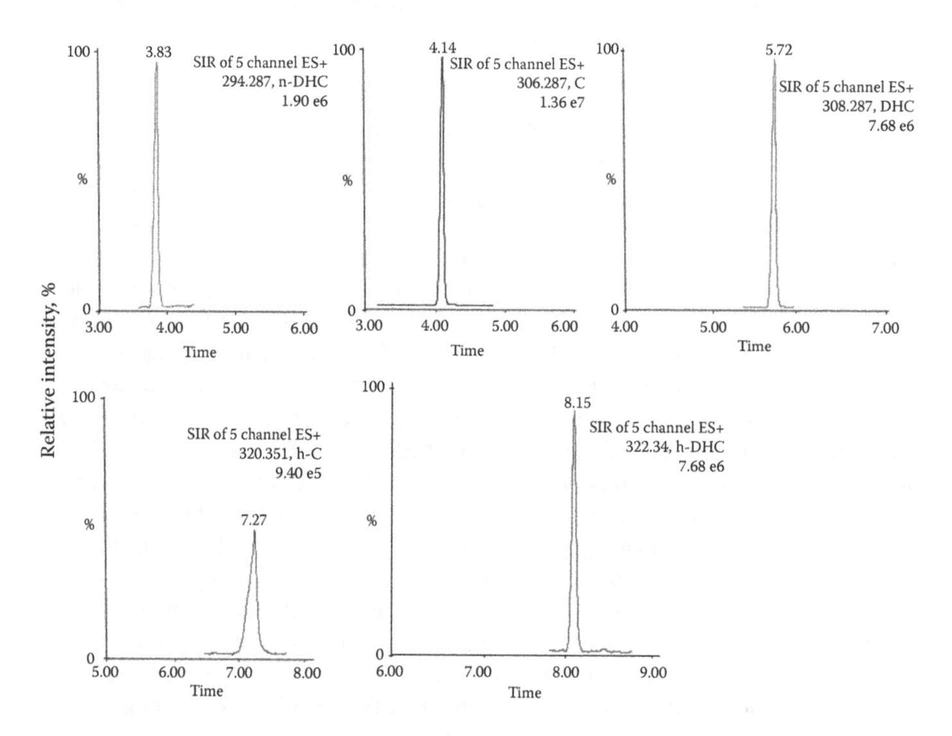

FIGURE 5.3 Chromatogram of the separation in standard mixture of capsaicinoids–SIR record is displayed from calibration level 50 μg/mL. Each chromatogram displays SIR transition for individual compound. (Adapted from Alothman, Z. A. et al. 2012. *J. Sep. Sci.* 35:2892–2896. With permission.)

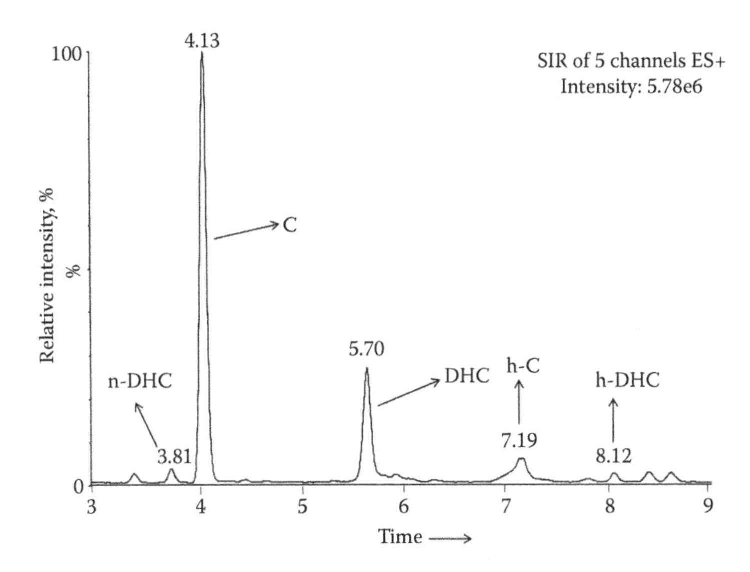

FIGURE 5.4 Chromatogram of the red chili extract obtained by using optimized UPLC–MS method. (Adapted from Alothman, Z. A. et al. 2012. *J. Sep. Sci.* 35:2892–2896. With permission.)

techniques was less than 9 min. The identifications of the compounds were confirmed by matching retention times as well as molecular mass of each component obtained under identical UPLC–MS conditions.

5.4 LEVELS OF CAPSAICINOIDS IN VARIOUS *CAPSICUM* SAMPLES

Alothman et al. applied the UPLC–MS method to quantify the major and minor capsaicinoids present in six varieties of peppers: hot chili, red chili, green chili, red pepper, green pepper, and yellow pepper [71]. The results for the concentration of capsaicinoids in the analyzed pepper samples obtained by Alothman et al. are shown in Table 5.5 [71]. The results concluded that C and DHC are the major capsaicinoids in all of the six varieties of analyzed peppers, which is in agreement with the reported results [51]. The corresponding pepper contents were calculated in microgram per gram and converted into SHUs in order to classify them with respect to their various pungency levels. There are five levels of pungency classified using SHU, which are nonpungent, mildly pungent, moderately pungent, highly pungent, and very highly pungent [24]. The analysis results were found to be that hot chili is the most pungent (76727.81 SHU) among the peppers studied, which are red chili (31058.38 SHU), green chili (10995.23 SHU), green pepper (3196.99 SHU), red pepper (2382.72 SHU), and yellow pepper (419.472 SHU) [71]. The pungency level of analyzed samples was decreased as follows: hot chili (very highly pungent), red chili (highly pungent), green chili (moderately pungent), green pepper (moderately pungent), red pepper (mildly pungent), and yellow pepper (nonpungent).

TABLE 5.5
Concentration (μg/g) of Capsaicinoids and Their Recovery in the Analyzed Pepper Samples ($n = 3$)

Sample	Hot Chili μg/g ± SD[a]	Recovery (%)	Green Chili μg/g ± SD[a]	Recovery (%)	Red Chili μg/g ± SD[a]	Recovery (%)	Red Pepper μg/g ± SD[a]	Recovery (%)	Green Pepper μg/g ± SD[a]	Recovery (%)	Yellow Pepper μg/g ± SD[a]	Recovery (%)
n-DHC	140.3 ± 3.5	88.2	12.2 ± 1.6	89.3	152.9 ± 3.7	90.2	4.0 ± 0.2	93.7	6.3 ± 0.3	92.2	1.1 ± 0.1	93.3
C	4795.5 ± 30.5	90.3	687.2 ± 9.5	89.7	1941.2 ± 20.6	90.7	148.9 ± 3.6	94.1	199.8 ± 4.3	93.1	26.2 ± 2.2	94.0
DHC	1399.3 ± 15.5	88.4	399.7 ± 7.1	88.9	1585.3 ± 16.6	90.1	41.6 ± 2.8	93.9	58.4 ± 3.1	92.7	7.8 ± 0.5	93.4
h-C	207.8 ± 4.1	89.2	527.4 ± 7.3	89.3	404.5 ± 7.0	90.0	2.2 ± 0.1	93.0	3.9 ± 0.2	92.4	nd	nd
h-DHC	139.4 ± 3.5	90.5	75.3 ± 2.5	89.0	118.4 ± 3.3	89.5	1.3 ± 0.1	93.2	6.1 ± 0.3	92.0	nd	nd

Source: Adapted from Alothman, Z. A. et al. 2012. *J. Sep. Sci.* 35:2892–2896. With permission.

Note: nd—not detected.

[a] Mean of three measurements.

5.5 SUMMARY AND CONCLUSIONS

Considering the importance of capsaicinoids found in capsicum species, the requirement of the potential knowledge for all aspects of identification and determination processes are highly demanded. Exposure to capsaicinoids is mainly caused by intentional and unintentional eating, inhaling, or touching of capsaicinoid products used in the daily lives of humans, animals, and plants. According to their vast applications in food and medicinal industry it is highly demanded to make people aware of their toxicological and nontoxicological effects to human body. These substances are widely used in foods since they produce characteristic sensations associated with the ingestion of spicy food. So, it is highly important to know how much too much consumption of these capsaicinoid materials. The antioxidant activity, the activation of vanilloid receptors that are responsible for inflammation and epithelial cell death in animal and human bodies and the characteristic pharmacological responses of capsaicinoids, including severe irritation, inflammation, erythema, and transient hyper- and hypoalgesia at exposed sites such as, eyes, skin, nose, tongue, and respiratory tract are briefly discussed in this chapter. Therefore, the only possibilities to minimize the exposure of these compounds are to obtain enough knowledge and make people aware of them.

These objectives have been achieved by applications of various extraction and determination methods to quantitate the compounds precisely and accurately. Therefore, the highly sensitive and accurate methods have been discussed with their merits and demerits in this book chapter. Mainly UPLC–MS technique using tandem quadruple mass spectrometer has been discussed in detail. The UPLC–MS has shown promising breakthroughs in the field of analytical chemistry. In addition, the incorporation of electro spray and/or atmospheric-pressure chemical ionization ion source has made UPLC–MS techniques more promising, especially with respect to easier operation, robustness of method, remarkably low detection limits, and applicability ranges for analysis. In this book chapter, the determinations of capsaicinoids using UPLC–MS techniques in different *Capsicum* species have been enlightened in detail. The various volatile compounds found in capsicum species are also reported here. The chapter also covers brief discussion on the pungency level of capsicums on the basis of SHU. According to the reported method the hot chili has shown a very high pungency level (76727.81 SHU) compared to red chili (31058.38 SHU), green chili (10995.23), green pepper (3196.99), red pepper (2382.72), and yellow pepper (419.472). The pungency level of analyzed capsicum samples was decreased as follows: hot chili (very highly pungent), red chili (highly pungent), green chili (moderately pungent), green pepper (moderately pungent), red pepper (mildly pungent), and yellow pepper (nonpungent) [71].

ABBREVIATIONS

C	Capsaicin
CE	Capillary electrophoresis
DHC	Dihydrocapsaicin
EPA	Environmental Protection Agency

GC	Gas chromatography
GC-FID	Gas chromatography-flame ionization detector
GC/LC–MS	Gas chromatography/liquid chromatography–mass spectrometry
GC/MS	Gas chromatography/mass spectrometry
h-C	Homocapsaicin
h-DHC	Homodihydrocapsaicin
HPLC–ESI/MS(TOF)	High-performance liquid chromatography–electrospray ionization/time-of-flight mass spectrometry
i.v.	Intravenous
LC	Liquid chromatography
LD50	Lethal dosage
MAE	Microwave-assisted extraction
MS	Mass spectrometry
n-DHC	Nordihydrocapsaicin
NMR	Nuclear magnetic resonance
P.O.	Oral therapy
RP-HPLC	Reversed-phase high-performance liquid chromatography
SHU	Scoville heat unit
SPME	Solid-phase micro extraction
UAE	Ultrasound-assisted extraction
u-HPLC	Ultra-high-performance liquid chromatography
UPLC–MS	Ultra-performance liquid chromatography combined with mass spectrometry
UV	Ultraviolet

ACKNOWLEDGMENTS

The author would like to extend his sincere appreciation to King Saud University, Deanship of Scientific Research, College of Science Research Center for its supporting of this book chapter.

REFERENCES

1. Palevitch, D. and Craker, L. E. 1995. Nutritional and medicinal importance of red pepper (*Capsicum* spp.). *J. Herbs Spices Med. Plants* 3:55–83.
2. Daood, H. G., Vinkler, M., Markus, F., Hebshi, E. A., and Biacs, P. A. 1996. Antioxidant vitamin content of spice red pepper (paprika) as affected by technological and varietal factors. *Food Chem.* 55:365–372.
3. Krinsky, N. I. 1994. The biological properties of carotenoids. *Pure Appl. Chem.* 66:1003–1010.
4. Krinsky, N. I. 2001. Carotenoids as antioxidants. *Nutrition* 17:815–817.
5. Perucka, I. and Materska, M. 2003. Antioxidant activity and contents of capsaicinoids isolated from paprika fruits. *Pol. J. Food Nutr. Sci.* 2:15–218.
6. Pruthi, J. S. 2003. In Capsicum, The genus *Capsicum*, In *Advances in Post-Harvest Processing Technologies of Capsicum*, ed. A. K. De, CRC Press, Boca Raton, Florida, USA.

7. Suzuki, T. and Iwai, K. 1984. Constituents of red pepper species: Chemistry, biochemistry, pharmcology, and food science of the pungent principle of *Capsicum* species. In *The Alkaloids: Chemistry and Pharmacology,* ed. A. Brossi, Academic Press, New York, USA, pp. 227–299.
8. Ayhan, T. and Feramuz, O. 2007. Assessment of carotenoids, capsaicinoids and ascorbic acid composition of some selected pepper cultivars (*Capsicum annuum* L.) grown in Turkey. *J. Food Comp. Anal.* 20:596–602.
9. Hoffman, P. G., Lego, M. C., and Galetto, W. G. 1983. Separation and quantitation of red pepper major heat principles by reverse high performance liquid chromatography. *J. Agric. Food Chem.* 31:1326–1330.
10. Henderson, D. E., Slickman, A. M., and Henderson, S. K. 1999. Quantitative HPLC determination of the antioxidant activity of capsaicin on the formation of lipid hydroperoxides of linoleic acid: a comparative study against BHT and melatonin. *J. Agric. Food Chem.* 47:2563–2570.
11. Bors, W., Heller, W., Michel, C., and Stettmaier, K. 1996. Flavonoids and polyphenols: Chemistry and biology. *Handb. Antioxid.* 26:409–466.
12. Halliwell, B. 1996. Antioxidants in human health and disease. *Annu. Rev. Nutr.* 16:39–50.
13. Murakami, K., Ito, M., Htay, H. H., Tsubouchi, R., and Yoshino, M. 2001. Antioxidant effect of capsaicinoids on the metal-catalyzed lipid peroxidation. *Biomed. Res. Tokyo* 22:15–17.
14. Srinivasan, K., Sambaiah, K., and Chandrasekhar, N. 1992. Loss of active principles of common spices during domestic cooking. *Food Chem.* 43:271–274.
15. Backonja, M. M., Malan, T. P., Vanhove, G. F., and Tobias, J. K. 2010. NGX-4010, a high-concentration capsaicin patch, for the treatment of postherpetic neuralgia: A randomized, double-blind, controlled study with an open-label extension. *Pain Med.* 11:600–608.
16. Tesfaye, S. 2009. Advances in the management of diabetic peripheral neuropathy. *Curr. Opin. Support. Palliat. Care* 3:136–143.
17. Maria, L. R. E., Edith G. G. M., and Erika V. T. 2011. Chemical and pharmacological aspectspa of capsaicin. *Molecules* 16:1253–1270.
18. Morris, G. C., Gibson, S. J., and Helme, R. D. 1995. Capsaicin-induced flare and vasodilatation in patients with post-herpetic neuralgia. *Pain* 63:93–101.
19. Laskaridou-Monnerville, A. 1999. Determination of capsaicin and dihydrocapsaicin by micellar electrokinetic capillary chromatography and its application to various species of *Capsicum*, Solanaceae. *J. Chromatogr. A.* 838:293–302.
20. Constant, H. L. and Cordell, G. A. 1996. Nonivamide, a constituent of capsicum oleoresin. *J. Nat. Prod.* 59:425–426.
21. Kaale, E., Van Schepdael, A., Roets, E., and Hoogmartens, J. 2002. Determination of capsaicinoids in topical cream by liquid–liquid extraction and liquid chromatography. *J. Pharm. Biomed. Anal.* 30:1331–1337.
22. Scoville, W. L. 1912, Note *Capsicum. J. Am. Pharm. Assoc.* 1:453–454.
23. Sanatombi, K. and Sharma, G. J. 2008. Capsaicin content and pungency of different *Capsicum* spp. cultivars. *Not. Bot. Hort. Agrobot. Cluj.* 36:89–90.
24. Weiss, E. A. 2002. *Spice Crops*, CABI Publishing International, New York, USA, pp. 411.
25. Pino, J., Gonzalez, M., Ceballos, L., Centurion-Yah, A. R., Trujillo-Aguirre, J., Latournerie-Moreno, L., and Sauri-Duch, E. 2007. Characterization of total capsaicinoids, colour and volatile compounds of Habanero chili pepper (*Capsicum chinense* Jack.) cultivars grown in Yucatan. *Food Chem.* 104:1682–1686.
26. Materska, M. and Perucka, I. 2005. Antioxidant activity of the main phenolic compounds isolated from hot pepper fruit (*Capsicum annuum* L.). *J. Agric. Food Chem.* 53:1750–1756.
27. Reilly, C. A., Taylor, J. L., Lanza, D. L., Carr, B. A., Crouch, D. J., and Yost, G. S. 2003. Capsaicinoids cause inflammation and epithelial cell death through activation of vanilloid receptors. *Toxicol. Sci.* 73:170–181.

28. Reilly, C. A., Veranth, J., Veronesi, B., and Yost, G. S. 2006. Vanilloid receptors in the respiratory tract. In *Target Organ Toxicity Series, No. 22*, ed. D. E. Gardner, CRC Press, Boca Raton, Florida, pp. 297–349.
29. Johansen, M. E., Reilly, C. A., and Yost, G. S. 2005. TRPV1 antagonists elevate cell surface populations of receptor protein and exacerbate TRPV1-mediated toxicities in human lung epithelial cells. *Toxicol. Sci.* 89:278–286.
30. Richardson, J. D. and Vasko, M. R. 2002. Cellular mechanisms of neurogenic inflammation. *J. Pharmacol. Exp. Ther.* 302:839–845.
31. Veronesi, B., Oortgiesen, M., Roy, J., Carter, J. D., Simon, S. A., and Gavett, S. H. 2000. Vanilloid (capsaicin) receptors influence inflammatory sensitivity in response to particulate matter. *Toxicol. Appl. Pharmacol.* 169:66–76.
32. Veronesi, B. and Oortgiesen, M. 2006. The TRPV1 receptor: Target of toxicants and therapeutics. *Toxicol. Sci.* 89:1–3.
33. Morre, D. J., Sun, E., Geilen, C., Wu, L. Y., de Cabo, R., Krasgakis, K., Orfanos, C. E., and Morre, D. E. 1996. Capsaicin inhibits plasma membrane NADH oxidase and growth of human and mouse melanoma lines. *Eur. J. Cancer* 32:1995–2003.
34. Takahata, K., Chen, X., Monobe, K., and Tada, M. 1999. Growth inhibitions of capsaicin on Hela cells are not mediated by intercellular calcium mobilization. *Life Sci.* 64:165–171.
35. Jung, M. Y., Kang, H. J., and Moon, A. 2001. Capsaicin-induced apoptosis in SK-Hep-1 hepatocarcinoma cells involves Bcl-2 down-regulation and caspase-3 activation. *Cancer Lett.* 165:139–145.
36. Kim, C. S., Park, W. H., Park, J. Y., Kang, J. H., Kim, M. O., Kawada, T., Yoo, H., Han, I. S., and Yu, R. 2004. Capsaicin, a spicy component of hot pepper, induces apoptosis by activation of the peroxisome proliferators-activated receptor gamma in HT-29 human colon cancer cells. *J. Med. Food* 7:267–273.
37. Zhang, J., Nagasaki, M., Tanaka, Y., and Morikawa, S. 2003. Capsaicin inhibits growth of adult T-cell leukemia cells. *Leuk. Res.* 27:275–283.
38. Jun, H. S., Park, T., Lee, C. K., Kang, M. K., Park, M. S., Kang, H. I., Surh, Y. J., and Kim, O. H. 2007. Capsaicin induced apoptosis of B16-F10 melanoma cells through down-regulation of Bcl-2. *Food Chem. Toxicol.* 45:708–715.
39. Lee, Y. S., Nam, D. H., and Kim, J. A. 2001. Induction of apoptosis by capsaicin in A172 human glioblastoma cells. *Cancer Lett.* 161:121–130.
40. Dorantes, L., Colmenero, R., Hernandez, H., Mota, L., and Jaramillo, M. E. 2000. Inhibition of growth of some food born epathogenic bacteria by *Capsicum annum* extracts. *Int. J. Food Microbiol.* 57:125–128.
41. Kurita, S., Kitagawa, E., Kim, C., Momose, Y., and Iwahashi, H. 2002. Studies on the antimicrobial mechanism of capsaicin using yeast DNA microarray. *Biosci. Biotechnol. Biochem.* 66:532–536.
42. Jones, N. L., Shabib, S., and Sherman, P. M. 1997. Capsaicinasan inhibitor of the growth of the gastric pathogen *Helicobacter pylori*. *FEMS Microbiol. Lett.*146:223–227.
43. Xing, F., Cheng, G., and Yi, K. 2006. Study on the antimicrobial activities of the capsaicin microcapsules. *J. Appl. Polym. Sci.* 102:1318–1321.
44. EPA (Environmental Protection Agency) 1992 Capsaicin: Reregistration Eligibility Document (RED). National Technical Information Service, Springfield, VA, EPA-738-F-92-016.
45. Glinsukon, T., Stitmunnaithum, V., Toskulkao, C., Buranawuti, T., and Tangkrisanavinont, V. 1980. Acute toxicity of capsaicin in several animal species. *Toxicon* 18:215–220.
46. Olajos, E. J. and Salem, H. 2001. Riot control agents: pharmacology, toxicology, biochemistry and chemistry. *J Appl. Toxicol.* 21:355–391.
47. Snyman, T., Stewart, M. J., and Steenkamp, V. 2001. A fatal case of pepper poisoning. *Forensic Sci. Int.* 124:43–46.

48. Govindarajan, V. S. 1985. Capsicum production, technology, chemistry, and quality. Part I: history, botany, cultivation, and primary processing. *CRC Crit. Rev. Food Sci. Nutr.* 22:109–176.
49. Huffman, V. L., Schadle, F. R., Villalon, B., and Burns, E. E. 1978. Volatile components and pungency in fresh and processed jalapeño peppers. *J. Food Sci.* 43:1809–1811.
50. Balbaa, S. L., Karawya, M. S., and Girgis, A. N. 1968. The capsaicin content of capsicum fruits at different stages of maturity. *Lloydia* 31:272–274.
51. AlOthman, Z. A., Yacine, B. H. A., Habila, M. A., and Ghafar, A. A. 2011. Determination of capsaicin and dihydrocapsaicin in capsicum fruit samples using high performance liquid chromatography. *Molecules* 16:8919–8929.
52. Collins, M. D., Mayer-Wasmund, L., and Bosland, P. W. 1995. Improved method for quantifying capsaicinoids in capsicum using high-performance liquid chromatography. *HortScience* 30:137–139.
53. Perucka, I. and Oleszek, W. 2000. Extraction and determination of capsaicinoids in fruit of hot pepper *Capsicum annuum* L. by spectrophotometry and high-performance liquid chromatography. *Food Chem.* 71:287–291.
54. Kirschbaum-Titze, P., Hiepler, C., Mueller-Seitz, E., and Petz, M. 2002. Pungency in paprika (*Capsicum annuum*). 1. Decrease of capsaicinoid content following cellular disruption. *J. Agric. Food Chem.* 50:1260–1263.
55. Contreras-Padilla, M. and Yahia, E. M. 1998. Changes in capsaicinoids during development, maturation, and senescence of chile peppers and relation with peroxidase activity. *J. Agric. Food Chem.* 46:2075–2079.
56. Korel, F., Bagdatlioglu, N., Balaban, M. O., and Hisil, Y. 2002. Ground red peppers: Capsaicinoids content, Scoville scores, and discrimination by an electronic nose. *J. Agric. Food Chem.* 50:3257–3261.
57. Krajewska, A. M. and Powers, J. J. 1987. Gas chromatography of methyl derivatives of naturally occurring capsaicinoids. *J. Chromatogr.* 409:223–233.
58. Barbero, G. F., Palma, M., and Barroso, C. G. 2006. Pressurized liquid extraction of capsaicinoids from peppers. *J. Agric. Food Chem.* 54:3231–3236.
59. Santamara, R. I., Reyes-Duarte, M. D., Barzana, E., Fernando, D., Gama, F. M., Mota, M., and Lopez-Mungua, A. 2000. Selective enzyme-mediated extraction of capsaicinoids and caratenoids from chili guajillo puya (*Capsicum annuum* L.) using ethanol as solvent. *J. Agric. Food Chem.* 48:3063–3067.
60. Luque de Castro, M. D. and Garćia-Ayuso, L. E. 1998. Soxhlet extraction of solid materials: An outdated technique with a promising innovative future. *Anal. Chim. Acta* 369:1–10.
61. Karnka, R., Rayanakorn, M., Watanesk, S., and Vaneesorn, Y. 2002. Optimization of high-performance liquid chromatographic parameters for the determination of capsaicinoid compounds using the simple method. *Anal. Sci.* 18:661–665.
62. Pan, X., Liu, H., Jia, G., and Shu, Y. Y. 2000. Microwave-assisted extraction of glycyrrhizic acid from licorice root. *Biochem. Eng. J.* 5:173–177.
63. Kwon, J. H., Belanger, J. M. R., and Pare, J. R. J. 2003. Optimization of microwave assisted extraction (MAP) for ginseng components by response surface methodology. *J. Agric. Food Chem.* 51:1807–1810.
64. Barbero, G. F., Palma, M., and Barroso, C. G. 2006. Determination of capsaicinoids in peppers by microwave-assisted extraction-highperformance liquid chromatography with fluorescence detection. *Anal. Chim. Acta* 578:227–233.
65. Daood, H. G., Illes, V., Gnayfeed, M. H., Meszaros, B., Horvath, G., and Biacs, P. A. 2002. Effects of particle size distribution, moisture content, and initial oil content on the supercritical fluid extraction of paprika. *J. Supercrit. Fluids* 23:143–152.
66. Wang, L. and Weller, C. L. 2006. Recent advances in extraction of nutraceuticals from plants. *Trends Food Sci. Technol.* 17:300–312.

67. Paniwnyk, L., Beaufoy, E., Lorimer, J. P., and Mason, T. J. 2001. The extraction of rutin from flower buds of *Sophora japonica*. *Ultrason. Sonochem.* 8:299–302.
68. Wu, J., Lin, L., and Chau, F. 2001. Ultrasound-assisted extraction of ginseng saponins from ginseng roots and cultured ginseng cells. *Ultrason. Sonochem.* 8:347–352.
69. Lambropoulou, D. A., Konstantinou, I. K., and Albanis, T. A. 2006. Sample pretreatment method for the determination of polychlorinated biphenyls in bird livers using ultrasonic extraction followed by headspace solid-phase microextraction and gas chromatography–mass spectrometry. *J. Chromatogr. A* 1124:97–105.
70. Barbero, G. F., Liazid, A., Palma, M., and Barroso, C. G. 2008. Ultrasound-assisted extraction of capsaicinoids from peppers. *Talanta* 75:1332–1337.
71. Alothman, Z. A., Wabaidur, S. M., Khan, M. R., Ghafar, A. A., Habila, M. A., and Yacine, B. H. A. 2012. Determination of capsaicinoids in *Capsicum* species using ultra performance liquid chromatography–mass spectrometry. *J. Sep. Sci.* 35:2892–2896.
72. Gibbs, H. and OGarro, L. O. 2004. Capsaicin content of West Indies hot pepper cultivars using colorimetric and chromatographic techniques. *HortScience* 39:132–135.
73. North, H. 1949. Colorimetric determination of capsaicin in oleoresin of *Capsicum*. *Anal. Chem.* 21:934–936.
74. Bajaj, K. L. 1980. Colorimetric determination of capsaicin in capsicum fruits. *J. Assoc. Offic. Anal. Chem.* 63:1314–1316.
75. Krajewska, A. M. and Powers, J. J. 1988. Pentafluorobenzylation of capsaicinoids for gas chromatography with electron capture detection. *J. Chromatogr.* 457:279–286.
76. Thomas, B. V., Schreiber, A. A., and Weisskopf, C. P. 1998. Simple method for quantitation of capsaicinoids in peppers using capillary gas chromatography. *J. Agric. Food Chem.* 46:2655–2663.
77. Hawer, W. S., Ha, J., Hwang, J., and Nam, Y. 1994. Effective separation and quantitative analysis of major heat principles in red pepper by capillary gas chromatography. *Food Chem.* 49:99–103.
78. Iwai, K., Suzuki, T., Fujiwake, H., and Oka, S. 1979. Simultaneous microdetermination of capsaicin and its four analogues by using high-performance liquid chromatography and gas chromatography–mass spectrometry. *J. Chromatogr.* 172:303–311.
79. Constant, H. L., Cordell, G. A., West, D. P., and Johnson, J. H. 1995. Separation and quantification of capsaicinoids using complexation chromatography. *J. Nat. Prod.* 58: 1925–1928.
80. Betts, T. A. 1999. Pungency quantitation of hot pepper sauces using HPLC. *J. Chem. Educ.* 76:240–244.
81. Chiang, G. H. 1986. HPLC analysis of capsaicins and simultaneous determination of capsaicins and piperine by HPLC–ECD and UV. *J. Food Sci.* 51:499–503.
82. Parrish, M. 1996. Liquid chromatographic method for determining capsaicinoids in capsicums and their extractives: collaborative study. *J. AOAC Int.* 79:738–745.
83. Maillard, M., Giampaoli, P., and Richard, H. M. J. 1997. Analysis of eleven capsaicinoids by reversed-phase high performance liquid chromatography. *Flav. Frag. J.* 12:409–413.
84. Yao, J., Nair, M. G., and Chandra, A. 1994. Supercritical carbon dioxide extraction of Scotch bonnet (*Capsicum annuum*) and quantification of capsaicin and dihydrocapsaicin. *J. Agric. Food Chem.* 42:1303–1305.
85. Reilly, C. A., Crouch, D. J., and Yost, G. S. 2001. Quantitative analysis of capsaicinoids in fresh peppers, oleoresin capsicum and pepper spray products. *J. Forensic Sci.* 46:502–509.
86. Nyberg, N. T., Baumann, H., and Kenne, L. 2001. Application of solidphase extraction coupled to an NMR flow-probe in the analysis of HPLC fractions. *Magn. Reson. Chem.* 39:236–240.
87. DiCecco, J. J. 1979. Spectrophotometric difference method for determination of capsaicin. *J. Assoc. Offic. Anal. Chem.* 62:998–1000.

88. Tice, L. F. 1933. A simplified and more efficient method for the extraction of capsaicin together with the colorimetric method for its determination in capsicum fruit and oleoresin. *Am. J. Pharm.* 105:320–325.
89. Nikolaeva, D. A. 1984. Spectrophotometric determination of capsaicin in peppers (*Capsicum annuum* L.). *Biokhim. Metody Analiza Plodov, Kishinev* 99–102.
90. Suzuki, J. I., Tausig, F., and Morse, R. E. 1957. A new method for the determination of pungency in red pepper. *Food Technol.* 11:100–104.
91. Pryakhin, O. R., Tkach, V. I., Golovkin, V. A., Gladyshev, V. V., and Kuleshova, N. D. 1992. Method for determination of the total amount of capsaicinoids in thick red pepper extract by amperometric titration. *U.S.S.R.* 90–4880330.
92. Khaled, M. Y., Anderson, M. R., and McNair, H. M. 1993. Micellar electrokinetic capillary chromatography of pungent compounds using simultaneous on-line ultraviolet and electrochemical detection. *J. Chromatogr. Sci.* 31:259–264.
93. Barbero, G. F., Liazid, A., Palma, M., and Barroso, C. G. 2008. Fast determination of capsaicinoids from peppers by high-performance liquid chromatography using a reversed phase monolithic column. *Food Chem.* 107:1276–1282.
94. Davis, C. B., Markey, C. E., Busch, M. A., and Busch, K. W. 2007. Determination of capsaicinoids in habanero peppers by chemometric analysis of UV spectral data. *J. Agric. Food Chem.* 55:5925–5933.
95. Kawada, T., Watanabe, T., Katsura, K., Takami, H., and Iwai, K. 1985. Formation and metabolism of pungent principle of capsicum fruits. XV. Microdetermination of capsaicin by liquid chromatography with electrochemical detection. *J. Chromatogr.* 329:99–105.
96. Reilly, C. A., Crouch, D. J., Yost, G. S., and Fatah, A. A. 2002. Determination of capsaicin, nonivamide, and dihydrocapsaicin in blood and tissue by liquid chromatography–tandem mass spectrometry. *J. Anal. Toxicol.* 26:313–319.
97. Thompson, R. Q., Phinney, K. W., Welch, M. J., and White, E. 2005. Quantitative determination of capsaicinoids by liquid chromatography-electrospray mass spectrometry. *Anal. Bioanal. Chem.* 381:1441–1451.
98. Garces-Claver, A., Arnedo-Andres, M. S., Abadia, J., Gil-Ortega, R., and Alvarez-Fernandez, A. 2006. Determination of capsaicin and dihydrocapsaicin in capsicum fruits by liquid chromatography–electrospray/time-of-flight mass spectrometry. *J. Agric. Food Chem.* 54:9303–9311.
99. Kozukue, N., Han, J. S., Kozukue, E., Lee, S. J., Kim, J. A., and Lee, K. R. 2005. Analysis of eight capsaicinoids in peppers and pepper-containing foods by highperformance liquid chromatography and liquid chromatography–mass spectrometry. *J. Agric. Food Chem.* 53:9172–9181.
100. Schweiggert, U., Carle, R., and Schieber, A. 2006. Characterization of major and minor capsaicinoids and related compounds in chili pods (*Capsicum frutescens* L.) by high-performance liquid chromatography/atmospheric pressure chemical ionization mass spectrometry. *Anal. Chim. Acta* 557:236–244.
101. Olga, C. P., Luis, W. T. T., Luis, C. G. P., Fernando, C. M., Tomás, G. E., and Sergio, R. P. S. 2007. Capsaicinoids quantification in chili peppers cultivated in the state of Yucatan, Mexico. *Anal. Nutri. Clin. Meth.* 104:1755–1760.
102. Liu, L., Chen, X., Liu, J., Deng, X., Duan, W., and Tan, S. 2010. Determination of capsaicin and dihydrocapsaicin in *Capsicum anuum* and related products by capillary electrophoresis with a mixed surfactant system. *Food Chem.* 119:1228–1232.
103. Estella-Hermoso, M. A., Campanero, M. A., Mollinedo, F., and Blanco-Prieto, M. J. 2009. Comparative study of a HPLC–MS assay versus an UHPLCMS/MS for anti-tumoral alkyl lysophospholipid edelfosine determination in both biological samples and in lipid nanoparticulate systems. *J. Chromatogr. B* 877:4035–4041.

104. Marin, J. M., Gracia-Lor, E., Sancho, J. V., Lopez, F. J., and Hernandez, F. 2009. Application of ultra-high-pressure liquid chromatography-tandem mass spectrometry to the determination of multi-class pesticides in environmental and wastewater samples study of matrix effects. *J. Chromatogr. A* 1216:1410–1420.
105. Saria, A., Lembeck, F., and Skofitsch, G. 1981. Determination of capsaicin analogues by high performance liquid chromatography. *J. Chromatogr.* 208:41–46.
106. Xiang, Y., Liu, Y., and Lee M. L. 2006. Ultrahigh pressure liquid chromatography using elevated temperature. *J. Chromatogr. A* 1104:198–202.
107. Horváth, C. S., Preiss, B. A., and Lipsky S. R. 1967. Fast liquid chromatography. Investigation of operating parameters and the separation of nucleotides on pellicular ion exchangers. *Anal. Chem.* 39:1422–1428.
108. Jerkovich, A. D., Mellors, J. S., and Jorgenson, J. W. 2003. The use of micron-sized particles in ultrahigh-pressure liquid chromatography. *LCGC* 21:600–610.
109. Wu, N., Lippert, J. A., and Lee, M. L. 2001. Practical aspects of ultrahigh pressure capillary liquid chromatography. *J. Chromotogr. A* 911:1–12.
110. Unger, K. K., Kumar, D., Grun, M., Buchel, G., Ludtke, S., Adam, T., Scumacher, K., and Renker, S. 2000. Synthesis of spherical porous silicas in the micron and submicron size range: challenges and opportunities for miniaturized high-resolution chromatographic and electrokinetic separations. *J. Chromatogr. A* 892:47–55.
111. Swartz, M. E. and Murphy, B. 2004. Ultra performance liquid chromatography: Tomorrow's HPLC technology today. *Lab Plus Int.* 18:6–9.
112. Swartz, M. E. and Murphy, B. 2004. Ultra performance liquid chromatography: Tomorrow's HPLC technology today. *Pharm. Formu. Qual.* 6:40.
113. Van Deemter, J. J., Zuiderweg, E. J., and Klinkenberg, A. 1956. Longitudinal diffusion and resistance to mass transfer as causes of non ideality in chromatography. *Chem. Eng. Sci.* 5:271–289.
114. MacNair, J. E., Lewis, K. C., and Jorgenson, J. W. 1997. Ultrahigh-pressure reversed-phase liquid chromatography in packed capillary columns. *Anal. Chem.* 69:983–989.
115. Swartz, M. E. 2005. Ultra Performance Liquid Chromatography (UPLC): An introduction, separation science re-defined. *LCGC Supp.* 12:8–13.
116. Srivastava, B., Sharma, B. K., Baghel, U. S., Yashwant, and Sethi, N. 2010. Ultra performance liquid chromatography (UPLC): A chromatography technique. *Int. J. Pharm. Qual. Assur.* 2:19–25.
117. Broske, P., Ricker, R., Permar, B., Chen, W., and Joseph, M. 2003. The influence of sub-two micron particles on HPLC performance. *Agilent Application Note* 5988–9251EN.
118. Choi, S. H., Suh, B. S., Kozukue, E., Kozukue, N., Levin, C. E., and Friedman, M. 2006. Analysis of the contents of pungent compounds in fresh korean red peppers and in pepper-containing foods. *J. Agric. Food Chem.*54:9024–9031.
119. Ha, J., Han, K. J., Kim, J. K., and Jeong, S. W. 2008. Gas chromatographic analysis of capsaicin in Gochujang. *J. AOAC Int.* 91:387–391.
120. Spiecer, O. and Almirall, J. R. 2005. Extraction of capsaicin in aerosol defense sprays from fabrics. *Talanta* 67:377–382.
121. Cooper, T. H., Guzinski, J. A., and Fisher, C. 1991. Improved high-performance liquid chromatography method for the determination of major capsaicinoids in capsicum oleoresins. *J. Agric. Food Chem.* 39:2253–2256.
122. Krajewska, A. M. and Power, J. J. 1986. Isolation of naturally occurring capsaicinoids by reversed phase low pressure liquid chromatography. *J. Chromatogra. A* 367:267–270.
123. Esaka, Y., Tanaka, K., Uno, B., and Goto, M. 1997. Sodium dodecyl sulfate Tween 20 mixed micellar electrokinetic chromatography for separation of hydrophobic cations: Application to adrenaline and its precursors. *Anal. Chem.* 69:1332–1338.

124. de Wasch, K., de Brabander, H. F., Impens, S., Okerman, L., and Kamel, C. H. 2001. Detection of the major components of capsicum oleoresin and zigerone by high performance liquid chromatography–tandem mass spectrometry. *Biol.-ActiVe Phytochem. Food* 269:134–139.
125. Ana, G. S. C., Ramiro, G. O., Ana, A. F., and Mariaa, S. A. A. 2007. Inheritance of capsaicin and dihydrocapsaicin, determined by HPLC–ESI/MS, in an intraspecific cross of *Capsicum annuum* L. *J. Agric. Food Chem.* 55:6951–6957.

6 Applications of UPLC–MS/MS for the Quantification of Folate Vitamers

Maria V. Chandra-Hioe, Jayashree Arcot, and Martin P. Bucknall

CONTENTS

6.1 INTRODUCTION

Vitamin B_9 or folate is an essential water-soluble B vitamin that exists in diverse forms (or vitamers). In general, the term folate encompasses both naturally occurring folate and synthetic folic acid. Folic acid or pteroylmonoglutamic acid is the fully oxidized vitamer and it is naturally present in foods at very low concentrations. It does not carry any carbon unit for folate metabolism [1] and is therefore biologically inactive [2]. Nevertheless, folic acid is widely used for the fortification of food and as a dietary supplement, owing to its stability and low manufacturing cost [3,4]. It is readily converted into tetrahydrofolate, the fully reduced form, following absorption [5]. Structurally, folate is composed of three distinct moieties: a heterocyclic pteridine ring, para-aminobenzoic acid, and glutamic acid, as shown in Figure 6.1.

The diversity of folate vitamers is attributed to different oxidation states of the pteridine ring, the number of glutamic acid residues, and the type of one-carbon unit attached at positions 5 and/or 10 relative to the nitrogen atom [5,6], as illustrated in Figure 6.2.

Dried peas and beans, leafy green vegetables, citrus fruits, and juice are some examples of folate-rich foods. Naturally occurring folate is also found in yeast

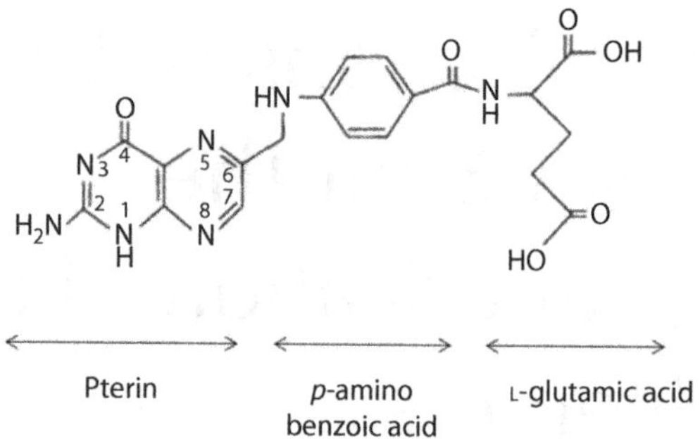

FIGURE 6.1 The structure of folate in the state of oxidation. (Adapted from Gerald F. Combs Jr, *The Vitamins: Properties of Vitamins*, 4th ed., Academic Press, San Diego, 2012. With permission.)

extracts, whole grains, nuts, avocado, and organ meats such as liver and kidney [7]. Folate found in food may contain mixtures of pteroylmonoglutamate and pteroylpolyglutamate with 2–7 glutamic acid residues [2]. This is because the glutamic acid residues are hydrolyzed by folypoly-γ-glutamate carboxypeptidase (FGCP) enzyme to monoglutamate vitamers. These vitamers are predominantly 5-methyltetrahydrofolate and formyltetrahydrofolate [8].

Folate is involved in one-carbon unit transfer reactions during DNA synthesis, DNA methylation, and amino acid metabolism. Evidence to date shows that maternal dietary intake of folic acid is inversely associated with the risk of neural tube defect-affected pregnancies [9,10]. Neural tube defects (a term which includes spina bifida) are anatomical birth anomalies affecting the brain and the spinal cord. As a result of these landmark findings, the first folic acid fortification program was introduced in the United States during 1998, in an attempt to reduce the prevalence

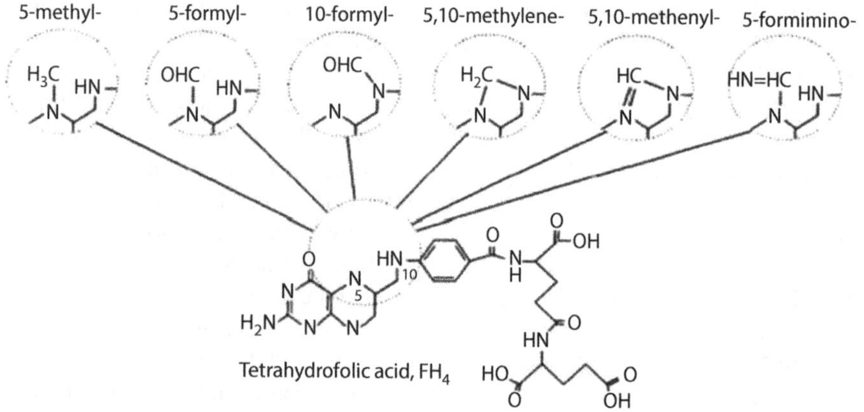

FIGURE 6.2 Various folate vitamers differ in the one-carbon unit that can be substituted at N-5, N-10, or N-5 and N-10 positions. (From Gerald F. Combs Jr, *The Vitamins: Folate*, 4th ed., Academic Press, San Diego, 2012. With permission.)

of neural tube defects *in utero*. Even though the underlying pathology of neural tube defects is not fully understood, approximately 60 countries have fortification programs in place [11].

Dietary folate intake determines folate nutritional status. It is therefore important to accurately measure folate in food using the best analytical platform currently available. In particular, folate analysis is confounded by several factors, such as the presence of many vitamers, low concentrations, physicochemical instability, and complex sample preparation [12]. It has been reported that the levels of endogenous 5-methyltetrahydrofolate and 5-formyltetrahydrofolate in canned chickpeas, a rich folate source, were 52 and 8 μg/100 g, respectively [13]. Folate-fortified foods contain higher total folate concentrations than those found in natural folate-rich foods. We recently determined folic acid in Australian fortified bread varieties that was approximately 100 μg/100 g [14].

6.2 QUANTITATIVE TECHNIQUES TO DETERMINE FOLATE IN FOOD

There are various analytical platforms used to measure folate concentrations in food, each with its advantages and drawbacks. Microbiological assay makes use of bacteria (e.g. *Lactobacillus casei*) that are very dependent on folate availability for their growth. This assay is commonly used to determine the total folate contents in food, which is expressed in terms of folic acid equivalents. The most important limitation of microbiological assay is that the various vitamers are not distinguishable [15]. The main folate vitamers occurring in fruits and vegetables are different from those found in animal and cereal products [16]. There is also evidence that the vitamers have different bioavailability [17] and stability [18]. A clear need existed for an analytical method that would be capable of independently quantifying each specific folate vitamer. Liquid chromatography (LC) seemed an ideal and logical technology for this purpose.

High-pressure liquid chromatography (HPLC) is able to separate compounds in complex mixtures and matrices. HPLC separation can be based on polarity (normal phase chromatography), hydrophobicity (reversed-phase chromatography), and charge (ion exchange chromatography), as well as other molecular characteristics. Folate vitamers show small differences in ionic character and hydrophobicity, and they are therefore candidates for separation using HPLC [19]. This approach was then applied to quantitatively determine natural folate vitamers in food [16,20,21] and for measuring folic acid in fortified food [22,23]. Folate was separated mainly by reversed-phase chromatography, except for one approach that used ion-pairing chromatography [24]. In many cases, HPLC was limited by inadequate specificity [15,25]. It is susceptible to matrix interferences [15], and chromatograms showed interfering peaks with similar retention times to those of the folate vitamer analytes [16]. In addition, sensitivity has been identified as another weakness of HPLC [25]. In spinach, 5-formyltetrahydrofolate was not detected using a fluorescence detector [26] and in other foods its level was close to the limit of detection, leading to inaccurate quantification [19]. It was not until the late 1990s that mass spectrometry found application as a highly specific detector for individual folate vitamers that can be readily coupled with reversed-phase HPLC. A mass spectrometer (MS) is a versatile tool for identifying sample analytes and determining their concentrations [27]. When interfaced with HPLC, MS provides more specificity and sensitivity for folate vitamers than any other HPLC detector currently available. Following electrostatic ionization of the compounds eluting from the HPLC column, the MS can be set up to detect only those ions with specific mass-to-charge ratios. Other species with different mass-to-charge ratios, originating from the same sample matrix are eliminated from detection.

The first liquid chromatography–mass spectrometry (LC–MS) method was developed by Stokes and Webb (1999). Here, folate was separated using reversed-phase chromatography, detected in a single ion monitoring (SIM) mode and quantified using external standards [25]. These authors successfully identified and determined folic acid, tetrahydrofolate, 5-methyltetrahydrofolate, and formyltetrahydrofolate in multivitamins and foods.

The quantitative accuracy of the LC–MS method was improved by adding stable isotope labeled ($^{13}C_5$) internal standards into various food matrices [28–31]. This method is known as isotope dilution mass spectrometry. As the internal standard is a stable isotope labeled analogue of the analyte, it has the same chemical properties, but different molecular weight when compared to the analyte. This characteristic is very advantageous for the MS, since it measures the mass-to-charge ratio (*m/z*) of molecular ions. For quantitative purposes, a stable isotope labeled analog of the analyte is an excellent choice for internal standards. This can be achieved by incorporating one or more atoms of deuterium, carbon-13 (^{13}C) or nitrogen-15 (^{15}N) to make the stable isotope labeled internal standards, which will chemically behave in an identical manner to the analyte [27]. Because of the strong bonds between carbon–carbon or carbon–nitrogen atoms, ^{13}C or ^{15}N internal standards are considered very stable in comparison to the more labile deuterium [32]. Internal standards compensate for any analyte loss during the sample preparation.

Even though addition of stable isotope labeled internal standards is useful for quantitation, in SIM mode both analytes and internal standards co-elute. Further, enhanced specificity and sensitivity were achieved by using a tandem mass spectrometer (MS/MS), operated in selected reaction monitoring (SRM) mode to monitor isolated folate-specific ion fragmentation [15,26,33].

6.3 FAST LIQUID CHROMATOGRAPHY

Fast LC emerged to meet the growing demand for high-throughput analysis with increased sensitivity, robustness, and high resolution [34]. The trend followed the advances in column sorbent technologies, such as monolithic, sub 2 μm particle, porous shell, as well as reversed-phase (C_8 and C_{18}), hydrophilic interaction liquid chromatography (HILIC) and fluorinated column [34]. Conventional HPLC columns containing stationary phase sorbents with 3 or 5 μm particle sizes are commonly used in LC–MS [31,35] and LC–MS/MS for separation of folate vitamers in food [13,26,36,37]. Nevertheless, HPLC is still advantageous in many types of analysis, especially those where time is required to perform multiple MS analyses, or accurate mass analysis on co-eluting peaks.

Ultra-performance liquid chromatography (UPLC) takes advantage of the column-packing materials with sub-2 μm particle size. The underlying principle for improved chromatographic performance is that smaller particles give greater separation efficiency for a given column length. According to the Van Deemter equation, reducing the particle size <2.5 μm leads to a significant gain in efficiency, even at increased flow rates or linear velocities [38]. The Van Deemter plot (Figure 6.3) shows the relationship between column efficiency and the linear velocity/flow rate [38]. Column efficiency is expressed in terms of the "height" of a theoretical plate (height equivalent to theoretical plate or HETP). The terminology has its roots in old-fashioned, gravity-fed, packed column chromatography, where the column was vertically mounted and mathematically modeled as a series of horizontal separating plates. When this height is a minimum, a column of given length has the maximum number of theoretical plates along its length and exhibits its highest separation efficiency.

From the Van Deemter plot, it is evident that smaller particle sizes produce a column with a higher optimal flow rate and that the gains in column efficiency are better maintained as flow rate is increased than with conventional 3–5 μm particles. This means that flow rates can be raised to approximately 4 times the optimal value with very little loss of column efficiency.

The increased column efficiency can be used either to increase chromatographic resolution, increase the run speed, or toward achieving a combination of both benefits. Normally, fast UPLC runs are accomplished using a shorter column and a linear velocity approximately 4 times higher than might be employed for the same

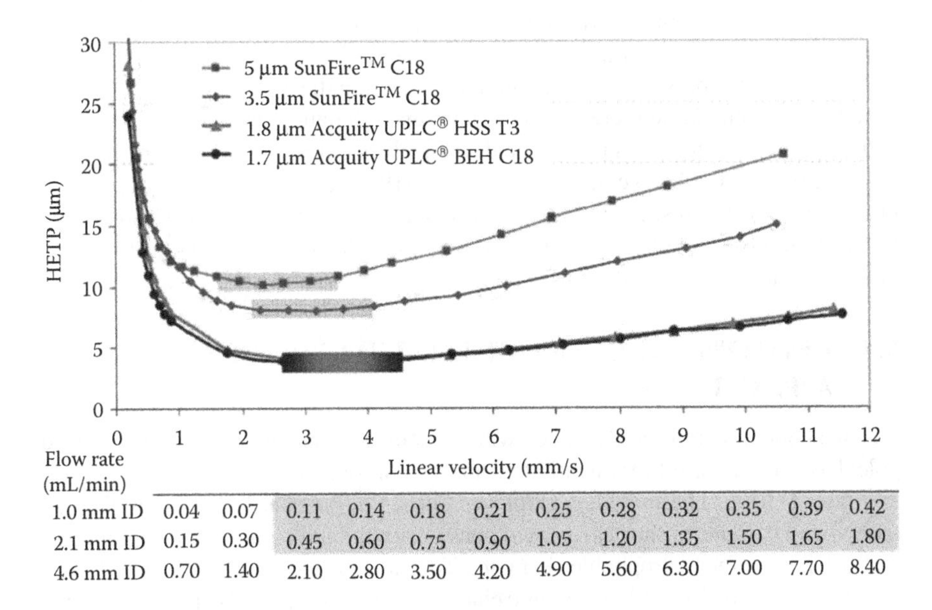

Flow rate (mL/min)												
1.0 mm ID	0.04	0.07	0.11	0.14	0.18	0.21	0.25	0.28	0.32	0.35	0.39	0.42
2.1 mm ID	0.15	0.30	0.45	0.60	0.75	0.90	1.05	1.20	1.35	1.50	1.65	1.80
4.6 mm ID	0.70	1.40	2.10	2.80	3.50	4.20	4.90	5.60	6.30	7.00	7.70	8.40

FIGURE 6.3 The Van Deemter plot describes the relationship between linear velocity and HETP. (From Waters Inc. *An Introduction to UPLC technology: Improve Productivity and Data Quality*, presented at the American Association of Pharmaceutical Scientists conference, 2007. With permission.)

analysis using conventional HPLC. Gradients are also increased by a factor of 4–8 times. This approach typically aims to reproduce conventional HPLC resolution levels and increase sample throughput by 4–8 times. Correspondingly, peak widths are also reduced by a factor of 4–8, though their areas are essentially the same as those obtained by conventional HPLC. This means that peak height is increased by 4–8 times, giving a signal-to-noise ratio and sensitivity gain, assuming baseline noise to be at similar levels to the conventional HPLC system. Columns of reduced internal diameter relative to those used for HPLC are employed in order to achieve the higher linear velocities without greatly increasing flow rate and solvent consumption.

If the objective is higher resolution, a UPLC column of similar length to that employed for the same analysis in conventional HPLC is employed with a linear velocity of 1.5–3 times that used in HPLC and a similar gradient profile and total run times. Resolution typically increases by a factor of 2–3, peak width decreases by the same factor and signal-to-noise (sensitivity) increases.

In migrating folate work from HPLC to UPLC, the authors have used a slightly shorter column, decreased column diameter by a factor of 2 and increased linear velocity by a factor of approximately 2.5 times. This has shortened the run times, increased resolution, and reduced the limit of detection.

Although UPLC offers a wider range of linear velocity, it creates a high back pressure. A system capable of withstanding a backpressure up to 1000 bar (15,000 psi) is required to deliver the mobile phase and elute analytes. Importantly, the system and the methods need to be optimized in order to benefit from the rapid analysis, resolution, and sensitivity. Thus, extra-column dispersion in the tubing, injector and detector should be minimized. The pump dwell time should be minimized to give the best results with fast gradients. Also, the injector has to have a pulse-free operation to tolerate the pressure fluctuation, thus protecting the column [38]. The detector sampling rate should be increased to acquire enough data points across a peak [39] and special "fast" detectors have been developed specifically for use with UPLC.

Jastrebova et al. (2011) compared the use of HPLC and UPLC, both coupled with ultra-violet and fluorescence detection, to determine folate in foods. In addition to reduced run times, the limit of detection was lower due to improved signal-to-noise ratio for folate vitamers [40].

6.4 DEVELOPMENT OF FOLATE-RELATED UPLC–MS/MS APPLICATIONS

Recent application of UPLC coupled with tandem mass spectrometry (triple quadrupole) has been reported for quantitative analysis of folate in rice [41], fortified bread and flour [14,42]. These studies employed sub-2 μm particle columns, as they are specialized variants of the C_{18} reversed phase that can bind polar organic compounds that still have some hydrophobic character. The aqueous mobile phase was 0.1% (v/v) formic acid in Milli-Q and the organic eluents were acetonitrile [14] and 0.1% formic acid in acetonitrile [41]. In both cases, the analytes were ionized and introduced into the MS using heated electrospray ionization (HESI).

Adding volatile formic acid is essential for retention of analytes, as the vitamers must be protonated rather than ionized during chromatographic separation [43]. The presence of formic acid also promotes the formation of protonated molecular ions during positive ion electrospray ionization (ESI). These are the most desirable ionic species for tandem MS analysis, as they readily fragment giving structure-specific product ions. ESI is a soft ionization technique involving the transfer of charges from aerosols within a high electric field without fragmentation of ions during the spray process [44]. The addition of a heater to the traditional ESI source (HESI) in order to promote solvent evaporation at higher flow rates is an important innovation that makes modern MS more compatible with UPLC flow rates.

In the method, pioneered by Stokes and Webb (1999), the ion source was operated in negative ESI mode [25]. Folate can actually be detected in either positive or negative ion ESI mode [45,46]. In the positive ESI mode, ionization of folate occurs via the addition of charge carriers, such as H^+, Na^+, K^+, or other cations. There is a limitation of the positive ion mode, which is the formation of sodium and potassium adducts, leading to a potential reduction in the abundance of the protonated molecular ions [47]. Sodium and potassium adducts require application of high collision energies in order to achieve any fragmentation, and any resulting fragments are often very small and lack a clear structural relationship to their precursors. Accordingly, dominant formation of these adducts is undesirable for ESI–MS/MS analysis.

In negative ion mode, deprotonated molecules are the only folate species detected; however, acidification of the mobile phase, which is essential for retention of folate on reversed-phase HPLC columns, suppresses the formation of deprotonated molecules [48]. As a consequence, negative ion ESI requires post column addition of organic bases, such as triethylamine [47,48], to improve ionization efficiency when low pH buffers are used in the mobile phase. This adds unnecessary complexity to the analytical system.

In a tandem MS/MS (triple quadrupoles), the first mass analyzer filters the precursor ions originating from the ion source, so that the ions leave the mass analyzer in a narrow *m/z* range. This is normally used to isolate the protonated molecular ions of the individual folate vitamers being measured. The second mass analyzer filters product ions and can be set to scan a given *m/z* range for product species or set to isolate and monitor specific target fragments. The monitoring of specific products that originate from specific isolated precursors is called SRM. This is used to selectively quantify target folate vitamers in complex matrices, free of interference from co-eluting molecules with similar molecular weights. The sensitivity of triple quadrupole is enhanced in SRM mode compared to scanning mode, because the mass analyzers are dedicated to the monitoring of single folate-related *m/z* values and do not spend large parts of their duty cycle scanning *m/z* values which are not relevant to folate determination. In positive (or negative) ESI mode, the fragmentation of the singly protonated (or deprotonated) precursor ions into product ions is performed during collision-induced dissociation (CID) by the neutral loss of a glutamic acid residue (147 Da) from folate vitamers. However, fragmentation of formyl-folate shows the most abundant product due to loss of glutamic acid (147 Da) and carbon monoxide (28 Da) [13].

The *m/z* values monitored in SRM mode for various folate vitamers are shown in Table 6.1. For quantitative purposes, commercially available $^{13}C_5$-labeled internal standards (Merck-Eprova, Schaffhausen, Switzerland) are added at the start of the sample extraction.

TABLE 6.1
Folate Detected and Monitored during SRM Based on the Specific *m/z*

Folate	*m/z* of the Precursor Ion	*m/z* of the Product Ion
Folic acid	442	295
5-Methyltetrahydrofolate	460	313
10-Formylpteroylmonoglutamate	470	295
Tetrahydrofolate	446	299
5,10-Methenyltetrahydrofolate	456	412
5-Formyltetrahydrofolate	474	327 (299)

De Brouwer et al. (2010) identified 5 folate vitamers in their study: 5-methyltetrahydrofolate (the predominant vitamer), 5,10-methenyltetrahydrofolate and tetrahydrofolate (the most labile vitamers), folic acid, and 10-formyl folic acid. This study reported that folate levels in wild-type rice were approximately 20 μg/100 g lower than in bio-fortified rice [41]. While separation of folate vitamers was obtained in 8 min run time, a previous LC–MS/MS method required 20 min [49]. Also, UPLC improved the sensitivity by reducing the limit of quantitation. Another study documented that endogenous 5-methyltetrahydrofolate in bread ranged between 1.3 and 3.4 μg/100 g [14]. Measurements of folic acid and 5-methyltetrahydrofolate in flour and infant milk formula have been successfully made using this method, described in the latter study.

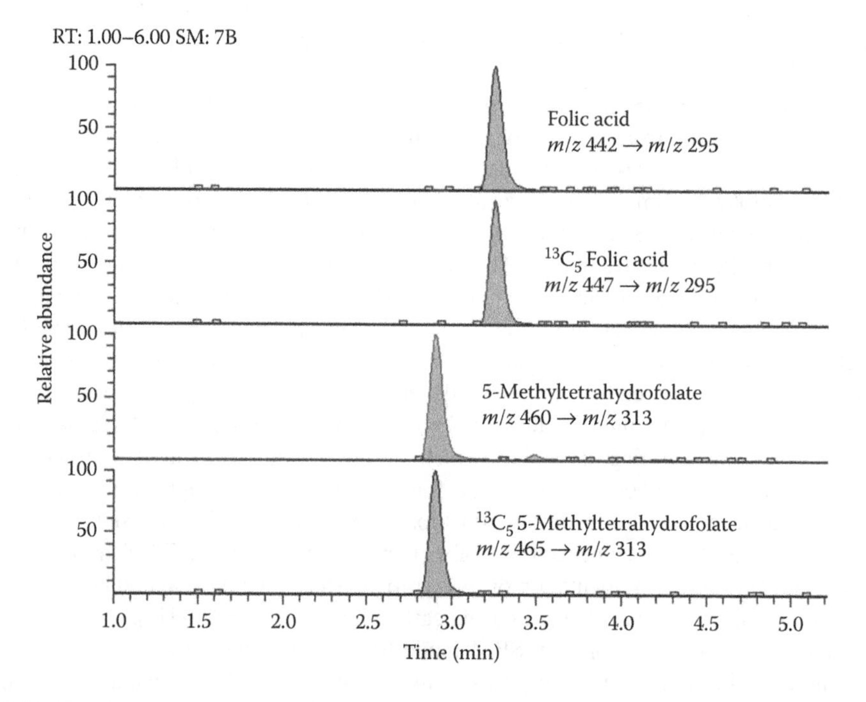

FIGURE 6.4 The SRM chromatogram of the adult/infant milk formula (standard reference material 1849a).

This method has a broad applicability and was fully validated for both matrices. The optimization of sample processing protocols for each food matrix was essential toward achieving good sensitivity and high quantitative precision. Figure 6.4 displays the SRM chromatogram of standard reference material 1849a (NIST Infant/Adult Nutritional formula), which was used to verify the accuracy of the method for determining the concentrations of folic acid and 5-methyltetrahydrofolate in milk formula.

6.5 SAMPLE PREPARATION AND FOLATE STABILITY

Besides a validated UPLC/MS–MS method, an optimized sample preparation is crucial for quantitative analysis of folate, because it governs the extent to which endogenous bound folate is released from the food matrix prior to measurement. Since endogenous folate occurs in considerably lower concentrations, folate degradation and loss need to be minimized. In the past, folate extraction was performed in two stages: heat treatment to release folate from its binding proteins and the deconjugation of polyglutamate chains [50]. Tri-enzyme treatment was proposed in the mid-1990s and then widely used. This procedure required treatments with α-amylase, protease, and deconjugase in order to extract folate completely [50]. However, it has been argued that the tri-enzyme treatment does not necessarily increase the folate release compared to deconjugase treatment alone [21,50]. It was then proposed that other factors, such as different food matrices and the stability of folate vitamers, are likely to play a role during sample extraction [21,50]. The suggestion is that folate extractions need to be experimentally optimized for each food matrix in order to maintain assay sensitivity across food groups [42]. The rice samples analyzed for UPLC–MS/MS were treated with the tri-enzyme treatments and filtration [41]. In other studies solid-phase extraction (SPE) was performed in addition to aqueous extraction from bread and flour samples [14,42]. SPE was carried out to selectively concentrate the analytes, reduce matrix interferences, and remove the matrix components. In particular, SPE using phenyl reversed-phase cartridges avoids the sodium concentrations used with ion-exchange sample preparation that might contaminate the UPLC column or quickly dirty the ion source of the MS system [14,42].

The other factor that needs to be taken into consideration during sample processing is the stability of folate vitamers. Stability of the vitamers differs with respect to the susceptibility to oxidative degradation, thermal, pH, and ultra-violet light [51]. Interconversion of folate vitamers occurs during sample preparation and analysis, for example, 5,10-methylenetetrahydrofolate dissociates from tetrahydrofolate and formaldehyde in the presence of mercaptoethanol (antioxidant) at low pH [49]. Furthermore, it has been argued that 5,10-methenyltetrahydrofolate is difficult to analyze on a reversed-phase (RP-18) HPLC column with a low pH mobile phase [13]. This is because interconversion of 10-formyltetrahydrofolate to 5,10-methenyltetrahydrofolate occurs at low pH [18].

6.6 SUMMARY AND CONCLUSION

UPLC takes advantage of sub-2 μm particles as the stationary phase to improve speed and resolution of a chromatographic separation. Sub-2 μm particles allow the use of higher chromatographic linear velocities while maintaining column efficiency, which in turn decreases analysis time.

Besides speed of analysis and increased resolution, UPLC can improve sensitivity by narrowing peak widths and increasing peak height. Enhanced specificity and sensitivity are further achieved by using a tandem MS in SRM mode. This enables the chromatographic peak to be better defined, reducing quantitative variability. For these reasons SRM and UPLC are very compatible technologies for the rapid and precise target quantification of small molecules with high sensitivity and specificity. The approach only works for analytes with known molecular weight and CID fragmentation characteristics. Addition of stable isotope-labeled internal standards improves accuracy for measuring individual folate vitamers in food. Isotope dilution (UPLC–MS/MS) is a robust, fast, specific, and sensitive method for quantitative analysis of folate vitamers.

ACKNOWLEDGMENT

The chromatogram shown in Figure 6.4 was obtained at the Bioanalytical Mass Spectrometry Facility within the Mark Wainwright Analytical Centre of the University of New South Wales. This work was undertaken using infrastructure provided by New South Wales Government co-investment in the National Collaborative Research Infrastructure Scheme (NCRIS). Subsidised access to this facility is gratefully acknowledged.

REFERENCES

1. Brouwer, I. A., Dusseldorp, M. V., West, C. E., and Steegers-Theunissen, R. P. M. 2001. Bioavailability and bioefficacy of folate and folic acid in man. *Nutr. Res. Rev.* 14:267–293.
2. Molloy, A. M. 2002. Folate bioavailability and health. *Int. J. Vitm. Nutr. Res.* 72:46–52.
3. Finglas, P. M., Wright, A. J. A., Wolfe, C. A., Hart, D. J. et al. 2003. Is there more to folates than neural-tube defects? *Pro. Nutr. Soc.* 62:591–598.
4. Wright, A. J. A., Dainty, J. R., and Finglas, P. M. 2007. Folic acid metabolism in human subjects revisited: Potential implications for proposed mandatory folic acid fortification in the UK. *Br. J. Nutr.* 98:667–675.
5. Combs Jr, G. F. 2012. Chapter 3—Properties of vitamins. In *The Vitamins (Fourth Edition)*, edited by G.F. Combs Jr, pp. 33–70. San Diego: Academic Press.
6. Blakley, R. L. 1969. *The Biochemistry of Folic Acid and Related Pteridines*. North-Holland: North-Holland Publishing Co.
7. Truswell, A. S. and Milne, R. 2003. The B vitamins. In *Essentials of Human Nutrition*, edited by J. Mann and A.S. Truswell, pp. 222–224. Melbourne: Oxford University Press.
8. Lucock, M. 2000. Folic acid: Nutritional biochemistry, molecular biology, and role in disease processes. *Mol. Genet. Metab.* 71:121–138.
9. Wald, N., Sneddon, J., Densem, J., Frost, C., and Stone, R. 1991. Prevention of neural tube defects: Results of the Medical Research Council vitamin study. *Lancet* 338:131–137.
10. Czeizel, A. E. and Dudas, I. 1992. Prevention of the first occurrence of neural-tube defects by periconceptional vitamin supplementation. *New Engl. J. Med.* 327:1832–1835.
11. Oakley Jr, G. P. 2010. Folic acid-preventable spina bifida: A good start but much to be done. *Am. J. Prev. Med.* 38:569–570.

12. Arcot, J. and Shrestha, A. 2005. Folate: Methods of analysis. *Tr. Food Sci. Technol.* 16:253–266.
13. Ringling, C. and Rychlik, M. 2013. Analysis of seven folates in food by LC-MS/MS to improve accuracy of total folate data. *Eur. Food Res. Technol.* 236:17–28.
14. Chandra-Hioe, M. V., Bucknall, M. P., and Arcot, J. 2011. Folate analysis in foods by UPLC-MS/MS: Development and validation of a novel, high throughput quantitative assay; folate levels determined in Australian fortified breads. *Anal. Bioanal. Chem.* 401:1035–1042.
15. Freisleben, A., Schieberle, P., and Rychlik, M. 2002. Syntheses of labeled vitamers of folic acid to be used as internal standards in stable isotope dilution assays. *J. Agr. Food Chem.* 50:4760–4768.
16. Ruggeri, S., Vahteristo, L. T., Aguzzi, A., Finglas, P., and Carnovale, E. 1999. Determination of folate vitamers in food and in Italian reference diet by high-performance liquid chromatography. *J. Chromatogr. A.* 855:237–245.
17. Gregory III, J. F., Bhandari, S. D., Bailey, L. B., Toth, J. P., Baumgartner, T.G., and Cerda, J.J. 1992. Relative bioavailability of deuterium-labeled monoglutamyl tetrahydrofolates and folic acid in human subjects. *Am. J. Clin. Nutr.* 55:1147–1153.
18. O'Broin, J. D., Temperley, I. J., Brown, J. P., and Scott, J. M. 1975. Nutritional stability of various naturally occurring monoglutamate derivatives of folic acid. *Am. J. Clin. Nutr.* 28:438–444.
19. Vahteristo, L. T., Ollilainen, V., Koivistoinen, P. E., and Varo, P. 1996. Improvements in the analysis of reduced folate monoglutamates and folic acid in food by high-performance liquid chromatography. *J. Agr. Food Chem.* 44:477–482.
20. Patring, J. D. M., Jastrebova, J. A., Hjortmo, S. B., Andlid, T. A., and Jägerstad, I.M. 2005. Development of a simplified method for the determination of folates in baker's yeast by HPLC with ultraviolet and fluorescence detection. *J. Agr. Food Chem.* 53:2406–2411.
21. Pfeiffer, C. M., Rogers, L. M., and Gregory, J. F. III. 1997. Determination of folate in cereal-grain food products using trienzyme extraction and combined affinity and reversed-phase liquid chromatography. *J. Agr. Food Chem.* 45:407–413.
22. Johansson, M., Witthöft, C. M., Bruce, A., and Jägerstad, M. 2002. Study of wheat breakfast rolls fortified with folic acid: The effect on folate status in women during a 3-month intervention. *Eur. J. Nutr.* 41:279–286.
23. Alaburda, J., Almeida, A. P., Shundo, L., Ruvieri, V., and Sabino, M. 2008. Determination of folic acid in fortified wheat flours. *J. Food Comp. Anal.* 21:336–342.
24. Nelson, B. C. 2007. The expanding role of mass spectrometry in folate research. *Curr. Anal. Chem.* 3:219–231.
25. Stokes, P. and Webb, K. 1999. Analysis of some folate monoglutamates by high-performance liquid chromatography-mass spectrometry. I. *J. Chromatogr. A.* 864:59–67.
26. Freisleben, A., Schieberle, P., and Rychlik, M. 2003. Specific and sensitive quantification of folate vitamers in foods by stable isotope dilution assays using high-performance liquid chromatography-tandem mass spectrometry. *Anal. Bioanal. Chem.* 376:149–156.
27. Skoog, D. A. 2004. *Fundamentals of Analytical Chemistry*. Belmont, CA: Thomson-Brooks/Cole.
28. Pawlosky, R. J., Flanagan, V. P., and Pfeiffer, C. M. 2001. Determination of 5-methyltetrahydrofolic acid in human serum by stable-isotope dilution high-performance liquid chromatography-mass spectrometry. *Anal. Biochem.* 298:299–305.
29. Pawlosky, R. J. and Flanagan, V. P. 2001. A quantitative stable-isotope LC-MS method for the determination of folic acid in fortified foods. *J. Agr. Food Chem.* 49:1282–1286.
30. Pawlosky, R. J., Hertrampf, E., Flanagan, V. P., and Thomas, P. M. 2003. Mass spectral determinations of the folic acid content of fortified breads from Chile. *J. Food Comp. Anal.* 16:281–286.

31. Pawlosky, R. J., Flanagan, V. P., and Doherty, R. F. 2003. A mass spectrometric validated high-performance liquid chromatography procedure for the determination of folates in foods. *J. Agr. Food Chem.* 51:3726–3730.
32. Rychlik, M. 2011. Stable isotope dilution assays in vitamin analysis–A review of principles and applications. In *Fortified Foods with Vitamins*, edited by M. Rychlik, pp. 1–19. Wiley-VCH Verlag GmbH & Co. KGaA.
33. Freisleben, A., Schieberle, P., and Rychlik, M. 2003. Comparison of folate quantification in foods by high-performance liquid chromatography-fluorescence detection to that by stable isotope dilution assays using high-performance liquid chromatography-tandem mass spectrometry. *Anal. Biochem.* 315:247–255.
34. Núñez, O., Gallart-Ayala, H., Martins, C. P. B., and Lucci, P. 2012. New trends in fast liquid chromatography for food and environmental analysis. *J. Chromatogr. A.* 1228:298–323.
35. Patring, J., Wandel, M., Jägerstad, M., and Frølich, W. 2009. Folate content of Norwegian and Swedish flours and bread analysed by use of liquid chromatography-mass spectrometry. *J. Food Comp. Anal.* 22:649–656.
36. Phillips, K. M., Ruggio, D. M., Ashraf-Khorassani, M., and Haytowitz, D. B. 2006. Difference in folate content of green and red sweet peppers (*Capsicum annuum*) determined by liquid chromatography-mass spectrometry. *J. Agr. Food Chem.* 54:9998–10002.
37. Vishnumohan, S., Arcot, J., and Pickford, R. 2011. Naturally-occurring folates in foods: Method development and analysis using liquid chromatography-tandem mass spectrometry (LC-MS/MS). *Food Chem.* 125:736–742.
38. Swartz, M. E. 2005. UPLC™: An introduction and review. *J. Liq. Chromatogr. Relat Technol.* 28:1253–1263.
39. Nguyen, D. T. T., Guillarme, D., Rudaz, S., and Veuthey, J. L. 2006. Fast analysis in liquid chromatography using small particle size and high pressure. *J. Sep. Sci.* 29:1836–1848.
40. Jastrebova, J., Strandler, H. S., Patring, J., and Wiklund, T. 2011. Comparison of UPLC and HPLC for analysis of dietary folates. *Chromatographia.* 73:219–225.
41. De Brouwer, V., Storozhenko, S., Stove, C. P., Van Daele J., Van Der Straeten, D., and Lambert, W. E. 2010. Ultra-performance liquid chromatography-tandem mass spectrometry (UPLC-MS/MS) for the sensitive determination of folates in rice. *J. Chromatogr. B* 878:509–513.
42. Chandra-Hioe, M. V., Bucknall, M. P., and Arcot, J. 2013. Folic acid-fortified flour: Optimised and fast sample preparation coupled with a validated high-speed mass spectrometry analysis suitable for a fortification monitoring program. *Food Anal. Methods.* DOI 10.1007/s12161-012-9559-3.
43. Nelson, B. C., K. E. Sharpless, and L. C. Sander. 2006. Quantitative determination of folic acid in multivitamin/multielement tablets using liquid chromatography/tandem mass spectrometry. *J. Chromatogr. A* 1135:203–211.
44. Stark, A. K., Meyer, C., Kraehling, T., Jestel, G., Marggraf, U., Schilling, M., Janasek, D., and Franzke, J. 2011. Electronic coupling and scaling effects during dielectric barrier electrospray ionization. *Anal. Bioanal. Chem.* 400:561–569.
45. Leporati, A., Catellani, D., Suman, M., Andreoli, R., Manini, P., and Niessen, W. M. A. 2005. Application of a liquid chromatography tandem mass spectrometry method to the analysis of water-soluble vitamins in Italian pasta. *Anal. Chim. Acta* 531:87–95.
46. Patring, J. D. M. and Jastrebova, J. A. 2007. Application of liquid chromatography-electrospray ionisation mass spectrometry for determination of dietary folates: Effects of buffer nature and mobile phase composition on sensitivity and selectivity. *J. Chromatogr. A.* 1143:72–82.
47. Kok, R. M., Smith, D. E. C., Dainty, J. R., Van Den Akker, J. T., Finglas, P. M., Smulders, Y. M., Jakobs, C., and De Meer, K. 2004. 5-Methyltetrahydrofolic acid and

folic acid measured in plasma with liquid chromatography tandem mass spectrometry: Applications to folate absorption and metabolism. *Anal. Biochem.* 326:129–138.
48. Garbis, S. D., Melse-Boonstra, A., West, C. E., and Van-Breemen, R. B. 2001. Determination of folates in human plasma using hydrophilic interaction chromatography-tandem mass spectrometry. *Anal. Chem.* 73:5358–5364.
49. De-Brouwer, V., Storozhenko, S., Van De Steene, J. C., Wille, S. M. R., Stove, C. P., Van Der Straeten, D., and Lambert, W. E. 2008. Optimisation and validation of a liquid chromatography-tandem mass spectrometry method for folates in rice. *J. Chromatogr. A.* 1215:125–132.
50. Hyun, T. H. and Tamura, T. 2005. Trienzyme extraction in combination with microbiologic assay in food folate analysis: An updated review. *Experimental Biology and Medicine.* 230: 444–454.
51. Eitenmiller, R. R. 2008. *Vitamin Analysis for the Health and Food Sciences.* Boca Raton, FL: Taylor & Francis.

7 UPLC–MS/MS Analysis of Heterocyclic Amines in Cooked Food

Mohammad Rizwan Khan and Mu. Naushad

CONTENTS

7.1 INTRODUCTION

7.1.1 History of Heterocyclic Amines

The presence of carcinogenic substances in foods was first reported in 1939 by a Swedish scientist, E. M. P. Widmark, at Lund University, with the finding that organic solvent extracts of roasted horse meat caused carcinogenic effect when repeatedly applied to mouse skin [1]. A short-term assay for the determination of mutagenic activity based on *Salmonella* strains was developed by an American scientist, Ames and his coworkers [2], and not long after, a Japanese scientist, Professor Sugimura, and his group utilized this assay to demonstrate the presence of high mutagenic activity in the charred surface of beef and fish, broiled over a naked flame or charcoal [3,4]. The formation of mutagenic activity in meat, cooking under normal domestic conditions, and even from boiling beef extracts, was reported shortly after [5]. These findings initiated the study on heterocyclic amines (HCAs) and since then, around 25 highly mutagenic HCAs have been isolated and identified from various cooked foods, mainly protein-rich foods of animal origin [6–8]. Some of them have been classified as *probable* or *possible human carcinogens* by IARC (International Agency for Research on Cancer) [9]. Recently, the National Toxicology Program (NTP) has also listed some of them as *reasonably anticipated human carcinogens* [10,11]. Intense research relating to their formation, metabolism, and carcinogenicity has been carried out and is ongoing to evaluate the relevance of HCAs to human cancer.

7.1.2 Chemistry of Heterocyclic Amines

Until now more than 25 HCAs have been characterized as potent mutagens in the Ames/*Salmonella* assay [7,12–14]. Figure 7.1 displays the HCAs structures and abbreviated names. Many of these mutagenic HCAs have been isolated and identified not only from various proteinaceous foods such as cooked meat and fish [7,15], but also from environmental components, for instance, airborne particles and diesel exhaust particles [16], river water [17], and mainstream cigarette smoke [18].

HCAs have at least one aromatic and one heterocyclic structure, which give them another name, heterocyclic aromatic amines. Most of them have an exocyclic amino group, except β-carbolines (harman and norharman) [19]. In the β-carbolines, it is the lack of this functional group that causes them to be nonmutagenic in the Ames/*Salmonella* mutagenicity test [20]. The amine groups or nitrogen atoms may have different pK_a values. This together with different positions and number of these ionizable moieties will, therefore, affect the behavior of HCAs during chromatographic separation. As an example, it has been found that PhIP (2-amino-1-methyl-6-phenylimidazo(4,5-*b*)pyridine) and AαC were strongly retained via electrostatic interactions between deprotonated silanols and protonated HCAs at higher pH (4.7), while at pH 2.8, late elution was due to their capability of having both N–H acceptor and N–H donor groups on the same part of the molecule to form hydrogen bonds with SiOH and SiO^- groups [21]. The fact that most HCAs are stronger bases than the components of the mobile phase suggests that protonation to form $[M + H]^+$ is their common mode of ionization using soft ionization techniques [22].

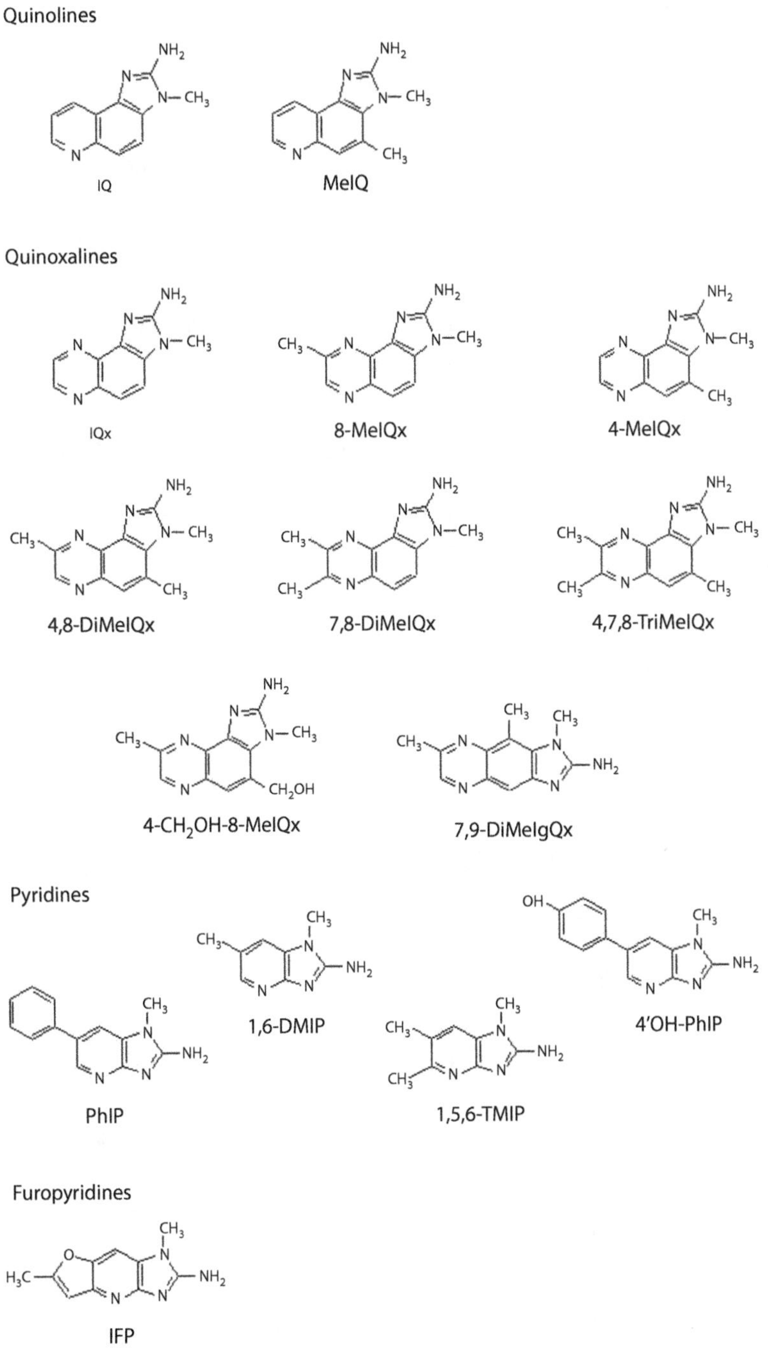

Quinolines

NH_2 N N–CH_3 N

IQ

NH_2 N N–CH_3 N CH_3

MeIQ

Quinoxalines

NH_2 N N–CH_3 N N

IQx

NH_2 N N–CH_3 CH_3 N N

8-MeIQx

NH_2 N N–CH_3 N N CH_3

4-MeIQx

NH_2 N N–CH_3 CH_3 N N CH_3

4,8-DiMeIQx

NH_2 N N–CH_3 CH_3 N CH_3 N

7,8-DiMeIQx

NH_2 N N–CH_3 CH_3 N CH_3 N CH_3

4,7,8-TriMeIQx

NH_2 N N–CH_3 CH_3 N N CH_2OH

4-CH_2OH-8-MeIQx

CH_3 CH_3 CH_3 N N NH_2 N N

7,9-DiMeIgQx

Pyridines

CH_3 CH_3 N NH_2 N N

1,6-DMIP

OH CH_3 N NH_2 N N

4′OH-PhIP

CH_3 N NH_2 N N

PhIP

CH_3 CH_3 N NH_2 CH_3 N N

1,5,6-TMIP

Furopyridines

CH_3 O N H_3C NH_2 N N

IFP

FIGURE 7.1 Structures and abbreviated names of HCAs.

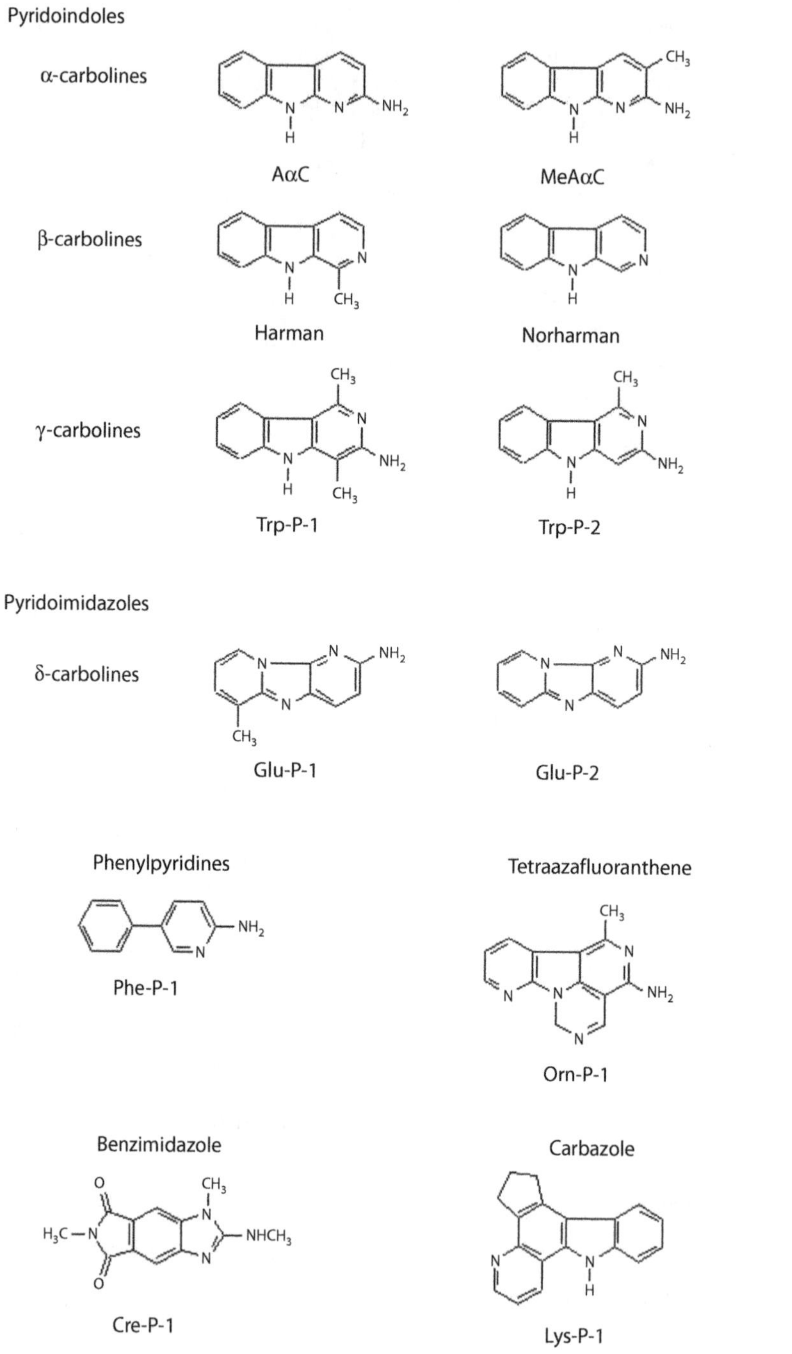

Pyridoindoles

α-carbolines

AαC

MeAαC

β-carbolines

Harman

Norharman

γ-carbolines

Trp-P-1

Trp-P-2

Pyridoimidazoles

δ-carbolines

Glu-P-1

Glu-P-2

Phenylpyridines

Phe-P-1

Tetraazafluoranthene

Orn-P-1

Benzimidazole

Cre-P-1

Carbazole

Lys-P-1

FIGURE 7.1 (continued) Structures and abbreviated names of HCAs.

For the amino-carbolines, the five-membered heterocyclic aromatic ring is sandwiched between two six-membered aromatic rings, one or both of which can be pyridine, in contrast to amino-imidazoazarenes (AIAs). All AIAs have an N-attached methyl group at the imidazole ring of the molecule, whereas for amino-carbolines, they either lack a methyl group or the methyl group is attached to a six-membered ring. Different number and positions of methyl groups largely increase the number of this class of mutagens and it is, therefore, likely that new mutagens will be identified [23–25].

7.1.3 Formation of Heterocyclic Amines

Heat treatment is used frequently to improve meat palatability and organoleptic properties. Baking, roasting, broiling, or frying of meat or fish bring about the formation of flavor compounds and brown pigments, which is sensorially perceived by the consumer as a golden brown crispy crust with good flavor. During heat treatment a temperature gradient develops in the meat since heat is transferred inwardly from the surface. Furthermore, mass transfer of water (as vapor and sometimes as drips) progresses during heat processing and a moisture gradient is established, with the highest content in the center of the meat. The rapid evaporation of water from the meat surface induces the formation of a crust with low free water content. Melted fat from the meat can readily absorb heat energy from the metal pan, raising the crust temperature above 100°C. The combination of high temperature, absence of water, and high concentrations of reactants is responsible for the rapid browning of the crust.

This nonenzymatic browning, also known as the Maillard reaction, mainly originated from the pyrolysis of carbonyl compounds, especially reducing sugars and compounds with free amino groups (amines, amino acids, and proteins). The Maillard reaction is also responsible for the high mutagenic activity in meat including the generation of HCAs. The general scheme of the Maillard reaction has been presented in Figure 7.2 [26].

In general, HCAs are formed during cooking of protein-rich foods such as meat and fish by the condensation of creatinine with amino acids. This formation is dependent upon the type of meat products, cooking temperature, duration of cooking process, and the concentrations of HCA precursors and compounds with enhancing or inhibiting effects [27–30]. Prolonged cooking time and high-temperature cooking surface produce the highest quantities of HCAs [27]. The level of HCAs formed in meats prepared by common household cooking practices are generally in the low parts-per-billion range, although concentrations in meats or poultry that are cooked well done [31], or the grilled pan scrapings often used for gravy, can be as high as 500 ng/g [31].

In view of their chemical structure, they can be classified into two main groups called "thermic HCAs," IQ-type or amino-imidazoazaarenes and "pyrolytic HCAs," non-IQ-type or amino-carbolines. Those of the first type are produced by Maillard reaction when mixtures of creat(ni)ne, amino acids, and sugars are heated at temperatures between 100°C and 300°C [32]. Those of the second type are mainly formed by the pyrolysis of amino acids and proteins at higher temperatures, above 300°C

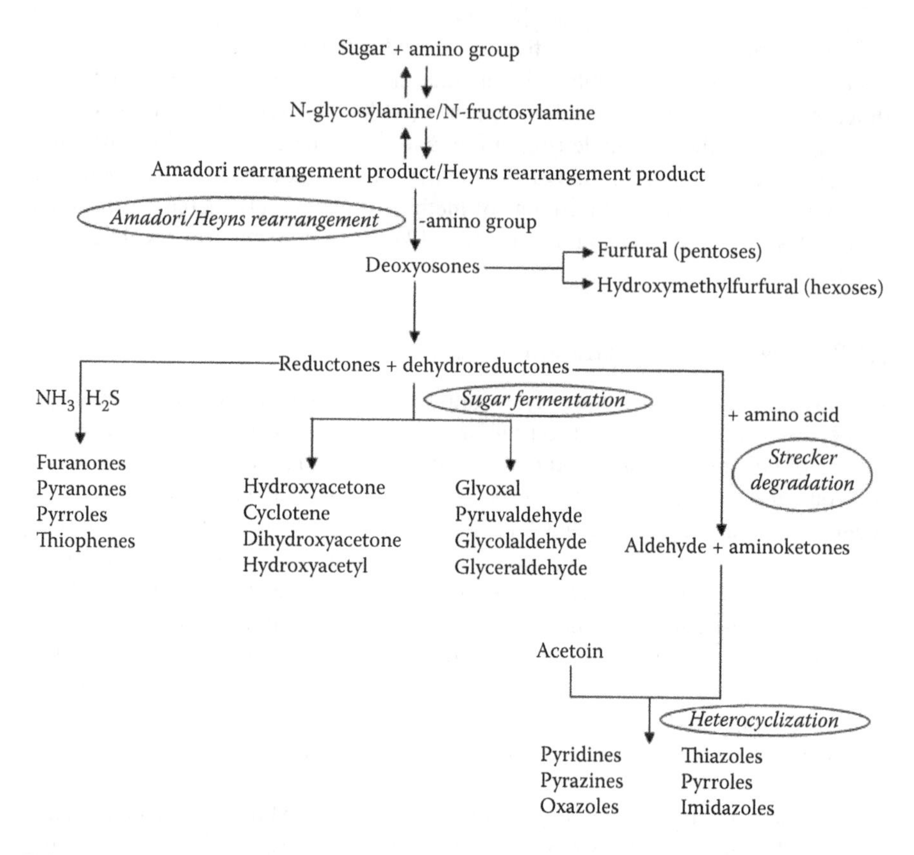

FIGURE 7.2 The general scheme of the Maillard reaction. (Adapted from Hodge, J. E. 1953. *J. Agric. Food Chem.* 1:928–943. With permission.)

[33]. Based on their apparent polarity, HCAs can be divided into polar HCAs, which are mainly of the quinolines (IQ), quinoxalines (IQx), and imidazopyrydine type, and less-polar HCAs, which have a common pyridoindole or dipyridoimidazole moiety. Table 7.1 illustrates the HCAs classifications and properties.

7.1.3.1 Amino-Imidazoazaarenes (IQ-Type)

Amino-imidazoazaarenes types of HCAs including IQ, IQx, pyridines, and furopyridines are generated at normal cooking temperatures (100–300°C). AIAs have been found to be responsible for most of the mutagenic activity in cooked foods, especially in Western diets [34]. Therefore, identification of precursors and elucidation of their mechanism of formation is needed for later development of strategies to inhibit their formation. For this purpose, model systems that simulate the chemical environment in the meat were developed, in which complex side reactions are reduced and reactions of other constituents of the meat that are not involved in the formation of HCAs are excluded [35].

IQ and IQx types of HCAs and their derivatives (methylated forms) are suggested to be formed from three endogenous constituents of meat: creatine (a compound

TABLE 7.1
HCAs Classification and Properties

Compound Name and Classification	Abbreviation	Molecular Formula	Molecular Mass	pK_a	Chemical Polarity
	Imidazopyridine derivatives				
2-Amino-1,6-dimethyl-imidazo[4,5-*b*]pyridine	DMIP	$C_8H_{10}N_4$	162	–	Polar
2-Amino-1,5,6-trimethyl-imidazo[4,5-*b*]pyridine	1,5,6-TMIP	$C_9H_{12}N_4$	176	–	Polar
2-Amino-1-methyl-6-phenyl-imidazo[4,5-*b*]pyridine	PhIP	$C_{13}H_{12}N_4$	224	5.6	Polar
2-Amino-1-methyl-6-(4′-hydroxyphenyl)-imidazo[4,5-*b*]pyridine	4′-OH-PhIP	$C_{13}H_{12}N_4O$	240	–	Polar
2-Amino-1,6-dimethyl-furo[3,2-*e*]imidazo[4,5-*b*]pyridine	IFP	$C_{10}H_{10}N_4O$	202	–	Polar
	Imidazoquinoline derivatives				
2-Amino-3-methyl-imidazo[4,5-*f*]quinoline	IQ	$C_{11}H_{10}N_4$	198	6.1	Polar
2-Amino-3,4-dimethyl-imidazo[4,5-*f*]quinoline	4-MeIQ	$C_{12}H_{12}N_4$	212	6.4	Polar
	Imidazoquinoxaline derivatives				
2-Amino-3-methyl-imidazo[4,5-*f*]quinoxaline	IQx	$C_{10}H_9N_5$	199	–	Polar
2-Amino-3,4-dimethyl-imidazo[4,5-*f*]quinoxaline	4-MeIQx	$C_{11}H_{11}N_5$	213	–	Polar
2-Amino-3,8-dimethyl-imidazo[4,5-*f*]quinoxaline	8-MeIQx	$C_{11}H_{11}N_5$	213	5.95	Polar
2-Amino-3,4,8-trimethyl-imidazo[4,5-*f*]quinoxaline	4,8-DiMeIQx	$C_{12}H_{13}N_5$	227	5.8	Polar
2-Amino-3,7,8-trimethyl-imidazo[4,5-*f*]quinoxaline	7,8-DiMeIQx	$C_{12}H_{13}N_5$	227	6.5	Polar
2-Amino-3,4,7,8-tetramethyl-imidazo[4,5-*f*]quinoxaline	4,7,8-TriMeIQx	$C_{13}H_{15}N_5$	241	6.0	Polar
2-Amino-4-hydroxymethyl-3,8-dimethyl-imidazo[4,5-*f*]quinoxaline	4-CH2OH-8-MeIQx	$C_{12}H_{13}N_5O$	243	–	Polar
2-Amino-1,7,9-trimethyl-imidazo[4,5-*g*]quinoxaline	7,9-DiMeIgQx	$C_{12}H_{13}N_5$	227	–	Polar
	Phenylpyridine derivatives				
2-Amino-5-phenylpyridine	Phe-P-1	$C_{11}H_{10}N_2$	170	–	Less-polar

continued

TABLE 7.1 (continued)
HCAs Classification and Properties

Compound Name and Classification	Abbreviation	Molecular Formula	Molecular Mass	pK_a	Chemical Polarity
	Pyridoindole derivatives (α-Carbolines)				
2-Amino-9H-pyrido[2,3-*b*]indole	AαC	$C_{11}H_9N_3$	183	4.4	Less-polar
2-Amino-3-methyl-9H-pyrido[2,3-*b*]indole	MeAαC	$C_{12}H_{11}N_3$	197		Less-polar
β-Carbolines					
1-Methyl-9H-pyrido[3,4-*b*]indole	Harman	$C_{12}H_{10}N_2$	182	–	Less-polar
9H-pyrido[3,4-*b*]indole	Norharman	$C_{12}H_{11}N_3$	168	6.8	Less-polar
γ-Carbolines					
3-Amino-1,4-dimethyl-5H-pyrido[4,3-*b*]indole	Trp-P-1	$C_{13}H_{13}N_3$	211	8.6	Less-polar
3-Amino-1-methyl-5H-pyrido[4,3-*b*]indole	Trp-P-2	$C_{12}H_{11}N_3$	197	8.5	Less-polar
	Pyridoimidazole derivatives (δ-carbolines)				
2-Amino-6-methyldipyrido[1,2-α:3´,2´-*d*]imidazole	Glu-P-1	$C_{11}H_{10}N_4$	198	6.0	Less-polar
2-Aminodipyrido[1,2-α:3´,2´-*d*]imidazole	Glu-P-2	$C_{10}H_8N_4$	184	5.9	Less-polar
	Benzimidazole derivatives				
4-Amino-1,6-dimethyl-2-methylamino-1H,6Hpyrrolo [3,4-*f*]benzimidazole-5,7-dione	Cre-P-1	$C_{13}H_{15}N_3O_2$	245	–	Less-polar
	Tetraazafluoranthene derivatives				
4-Amino-6-methyl-1H-2,5,10,10b-tetraazafluoranthene	Orn-P-1	$C_{13}H_{11}N_2$	237	–	Less-polar
	Carbazole derivatives				
3,4-Cyclopenteno-pyrido[3,2-*a*]carbazole	Lys-P-1	$C_{18}H_{14}N_2$	258	–	Less-polar

present only in muscle in the form of creatine phosphate, which is transformed into free creatine within 24 h of animal killing), free amino acids, and reducing sugars, through Maillard reaction and Strecker degradation upon heating [32,36] (Figure 7.2) [26]. Weisburger et al. found that the addition of extra creatinine to the surface of meat prior to frying increased the yield of HCAs, indicating the important role of creatinine [37]. On the other hand, as several amino acids may give rise to a common HCA, though likely with different yields, no definite amino acid precursor HCA relationship has been drawn. Findings of studies intending to identify the limiting factor have also been inconsistent. Other studies found that HCAs can also be formed in dry-heated mixtures of amino acids and creatine [38]. It appears that HCA formation proceeds via different reaction kinetics under the above different sets of parameters, the variation of which has strong influence on the rate-limiting step.

It was postulated and later demonstrated that creatinine, which is produced from creatine through cyclization and water elimination, forms the amino-imidazo part of the IQ and IQx, while the pyridines or pyrazines contribute to the remaining part of the molecule. Aldol condensation was thought to link the two parts together via a Strecker aldehyde, while the Strecker degradation products such as pyridines or pyrazines formed in the Maillard reaction between amino acids and hexose contributed to the remaining part of the molecule, probably via aldol condensation [19,39] (Figure 7.3). This reaction is especially favored at temperatures above 100°C [40]. Nevertheless, yield of pyrazines and pyridines from the Maillard reaction and Strecker degradation is still low and Arvidsson et al. proposed this as a causal factor for low yield of HCA in foods or model systems [41]. Despite evident involvement of the Maillard reaction, a previous study has shown suppression of IQ compound formation in model systems by addition of preformed Maillard reaction products [42].

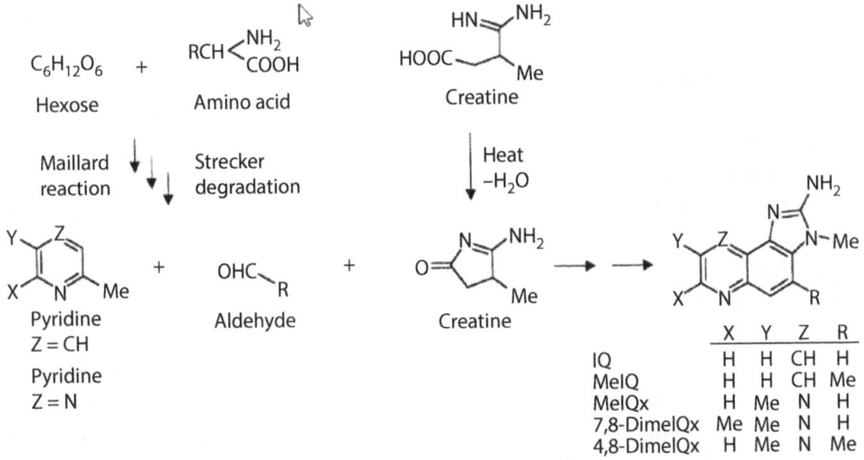

	X	Y	Z	R
IQ	H	H	CH	H
MeIQ	H	H	CH	Me
MeIQx	H	Me	N	H
7,8-DimeIQx	Me	Me	N	H
4,8-DimeIQx	H	Me	N	Me

FIGURE 7.3 Suggested pathway for the formation of imidazo-quinolines and imidazo-quinoxalines. (Adapted from Jägerstad, M. et al. 1998. *Z. Lebensm. Unters. Forsch. A* 207:419–427. With permission.)

7.1.3.2 Amino-Carbolines (Non-IQ Type)

These types of HCAs are generated through pyrolysis of amino acids or proteins at cooking temperatures above 300°C, with the exception of harman and norharman, which are both formed at lower temperature. The amino-carbolines are generated in lower amounts in cooked meat or fish because of the high temperatures required for their formation, therefore little investigation has been carried out to elucidate their mechanism of formation. However, as addition of excess creatinine to a model system was found to increase harman and norharman formation [19], it is probable that creatinine may serve as an extra source of structural fragments. The most popular hypothesis for the formation of amino-carbolines under such drastic thermal environment has been a pathway via free radical reactions, which produce many reactive fragments [29]. These fragments may then condense to form new structures [29]. However, due to their much lower occurrence in normally cooked foods and greater difficulty in manipulation of the high temperatures required in experimental set up, relatively little investigation has been carried out to verify the above hypothesis compared with the AIAs.

7.1.3.3 Occurrence of HCAs in Cooked Foods

Even though HCAs are ubiquitous, it is only the uptake of thermally processed meat and fish products that contributes significantly to the exposure, since the concentrations in other foods or in the environment are extremely low. Tables 7.2 through 7.4 present a brief summary of quantitative data from the literature on the amounts of five highly mutagenic HCAs in cooked meat, poultry, fish and offal products. The HCAs have been reported in all kinds of meat and fish products, generally those cooked by frying, grilling, broiling, barbecuing, and even smoked foods. From the literature survey, the red meats (nearly 60%) followed by poultry, fish, and offal products are the most frequently consumed throughout the world. The most common cooking method, pan-frying (nearly 80%) was used for all kinds of meat products. Frying followed by broiling and barbecuing produce the largest amounts of HCAs because the meats are cooked at very high temperatures. One study conducted by researchers showed a threefold increase in the content of HCAs when the cooking temperature was increased from 200°C to 250°C. Oven roasting and baking are done at lower temperatures, so lower levels of HCAs are likely to form; however, gravy made from meat drippings does contain substantial amounts of HCAs. Stewing or boiling is done at or below 100°C; cooking at this low temperature creates negligible amounts of the HCAs. Foods cooked a long time ("well done" instead of "medium") by other methods will also form slightly more of the chemicals. Meats that are partially cooked in the microwave oven before cooking by other methods also have lower levels of HCAs.

HCAs are primarily found in the crust of cooked meat and fish with minor amounts left in the inner parts of fried meat [28]. In the few studies where co-mutagens harman and norharman have been analyzed, these β-carbolines have been found in the highest range [15,67]. Usually, PhIP is the most abundant HCA, detected in amounts of up to almost 500 ng/g. PhIP seems to form more easily in chicken (up to 40 ng/g) than in beef, pork, fish, or offal products during cooking, while the amount of, for example, MeIQx is generally lower in cooked chicken than in cooked beef and pork

TABLE 7.2
Content of HCAs (ng/g) in Cooked Beef, Lamb, Pork, and Reindeer

Food Type	Cooking Method	Temp (°C)	Time[a] (min)	IQ	MeIQ	MeIQx	4,8-DiMeIQx	PhIP	References
Bacon	Fried	150	5–10	3.8–10.5	nd-1.7	2.5–2.8	1.0–3.4	0.2–1.0	[43]
Bacon, pan r	Fried	150	5–10	nd	nd	0.2–5.9	0.2–1.7	nd	[43]
Bacon	Fried	150–225	4–8	nd	nd	nd-23.7	0.2–1.4	0.3–4.5	[28]
Bacon	Grilled	230–260	4	nd	nd	nd	nd	16.5–248.8	[44]
Beef	Fried	200–250	12	nd-1.0	na	nd-5.1	0.1–1.2	0.7–13.2	[25]
Beef	Fried	277	7	na	na	16.4	4.5	67.5	[45]
Beef	Fried	–	–	nd	0.09	0.19	nd	0.63	[46]
Beef	Baked	275	30	na	na	1.43	0.20	1.2	[47]
Beef	Fried	165–200	10–6	na	na	0.2–1.6	nd-0.4	0.08–1.5	[48]
Beef pan r	Fried	165–200	10–6	nd	nd	0.8–4.3	0.4–1.3	0.4–13.3	[48]
Beef	Fried	190	24	0.04	na	3.7	0.58	0.4	[24]
Beef scraping	Fried	150–180	10	1.9	na	62.6	15.0	7.6	[24]
Beef	Grilled	–	–	na	na	6	1.2	14	[49]
Beef	Barbecued	230–300	20	0.03	na	0.5	0.1	2.2	[24]
Beef	Cooked	100/90	20/20	na	na	nd	nd	nd	[50]
Beef	Com.	–	–	nd	nd	nd	nd	0.3	[51]
Beef hamburger	Grilled	–	–	na	na	0.02	0.03	0.06	[52]
Beef kebab	Fried	–	–	na	na	0.14	0.22	0.22	[52]
Beef extract	Com.	—	–	nd	5.77	5.07	nd	na	[53]
Beef stock cube	–	–	–	na	na	0.6	0.3	0.3	[54]
Beef, min steak	Fried	150–225	2.5–4	nd	nd	nd-6.2	nd-2.7	0.02–12.7	[28]
Beef, min steak, pan r	Fried	150–225	2.5–4	0.1	nd	0.1–23.3	0.1–4.1	0.2–82.4	[28]
Beef sirloin steak	Fried	150–225	7	nd-0.04	nd	0.02–1.6	0.02–0.6	0.06–1.8	[28]
Beef steak	Fried	180–200	12	na	na	0.8	0.7	1.1	[55]
Beef steak	Grilled	225	12	na	na	0.5	0.1	0.6	[54]
Beef liver	Fried	210–225	8	nd	nd	nd	nd	nd	[56]
Beef tongue	Fried	210–225	8	nd	nd	0.21	0.02	0.02	[56]

continued

TABLE 7.2 (continued)
Content of HCAs (ng/g) in Cooked Beef, Lamb, Pork, and Reindeer

Food Type	Cooking Method	Temp (°C)	Time[a] (min)	IQ	MeIQ	MeIQx	4,8-DiMeIQx	PhIP	References
Black pudding	Fried	150	8	0.2	nd	0.5	0.5	nd	[43]
Bouillion cubes	Com.	–	–	nd	nd	nd	nd	0.8	[51]
Lamb chops	Fried	150–225	9	nd	nd	nd-0.4	nd-0.6	nd-1.5	[15]
Lamb chops, pan r	Fried	150–225	9	nd	nd	0.08–0.6	0.04–0.3	0.01–2.3	[15]
Lamb	Grilled	–	–	na	na	1.6	nd	11	[57]
Lamb kidney	Fried	210–225	8	nd	nd	0.08	< 0.01	0.11	[56]
Meatballs	Fried	150–225	6.5–9	nd	nd	nd-0.8	nd-0.3	nd-0.1	[28]
Meatballs, pan r	Fried	150–225	6.5–9	0.05	nd	0.02–0.7	0.02–0.1	0.03–0.5	[28]
Meat cuts	Com.	–	–	nd	nd	nd-0.4	nd	nd	[58]
Meat extract	–	–	–	na	na	29.0–46.0	4.8–6.2	nd-7.5	[49]
Meat sauce	Fried	150–225	6	nd	nd	nd-1.1	nd-0.4	0.07–2.1	[15]
Meatloaf	Roasted	150	55	nd	nd	0.1	nd	0.3	[15]
Meatloaf pan r	Roasted	150	55	nd	nd	0.04	0.03	nd	[15]
Pork loin	Grilled	230–260	8	nd	nd	nd-13.2	nd-1.3	6.4–569.8	[44]
Pork lion	Boiled	100	15	nd	nd	nd	nd	0.8	[44]
Pork	Barbecued	–	–	na	na	0.4	0.1	4.2	[54]
Pork	Fried	–	–	nd	0.08	0.20	nd	0.85	[46]
Pork loin	Baked	275	30	na	na	3.5	0.4	4.7	[47]
Pork chop, pan r	Fried	150–225	8–9.5	0.1	nd	nd-1.9	nd-0.5	0.02–3.8	[28]
Pork, ham	Grilled	–	–	na	na	2.3	0.2	1.5	[57]
Pork, fillet	Fried	150–225	7	nd	nd	nd-4.6	nd-3.3	nd-13.4	[15]
Pork, fillet, pan r	Fried	150–225	7	0.1	0.1	0.06–5.6	0.08–4.2	0.3–32.0	[15]
Reindeer	Fried	150–225	10	nd	nd	nd-1.0	nd	0.4–5.8	[15]
Reindeer, pan r	Fried	150–225	10	nd	nd	0.1–0.8	0.03–0.6	nd-3.5	[15]

Note: pan r = pan residue; nd = not detected; na = not analyzed; Com. = commercially cooked food; – not described.

[a] Total cooking time.

TABLE 7.3
Content of HCAs (ng/g) in Cooked Poultry Products

Food Type	Cooking Method	Temp (°C)	Time[a] (min)	IQ	MeIQ	MeIQx	4,8-DiMeIQx	PhIP	Reference
Chicken breast	Fried	150–225	30	nd	nd	0.4–0.5	0.2–0.5	0.5–10	[15]
Chicken breast, pan r	Fried	150–225	30	nd	0.3	0.08–0.6	nd-0.3	0.02–1	[15]
Chicken breast	Fried	80[b]	14	na	na	26.4	16.6	1079.8	[59]
Chicken patty	Fried	–	–	nd-4.5	nd	nd-5.4	nd	2.1–17.8	[44]
Chicken breast	Baked	275	30	na	na	0.5	0.2	37.5	[47]
Chicken thigh	Baked	275	30	na	na	0.02	0.05	8.0	[47]
Chicken	Grilled	150–250	8–20	nd	0.14	0.07	nd	0.28	[46]
Chicken	Fried	190.1	7.19	na	na	0.18	0.11	2.37	[60]
Chicken liver	Fried	190	9	na	na	nd	nd	nd	[61]
Chicken	Stewed	100	–	nd	nd	nd	nd	nd	[62]
Chicken breast	Baked	250	14	na	na	23.3	2.9	427.5	[59]
Chicken legs, with skin	Microwaved	–	5–15	nd	nd	nd	nd	nd	[63]
Chicken legs, skinless	Microwaved	–	5–15	nd	nd	nd	nd	nd	[63]
Chicken breast	Broiled	180	43	nd	na	nd	nd	131	[62]
Chicken breast	Barbecued	177–260	10–43	nd	na	nd-9	nd-2	27–480	[62]
Turkey breast	Baked	275	30	na	na	3.5	0.4	4.7	[47]
Turkey hotdog	Fried	–	–	0.51	na	4.2	na	4.4	[64]

Note: pan r = pan residue; nd = not detected; na = not analyzed; Com. = commercially cooked food; – not described.

[a] Total cooking time.

[b] Internal temperature.

TABLE 7.4
Content of HCAs (ng/g) in Cooked Fish Products

Fish Type	Cooking Method	Temp (°C)	Time[a] (min)	IQ	MeIQ	MeIQx	4,8-DiMeIQx	PhIP	References
Cod	Fried	–	–	0.16	0.03	6.44	0.10	69.2	[65]
Cod	Fried	150–225	10	nd	nd	nd-0.9	nd	0.02–2.2	[15]
Baltic herring	Fried	150–225	4	nd	nd	nd-0.2	nd	0.06–0.3	[15]
Baltic herring	Com.	–	–	0.2	0.1	0.6	0.3	nd	[43]
Cod	Baked	275	30	na	na	nd	nd	3.2	[47]
Salmon	Fried	150	18	0.6	1.3	0.6	0.2	3.0	[43]
Salmon	Smoked	–	–	0.3	nd	1.3	nd	nd	[43]
Red snapper with skin	Fried	222.5	8.24	na	na	0.10	0.03	1.37	[60]
Pacific saury (flesh)	Grilled	150–250	8–20	nd	nd	0.2	nd	0.3	[46]
Horse mackerel (flesh)	Grilled	150–250	8–20	nd	nd	0.02	nd	0.16	[46]
Mackerel (flesh)	Grilled	150–250	8–20	nd	0.06	nd	nd	0.23	[46]
Atka mackerel (flesh)	Grilled	150–250	8–20	nd	0.09	nd	nd	0.65	[46]
Brown trout	Barbecued	–	20	0.12	na	na	0.02	na	[66]
Rainbow trout	Grilled	180	8	na	na	na	0.02	na	[66]

Note: nd = not detected; ; na = not analyzed; Com. = commercially cooked food; – not described.

[a] Total cooking time.

[31]. Other HCAs, IQ, IQx, MeIQ, Trp-P-1, Trp-P-2, AαC, and MeAαC, are generally found at levels not detectable to below 1 ng/g in all kinds of food. The amounts of HCAs are generally higher in cooked meats than fish and in pure meat than in mixed meat products, for example, meatballs or sausage. Commercially cooked food and offal products generally contain very low or undetectable amounts of HCAs [31], with a few exceptions. The levels of HCAs in pan residues after frying different meat and fish products are generally the same as in the corresponding food products, but in some cases the amounts are considerably higher [31]. In some countries pan residues are used to make gravy, which may result in a substantial contribution of HCAs to the diet. Bouillon cubes contain very low or undetectable amounts of HCAs [31] and the dietary intake of HCAs may thus be reduced simply by discarding the pan residue after frying and preparing gravy and sauces using commercial products such as bouillon cubes.

7.1.3.4 Intake of HCAs

To approximate the intake of HCAs by human beings, a number of methods are available ranging from dietary questionnaires to determination of biomarkers. However, all of these methods have their limitations so that only a combination can be used to calculate the exposure to HCAs and relate these data to epidemiological studies. Especially by using questionnaires for HCA intake there are several errors including inconsistent reporting, difficulty in quantifying cooking doneness, and the day-to-day variation in the diet [68]. In order to overcome some of the problems related to the use of questionnaires in the assessment of HCA exposure, the development of biomarkers is an important concern. Since the exposure to HCAs may vary to a great extent from day to day, a biomarker should ideally integrate the exposure of a certain time period. A detailed overview is given by Alexander et al. [69], where the available methods are discussed. Because HCAs are known for their extensive metabolism as well as fast urinary excretion, not only the parent compounds but also metabolites can be found in urine as was shown for rodents [70]. It was shown, using radioactive labeled AαC, that within 72 h, 55% of the radioactivity was excreted in urine and feces [71]. In humans, conjugated HCAs (glucuronides and sulfates) [72] and hydroxylation products are found as well [71]. Adduct formation with proteins was shown to be another possibility but the measured levels were very low [73]. Another possibility to estimate the HCA intake is to measure the accumulation of these basic compounds in hair as is done with drugs, narcotics, or nicotine. However, the amounts of HCAs in hair are at least three orders of magnitude lower compared to these compounds. Alexander et al. showed that PhIP accumulates in the hair of exposed mice as well as humans consuming an ordinary diet [74,75].

To estimate the uptake of HCAs, Keating and Bogen [76] applied a method that combines laboratory data to predict HCA concentrations from meat type, cooking method, and meat doneness with national dietary data to estimate daily HCA intake for segments of the U.S. population. PhIP was found to comprise 70% of U.S. mean dietary intake of total HCAs, with pan-frying and chicken being the single cooking method and meat type contributing the greatest to total estimated HCA exposures. This analysis demonstrated significantly higher concentrations in grilled/barbecued meats than in other cooked meats. The daily intake of PhIP, MeIQx, and DiMeIQx for an adult was in the range of 11.0–19.9 ng/kg per day. Earlier estimates for the U.S. population ranged from 6.3 to 20.1 ng/kg per day [77,78], whereas estimates for European populations range from 2.3 to 6.6 ng/kg per day [79,80]. In a recent study the intake of the Spanish population was estimated at 606 ng of HCAs *per capita* per day with DMIP and PhIP being the main contributors [81]. In a similar study in Poland the intake of the sum of IQ, MeIQ, MeIQx, 4,8-DiMeIQx, and PhIP was estimated from 200 to 7000 ng of HCAs *per capita* per day [82].

7.1.4 Exposure of Heterocyclic Amines

Exposure scenarios for genotoxic carcinogens in food and diet will almost always relate to naturally occurring substances, illegal additives, or substances formed during cooking or other processing. In any food chemical risk assessment, exposure assessment is often a source of great uncertainty. This situation may be even more pronounced in the case of genotoxic carcinogens in food due to the low level in food, variability in concentrations present, and variability in food consumed.

The exposure of HCAs in human beings derived from cooked meats in the diet has become a subject of increasing concern during the past two decades. The relationship between HCAs in the diet and associated health risks is reflected in the number of epidemiological and dietary studies that have been conducted in recent years [7]. The HCAs are present in ng/g quantities in cooked meat; nevertheless, the estimated total dietary intake of HCAs is low but chronic since meat is consumed daily over a lifetime. Thus, depending upon dietary preferences, an individual's daily exposure to HCAs is likely to range from microgram quantities to essentially zero in the case of vegetarians. Significantly, the incidence of cancer is approximately 60% lower in vegetarians than in nonvegetarians [83].

With epidemiological data indicating a link with cancers [77], several studies have been carried out in the analysis of HCAs and/or metabolites in biological samples as a means of assessing human exposure to these compounds. In one such study, fried beef containing known quantities of HCAs was fed to healthy volunteers on each of four separate occasions, and urinary excretion of unchanged HCAs was examined. Irrespective of the dose, about 2% of the ingested MeIQx and between 0.5% and 1% of the PhIP was excreted unchanged in the urine within 24 h of the test meal, the majority being eliminated within 8 h [84]. Although there was up to a fivefold interindividual variation in the excretion of unchanged amine, within an individual the percentage excreted on different occasions remained remarkably consistent [84]. Excretion of unchanged amine in urine is a function of the extent of absorption and clearance due to metabolism. From animal studies, HCA absorption seems to be almost complete; thus, variations in urinary excretion should provide a measure of interindividual variability in the metabolism of these compounds.

7.1.5 Regulations and Guidelines

No regulations or guidelines relevant to the prevention of exposure to HCAs were identified. Nevertheless, the International Agency for Research on Cancer [9] has classified some of them as *probably carcinogenic to humans* (Group 2A) and *possibly carcinogenic to humans* (Group 2B). The Group 2A includes IQ, whereas, the Group 2B includes MeIQ, MeIQx, and PhIP.

Recently, the NTP has listed four individual HCAs in the Report on Carcinogens as *reasonably anticipated to be a human carcinogen*. IQ was first listed in the *Tenth Report on Carcinogens* [10], and three other HCAs, MeIQ, MeIQx, and PhIP, were listed for the first time in the *Eleventh Report on Carcinogens* [11]. Studies have reliably displayed that MeIQ, MeIQx, IQ, and PhIP cause mutations in most test systems, including rodents exposed *in vivo*, bacteria, and cultured human cells.

7.2 ANALYTICAL METHODOLOGIES

To fully assess the risk to human health posed by the daily consumption of foods containing these mutagens/carcinogens, it is essential to quantify the amount of carcinogen to which man is chronically exposed. Nevertheless, the quantitative determination of HCAs in food samples is mainly hindered by the low level of concentration of these microcomponents and the high complexity of the matrix. Therefore, the development of sensitive and selective analytical approach is mandatory.

7.2.1 Extraction Methods

Owing to very low levels (ng/g) of HCAs in cooked foods and a high number of matrix interferences, to date various multi-step extraction methods—for instance liquid–liquid extraction (LLE), solid-phase extraction (SPE) with disposable columns, solid-phase microextraction (SPME), supercritical fluid extraction (SFE), tandem extraction procedures consisting of the coupling of LLE and SPE, or pressurized liquid extraction (PLE)—have been employed for the extraction and purification of HCAs from food samples. The analysis of HCAs is commonly carried out by means of chromatographic or electrophoretic methods using different detection systems. The sample matrix greatly influences the clean-up procedures and many peaks with the same retention times as those of HCAs are often present in the chromatograms of real samples. Thus, it is not too much to say that the clean-up procedures for the complex sample matrix greatly influence the reliable and accurate analysis of these compounds. The first step for the sample preparation usually consists of a solubilization step, where the sample is homogenized and dispersed using different solvents. The solvents used are organic (such as acetone, dichloromethane, ethyl acetate, methanol, hydro-alcoholic mixtures) or aqueous solvents (such as water, hydrochloric acid or, more frequently, sodium hydroxide). After solubilization, the solvent is used to remove proteins by precipitation using traditional procedures and to make their separation by centrifugation or filtration. In order to remove interferences and to preconcentrate HCAs, the extraction and clean-up of the sample has been carried out using a number of different purification techniques as named above. The details of the most important applied methods are described in the following.

7.2.1.1 Liquid–Liquid Extraction

LLE extraction method has been preferred by most of the authors for the first step in the isolation of the HCAs from the food matrix. In the LLE, if an organic solvent has been used to homogenize the sample, the analytes must be extracted with HCl. But when an aqueous solvent is used, the acidic solution obtained is directly extracted with an organic solvent, such as dichloromethane [54,85], ethyl acetate [64], or diethyl ether [86] in order to remove acidic or neutral interferences. If the obtained solution is basic, analytes can be extracted in their neutral form with dichloromethane [87]. In most cases, further purification is carried out by consecutive acid–base partition processes with dichloromethane [85] or by combination with extraction using sorbents, such as Kieselgur, Extrelut NT, diatomaceous earth, or with Blue Rayon [88]. These materials can be added to the liquid in the batch mode or, more frequently, as a support in a chromatographic column. The method using diatomaceous earth is usually referred to as LLE, in contrast to SPE, which is usually coupled online to the column that contains the solid material, which will be described in following.

7.2.1.2 Solid-Phase Extraction

The SPE procedure can be considered a special case of liquid chromatography (LC). In general, the SPE procedure is a highly selective technique, where the extraction of the analyte is performed using disposable commercial cartridges, which typically contain from 100 to 500 mg of a solid sorbent as the stationary phase. Most of the SPE procedures allow working at micro-analytical scale. Thus, most of the sample preparation procedures apply this separation technique for the analysis of HCAs, which allows one to obtain extracts, purified enough to prevent interferences and a high-throughput analysis.

Other SPE methods are based on affinity mechanisms using different sorbents, such as Blue Rayon [89], Blue Cotton [90], or Blue Chitin [91]. Using Blue Cotton, some HCAs that are not commonly found in fried foods have been identified, for example, 4-OH-PhIP, 4-CH_2OH-8-MeIQx, and 7,9-DiMeIgQx. Blue Cotton is cotton-bearing, covalently linked copper phthalocyanine trisulfonate as ligand. It can adsorb selectively aromatic compounds having three or more fused rings. The adsorption occurs in aqueous media, involving a 1:1 complex formation linking the ligand and the aromatic compound. Desorption can be completed by elution with organic solvents, while a treatment with methanol containing ammonia is generally more efficient.

An enhanced method uses Blue Rayon as the supporting material in place of cotton. Blue Rayon is, likewise, rayon-bearing, covalently bound copper phthalocyanine trisulfonate, but it can contain two to three times more blue pigment, making Blue Rayon a more efficient adsorbent than Blue Cotton [92].

The extraction method using Blue Rayon is the same as for Blue Cotton. Blue Rayon has mainly been used for the adsorption of HCAs from river water [93]. Both Blue Cotton and Blue Rayon can be packed into glass or plastic columns, which assist the extraction and purification procedure [88]. Chitin (poly-*N*-acetylglucosamine) can also covalently link copper phthalocyanine trisulfonate as ligand. By using chitin powder as the supporting material, the content of the blue pigment can be doubled when compared with rayon and increased by four times compared with cotton [94]. Methods based on Blue Chitin columns are simpler and less-time consuming than methods based on Blue Cotton or Blue Rayon, and allow us to obtain higher HCAs recoveries for compounds having more than three rings. Nevertheless, compounds with two- or one-ring structures gave little or no adsorptions [94]. In order to eliminate interferences, to preconcentrate the HCAs, and to separate the analytes in diverse fractions, column chromatography on XAD-2 resin [88], preparative HPLC [95], or in-tube SPME [89], have also been used.

7.2.1.3 Solid-Phase Microextraction

Conventional extraction techniques such as LLE and, in particular, SPE are, however, characterized by intrinsic disadvantages like the use of toxic solvents and plugging of the cartridges. These drawbacks can be avoided by using SPME technique. It enables simultaneous extraction and preconcentration of analytes from gaseous, aqueous, and solid matrices. The principle of SPME is equilibration of the analytes between the sample matrix and an organic polymeric phase usually coating a fused-silica fiber; the amount of the analyte absorbed by the fiber is proportional to the initial concentration. In order to apply to nonvolatile or thermally unstable compounds, SPME can be performed in combination with HPLC [96] or capillary electrophoresis. The difference between SPME-GC and SPME-HPLC is the desorption step. Four kinds of fiber coatings are compared for the extraction efficiency of HCAs from beef extracts. The most polar fiber studied (CW-TPR) exhibits better extracting efficiency and is recommended [97]. Factorial designs were used to optimize variables affecting the microextraction process [98]. The high fat content of the samples used led to low recoveries, probably due to the fiber coating poisoning. To minimize the fat content in the extract, it was frozen between −18°C and −20°C for 1 h [96]. Besides the simplification of the clean-up step, this method eliminates different SPE stages required in the analysis of HCAs, reducing the time and the amounts of organic solvents needed [96].

The in-tube SPME method is suitable for the extraction of less volatile or thermally labile HCAs compounds [89]. The food sample is treated with HCl followed by centrifugation, the sample supernatant is neutralized with NaOH, and the HCAs are extracted by the Blue Rayon adsorption method. This method can selectively adsorb compounds having polycyclic planar molecular structures, such as HCAs, in order to concentrate them from the aqueous solution. The extract is passed through a syringe micro filter, and a capillary column is used as a SPME device. This column is placed between the injection loop and the injection needle of the auto sampler. The method is simple, rapid, automatic, and gives 3–20 times higher sensitivity in comparison with the direct liquid injection method [89].

7.2.1.4 Tandem Extraction Procedures

The analytical performance can be optimized by combining different sorbents and eluents, or by coupling different sorbents in tandem [56,99,106]. A number of methods are based on the HCAs extracted by sample alkalinization and subsequent extraction with diatomaceous earth or with Kieselgur using different extraction solvents as previously stated. Then, the extract is passed to purification on a Bond-Elut propylsulfonyl silica gel (PRS) cartridge of 200–500 mg followed by octadecylsilane (C_{18}) [56], benzene sulfonic acid silica (SCX) [101], Oasis MCX LP SPE extraction cartridge [66], or carboxypropyl silica (CBA) [102] columns. The elution from diatomaceous earth seems to get better when solvents (toluene or phenol) are added to dichloromethane [103]. In addition, ethyl acetate improves slightly the recoveries of some HCAs in meat samples [66].

The most popular tandem method is proposed by Gross et al. [99], who separated a series of HCAs into a polar group and less-polar group by the optimization of the PRS (a cationic exchanger column) step in the SPE with Extrelut-PRS-C_{18} coupled cartridges. One of the main advantages of this technique is that it allows the elution of all the fluorescent compounds in the same fraction. However, although the method worked well for some process flavors, it was insufficient for the analysis of the more complex ones, such as those formed at elevated temperatures [49]. Recoveries of PhIP were inconsistent and too low [67]. Gross et al. [67] have dealt with these clean-up problems and recommended an extra clean up on TSK gel, but this material is not available in pre-packed columns. A study was made to modify the gross procedure [67] to improve its reliability, to determine PhIP, and to facilitate the analysis of very complex process flavors.

7.2.1.5 Supercritical Fluid Extraction

This technique has only been applied to extract HCAs from cooking fumes [45]. Supercritical CO_2 was ineffective in extracting analytes spiked onto a solid matrix, while supercritical CO_2/10% methanol at 55°C and 6000 psi resulted in good recoveries of quinolines and quinoxalines types HCAs. Supercritical fluid extraction (SFE) is an efficient and reliable technique that has several advantages. It allows the extraction and concentration of volatile compounds in one step, minimizing potential loss of the compounds, and providing a methanolic extract that can directly be analyzed by GC–MS. One disadvantage is that the flow restrictor is subject to plugging when the samples are wet or contain high amounts of extractable material and particulate matter. Thus, the bead trap condensate and the filters were extracted with traditional liquid solvents.

7.2.1.6 Pressurized Liquid Extraction

In terms of sample preparation, the analytical methodologies already described for the determination of HCAs in food samples are usually multistep procedures typically based on exhaustive extraction from the matrix and the subsequent removal of coextracted material by successive clean-up steps prior to the actual analysis. Such sample preparation involves a large amount of high-quality organic solvents and requires much manual handling of the extracts. That is, these methods are expensive in terms of time and material consumption, and sample throughput is too low to meet the challenges of modern food analysis. Developing faster, more cost-effective and environmental friendly procedures is, therefore, a pressing demand.

As an answer to this demand, in recent years, a new PLE technique has been developed in an attempt to overcome the main limitations of the conventional methods [104]. In general, this alternative technique allow a more efficient extraction of the analyte from the matrix by improving the contact of the target compounds with the extraction solvent, which allows a reduction of both the extraction time and the organic solvent consumption and increases sample throughput. In PLE, enhanced extraction efficiency can be achieved by solvents at high pressures and temperatures being near their supercritical region. Originally the use of the PLE was mainly focused on the extraction of the environmental pollutants present in soil matrices, sediments, and sewage sludge. However, more recently the distinct advantages of this technique are being exploited in diverse areas, including biology and thepharmaceutical and food industries. Recently, analytical application of the PLE technique in the extraction of HCAs in food samples has been explored, placing special emphasis on the strategies followed to obtain a rapid, selective, efficient, and reliable extraction process.

7.2.2 Determination Methods

The analysis of HCAs is quite challenging for food researchers because they are present at very low concentrations in heat-treated foods. For quantitative–qualitative analysis of HCAs, high-performance liquid chromatography (HPLC) techniques using fluorescence, electrochemical detection (ED) diode array detection (DAD), and mass spectrometry (MS) have been commonly applied [60,105–108]. Gas chromatography (GC) has been rarely applied due to need of the derivatization of HCAs into their less polar compounds [46]. Recently, an ultra-performance liquid chromatography–mass spectrometry (UPLC–MS) method has been [109,110] used for the analysis of HCAs, which is known to be the fastest separation technique. At present, the most frequently used techniques are liquid chromatography–mass spectrometry (LC–MS), LC–MS/MS, UPLC–MS, and UPLC–MS/MS. They allow the separation and unambiguous qualitative determination of HCAs as well as quantitative determination of a great number of compounds at the low ppb concentration in highly complex food matrix.

7.2.2.1 High-Performance Liquid Chromatography

HPLC is one of the most well-known analytical methods for HCAs. All HCAs have characteristic ultraviolet (UV) spectra and high extinction coefficients, and they are also electrochemically oxidizable. Thus, these highly polar, nonvolatile, and thermally unstable compounds can be measured with HPLC in combination with UV, electrochemical, and fluorescence detectors. HPLC with fluorescence or electrochemical detection presents high selectivity and sensitivity, although these detectors are restricted to the determination of selected groups. Sensitivity of HPLC with UV detection is not so high, around 100–400-fold lower than fluorescence, but fluorescence detection does not allow confirmation of the chromatographic peaks, and for that reason, the detection method most commonly used is DAD detection, which allows the online identification of the analytes by spectral library matching. Usually fluorescence detection is used as a complement to DAD in order to eliminate interferences produced when using UV detection, or to confirm the peaks obtained. The figures of merit for the determination of HCAs by HPLC are shown in Table 7.5.

7.2.2.2 Gas Chromatography

Usually, HCAs are polar and nonvolatile, and tend to elute as broad tailing peaks due to their strong adsorption to the column and injector during gas chromatography analysis. Thus, they cannot be detected in low levels without derivatization. Derivatization of HCAs may be employed not only to reduce the polarity but also to improve the volatility, selectivity, sensitivity, and separation [115].

TABLE 7.5
Figures of Merit for the Determination of HCAs by LC

Method	Pyridines	Quinolines	Quinoxalines	Pyridoindoles	References
HPLC-UV	PhIP D.L. (1 ng/g)	IQ D.L. (70 ng/g)	MeIQx, 4,8-DiMeIQx D.L. (8–90 ng/g)		[111]
HPLC-UV-FD	PhIP D.L. (0.4 ng/g) Recovery (20%)	IQ, MeIQ D.L. (4–10 ng/g) Recovery (63–66%)	MeIQx, 4,8-DiMeIQx D.L. (2–4 ng/g) Recovery (68–72%)	Harman, Norharman, Trp-P-1, Trp-P-2, AαC, MeAαC D.L. (0.3–5 ng/g) Recovery (7–66%)	[15]
HPLC-FD-DAD	PhIP Recovery (68%)	IQ, MeIQ Recovery (72–75%)	IQx, MeIQx, 4,8-DiMeIQx, 7,8-DiMeIQx Recovery (55–75%)	Harman, Norharman, Trp-P-1, Trp-P-2, AαC, MeAαC Recovery (55–75%)	[112]
HPLC-UV	PhIP D.L. (0.1 ng/g) RSD (15%) Recovery (64.4%)	IQ, MeIQ D.L. (0.1–0.1 ng/g) RSD (6–15%) Recovery (104.5–123.6%)	MeIQx, 4,8-DiMeIQx, 7,8-DiMeIQx D.L. (0.1–0.1 ng/g) RSD (6–15%) Recovery (99.4–113.1%)		[80]
HPLC-DAD	PhIP D.L. (9.1 ng/g) RSD (3.1%) Recovery (91.1%)	IQ, MeIQ D.L. (4.3–4.6 ng/g) RSD (4.1–4.2%) Recovery (62.3–64.5%)	MeIQx, 4,8-DiMeIQx, 7,8-DiMeIQx, 4,7,8-TriMeIQx D.L. (2.8–3.4 ng/g) RSD (2.9–3.4%) Recovery (83.6–95.1%)	Harman, Norharman, MeAαC D.L. (3.8–5.3 ng/g) RSD (3.7–5.2%) Recovery (24.7–78.9%)	[113]
HPLC-DAD	PhIP D.L. (2.0 ng/g) RSD (10%) Recovery (61.7%)	IQ, MeIQ D.L. (0.4–0.4 ng/g) RSD (9–11%) Recovery (77.5–83.7%)	MeIQx, 4,8-DiMeIQx D.L. (0.8–1.0 ng/g) RSD (6–12%) Recovery (60.6–77.0%)		[82]

HPLC-DAD	PhIP D.L. (23.8 ng/g) RSD (9.65%) Recovery (57.2%)	IQ, MeIQ D.L. (9.81–11.3 ng/g) RSD (9.81–11.3%) Recovery (5.78–16.6%)	MeIQx D.L. (16.8 ng/g) RSD (28.3%) Recovery (115.4%)	Harman, Norharman, Trp-P-1, Trp-P-2, AαC, MeAαC D.L. (1.58–7.47 ng/g) RSD (2.21–8.65%) Recovery (41.7–111.7%)	[98]
HPLC-DAD	DMIP, PhIP D.L. (2.40–4.59 ng/g) Recovery (18–25%)	MeIQ D.L. (9.28 ng/g) Recovery (20%)	MeIQx, 4,8-DiMeIQx D.L. (1.74–4.58 ng/g) Recovery (76–100%)		[106]
HPLC-ECD	PhIP D.L. (0.98 ng/g) RSD (6.1%) Recovery (77.3%)	IQ, MeIQ D.L. (1.15–1.66 ng/g) RSD (6.9–10.4%) Recovery (77.2–83.3%)	MeIQx D.L. (0.83 ng/g) RSD (12.8%) Recovery (101.3%)	Harman, Norharman, Trp-P-1, Trp-P-2, AαC, MeAαC D.L. (0.41–2.68 ng/g) RSD (7.7–12.5%) Recovery (57.5–109.5%)	[96]
HPLC-ECD/ FD-DAD	PhIP Recovery (55%)	IQ, MeIQ Recovery (82–99%)	MeIQx, 4,8-DiMeIQx Recovery (78–87%)	Harman, Norharman, Trp-P-1, Trp-P-2, AαC, MeAαC Recovery (68–91%)	[100]
HPLC-ECD	DMIP, PhIP DL (0.6–2 ng/g) RSD (1.2–6.1%)	IQ, MeIQ D.L. (1.6–2 ng/g) RSD (4.7–6.2%)	MeIQx, 4,8-DiMeIQx, 7,8-DiMeIQx D.L. (0.2–2 ng/g) RSD (1.2–7.6%)	Harman, Norharman,Trp-P-1, Trp-P-2, AαC, MeAαC D.L. (0.4–2.4 ng/g) RSD (0.9–3.5%)	[102]
HPLC-UV-FLD	PhIP D.L. (3.6 ng/g) RSD (21.3%)		MeIQx, 4,8-DiMeIQx D.L. (1.3–4.8 ng/g) RSD (9.2–25.1%)	Harman, Norharman D.L. (8.9–10.4 ng/g) RSD (8.2–11.5%)	[114]

Note: D.L. = detection limit; RSD = relative standard deviation.

Formerly, a GC method with nitrogen-phosphorus selective detector (NPD) was developed for the determination of HCAs with the advantage of the high response of these compounds in the detector due to the nitrogen atoms present in the structure of the HCAs. Kataoka and Kijima developed a simple and rapid derivatization method for GC analysis of HCAs. Ten HCAs were converted into their *N*-dimethylaminomethylene derivatives with *N,N*-dimethylformamide dimethylacetal and measured by GC with NPD using two connected fused silica capillary columns in order to improve the separation of HCAs [115]. The structures of the HCAs derivatives were confirmed by GC–MS.

7.2.2.3 Gas Chromatography–Mass Spectrometry

For nonpolar and volatile compounds, GC–MS is indeed one of the best online detection techniques, which ideally combines the advantages of the high separation efficiency of capillary GC with high sensitivity and selectivity of the MS detector. GC–NPD and GC–MS techniques have been applied to determine a few HCAs, which require a derivatization step prior to the analysis. A number of derivatizing agents, such as acetic, trifluoroacetic, pentafluoro-propionic, and heptafluoro-butyric anhydrides, pentafluoro-benzyl bromide, 3,5-bistrifluoro-methylbenzyl bromide, and 3,5 bistrifluoro-methylbenzoyl chloride have been tested for the analysis of some HCAs [116].

The GC–MS can usually be operated in two modes, total ion scanning and single ion monitoring (SIM). For SIM, only the base peaks are chosen to obtain the highest possible sensitivity [115]. GC–MS negative ion mode-SIM offers high chromatographic efficiency and provides an alternative method of analyzing less-polar HCAs in complex samples [117]. Nevertheless, it causes contamination of the ion source through the deposition of nonvolatile material. Trp-P-1, Trp-P-2, AαC, MeAαC, harman, and norharman, due to their low polarity can be directly analyzed without prior derivatization.

Silylation is probably the most versatile GC derivatization technique. Besides improving volatility and stability, the introduction of the silyl group can also serve to enhance mass spectrometric properties. Consequently, a derivatization method in a one-step reaction with *N*-methyl-*N*-(*tert*-butyldimethyl-silyl) trifluoroacetamide for the analysis of 12 HCAs by GC–EI–MS analysis with SIM quantification was developed [118]. The derivatives are characterized by easy-to-interpret mass spectra due to the prominent ion $[M\text{-}57]^+$ by loss of a *tert*-butyl-dimethyl-silyl group. The derivatization of the pyridoimidazoles Glu-P-1 Glu-P-2 and the β-carboline harman is incomplete for all the temperatures tested, and a tailed peak due to the underivatized compound is observed. The procedure is simple, rapid, and accurate. However, the instability of the imidazo-quinoline and imidazo-quinoxaline derivatives, requiring their injection on the same working day, is a further disadvantage. In Table 7.6 the figures of merit for the determination of HCAs by gas chromatography with NPD and MS detectors have been presented.

7.2.2.4 Liquid Chromatography–Mass Spectrometry

Liquid chromatography–mass spectrometry (LC–MS) is one of the best online identification systems, because of its high sensitivity and selectivity. LC–MS is capable of

TABLE 7.6
Figures of Merit for the Determination of HCAs by GC

Method	Pyridines	Quinolines	Quinoxalines	Pyridoindoles	Pyridoimidazoles	References
GC–NPD	PhIP D.L. (15 pg/g)	IQ, MeIQ D.L. (2–4 pg/g)	MeIQx, 4,8-DiMeIQx D.L. (8–10 pg/g)	Trp-P-1, Trp-P-2, AαC D.L. (3–14 pg/g)	Glu-P-1, Glu-P-2, D.L. (8–14 pg/g)	[115]
GC–EI–MS	PhIP D.L. (6 ng/g)	IQ, MeIQ D.L. (50 ng/g)	MeIQx, 4,8-DiMeIQx D.L. (6.0–7.5 ng/g)			[82]
GC–EI–MS–SIM	PhIP D.L. (0.12 ng/g) RSD (20.7%)	IQ, MeIQ D.L. (0.09 ng/g) RSD (8.8–7.1%)	IQx, MeIQx, 4,8-DiMeIQx, 7,8-DiMeIQx D.L. (0.05–0.08 ng/g) RSD (10.2–12.6%)	Norharman, Trp-P-1, Trp-P-2, AαC, MeAαC D.L. (0.02–0.35 ng/g) RSD (5.0–15.0%)		[118]
GC–MS–MS triple quadrupole	DMIP, PhIP D.L (3.6–7.5 pg/g) Quantification limit (0.32–0.58 ng/g) RSD (4.2–4.7%) Recovery (34.2–85.3%)	IQ, MeIQ D.L (1.1–1.6 pg/g) Quantification limit (0.12–0.16 ng/g) RSD (3.3–9.7%) Recovery (57.5–61.0%)	MeIQx, 4,8-DiMeIQx, 7,8-DiMeIQx Quantification limit (0.38–0.58 ng/g) D.L (4.2–6.9 pg/g) RSD (4.7–5.2%) Recovery (71.5–87.2%)	Trp-P-1, Trp-P-2, AαC, MeAαC Quantification limit (0.26–0.48 ng/g) D.L (1.1–6.8 pg/g) RSD (6.9–9.6%) Recovery (32.7–57.7%)	Glu-P-1, Glu-P-2 D.L (1.3–2.2 pg/g) Quantification limit (0.14–0.21 ng/g) RSD (6.5–12.9%) Recovery (36.5–46.1%)	[119]

Note: D.L. = detection limit; RSD = relative standard deviation.

simultaneously measuring retention times and molecular mass, and can identify and quantify HCAs in highly complex samples without derivatization, that is, required in GC–MS systems. The ionization techniques that have been widely used are electrospray ionization (ESI) followed by atmospheric pressure chemical ionization (APCI) and thermospray ionization (TSI). The sensitivity of MS can be increased if only a few selected ions are monitored instead of full spectra, as it occurs when the single ion monitoring technique is applied [22]. Another procedure, selected reaction monitoring (SRM), achieves high selectivity and extreme sensitivity as compared with SIM mode.

The thermospray LC–MS can work with conventional size LC columns and with reversed-phase columns. The ionization process for HCAs produces abundant pseudo molecular ions and the base peaks in the mass spectra are detected as $[M + H]^+$. These amines are stable toward the ionization process and do not undergo notable fragmentation. Single ion monitoring of the $[M + H]^+$ ion of the respective HCAs, such as Trp-P-1, Trp-P-2, IQ, MeIQ, and MeIQx can be used for analysis in complex matrices [120]. However, this technique has been replaced by atmospheric pressure ionization (API) techniques because of their higher sensitivity. LC–MS appears to be the main technique able to screen most of the known HCAs simultaneously, either using a thermospray [120] or an electrospray [48] interface [49].

Two powerful and promising interface methods based on API sources are ESI and APCI. In ESI, droplet formation and charging take place simultaneously, while in APCI droplets are formed prior to ionization. Both API techniques involve soft ionization and, therefore, the unfragmented ions obtained, quasi molecular ions, provide information on molecular mass, but little structural information. The application of higher voltage difference between different regions of an API source generally induces more fragmentation of the formed ions. This procedure is designed as in-source fragmentation or pre-analyzer collision-induced dissociation and allows induction of fragmentation before entering the quadrupole in LC–MS [22], or between the two quadrupoles when LC–MS/MS is used [121]. The application of in-source fragmentation provides an easier and less expensive method than tandem-MS for confirmation of the HCAs. In this way, the lowest extraction potential applied (+100 V) is used for quantification purposes, because higher responses are obtained. The highest potential (+150 V) induces fragmentation of the primarily formed ions $[M + H]^+$, and allows the confirmation of the detected peaks [47]. It is a selective and highly specific technique and can be considered one of the best online identification methods, which is an important requisite when working in the analysis of HCAs in complex matrices, such as processed food samples. Both methods (ESI and APCI) are more sensitive than the usually used LC–UV method, and give similar results to those obtained using HPLC with electrochemical detection, and they have the advantage of being more stable than the latter [122]. The chromatograms are almost free of interfering peaks due to the high selectivity and specificity of this technique. Galceran et al. [22] have determined simultaneously harman, norharman, and other HCAs in processed food samples, with a triple quadrupole mass spectrometer using APCI or ESI with positive ionization. Measurements were performed by SIM of the protonated molecular ions [22]. In this most sensitive single-ion mode, frequently no abundant secondary ions are present for confirmation of the base peak. Also, the LC–MS measurements were performed by multiple ion detection (MID) of the most important masses for each HCAs [122]. The potential and the limitations of LC–APCI–MS/MS and of LC–ESI–MS/MS techniques applied to HCAs have been discussed [87].

As low-flow rates are usually needed for the electrospray LC–MS technique, microbore or semi-microbore columns must be used. Columns with smaller diameter have the advantages of low solvent consumption, higher sensitivity, and good separation at low flow rates. This last characteristic makes micro-columns and capillary columns suitable for LC–ESI–MS techniques. The sensitivity increases because micro-columns elute analytes at higher concentrations than traditional columns. Different narrow-bore reversed-phase columns employed in LC–ESI–MS were studied [123].

The electrospray LC–MS using soft ionization interface is a powerful technique for the analysis of low molecular weight trace constituents in complex matrices. ESI source requires analytes to be ionized in the liquid phase, so for HCAs analysis the pH of the mobile phase should be lower than pK_a of the HCAs to protonate the amino group. As HCAs are stronger bases than the components of the mobile phase, this ionization process can transform the HCAs from solution to protonated ions in the gas phase. As a result, the HCAs give a simple mass spectrum in which the only peak is due to $[M + H]^+$, the abundant protonated molecular ion. These compounds are stable toward the ionization process and do not undergo notable fragmentation except for IQ and 4,7,8-TriMeIQx, which show the $[MH\text{-}15]^+$ fragment [122]. When higher extraction voltages were used, more fragmentation was observed and a decrease in the intensity of the protonated molecule $[M + H]^+$ occurred [122]. $[M + NH_4\text{-}H_2O]^+$ ions were observed from Glu-P-1 and Glu-P-2 [89]. The loss of CH_3 from protonated molecules and the loss of the aminoimidazyl moiety ($\text{-}CH_3\text{-}HCN$ and $\text{-}C_3H_4N_2$) are the common route of fragmentation for these compounds.

Stavric et al. applied LC–APCI–MS to the determination of HCAs [58]. A dual channel with UV detector was installed after the LC column but before the LC–MS interface, which was attached to the APCI source of the triple quadrupole MS, operated in the single quadrupole mode. The mass spectrometer was operated in SIM mode and the resolution was set at around 1–1.2 mass units at the base line. Although additional clean-up procedures were used, interferences were still observed even with trideuterated standards. Therefore, for samples where some interference was observed a second LC column, TSK gel ODS, was used. All the studied HCAs were quantified and the minimum detection limits were 1–3 ng/g [58]. The problems derived from a less exhaustive purification of the extract have been resolved by using LC–APCI provided as an ion trap (IT) mass analyzer [124], but with this simplification of the clean-up, detection limits in the meat extract analyzed were higher than expected [124]. Comparison of different commercial SPE cartridges to extract HCAs was made [125] by this simplified purification procedure. A liquid chromatography–electrospray ionization–ion trap mass spectrometry (LC–ESI–IT–MS) method has been developed [126] to study the metabolism of PhIP by the human liver microsomes and prostate tissue. A mixture of ammonium acetate buffer and acetonitrile was used for elution from the SPE cartridges. To improve the recovery, dimethyl sulfoxide was added because it is a very good solvent for PhIP and its metabolite

2-hydroxyamino-1-methyl-6-phenylimidazo[4,5-*b*]pyridine. Although the recovery was better, the evaporation of dimethyl sulfoxide was difficult and therefore unsuitable for larger sample volumes or greater number of samples [126].

7.2.2.5 Tandem MS Techniques

The ESI can also be used in combination with tandem mass spectrometry (MS/MS) to enhance the sensitivity of the detection. MS/MS technique provides a high degree of selectivity, leading to chromatograms that are almost free of interfering peaks. Moreover, false peak identification was avoided by comparing the product ion full-scan mass spectra of the sample with those of the standards [127]. Richling et al. developed a sensitive and selective method for the simultaneous analysis of the 10–16 most abundant HCAs in several food samples by LC–ESI MS/MS using triple quadropole in combination with SRM [121]. The ionization of analytes in LC–ESI–MS/MS is influenced by different factors, in particular, the sample matrix, thus requiring the use of deuterated standards. Separation of the polar and less polar amines was achieved by means of two different LC gradients with trifluoroacetic/H_2O and CH_3OH/acetonitrile as solvents [87].

As the solvent composition affects the LC–ESI–MS systems, the influence of the concentration of a volatile ion-pairing reagent has been studied [21]. The chromatographic behavior of the HCAs using a formic acid/ammonium formate (pH 2.8, 3.7, and 4.7) was compared with that observed using acetic acid/ammonium acetate buffer (pH 4.0) in the mobile phase. Reversed-phase ion-pair chromatography in tandem MS/MS using ESI and selected reaction monitoring (IP-LC/ESI–MS/MS-SRM) was carried out with formate, or acetate, as counter ion in an aqueous eluent with acetonitrile as organic modifier. Higher detectability was obtained with a formate buffer at pH 2.8 [21]. It was observed that under isocratic conditions, pH values higher than 3.7 produced broad peaks of all the HCAs [21].

An ion-pair LC–ESI tandem MS with SRM for identification is also reported for determining HCAs in meat-based infant foods. Mean recoveries ranged between $78 \pm 4\%$ and $98 \pm 2\%$ for IQ, MeIQ, MeIQx, PhIP, AαC, harman, and norharman and, LOQ generally are <8 ng/g. Some factors are identified as statistically significant in influencing chromatographic separation and response: The mobile-phase pH turned out to be a critical parameter for the capacity factor (k') of IQ, MeIQ, and norharman, whereas the mobile-phase flow rate was statistically significant for k' values of all analytes, except AαC peak [128].

Three LC–ESI–MS systems equipped with an electrospray as ionization source and different analyzers, using the same chromatographic conditions, were evaluated for the determination of 16 HCAs [107]. The analyzers were: (a) an ion trap, (b) a single quadrupole, and (c) a triple quadrupole. The (b) and (c) systems were equipped with a Turbo Ionspray as ionization source. Selected ion monitoring was used as data-acquisition mode for the systems (b) and (c). The systems (c) and (a) used an SRM and a product ion scanning, respectively, using as precursor ion the protonated molecular ions $[M + H]^+$. Post column addition of formic acid–acetonitrile was needed to increase ionization efficiency when using the ion trap analyzer. In contrast with the observed when using the ionization source of the ion-trap instrument, no post column addition was needed in either single or triple quadrupole instruments.

This fact can be explained by the higher ESI efficiency that provides the Turbo Ionspray compared with the ion trap system. The best RSD values were obtained when using triple quadrupole with SRM acquisition. In addition, triple quadrupole provided lower LOD than the other systems. Because of this, its linearity range was generally two orders of magnitude larger. The results obtained with all the instruments and acquisition modes are in agreement, although the most precise ones were obtained with the triple quadrupole instrument. Nevertheless, the results achieved with the ion trap were also good and the trap has the additional advantage of providing spectral information for false positive peak identification [107].

LC–APCI–MS/MS has also been applied [129]. To improve the detection and quantification limits, a 2-mm internal diameter LC column was used instead of the conventional 4.6-mm one [129]. When IQ, 8-MeIQx, 4,8-DiMeIQx, and 7,8-DiMeIQx were treated under higher collision-induced dissociation conditions, the found data imply that the pyrazine moiety has been lost, with retention of the charge on the benzo-imidazo-2-yl-amine moiety of 8-MeIQx and its homo homologues. Thus, MS/MS analysis of $[M + H\text{-}15]^+$ in the constant neutral acquisition mode enabled the identification of two other HCAs (IQx and 7,9-DiMeIgQx), which have rarely been reported in cooked meats [129]. Recently, two unidentified chromatographic peaks with product ion spectra and retention time very similar to those of 8-MeIQx and 4,8-DiMeIQx, have been found in the griddled beef samples. Turesky et al. have recently mentioned the presence of these chromatographic peaks among others in fried or barbecued chicken and beef, and have proposed them to be isomers of 8-MeIQx and DiMeIQx [24]. LC–MS/MS is used as quantification technique, and product ion scan mass spectra provided by the ion trap mass analyzer is used to confirm the identity of the analytes [108].

A method based on LC–APCI–IT–MS/MS for the analysis of 16 HCAs has been described [127]. The fragmentation patterns of the AIAs observed are consistent with those obtained by other authors [129] using triple quadrupole instruments [129]. In addition to that, some ion molecule reactions were observed into the trap [22] using LC–APCI–MS/MS [107]. These reactions occurred only for carbolines (Trp-P-1, Trp-P-2, AαC, MeAαC, norharman, harman, Glu-P-1, and Glu-P-2) by recombination of the product ion $[M + H\text{-}NH_3]^+$ with neutral molecules present in the ion-trap, such as water or acetonitrile. Also, adducts of *m/z* higher than parent ion were obtained [107]. The abundance of these product ions is highly dependent on small changes in experimental conditions. As these ions had a very high signal, they must be added to the base peak to carry out the quantification of carbolines by MS/MS in order to obtain reproducible results [107]. On the other hand, the MS/MS spectra obtained with the ion trap and the triple quadrupole systems were very similar in both fragment ions and relative abundances, except for carbolines that showed adduct formation in the ion trap [107]. These adducts observed in the ion trap spectra were not present in the MS/MS spectra obtained with the triple quadrupole instrument. This fact can be explained by the absence of neutral molecules from the mobile phase inside the collision cell [107].

Time required for extraction with SPE cartridges, chromatographic separation and LC–MS/MS determinations are too long. Therefore, a method using selective ionization of metastable atom bombardment (MAB) has been developed in order to detect HCAs in nonpurified meat extracts, thus avoiding purification and concentration steps and reducing analysis time around 18-fold [59]. MAB ionization forms radical ions by electron transfer from a molecule to a species (noble gas or nitrogen) excited in a metastable state. By selecting metastable gas for ionization, it is possible to precisely control the available ionization energy in the gas phase. This allows one to control the fragmentation extent of the studied species and to selectively ionize some molecules in a mixture depending on their ionization potentials. Metastable nitrogen was selected as the best MAB gas for the analysis of HCAs. The MAB ionization source was coupled to a pyrolyzer, which allows analysis by direct introduction of the sample into the mass spectrometer, therefore, constituting a fast analytical technique. The pyrolysis probe was not designed to achieve a real pyrolysis process involving thermal degradation of molecules, but rather was used to rapidly transfer molecules into the gas phase. Detection of HCAs is completed in 27 s. However, relative standard deviations are quite large and are due to the manual introduction of the sample into the pyrolysis probe. This pyrolysis-metastable atom bombardment ionization-time of flight-mass spectrometry (Py-MAB-ToF-MS) method was in good agreement with a LC–APCI–MS/MS method [59]. Analytical properties for the determination of HCAs by LC–MS are displayed in Table 7.7.

7.2.2.6 Ultra-Performance LC–MS

Ultra-performance LC is a new class of analytical separation techniques that retains the principles and pragmatism of HPLC while increasing the whole interlaced attributes of sensitivity, resolution, and speed.

HPLC is a well-established analytical technique that has been used in laboratories globally over the past 35 years. One of the major drivers for the development of this method has been the evolution of packing materials used to effect the separation. The fundamental principles of this evolution are governed by the van Deemter equation, which is an empirical formula that defines the correlation between linear velocity (flow rate) and plate height (HETP or column efficiency). Since particle size is one of the variables, a van Deemter curve can be used to investigate chromatographic performance.

Recently, column packings in the range of 1–2 μm in size have been introduced as a reasonable approach to improve resolution in LC separations.

According to the van Deemter equation, as the particle size decreases to <2 μm, not only is there a significant gain in efficiency, but the efficiency does not diminish at increased flow rates or linear velocities, providing both resolution and speed. However, the pressure required for pumping mobile phase through a relatively long column packed with such small particles is prohibitive for standard HPLC hardware, which is limited to pressures less than 6000 psi. Nevertheless, UPLC has been developed to operate at very high pressure, 15,000 psi [133]. The benefit of UPLC is that quantitative response improves and the analysis time can be shortened by raising the flow rates to allow a higher sample throughput. However, in order to address the very narrow peaks produced by UPLC, a high data capture rate detector is required. MS systems based on triple quadrupole [134] or quadrupole TOF [134] analyzers to achieve scans at very fast rates have typically been used for UPLC–MS/MS combination.

TABLE 7.7
Figures of Merit for the Determination of HCAs by LC/UPLC–MS

Method	Pyridines	Quinolines	Quinoxalines	Pyridoindoles	Pyridoimidazoles	References
LC–MS–TSI–SIM		IQ D.L. (5.6 ng/g)	MeIQx, 4,8-DiMeIQx, 4,7,8-TriMeIQx D.L. (3.2–29.2 ng/g)			[130]
LC–APCI–MS–SIM	PhIP D.L. (0.6 ng/g)	IQ, MeIQ D.L. (0.6–0.6 ng/g)	MeIQx, 7,8-DiMeIQx, 4,7,8-TriMeIQx D.L. (0.4–1.1 ng/g)	Trp-P-1, Trp-P-2 D.L. (0.6–1.1 ng/g)		[131]
LC–APCI–MS	PhIP D.L. (0.4 ng/g) RSD (3.6%) Recovery (50.3%)	IQ, MeIQ D.L. (0.2–0.2 ng/g) RSD (3.7–3.9%) Recovery (67.7–72.2%)	MeIQx, 4,8-DiMeIQx D.L. (1.0–1.4 ng/g) RSD (3.3–4.6%) Recovery (83.1–84.9%)	Harman, Norharman, Trp-P-1, Trp-P-2, AαC, MeAαC D.L. (0.08–0.8 ng/g) RSD (3.3–4.1%) Recovery (58.4–73.4%)	Glu-P-1 D.L. (0.4 ng/g) RSD (4.2%) Recovery (74.0%)	[132]
LC–APCI–IT–MS	PhIP D.L. (1.5 ng/g) RSD (2.9%)	IQ, MeIQ D.L. (4.9–10.1 ng/g) RSD (2.7–4.2%)	MeIQx, 4,8-DiMeIQx, 7,8-DiMeIQx D.L. (2.7–5.3 ng/g) RSD (3.1–4.8%)	Harman, Norharman, Trp-P-1, Trp-P-2, AαC, MeAαC D.L. (0.8–2. ng/g) RSD (2.1–5.1%)	Glu-P-1, Glu-P-2 D.L. (7.9–9.0 ng/g) RSD (2.2–3.8%)	[124]
LC–AP–ESI–MS–SIM	PhIP D.L. (0.21 ng/g)	IQ, MeIQ D.L. (0.39–0.55 ng/g)	MeIQx, 4,8-DiMeIQx, 7,8-DiMeIQx D.L. (0.53–1.33 ng/g)	Trp-P-1, Trp-P-2, AαC D.L. (0.95–1.57 ng/g)	Glu-P-1, Glu-P-2 D.L. (2.71–3. ng/g)	[89]

continued

TABLE 7.7 (continued)
Figures of Merit for the Determination of HCAs by LC/UPLC–MS

Method	Pyridines	Quinolines	Quinoxalines	Pyridoindoles	Pyridoimidazoles	References
LC–ESI–MS	PhIP D.L. (0.3 ng/g) Recovery (54%)	IQ, MeIQ D.L. (0.2–0.3 ng/g) Recovery (74–80%)	MeIQx, 4,8-DiMeIQx D.L. (0.2–1.1 ng/g) Recovery (82–89%)	Harman, Norharman, Trp-P-1, Trp-P-2, AαC, MeAαC D.L. (0.1–2.3 ng/g) Recovery (59–105%)	Glu-P-1 D.L. (2.3 ng/g) Recovery (79%)	[132]
HPLC–ESI–MS–SIM	PhIP D.L. (0.2 ng/g) Quantification limit (0.8 ng/g) Recovery (54.29%)	IQ, MeIQ D.L. (1.2–2.9 ng/g) Quantification limit (4.0–9.7 ng/g) Recovery (51.62–66.30%)	MeIQx, 4,8-DiMeIQx, 7,8-DiMeIQx, 4,7,8-TriMeIQx D.L. (0.6–1.0 ng/g) Quantification limit (2.0–3.2 ng/g) Recovery (15.70–64.01%)	Harman, Norharman, Trp-P-1, Trp-P-2, AαC, MeAαC D.L. (0.6–7.2 ng/g) Quantification limit (2.1–23.9 ng/g) Recovery (25.01–74.70%)	Glu-P-1, Glu-P-2 D.L. (0.4–1.3 ng/g) Quantification limit (1.4–4.2 ng/g) Recovery (48.67–68.69%)	[44]
LC–ESI–IT–MS	DMIP, PhIP D.L. (5–13 pg/g)	IQ, MeIQ D.L. (2–5 pg/g)	MeIQx, 4,8-DiMeIQx, 7,8-DiMeIQx D.L. (3–6 pg/g)	Harman, Norharman, Trp-P-1, Trp-P-2, AαC, MeAαC D.L. (3–8 pg/g)	Glu-P-1, Glu-P-2 D.L. (2–3 pg/g)	[107]
LC–APCI–IT–MS/MS	DMIP, PhIP D.L. (0.7–10. ng/g) RSD (3–4%) Recovery (14–87%)	IQ, MeIQ D.L. (1.0–1.1 ng/g) RSD (3–3%) Recovery 87–93(%)	MeIQx, 4,8-DiMeIQx, 7,8-DiMeIQx D.L. (0.9–1.2 ng/g) RSD (4–5%) Recovery (78–87%	Harman, Norharman, Trp-P-1, Trp-P-2, AαC, MeAαC D.L. (0.8–3.1 ng/g) RSD (4–16%) Recovery (46–72%)	Glu-P-1, Glu-P-2 D.L. (2.7–4.6 ng/g) RSD (5–7%) Recovery (74–83%)	[127]

HPLC–ESI–MS/MS–SRM	PhIP RSD (2.3%) Recovery (31%)	IQ RSD (3.1%) Recovery (54%)	IQx,MeIQx,4,8-DiMeIQx,7,9-DiMeIgQx RSD (2.7–10.2%) Recovery (20–50%)	AαC, MeAαC RSD (%) Recovery (%)		[103]
LC–ESI–IT–MS/MS	DMIP, PhIP D.L. (1.3–3.6 ng/g) RSD (6–8%)	IQ, MeIQ D.L. (0.6–0.7 ng/g) RSD (5–8%)	MeIQx, 4,8-DiMeIQx, 7,8-DiMeIQx D.L. (1.1–2.9 ng/g) RSD (7–9%)	Harman, Norharman, Trp-P-1, Trp-P-2, AαC, MeAαC D.L. (0.1–1.7 ng/g) RSD (5–9%)	Glu-P-1, Glu-P-2 D.L. (1.4–1.9 ng/g) RSD (2–8%)	[107]
LC–ESI–MS–SIM single quadrupole	DMIP, PhIP D.L. (0.5–0.8 ng/g) RSD (3–5%)	IQ, MeIQ D.L. (0.5–1.2 ng/g) RSD (5–8%)	MeIQx, 4,8-DiMeIQx, 7,8-DiMeIQx D.L. (0.1–1.7 ng/g) RSD (5–8%	Harman, Norharman, Trp-P-1, Trp-P-2, AαC, MeAαC D.L. (0.3–0.9 ng/g) RSD (2–7%)	Glu-P-1, Glu-P-2 D.L. (0.3–0.6 ng/g) RSD (7–7%)	[107]
LC–ESI–MS/MS–SIM triple quadrupole	DMIP, PhIP D.L. (0.01–0.1 ng/g) RSD (0.4–3%)	IQ, MeIQ D.L. (0.04–0.04 ng/g) RSD (2–3%)	MeIQx, 4,8-DiMeIQx, 7,8-DiMeIQx D.L. (0.1–0.5 ng/g) RSD (3–5%	Harman, Norharman, Trp-P-1, Trp-P-2, AαC, MeAαC D.L. (0.02–0.1 ng/g) RSD (1–4%)	Glu-P-1, Glu-P-2 D.L. (0.1–0.1 ng/g) RSD (3–4%)	[107]
LC–ESI–TOF[a]	DMIP, PhIP D.L. (0.6–1.4 ng/g) RSD (6.1–7.6%)	IQ, MeIQ D.L. (1.9–1.9 ng/g) RSD (5.8–7.4%)	MeIQx, 4,8-DiMeIQx, 7,8-DiMeIQx D.L. (0.4–0.5 ng/g) RSD (7.1–7.9%)	Harman, Norharman, Trp-P-1, Trp-P-2, AαC, MeAαC D.L. (0.2–2.9 ng/g) RSD (4.1–8.0%)	Glu-P-1, Glu-P-2 D.L. (0.8–1.1 ng/g) RSD (6.5–6.9%)	[133]

continued

TABLE 7.7 (continued)
Figures of Merit for the Determination of HCAs by LC/UPLC–MS

Method	Pyridines	Quinolines	Quinoxalines	Pyridoindoles	Pyridoimidazoles	References
Py–MAB–TOF–MS[b]	PhIP D.L. (0.9 ng/g) RSD (11.8%) Recovery (100.3%)	IQ, MeIQ D.L. (0.3 ng/g) RSD (22.6–23.1%) Recovery (96.3%)	IQx, MeIQx, 4,8-DiMeIQx D.L. (0.15–1.5 ng/g) RSD (12.2–20.7%) Recovery (81.3–85-.6%)	Trp-P-1, Trp-P-2, AαC, MeAαC D.L. (0.3–1.2 ng/g) RSD (10.5–21.6%) Recovery (78.3–94.8%)	Glu-P-2 D.L. (0.5 ng/g) RSD (18.1%) Recovery (82.5%)	[59]
UPLC–MS/MS–SRM	DMIP, PhIP D.L. (0.07–0.08 pg injected) RSD (4.4–6.5%)	IQ, MeIQ D.L. (0.06–0.08 pg injected) RSD (3.9–7.2%)	MeIQx, 4,8-DiMeIQx, 7,8- DiMeIQx D.L. (0.06–0.10 pg injected) RSD (6.5–7.9%)	Harman, Norharman, Trp-P-1, Trp-P-2, AαC, MeAαC D.L. (0.11–0.19 pg injected) RSD (3.8–8.2%)	Glu-P-1, Glu-P-2 D.L. (0.08–0.09 pg injected) RSD (4.9–6.8%)	[109]

Note: D.L. = detection limit; RSD = relative standard deviation.
[a] Liquid chromatography–electrospray ionization–time-of-flight–mass spectrometry.
[b] Pyrolysis-metastable atom bombardment–time-of-flight–mass spectrometry.

A new method based on ultra-performance liquid chromatography–tandem mass spectrometry (UPLC–MS/MS) has been developed to allow the analysis of 16 HCAs in less than 2 min (Figure 7.4) [109]. Nevertheless, the LC–MS/MS accomplishes the same separation in excess of 22 min (Figure 7.5) [107]. UPLC operates at much higher pressure, and to address the very narrow peaks produced, a high data capture rate detector is necessary. Argon was used as collision gas instead of N_2, because argon seems to need relatively low collision energy for fragmentation. The linearity range was established over three orders of magnitude. In addition to the reduction in analysis time, the detection limits obtained was up to 10 times lower (Table 7.8) than those obtained using similar triple quadrupole instruments but with a traditional LC system (Table 7.9) [107]. The precision values (Table 7.8) were nearly similar to those obtained previously using other HPLC–MS/MS systems (Table 7.9) [107]. Thus, besides the dramatic decrease in analysis time when

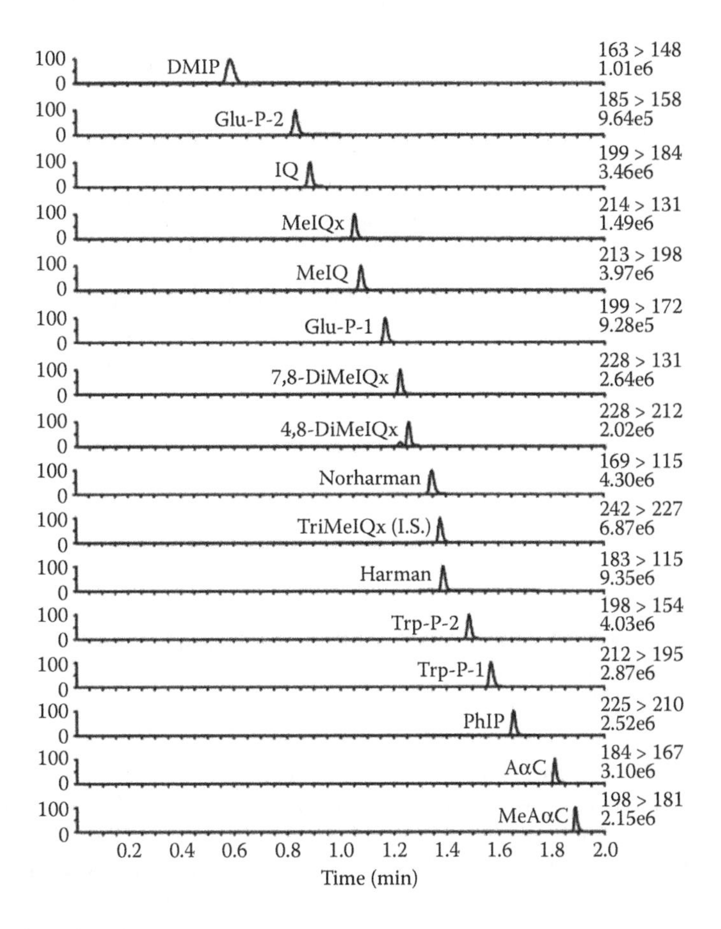

FIGURE 7.4 UPLC–MS/MS chromatograms corresponding to a HCAs standard solution at concentration of 60 ng/g in SRM acquisition method. (Adapted from Barcelò-Barrachina, E. et al. 2006. *J. Chromatogr. A* 1125:195–203. With permission.)

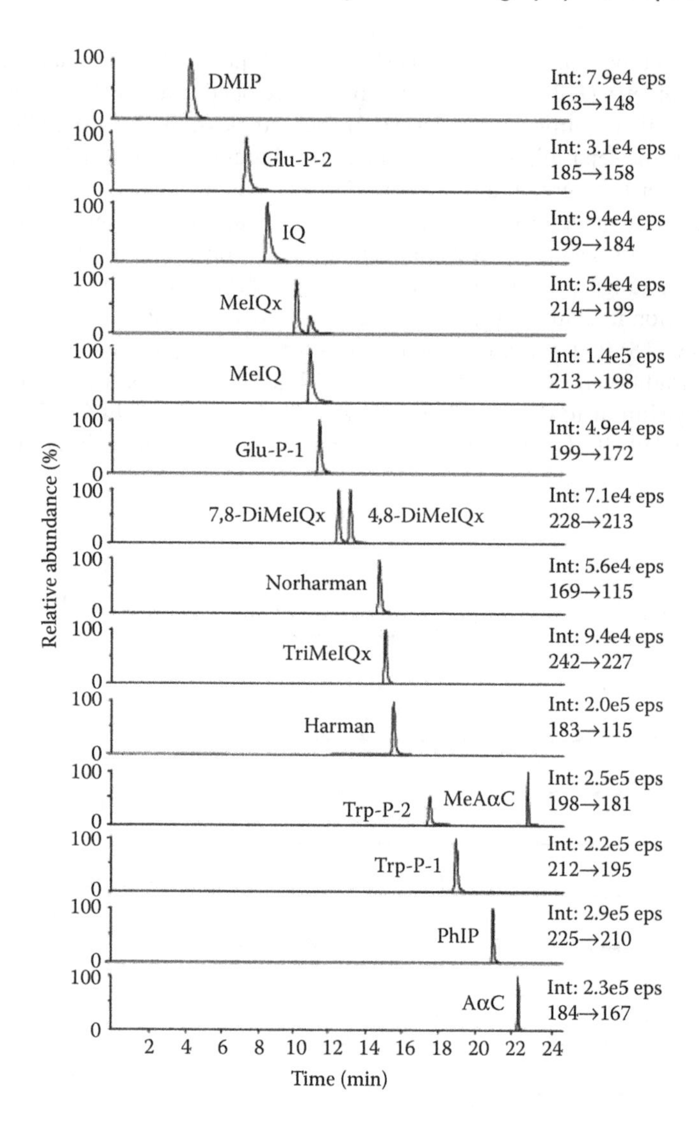

FIGURE 7.5 LC–MS/MS chromatograms of HCAs standard solution at a concentration of 0.5 μg/g in SRM acquisition method. (Adapted from Toribio, F. et al. 2007. *Food Chem. Toxicol.* 45:667–675. With permission.)

using the UPLC procedure, the narrower peaks produced a significant increase in sensitivity. Analytical properties for the analysis of HCAs by UPLC–MS/MS are displayed in Table 7.7.

The applicability of the UPLC–MS/MS method for the analysis of HCAs has been carried out in cooked meat samples; these values are lower (~five times) than those obtained with triple quadrupole instruments using a traditional narrow bore column [107], due to the high peak efficiency and the high sensitivity provided by the UPLC–MS/MS system.

TABLE 7.8
Quality Parameters of the UPLC–ESI–MS/MS Method

	LODs				
	Standards	Meat Extract		Run-to-Run (RSD%) (n = 6)	
HCAs	pg Inj.	pg Inj.	pg/g Meat	Medium[a] Concentration	Low[b] Concentration
DMIP	0.07	1.4	4.2	4.2	4.4
Glu-P-2	0.9	0.3	9.0	4.1	4.9
IQ	0.06	0.8	25	1.5	3.9
MeIQx	0.10	0.3	9.0	5.4	7.9
MeIQ	0.08	0.6	17.0	5.5	7.2
Glu-P-1	0.08	0.6	17.0	4.2	6.8
7,8-DiMeIQx	0.06	1.1	34.0	6.0	6.5
4,8-DiMeIQx	0.09	0.8	24	5.7	7.6
Norharman	0.19	1.7	51	3.7	7.5
Harman	0.18	1.5	45	4.7	8.1
Trp-P-2	0.23	0.9	28	6.1	9.1
Trp-P-1	0.16	0.2	5	4.4	8.2
PhIP	0.08	0.5	15	5.7	6.5
AαC	0.11	0.8	23	4.1	5.6
MeAαC	0.17	0.5	14	2.5	3.8

Source: Adapted from Barcelò-Barrachina, E. et al. 2006. *J. Chromatogr. A* 1125:195–203. With permission.

[a] 0.3 μg/g.

[b] 0.02 μg/g.

7.3 SUMMARY AND CONCLUDING REMARKS

To evaluate the threat to human health posed by the daily consumption of foods containing mutagenic/carcinogenic HCAs, it is necessary to quantify the amount of the carcinogen to which man is chronically exposed. Nonetheless, the quantitative determination of HCAs in food samples is mainly hindered by the low concentration of these microcomponents and the high complexity of the matrix. Thus, the development of sensitive and selective analytical approach is mandatory. LC coupled with MS and tandem MS is well established as the main analytical techniques capable of providing main information and throughput required for the HCA analysis in cooked food. UPLC using column particle size in the range of 1–2 μm and a properly holistically designed system provide significantly more resolution while reducing analysis time, and improve the sensitivity for the analysis of many compound types. The utility of UPLC–MS/MS for the rapid, accurate and precise analysis of HCAs in food samples was established. Chromatographic separation of 16 HCAs was achieved in <2 min, providing narrow peaks with good peak symmetry. The linearity range was

TABLE 7.9
Limits of Detection in Standards (pg Injected) and Lyophilized Meat Extract (pg Injected and μg/kg)

	Standards		Meat Extract				Standard			
	Triple Quadrupole		Triple Quadrupole				Triple Quadrupole			
	SIM	MRM	SIM		MRM		MRM			
							Low Level		Medium Level	
HCAs	pg	pg	pg	μg/kg	pg	μg/kg	Run-to-Run	Day-to-Day	Run-to-Run	Day-to-Day
DMIP	5	0.1	18	0.4	3	0.1	0.4	7	1	8
Glu-P-2	8	4	40	0.8	5	0.1	3	5	2	4
IQ	7	2	37	0.7	2	0.04	3	5	2	2
MeIQx	7	0.5	56	1.1	4	0.1	3	5	1	2
MeIQ	2	1	21	0.4	2	0.04	2	9	0.4	2
Glu-P-1	7	3	32	0.6	5	0.1	4	6	4	4
7,8-DiMeIQx	2	0.3	25	0.5	3	0.1	5	8	1	1
4,8-DiMeIQx	6	0.4	26	0.5	3	0.1	4	8	1	1
Norharman	7	1	10	0.2	2	0.04	3	5	3	3
Harman	9	3	21	0.4	4	0.1	4	6	2	3
Trp-P-2	0.2	0.3	9	0.2	1	0.02	1	7	0.5	1
Trp-P-1	0.1	0.3	2	0.04	1	0.02	1	6	1	1
PhIP	0.6	0.1	18	0.4	0.5	0.01	3	6	0.4	0.5
AαC	1	0.3	11	0.2	1	0.02	3	4	1	2
MeAαC	0.2	0.3	10	0.2	1	0.02	4	5	0.5	1

Source: Adapted from Barcelò-Barrachina, E. et al. 2004b. *J. Chromatogr. A* 1023:67–78. With permission.

over three orders of magnitude, and the instrumental LODs were 10-fold lower and excellent precision values than those previously obtained with LC–MS/MS [107]. For the analysis of HCAs in food samples, UPLC fulfills the promise of increased resolution, speed, and sensitivity.

ACKNOWLEDGMENTS

The authors would like to extend their sincere appreciation to King Saud University, Deanship of Scientific Research, College of Science Research Center for its supporting of this book chapter.

REFERENCES

1. Widmark, E. M. P. 1939. Presence of cancer-producing substances in roasted food. *Nature* 143:984–984.
2. Ames, B. N., McCann, J., and Yamasaki, E. 1975. Methods for detecting carcinogens and mutagens with the Salmonella/mammalian-microsomal mutagenicity test. *Mutat. Res.* 31:347–364.
3. Sugimura, T., Nagao, M., Kawachi et al. 1977. Mutagens carcinogens in food, with special reference to highly mutagenic pyrolytic products in broiled food, In *Origins of Human Cancer*, ed. H. H. Hiatt, J. D.Watson, and J. A. Winsten, 1561–1577, Cold Spring Harbor Laboratory, Cold Spring Harbor, New York.
4. Nagao, M., Honda, M., Seino, Y., Yahagi, T., and Sugimura, T. 1977. Mutagenicities of smoke condensates and the charred surface of fish and meat. *Cancer Lett.* 2:221–226.
5. Commoner, B., Vithayathil, A. J., Dolara, P., Nair, S., Madyasta, P., and Cuca, G. C. 1978. Formation of mutagens in beef and beef extract during cooking. *Science* 201:913–916.
6. Felton, J. S. and Knize, M. G. 1990. Heterocyclic amine mutagens/carcinogens in foods. In *Chemical Carcinogenesis and Mutagenesis I*, ed. C. S. Cooper and P. L. Grover, 471–502, Berlin: Springer-Verlag.
7. Wakabayashi, K., Nagao, M., Esumi, H., and Sugimura, T. 1992. Food-derived mutagens and carcinogens. *Cancer Res.* 52:2092–2098.
8. Sugimura, T., Wakabayashi, K., Nagao, M., and Esumi, H. 1993. A new class of carcinogens: Heterocyclic amines in cooked food. In *Food, Nutrition and Chemical Toxicity*, ed. D. V., Parke., C., Ioannides, R.,Walker, and G. Smith, 259–276, London, United Kingdom: Nishimura Ltd.
9. IARC. 1993. *Some Naturally Occurring Substances: Food Items and Constituents, Heterocyclic Aromatic Amines, and Mycotoxins. IARC Monographs on the Evaluation of Carcinogenic Risk of Chemicals to Humans, vol. 56.* Lyon, France: International Agency for Research on Cancer. 571 pp.
10. NTP. 1999. Report on Carcinogens Background Document for 2-Amino-3-Methylimidazo[4,5-f]Quinoline (IQ). National Toxicology Program. http://ntp-server.niehs.nih.gov/newhomeroc/roc10/IQ.pdf
11. NTP. 2002. Report on Carcinogens Background Document for Heterocyclic Amines: PhIP, MeIQ and MeIQx. National Toxicology Program. http://ntp-server.niehs.nih.gov/newhomeroc/roc11/HCAsPub.pdf
12. Sugimura, T. 1992. Multistep carcinogenesis: A 1992 perspective. *Science* 258:603–607.
13. Felton, J. S., Knize, M. G., Shen, N. H., Lewis, P. R., Andresen, B. D., Happe, J., and Hatch, F. T. 1986. The isolation and identification of a new mutagen from fried ground beef: 2-Amino-1-methyl-6-phenylimidazo[4,5-b]pyridine (PhIP). *Carcinogenesis* 7:1081–1086.

14. Becher, G., Knize, M. G., Nes, I. F., and Felton, J. S. 1988. Isolation and identification of mutagens from a fried Norwegian meat product. *Carcinogenesis* 9:247–253.
15. Skog, K., Augustsson, K., Steineck, G., Stenberg, M., and Jägerstad, M. 1997. Polar and nonpolar heterocyclic amines in cooked fish and meat products and their corresponding pan residues. *Food Chem. Toxicol.* 35:555–565.
16. Manabe, S., Kurihara, N., Wada, O., Izumikawa, S., Asakuno, K., and Morita, M. 1993a. Detection of a carcinogen, 2-amino-1-methyl-6-phenylimidazo (4,5-B)pyridine, in airborne particles and diesel-exhaust particles. *Environ. Pollut.* 80:281–286.
17. Ono, Y., Somiya, I., and Oda, Y. 2000. Identification of a carcinogenic heterocyclic amine in river water. *Water Res.* 34:890–894.
18. Saski, T. A., Wilkins, J. M., Forehand, J. B., and Moldoveanu, S. C. 2001. Analysis of heterocyclic amines in mainstream cigarette smoke using a new NCI GC–MS technique. *Anal. Lett.* 34:1749–1761.
19. Jägerstad, M., Skog, K., Arvidsson, P., and Solyakov, A. 1998. Chemistry, formation and occurrence of genotoxic heterocyclic amines identified in model systems and cooked foods. *Z. Lebensm. Unters. Forsch. A* 207:419–427.
20. Sugimura, T., Nagao, M., and Wakabayashi K. 1982. Metabolic aspects of the comutagenic action of norharman. *Adv. Exp. Med. Biol.* 136B:1011–1025.
21. Bianchi, F., Careri, M., Corradini, C., Elviri, L., Mangia, A., and Zagnoni, I. 2005. Investigation of the separation of heterocyclic aromatic amines by reversed phase ion-pair liquid chromatography coupled with tandem mass spectrometry: The role of ion pair reagents on LC-MS/MS sensitivity. *J. Chromatogr. B* 8251:93–200.
22. Galceran, M. T., Pais, P., and Puignou, L. 1996b. Isolation by solid-phase extraction and liquid-chromatographic determination of mutagenic amines in beef extracts. *J. Chromatogr. A* 719:203–212.
23. Wakabayashi, K., Kim, I. S., Kurosaka, R., and Yamaizumi, Z. 1995. Heterocyclic amines in cooked foods, In *Possible Human Carcinogens*, ed. R. H. Adamson, J. A. E. Gustavsson, N. Ito, and M. Nagao, 197–206. Princeton Scientific Publishing Co., Princeton, NJ.
24. Turesky, R. J., Taylor, J., Schnackenberg, L., Freeman, J. P., and Holland, R. D. 2005. Quantitation of carcinogenic heterocyclic aromatic amines and detection of novel heterocyclic aromatic amines in cooked meats and grill scrapings. *J. Agric. Food Chem.* 53:3248–3258.
25. Felton, J. S., Fultz, E., Dolbeare, F. A., and Knize, M. G. 1994. Effect of microwave pretreatment on heterocyclic aromatic amine mutagens/carcinogens in fried beef patties. *Food Chem. Toxicol.* 32:897–903.
26. Hodge, J. E. 1953. Dehydrated foods: Chemistry of Browning reactions in model systems. *J. Agric. Food Chem.* 1:928–943.
27. Knize, M. G., Dolbeare, F. A., Carroll, K. L., Moore, D. H., and Felton, J. S. 1994. Effect of cooking time and temperature on the heterocyclic amine content of fried beef patties. *Food Chem. Toxicol.* 32:595–603.
28. Skog, K., Steineck, G., Augustsson, K., and Jägerstad, M. 1995. Effect of cooking temperature on the formation of heterocyclic amines in fried meat-products and pan residues. *Carcinogenesis* 16:861–867.
29. Skog, K., Solyakov, A., and Jägerstad, M. 2000. Effects of heating conditions and additives on the formation of heterocyclic amines with reference to amino-carbolines in a meat juice model system. *Food Chem.* 68:299–308.
30. Persson, E., Sjöholm, I., and Skog, K. 2002. Heat and mass transfer in chicken breasts—Effect on PhIP formation. *Eur. Food Res. Technol.* 214:455–459.
31. Skog, K., Johansson, M. A. E., and Jägerstad, M. I. 1998a. Carcinogenic heterocyclic amines in model systems and cooked foods—A review on formation, occurrence and intake. *Food Chem. Toxicol.* 36:879–896.

32. Jagerstad, M., Laser-Reuterswärd, A., Olsson, R., Grivas, S., Nyhammar, T., Olsson, K., and Dahlqvist, A. 1983a. Creatin(in)e and Maillard reaction products as precursors of mutagenic compounds: Effects of various amino acids. *Food Chem.* 12:255–264.
33. Sugimura, T. 2000. Nutrition and dietary carcinogens. *Carcinogenesis* 21:387–395.
34. Felton, J. S. and Knize, M. G. 1991b. Occurrence, identification, and bacterial mutagenicity of heterocyclic amines in cooked food. *Mutat. Res.* 259:205–217.
35. Murkovic, M. 2004. Formation of heterocyclic aromatic amines in model system. *J. Chromatogr. B* 802:3–10.
36. Felton, J. S., Jägerstad, M., Knize, M. G., Skog, K., and Wakabayashi, K. 2000. Contents in foods, beverages and tobacco, In *Food Borne Carcinogens: Heterocyclic Amines*, ed. Nagao, M., and Sugimura, T. 31–71. John Wiley & Sons, Chichester.
37. Weisburger, J. H. 1994. Specific Maillard reactions yield powerful mutagens and carcinogens. In: *Maillard Reactions in Chemistry, Food, and Health*, ed. T. P. Labuza, G. A., Reineccius, V. M., Monnier, J., O'Brien and J. W. Baynes, 335–340, The Royal Society of Chemistry, Cambridge.
38. Overvik, E., Kleman, M., Berg, I., and Gustavsson, J. A. 1989. Influence of creatine, amino acids and water on the formation of the mutagenic heterocyclic amines found in cooked meat. *Carcinogenesis* 10:2293–2301.
39. Milic, B. L., Djilas, S. M., and Canadanovic-Brunet, J. M. 1993. Synthesis of some heterocyclic amino imidazoazarenes. *Food Chem.* 46:273–276.
40. Abdulkarim, B. G. and Smith, J. S. 1998 Heterocyclic amines in fresh and processed products. *J. Agric. Food Chem.* 46:4680–4687.
41. Arvidsson, P., Van-Boekel, M. A. J. S., Skog, K., and Jägerstad, M. 1997. Kinetics of formation of polar heterocyclic amines in a meat model system. *J. Food Sci.* 62:911–916.
42. Yen, G. C. and Chau, C. F. 1993. Inhibition by xylose-lysine Maillard reaction products of the formation of MeIQx in a heated creatinine, glycine, and glucose model system. *Biosci. Biotech. Biochem.* 57:664–665.
43. Johansson, M. and Jägerstad, M. 1994. Occurrence of mutagenic/carcinogenic heterocyclic amines in meat and fish products, including pan residues, prepared under domestic conditions. *Carcinogenesis* 15:1511–1518.
44. Back, Y. M., Lee, J. H., Shin, H. S., and Lee, K. G. 2009. Analysis of heterocyclic amines and β-carbolines by liquid chromatography-mass spectrometry in cooked meats commonly consumed in Korea. *Food Addit. Contam.* 26:298–305.
45. Thiebaud, H. P., Knize, M. G., Kuzmicky, P. A., Felton, J. S., and Hsieh, D. P. 1994. Mutagenicity and chemical analysis of fumes from cooking meat. *J. Agric. Food Chem.* 42:1502–1510.
46. Kataoka, H., Nishioka, S., Kobayashi, M., Hanaoka, T., and Tsugane, S. 2002. Analysis of mutagenic heterocyclic amines in cooked food samples by gas chromatography with nitrogen-phosphorus detector. *Bull. Environ. Contam. Toxicol.* 69:682–689.
47. Pais, P., Salmon, C. P., Knize, M. G., and Felton, J. S. 1999. Formation of mutagenic/carcinogenic heterocyclic amines in dry-heated model systems, meats, and meat drippings. *J. Agric. Food Chem.* 47:1098–1108.
48. Johansson, M., Fredholm, L., Bjerna, I., and Jägerstad, M. 1995b. Infuence of frying fat on the formation of heterocyclic amines in fried beefburgers and pan residues. *Food Chem. Toxicol.* 33:993–1004.
49. Fay, L. B., Ali, S., and Gross, G. 1997. Determination of heterocyclic aromatic amines in food products: Automation of the sample preparation method prior to HPLC and HPLC-MS quantification. *Mutat. Res.* 376:29–35.
50. Sinha, R., Rothman, N., Brown, E. D., Mark, S. D., Hoover, R. N., Caporaso, N. E., Levander, O. A., Knize, M. G., Lang, N. P., and Kadlubar, F. 1994. Pan-fried meat containing high levels of heterocyclic amines but low levels of polycyclic aromatic hydrocarbons induces cytochromP450IA2 activity in humans. *Cancer Res.* 54:6154–4159.

51. Holder, C. L., Cooper, W. M., Churchwell, M. I., Doerge, D. R., and Thompson, H. C. 1996. Multi-residue determination and confirmation of ten heterocyclic amines in cooked meats. *J. Muscle Foods* 7:281–290.
52. Borgen, E. and Skog, K. 2004. Heterocyclic amines in some Swedish cooked foods industrially prepared or from fast food outlets and restaurants. *Mol. Nutr. Food Res.* 48:292–298.
53. Galceran, M. T., Pais, P., and Puignou, L. 1993. High-performance liquid-chromatographic determination of 10 heterocyclic aromatic-amines with electrochemical detection. *J. Chromatogr. A* 655:101–110.
54. Murray, S., Lynch, A. M., Knize, M. G., and Gooderham, N. J. 1993. Quantification of the carcinogens 2-amino-3,8-dimethyl- and 2-amino-3,4,8-trimethylimidazo[4,5-f]quinoxaline and 2-amino-1-methyl-6-phenylimidazo[4,5-b]pyridine in food using a combined assay based on gas chromatography-negative ion mass spectrometry. *J. Chromatogr.* 616:211–219.
55. Busquets, R., Mitjans, D., Puignou, L., and Galceran, M. T. 2008. Quantification of heterocyclic amines from thermally processed meats selected from a small-scale population-based study. *Mol. Nutr. Food Res.* 52:1408–1420.
56. Khan, M. R., Bertus, L. M., Busquets, R., and Puignou, L. 2009. Mutagenic heterocyclic amine content in thermally processed offal products. *Food Chem.* 112:838–843.
57. Knize, M. G., Salmon, C. P., Mehta, S. S., and Felton, J. S. 1997b. Analysis of cooked muscle foods for heterocyclic aromatic amine carcinogens. *Mutat. Res.* 376:129–134.
58. Stavric, B., Lau, B. P. Y., Matula, T. I., Klassen, R., Lewis, D., and Downie, R. H. 1997b. Heterocyclic aromatic amine content in pre-processed meat cuts produced in Canada. *Food Chem. Toxicol.* 35:199–206.
59. Jamin, E., Chevolleau, S., Touzet, C., Tulliez, J., and Debrauwer, L. 2007. Assessment of metastable atom bombardment (MAB) ionization mass spectrometry for the fast determination of heterocyclic aromatic amines in cooked meat. *Anal. Bioanal. Chem.* 387:2931–2941.
60. Salmon, C. P., Knize, M. G., Felton, J. S., Zhao, B., and Seow, A. 2006. Heterocyclic aromatic amines in domestically prepared chicken and fish from Singapore Chinese households. *Food Chem. Toxicol.* 44:484–492.
61. Solyakov, A. and Skog, K. 2002. Screening for heterocyclic amines in chicken cooked in various ways. *Food Chem. Toxicol.* 40:1207–1212.
62. Sinha, R., Rothman, N., Brown, E. D., Salmon, C. P., Knize, M. G., Swanson, C. A., Rossi, S. C., Mark, S. D., Levander, O. A., and Felton, J. S. 1995. High concentrations of the carcinogen 2-amino-1-methyl-6-phenyl-imidazo[4,5-b]pyridine (PhIP) occur in chicken but are dependent on the cooking. *Cancer Res.* 55:4516–4519.
63. Chiu, C. P., Yang, D. Y., and Chen, B. H. 1998. Formation of heterocyclic amines in cooked chicken legs. *J. Food Protect.* 61:712–719.
64. Holder, C. L., Preece, S. W., Conway, S. C., Pu, Y. M., and Doerge, D. R. 1997. Quantification of heterocyclic amine carcinogens in cooked meats using isotope dilution liquid chromatography/atmospheric pressure chemical ionization tandem mass spectrometry. *Rapid Commun. Mass Spectrom.* 11:1667–1672.
65. Wakabayashi, K., Ushiyama, H., Takahashi, M., Nukaya, H., Kim, S. B., Hirose, M., Ochiai, M., Sugimura, T., and Nagao, M. 1993. Exposure to heterocyclic amines. *Environ. Health Perspect.* 99:129–133.
66. Oz, F., Kaban, G., and Kaya, M. 2007. Effects of cooking methods on the formation of heterocyclic aromatic amines of two different species trout. *Food Chem.* 104:67–72.
67. Gross, G. and Grüter, A. 1992.Quantitation of mutagenic HAAs in food products. *J. Chromatogr.* 592:271–278.
68. Skog, K. and Solyakov, A. 2002. Heterocyclic amines in poultry products. *Food Chem. Toxicol.* 40:1213–1221.

69. Alexander, J., Reistad, R., Hegstad, S., Frandsen, H., Ingebrigtsen, K., Paulsen, J. E., and Becher, G. 2002. Biomarkers of exposure to heterocyclic amines: Approaches to improve the exposure assessment. *Food Chem. Toxicol.* 40:1131–1137.
70. Alexander, J. and Wallin, H. 1991. Metabolic fate of heterocyclic amines in cooked food, In *Mutagens in Food: Detection and Prevention*, ed. Hayatsu H. 143–156, CRC Press, Boca Raton, FL.
71. Frederiksen, H. and Frandsen, H. 2004. Excretion of metabolites in urine and faeces from rats dosed with the heterocyclic amine, 2-amino-9H-pyrido[2,3-b]indole. *Food Chem Toxicol.* 42:879–85.
72. Kulp, K. S., Knize, M. G., Fowler, N. D., Salmon, C. P., and Felton, J. S. 2004. PhIP metabolites in human urine after consumption of well-cooked chicken. *J. Chromatogr. B* 802:143–153.
73. Magagnotti, C., Orsi, F., Bagnati, R., Celli, N., Rotilio, D., Fanelli, R., and Airoldi, L. 2000. Effect of diet on serum albumin and haemoglobin adducts of 2-amino-1-methyl-6-phenylimidazo[4,5-b]pyridine (PhIP) in humans. *Int. J. Cancer* 88:1–6.
74. Hegstad, S., Ingebrigtsen, K., Reistad, R., Paulsen, J. E., and Alexander, J. 2000a. Incorporation of the food mutagen 2-amino-1-methyl-6-phenylimidazo[4, 5-b]pyridine (PhIP) into hair of mice. *Biomarkers* 5:24–32.
75. Hegstad, S., Lundanes, E., Reistad, R., Haug, L. S., Becher, G., and Alexander, J. 2000b. Determination of the food mutagen carcinogen 2-amino-1-methyl-6-phenylimidazo[4, 5-b]pyridine (PhIP) in human hair by solid-phase extraction and gas chromatography-mass spectrometry. *Chromatographia* 52:499–504.
76. Keating, G. A. and Bogen, K. T. 2004. Estimates of heterocyclic amine intake in the US population. *J. Chromatogr. B* 802:127–133.
77. Byrne, C., Sinha, R., Platz, E. A., Giovannucci, E., Colditz, G. A., Hunter, D. J., Speizer, F. E., and Willett, W. C. 1998. Predictors of dietary heterocyclic amine intake in three prospective cohorts. *Cancer Epidemiol. Biomarkers Prev.* 7:523–529.
78. Layton, D. W., Bogen, K. T., Knize, M. G., Hatch, F. T., Johnson, V. M., and Felton, J. S. 1995. Cancer risk of heterocyclic amines in cooked foods: An analysis and implications for research. *Carcinogenesis* 16:39–52.
79. Augustsson, K., Skog, K., Jägerstad, M., and Steineck, G. 1997. Assessment of the human exposure to heterocyclic amines. *Carcinogenesis* 18:1931–1935.
80. Zimmerli, B., Rhyn, P., Zoller, O., and Schlatter J. 2001. Occurrence of heterocyclic aromatic amines in the Swiss diet: Analytical method, exposure estimation and risk assessment. *Food Addit. Contam.* 18:533–551.
81. Busquets, R., Bordas, M., Toribio, F., Puignou, L., and Galceran, M. T. 2004. Occurrence of heterocyclic amines in several home-cooked meat dishes of the Spanish diet. *J. Chromatogr. B* 802:79–86.
82. Warzecha, L., Janoszka, B., Błaszczyk, U., Stróżyk, M., Bodzek, D., and Dobosz C. 2004. Determination of heterocyclic aromatic amines (HAs) content in samples of household-prepared meat dishes. *J. Chromatogr. B* 802:95–106.
83. Thorogood, M., Mann, J., Appleby, P., and McPherson, K. 1994. Risk of death from cancer and ischaemic heart disease in meat and non-meat eaters. *Br. Med. J.* 308:1667–1670.
84. Lynch, A. M., Knize, M. G., Boobis, A. R., Gooderham, N. J., Davies, D. S., and Murray, S. 1992. Intra- and inter-individual variability in systemic exposure in man to MeIQx and PhIP, carcinogen present in cooked beef. *Cancer Res.* 52:6216–6223.
85. Gu, Y. S., Kim, I.S., Ahn, J. K., Park, D. C., Yeum, D. M., Ji, C. I., and Kim, S. B. 2002. Mutagenic and carcinogenic heterocyclic amines as affected by muscle types/skin and cooking in pan-roasted mackerel. *Mutat. Res.* 515:189–195.
86. Kasai, H., Nishimura, S., Nagao, M., Takahashi, Y., and Sugimura, T. 1979. Fractionation of a mutagenic principle from broiled fish by high-pressure liquid chromatography. *Cancer Lett.* 7:343–348.

87. Richling, E., Decker, C., Haring, D., Herderich, M., and Schraier P. 1997. Analysis of heterocyclic aromatic amines in wine by high-performance liquid chromatography-electrospray tandem mass spectrometry. *J. Chromatogr. A* 791:71–77.
88. Janoszka, B., Blaszczyk, U., Warzecha, L., Strozyk, M., Damasiewicz-Bodzek, A., and Bodzek D. 2001. Clean-up procedures for the analysis of heterocyclic aromatic amines (aminoazaarenes) from heat-treated meat samples. *J. Chromatogr. A* 938:155–165.
89. Kataoka, H. and Pawliszyn, J. 1999. Development of In-tube solid-phase microextraction/liquid chromatography/electrospray ionization mass spectrometry for the analysis of mutagenic heterocyclic amines. *Chromatographia* 50:532–538.
90. Murkovic, M., Friedrich, M., and Pfannhauser, W. 1997. Heterocyclic aromatic amines in fried poultry meat. *Z. Lebensm. Unters. Forsch. A* 205:347–350.
91. Bang, J., Frandsen, H., and Skog, K. 2004. Blue chitin columns for the extraction of heterocyclic amines from urine samples. *Chromatographia* 60:651–655.
92. Hayatsu, H. 1992. Cellulose bearing covalently linked copper phthalocyanine trisulphonate as an adsorbent selective for polycyclic compounds and its use in studies of environmental mutagens and carcinogens. *J. Chromatogr.* 597:37–56.
93. Ohe, T. 1997. Quantification of mutagenic/carcinogenic heterocyclic amines, MeIQx, Trp-P-1, Trp-P-2 and PhIP, contributing highly to genotoxicity of river water. *Mutat. Res.* 393:73–79.
94. Hayatsu, H., Hayatsu, T., Arimoto, S., and Sakamoto, H. 1996. A short-column technique for concentrating mutagens/carcinogens having polycyclic structure. *Anal. Biochem.* 235:185–190.
95. Nukaya, H., Koyota, S., Jinno, F., Ishida, H., Wakabayashi, K., Kurosaka, R., Kim, I. S., Yamaizumi, Z., Ushiyama, H., Sugimura, T., Nagao, M., and Tsuji, K. 1994. Structural determination of a new mutagenic heterocydic amine, 2-amino-1,7,9-trimethylimidazo[4,5-g]quinoxaline (7,9-DiMeIgQx), present in beef extract. *Carcinogenesis* 15:1151–1154.
96. Martín-Calero, A., Ayala, J. H., González, V., and Afonso, A. M. 2007a. Determination of less polar heterocyclic amines in meat extracts: Fast sample preparation method using solid-phase microextraction prior to high-performance liquid chromatography-fluorescence quantification. *Anal. Chim. Acta* 582:259–266.
97. Cárdenes, L., Ayala, J. H., Afonso, A. M., and González, V. 2004. Solid-phase microextraction coupled with high-performance liquid chromatography for the analysis of heterocyclic aromatic amines. *J. Chromatogr. A* 1030:87–93.
98. Cárdenes, L., Martín-Calero, A., Ayala, J. H., González, V., and Afonso, A. M. 2006. Experimental design optimization of solid-phase microextraction conditions for the determination of heterocyclic aromatic amines by high-performance liquid chromatography. *Anal. Lett.* 39:405–423.
99. Gross, G. A. 1990. Simple methods for quantifying mutagenic heterocyclic aromatic amines in food products. *Carcinogenesis* 11:1597–1603.
100. Galceran, M. T., Pais, P., and Puignou, L. 1996a. Isolation by solid-phase extraction and liquid-chromatographic determination of mutagenic amines in beef extracts. *J. Chromatogr. A* 719:203–212.
101. Fei, X. Q., Li, C., Yu, X. D., and Chen, H. Y. 2007. Determination of heterocyclic amines by capillary electrophoresis with UV-DAD detection using on-line preconcentration. *J. Chromatogr. B* 854:224–229.
102. Bermudo, E., Ruiz-Calero, V., Puignou, L., and Galceran, M. T. 2005. analysis of heterocyclic amines in chicken by liquid chromatography with electrochemical detection. *Anal. Chim. Acta* 536:83–90.
103. Turesky, R. J., Goodenough, A. K., Ni, W. J., McNaughton, L., LeMaster, D. M., Holland, R. D., Wu, R. W., and Felton, J. S. 2007a. Identification of 2-amino-1,7-dimethylimidazo[4,5-g]

quinoxaline: An abundant mutagenic heterocyclic aromatic amine formed in cooked beef. *Chem. Res. Toxicol.* 20:520–530.

104. Khan, M. R., Busquets, R., Santos, F. J., and Puignou, L. 2008. New method for the analysis of heterocyclic amines in meat extracts using pressurised liquid extraction and liquid chromatography–tandem mass spectrometry. *J. Chromatogr. A* 1194:155–160.
105. Gerbl, U., Cichna, M., Zsivkovits, M., Knasmuller, S., and Sontag, G. 2004. Determination of heterocyclic aromatic amines in beef extract, cooked meat and rat urine by liquid chromatography with coulometric electrode array detection. *J. Chromatogr.* 802:107–113.
106. Janoszka, B., Błaszczyk, U., Damasiewicz-Bodzek, A., and Sajewicz, M. 2009. Analysis of heterocyclic amines (HAs) in pan-fried pork meat and its gravy by liquid chromatography with diode array detection. *Food Chem.* 113:1188–1196.
107. Barcelò-Barrachina, E., Moyano, E., Puignou, L., and Galceran, M. T. 2004b. Evaluation of different liquid chromatography–electrospray mass spectrometry systems for the analysis of heterocyclic amines. *J. Chromatogr. A* 1023:67–78.
108. Toribio, F., Busquets, R., Puignou, L., and Galceran, M. T. 2007. Heterocyclic amines in griddled beef steak analysed using a single extract clean-up procedure. *Food Chem. Toxicol.* 45:667–675.
109. Barcelò-Barrachina, E., Moyano, E., Galceran, M. T., Lliberia, J. L., Bagò, B., and Cortes, M. A. 2006. Ultra-performance liquid chromatography-tandem mass spectrometry for the analysis of heterocyclic amines in food. *J. Chromatogr. A* 1125:195–203.
110. Fatih, O. 2010. Quantitation of heterocyclic aromatic amines in ready to eat meatballs by ultra fast liquid chromatography. *Food Chem.* 126:2010–2016.
111. Gross, G., Philipposian, and Aeschbacher, 1989. An efficient and convenient method for the purification of HAAs. *Carcinogenesis* 10:1175–1182.
112. Solyakov, A., Skog, K., and Jägerstad, M. 1999. Heterocyclic amines in-process flavors, process flavor ingredients, Bouillon concentrates and a pan residue. *Food Chem. Toxicol.* 37:1–11.
113. Vollenbroker, M., and Eichner, K. 2000. A new quick solid-phase extraction method for the quantification of heterocyclic aromatic-amines. *Eur. Food Res. Technol.* 212:122–125.
114. Jautz, U., Gibis, M., and Morlock, G. E. 2008. Quantification of heterocyclic aromatic amines in fried meat by HPTLC/UV-FLD and HPLC/UV-FLD: A comparison of two methods. *J. Agric. Food Chem.* 56:4311–4319.
115. Kataoka, H. and Kijima, K. 1997. Analysis of heterocyclic amines as their N-dimethylaminomethylene derivatives by gas chromatography with nitrogen-phosphorus selective detection. *J. Chromatogr. A* 767:187–194.
116. Janoszka, B., Blaszczyk, U., Warzecha, L., Luks-Betlej, K., and Strozyk, M. 2003. The analysis of aminoazaarenes as their derivatives with GC-MS method in the heat-processed meat samples. *Anal. Chem. (Warsaw)* 48:707–707.
117. Richling, E., Kleinschnitz, M., and Schreier, P. 1999. Analysis of heterocyclic aromatic-amines by high-resolution gas chromatography-mass spectrometry—A suitable technique for the routine control of food and process flavors. *Eur. Food Res. Technol.* 210:68–72.
118. Casal, S., Mendes, E., Fernandes, J. O., Oliveira, M. B. P. P., and Ferreira, M. A. 2004. Analysis of heterocyclic aromatic amines in foods by gas chromatography-mass spectrometry as their tert.-butyldimethylsilyl derivatives. *J. Chromatogr. A* 1040:105–114.
119. Zhang, F., Chu, X., Sun, Li., Zhao, Y., Ling, Y., Wang, X., Yong, W., Yang, M., and Li, X. 2008. Determination of trace food-derived hazardous compounds in Chinese cooked foods using solid-phase extraction and gas chromatography coupled to triple quadrupole mass spectrometry. *J. Chromatogr. A* 1209:220–229.
120. Gross, G. A., Turesky, R. J., Fay, L. B., Stillwell, W. G., Skipper, P. L., and Tannenbaum, S. R. 1993. Heterocyclic aromatic amine formation in grilled bacon, beef and fish and in grill scrapings. *Carcinogenesis* 14:2313–2318.

121. Richling, E., Häring, D., Herderich, M., and Schreier, P. 1998b. Determination of heterocyclic aromatic amines (HAA) in commercially available meat products and fish by high performance liquid chromatography electrospray tandem-mass spectrometry (HPLC-ESIMS-MS). *Chromatographia* 48:258–262.
122. Pais, P., Moyano, E., Puignou, L., and Galceran, M. T. 1997a. Liquid chromatography-electrospray mass spectrometry with in-source fragmentation for the identification and quantification of fourteen mutagenic amines in beef extracts. *J. Chromatogr. A* 775:125–136.
123. Barceló-Barrachina, E., Moyano, E., Puignou, L., and Galceran, M. T. 2004a. Evaluation of reversed-phase columns for the analysis of heterocyclic aromatic amines by liquid chromatography-electrospray mass spectrometry. *J. Chromatogr. B* 802:45–59.
124. Toribio, F., Moyano, E., Puignou, L., and Galceran, M. T. 2000b. Determination of heterocyclic aromatic amines in meat extracts by liquid chromatography-ion-trap atmospheric pressure chemical ionization mass spectrometry. *J. Chromatogr. A* 869:307–317.
125. Toribio, F., Moyano, E., Puignou, L., and Galceran, M. T. 2000a. Comparison of different commercial solid-phase extraction cartridges used to extract heterocyclic amines from a lyophilised meat extract. *J. Chromatogr. A* 880:101–112.
126. Prabhu, S., Lee, M. J., Hu, W. Y., Winnik, B., Yang, I., Buckley, B., and Hong, J. Y. 2001. Determination of 2-amino-1-methyl-6-phenylimidazo(4,5-B)pyridine (PhIP) and its metabolite 2-hydroxyamino-PhIP by liquid chromatography/electrospray ionization-ion trap mass-spectrometry. *Anal. Biochem.* 298:306–313.
127. Toribio, F., Moyano, E., Puignou, L., and Galceran, M. T. 2002. Ion-trap tandem mass spectrometry for the determination of heterocyclic amines in food. *J. Chromatogr. A* 948:267–281.
128. Calbiani, F., Careri, M., Elviri, L., Mangia, A., and Zagnoni, I. 2007. Validation of an ion-pair liquid chromatography-electrospray-tandem mass spectrometry method for the determination of heterocyclic aromatic amines in meat-based infant foods. *Food Addit. Contam.* 24:833–841.
129. Guy, P. A., Gremaud, E., Richoz, J., and Turesky, R. J. 2000. Quantitative analysis of mutagenic heterocyclic aromatic amines in cooked meat using liquid chromatography-atmospheric pressure chemical ionisation tandem mass spectrometry. *J. Chromatogr. A* 883:89–102.
130. Turesky, R. J., Bur, H., Huynh-Ba, T., Aeschbacher, H. U., and Milon, H. 1988. Analysis of mutagenic heterocyclic amines in cooked beef products by high-performance liquid chromatography in combination with mass spectrometry. *Food Chem. Toxicol.* 26:501–509.
131. Kim, I. S., Wakabayashi, K., Kurosawa, R., Yamaizumi, Z., Jinno, F., Koyota, S., Tada, A., Nukaya, H., Takahashi, M., Sugimura, T., and Nagao, M. 1994. Isolation and identification of a new mutagen, 2-amino-4-hydroxy-methyl-3, 8-dimethylimidazo[4,5-f] quinoxaline (4-CH_2OH-8-MeIQx), from beef extract. *Carcinogenesis* 15:21–26.
132. Pais, P., Moyano, E., Puignou, L., and Galceran, M. T. 1997b. Liquid chromatography-atmospheric-pressure chemical ionization mass spectrometry as a routine method for the analysis of mutagenic amines in beef extracts. *J. Chromatogr. A* 778:207–218.
133. Churchwell, M. I., Twaddle, N. C., Meeker, L. R., and Doerge, D. R. 2005. Improving LC-MS sensitivity through increases in chromatographic performance: Comparisons of UPLC-ES/MS/MS to HPLC-ES/MS/MS. *J. Chromatogr. B* 825:134–143.
134. Toshimasa, T. 2008. Determination of methods for biologically active compounds by ultra-performance liquid chromatography coupled with mass spectrometry: Application to the analyses of pharmaceuticals, foods, plants, environments, metabonomics, and metabolomics. *J. Chromatogr. Sci.* 46:233–247.

8 Determination of Avocado Fruit Metabolites by UHPLC–MS

Complementarity with Other Analytical Platforms

Elena Hurtado-Fernández, María Gómez-Romero, Tiziana Pacchiarotta, Oleg A. Mayboroda, Alegría Carrasco-Pancorbo, and Alberto Fernández-Gutiérrez

CONTENTS

8.1 INTRODUCTION

Botanically, avocado (*Persea americana*) is a berry that consists of a large central seed and pericarp, which is the sum of the skin, the edible portion, and the inner layer surrounding the seed [1]. It is a fruit of an evergreen tree that belongs to the *Lauraceae* family, which is typically from tropical or subtropical climates. Although the origin of this crop is Central America and Mexico, it is widely cultivated throughout the world [2,3]. In 2011, the world production of avocado was 4,434,425 tons, increasing by 10% over the previous year. The main producers are Mexico (28.5%), Chile (8.3%), Dominican Republic (6.7%), and Indonesia (6.2%), and the principal importers of this tropical fruit are the United States (8.6%), The Netherlands (2.6%), France (2.3%), and Japan (1.1%) [4]. The increased production of avocado over the last few years has been driven by a higher demand, as this fruit has distinctive and pleasant sensory attributes and is perceived by buyers as beneficial to health [5,6].

Numerous researchers have drawn attention to the beneficial properties that avocado could provide to human health. Some of these healthy benefits have been summarized in Figure 8.1. Wilson Grant, in 1960, published the first clinical study where avocado intake was associated with the maintenance of normal serum cholesterol levels, or even with their reduction [7]. Later on, in the 90s and beginning of 2000 there was an increase in the number of works that studied the relationship between avocado fat and its effect on cardiovascular diseases, cholesterol, lipid profile, weight control, and diabetes [8–16]. While the principal healthy benefits attributed to avocado fruit are the aforementioned, some others have also been evaluated, such as: prevention and treatment of osteoarthritis [17,18], anticancer properties [19–24], protective activity against liver injury [25], skin protection [26–28], reduction of risk of macular degeneration [29], influence on short-term memory [30], antioxidant activity [31–37], reduction of metabolic syndrome risk [38,39], and anti-inflammatory effects [31,32,40].

All these beneficial properties have been logically associated with some of the substances present in avocado fruit. Table 8.1 presents the families of compounds that have been identified in this matrix since 1969.

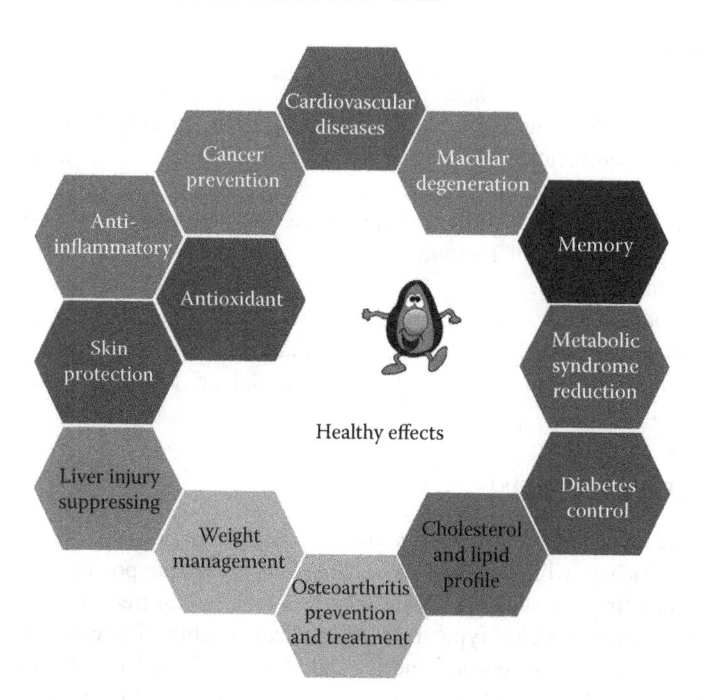

FIGURE 8.1 Principal healthy effects attributed to avocado consumption.

TABLE 8.1
Applications Where Avocado Fruit was the Matrix under Study, Including the Determined Metabolites, Techniques, Part of the Fruit Used, and Year of Publication

Metabolites	Technique	Avocado Part	Year	Reference
Aliphatic compounds	IR NMR MS	Pulp	1969	[41]
Carotenoids	TLC MS	Pulp, peel, leaves	1973	[42]
Fatty acids/vitamins/minerals	AOAC methods	Pulp	1975	[43]
Triglycerides/fatty acids	HPLC-RI GC-FID	Pulp	1987	[44]
Phenolic acids	HPLC-UV/Vis	Pulp	1987	[45]
Lipids	GC–MS NMR	Pulp	1991	[46]
Triglycerides	HPLC-LSD	Oil	1992	[47]
Volatiles	GC-MS	Pulp	1998	[15]
Sugars	HPLC-RI	Pulp	1999	[48]
Flavanols	HPLC-UV	Pulp	2000	[49]
Antifungal compounds	HPLC–MS HPLC-UV-RI NMR IR	Idioblast cells of mesocarp	2000	[50]
Fatty acids derivatives	NMR HR-FAB-MS HR-EI-MS IR HPLC GC	Pulp	2001	[25]
Sugars	HPLC-RI NIR	Honey	2002	[51]
Carbohydrates	HPLC-RI	Pulp	2002	[52]
Sterols	GC–MS	Pulp	2003	[53]
Fatty acids/volatiles	GC-FID GC–MS	Oil	2003	[54]
Flavanols	HPLC	Pulp	2004	[55]
Fatty acids	GC-FID	Pulp	2004	[56]
Minerals	ICP-OES	Honey	2004 2005	[57] [58]
Pigments	HPLC-PDA-FL	Oil, pulp, peel	2006	[29]
Vitamin E	HPLC-FL	Pulp	2006	[59]

continued

TABLE 8.1 (continued)
Applications Where Avocado Fruit was the Matrix under Study, Including the Determined Metabolites, Techniques, Part of the Fruit Used, and Year of Publication

Metabolites	Technique	Avocado Part	Year	Reference
Phenolic compounds/volatiles/ fatty acids	HPLC-DAD GC-FID-MS GC-FID	Oil	2007	[60]
Fatty acids/glycolipids/ phospholipids	GC-FID HPLC–MS–MS	Pulp	2007	[61]
Fatty acids/sugars	GC-FID HPLC-ELSD	Pulp	2008	[62]
Lipids/fatty acids	TLC GC-FID	Pulp	2008	[63]
Water content	NIR	Pulp	2009	[64]
Fatty acids/carbohydrates	GC-FID HPLC-ELSD	Pulp	2009	[65]
Fatty acids/carotenoids/ tocopherol	GC-FID HPLC-UV LC–MS	Pulp	2009	[66]
Fatty acids/sterols	GC–MS	Oil, pulp	2009	[67]
Fatty acids	GC NMR	Oil	2009	[68]
Aminoacids	HPLC-UV/Vis	Oil	2010	[69]
Phenolic compounds	UV/Vis HPLC-UV	Pulp	2010	[70]
Fatty acids/sugars	GC-FID HPLC-ELSD	Pulp	2010	[71]
Phenolic compounds	HPLC-3D-FL	Pulp	2010	[72]
Phenolic compounds	FT-IR FL	Pulp	2010	[33]
Fatty acids	GC–MS	Pulp	2010	[34]
Procyanidins/pigments	HPLC–MS^n UV-Vis	Pulp, peel, seed	2010	[35]
Phenolic compounds	UHPLC-UV-ESI-TOF MS	Pulp	2011	[73]
Phenolic compounds	UHPLC-PDA-FL-MS	Peel, pulp, seed	2011	[36]
Phenolic compounds	LC-DAD-MS	Pulp	2011	[5]
Fatty acids	CE-UV	Oil	2011	[74]
Fatty acids/sterols/tocopherols	GC-FID GC–MS HPLC-UV	Oil	2012	[75]
Proteins	Nano-LC–MS/MS	Pulp	2012	[76]

TABLE 8.1 (continued)
Applications Where Avocado Fruit was the Matrix under Study, Including the Determined Metabolites, Techniques, Part of the Fruit Used, and Year of Publication

Metabolites	Technique	Avocado Part	Year	Reference
Phenolic compounds	UV/Vis HPLC-DAD-MS	Peel, seed	2012	[77]
Fatty alcohol derivatives/alkanols/fatty acids/sterols/coumarins	FT-IR NMR HPLC–MS	Unripe pulp	2012	[78]
Chlorophyll/carotenoids	UV-Vis	Pulp	2012	[79]
Lipids	HA-LAESI MS	Pulp	2013	[80]
Carbohydrates/flavonoids/organic acids/phenolic acids/phytohormones/vitamins	CE-MS	Pulp	2013	[81]
Phenolic acids	CE-UV	Pulp	2013	[82]

The composition of avocado, in terms of macronutrients, has been widely studied and is compiled in different tables of food composition. According to the United States Department of Agriculture (USDA) National Nutrient Database for Standard Reference [83], 100 g of avocado are an important source of energy (160 kcal) and contain 73.23 g of water, 2 g of proteins, 14.66 g of total fat (67% monounsaturated, 12% polyunsaturated, and 14% saturated fatty acids), and 8.53 g of carbohydrates, of which 6.70 g are total dietary fiber and 0.66 g are sugars. As it can be seen, one of the main components of avocado is the fat, and thus it is not surprising that it is also known as "butter fruit." Besides, some of the principal health benefits of avocado have been attributed to its high monounsaturated fatty acid content. These facts make lipids one of the most studied families of compounds in avocado.

Some authors have noted that the existing synergy among some of the metabolites present in the avocado fruit may be responsible of the beneficial properties that this matrix possesses, rather than the effect caused by each individual compound [1]. For this reason, new trends in food science are based on "*omics*" approaches, which could allow linking food and health from a global point of view [84]. In this sense, a new concept has been defined: *foodomics* [85]. Foodomics is an emerging discipline that combines all of the modern "*omics*" technologies for the profiling of food compounds with other tools such as bioinformatics (detection of biomarkers for food quality and safety), toxicity assays (food contaminants), *in-vitro/vivo* assays (bioactivity), or clinical trials to determine the possible health effects [86,87].

All of the metabolomics experiments tend to follow a similar workflow (Figure 8.2), which includes different steps to reach a final interpretation for an initial biological question. Principally, three different parts can be distinguished: sample preparation, sample analysis, and data treatment.

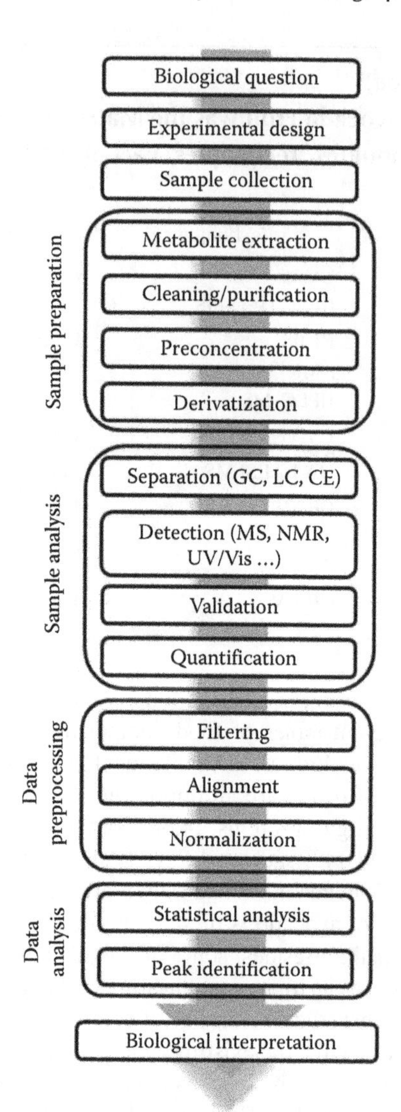

FIGURE 8.2 Diagram of the workflow involved in a metabolomics experiment. The diagram includes all the possible steps to be followed in a metabolomics study; it does not mean that all of them are absolutely necessary in all the cases.

Usually, sample analysis combines the use of different analytical techniques, such as spectroscopic (MS (mass spectrometry), NMR (nuclear magnetic resonance), IR (infra-red), FL, among others), electrochemical (voltammetry, conductimetry), separation (CE [capillary electrophoresis], GC [gas chromatography], LC [liquid chromatography]), or hyphenated techniques. All of them have been applied, to a greater or lesser extent, for food analysis [88]. As can be seen in Table 8.1, the use of separation and spectroscopic techniques, or the coupling of both, has been extensively applied for avocado metabolic profiling, since they are very useful and powerful platforms that provide a large amount of information of the food sample under study.

Among the separation techniques, liquid chromatography (LC) is one of the most used for plant metabolic profiling since the first application developed by Tswett in 1906. It has undergone several changes that have transformed the classical LC into a modern one, emerging in the 1960s high-pressure liquid chromatography (HPLC) [89]. It is characterized by higher resolution and appropriate separation times, thanks to the use of packed columns with small particle sizes and pumping systems that maintain a constant flow [90]. HPLC has become a routine technique in laboratories of analysis, because it is robust, reproducible, and sensitive, and it can be reasonably fast. However, nowadays the use of ultra-high-performance liquid chromatography (UHPLC) is gaining interest over conventional HPLC, because HPLC column chemistry and dimensions affect resolution and sensitivity, whereas sub-2 μm particle stationary phases combined with ultra-high-pressure pumps, which overcome the high backpressure associated with such columns, have helped produce narrower peaks, rapid analysis times, and lower detection limits. In addition, the solvent volume consumed is lower, decreasing the volumes of waste [91]. For these reasons, this book is focused on the use of this powerful technique in a very important field as food analysis. Specifically, the aim of this chapter is to summarize the applications described in literature for avocado studies by using UHPLC–MS. The first application of UHPLC–MS for the metabolome characterization of avocado fruit is explained in detail. Both qualitative and quantitative results achieved are commented and compared with those presented in similar works. In addition, a special section is dedicated to the data treatment, because of the importance that statistical tools have on the proper interpretation of the final results.

8.2 METABOLIC PROFILING OF AVOCADO FRUIT (*PERSEA AMERICANA*) BY UHPLC–MS

Bearing in mind the information provided by Table 8.1, UHPLC has not been a widely used technique for avocado study. Actually, to date only two publications are available [36,73], both of them in 2011. Rodríguez-Carpena et al. [36] studied the possible protective effect of the phenolic compounds present in different avocado extracts against meat lipid and protein oxidation. They performed the analysis of the mentioned analytes in extracts of peel, seed, and pulp of Hass and Fuerte varieties by using UHPLC–PDA–FL (ultra-high-performance liquid chromatography–photodiode array–fluorescence), studying also their *in vitro* antioxidant activity and antimicrobial potential. They were able to determine five families of phenolic compounds: catechins, hydroxybenzoic acids, hydroxycinnamic acids, flavonols, and procyanidins, which were quantified as equivalents of a representative compound of each group: gallic acid (hydroxybenzoic acids), chlorogenic acid (hydroxycinnamic acids), rutin (flavonols), and catechin (catechins and procyanidins). In general, peels and seeds showed the highest amounts of these compounds, although avocado pulp contained significant concentrations of procyanidins, hydroxycinnamic, and hydroxybenzoic acids. The values obtained for antioxidant and antimicrobial activities were also higher for avocado peels and seeds than for pulp. Finally, the *in vitro* antioxidant effect of avocado extracts was confirmed in real food products. The main objective of this study was to demonstrate the potential application of avocado subproduct extracts as natural food additives, using UHPLC just to have a general idea about the composition of avocado extracts in terms of phenolic compounds.

On the other hand, Hurtado et al. [73] developed the first method for studying the metabolome of different avocado varieties by using UHPLC as a separation technique coupled to two different detectors: ultraviolet (UV) and time-of-flight (TOF) mass spectrometry (MS) using electrospray ionization (ESI) as an ionization technique. The varieties included in this study were 13: Colin V 33, Gem, Harvest, Hass, Hass Motril, Jiménez 1, Jiménez 2, Lamb Hass, Marvel, Nobel, Pinkerton, Sir Prize, and Tacambaro, which were cultivated under identical conditions of soil, rain, light, among others, making feasible to carry out an exhaustive comparison between them. Four of these varieties come from putative mutations of Hass variety [2], specifically Hass Motril, Jiménez 1, Jiménez 2, and Tacambaro, therefore they present the same genetic profile as Hass. The analysis of these mutants could be interesting to check if there are similarities or differences in their metabolic profiles being so related genetically. All of the mentioned varieties were analyzed at two different ripening degrees: unripe (when the fruit was just harvested) and ripe (when the fruit was ready for human consumption). According to the authors, the interest in the comparison of 13 varieties in different ripening stages is double: on one side, to extend the information available about some fractions of avocado fruit that had not been studied in detail previously; and on the other, to assess how the ripening process changes the concentration of specific compounds, since it is well-known that harvesting time and maturity affects the color, size, dry matter, oil content, and rheological properties of the fruit [56,65], as well as fatty acid and carbohydrate content [44,52], and to a lesser extent, pigments, sterols, and antioxidant activity [29,34,67].

To perform that work, samples were subjected to a nonselective preparation, trying to extract the highest number of different compounds, since a metabolomics approach was pursued. The protocol followed by Hurtado and her collaborators [73] was quite simple: a solid–liquid extraction with methanol, starting from lyophilized and homogenized fruit pulp; followed by shaking in a vortex, centrifugation, evaporation to dryness, and reconstitution of the extract.

In an attempt to detect as many metabolites as possible in a single run, the authors optimized both chromatographic and detection conditions of the method, taking into account analysis time, peak shape, sensitivity, and the behavior of the analytes, among other conditions. Once the method was optimized, it was applied to the analysis of the 26 samples under study.

8.2.1 Identification of Compounds

Hurtado et al. [73] performed the identification of the metabolites considering retention time, spectral data, and the information provided by ESI–TOF MS for

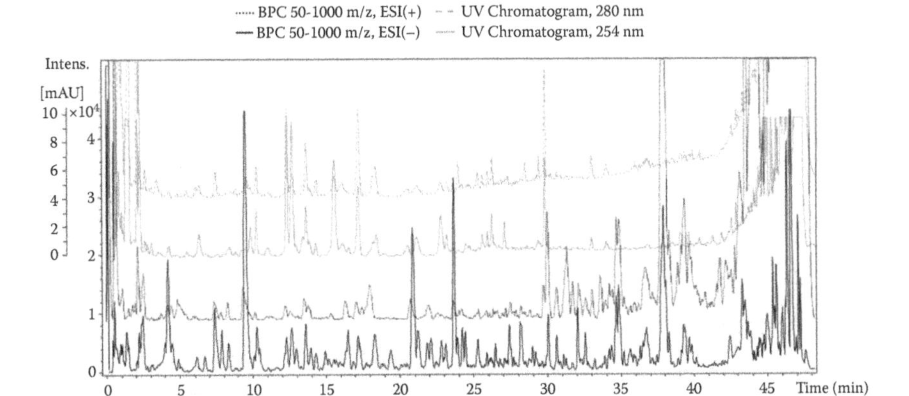

FIGURE 8.3 Profiles obtained (TOF MS in positive and negative modes, and UV at 254 and 280 nm) after the analysis of Colin V33 avocado variety at the second ripening degree under the optimum conditions.

both the avocado samples and the standards (32 standard solutions of pure compounds were used in that study), and also by using spiked samples at different concentrations. Taking advantage of the information provided by a TOF analyzer, a list of possible molecular formulas was generated by the software used, considering the accurate mass of the *m/z* signals detected in the spectrum and their true isotopic patterns. These elemental formulae made it possible to identify tentatively some other metabolites (which were not available as pure commercial standards) looking at open source databases (Kegg-Ligand [92], Pubchem [93], etc.) and literature.

Figure 8.3 shows the base peak chromatograms (BPCs) of one of the varieties of avocado fruit obtained for MS (in positive and negative modes) and also for UV at 254 and 280 nm. Among the detected signals, more than 200 *m/z* values were selected for an in-depth study, some of which were tentatively identified and some others corroborated by standards. The identified metabolites belonged to diverse chemical families, such as fatty acids, sugars, amino acids, vitamins, phenolic acids and related compounds, flavonoids, nucleosides, phytohormones, alkanols, among others.

Table 8.2 includes 60 of the compounds identified using all of the information acquired during analysis: retention time, observed signals in both ionization modes, molecular formula generated, and avocado samples containing the metabolite.

It is interesting to highlight the fact that the majority of the compounds were found in the totality of the samples. However, linolenic acid was only present in Gem and Marvel varieties at the second ripening degree. Some flavonoids (laricitrin, naringenin, chrysin, and kaempferide) were detected in ripe fruits from Colin V 33, Pinkerton, and Sir Prize. Ferulic and *p*-coumaric acids were determined in all of the samples ready for consumption and in just one unripe, Colin V 33. A similar situation is appreciated for syringic acid, which has been identified in Colin V 33 and Sir

TABLE 8.2
Peak Assignment of the Metabolites Present in Different Avocado Samples Using UHPLC–ESI–TOF MS

Retention Time (min)[a]	ESI(+) TOF MS[b]	ESI(–) TOF MS[b]	Molecular Formula [M] Generated	Assignment[c]	Avocado Samples Containing the Compound[d]
0.56	–	211.0817 [M-H]$^-$	$C_7H_{16}O_7$	*Perseitol*	ALL
0.61	193.0699 [M + H]$^+$ 215.0222 [M + Na]$^+$ 231.0617 [M + K]$^+$	191.0551 [M-H]$^-$	$C_7H_{12}O_6$	*Quinic acid**	ALL
0.62	–	341.1096 [M-H]$^-$	$C_{12}H_{22}O_{11}$	*Sucrose*	ALL
0.64	–	179.0557 [M-H]$^-$	$C_6H_{12}O_6$	*Glucose/fructose*	ALL
1.06	193.0649 [M + H]$^+$ 215.1195 [M + Na]$^+$ 230.9716 [M + K]$^+$	191.0191 [M-H]$^-$ 111.0099[M-H-44-18-18]$^-$	$C_6H_8O_7$	*Citric acid/isocitric acid*	G1, G2, HT1, HM1, HM2, HS1, HS2, J1-1, J1-2, J2-1, J2-2, LH1, LH2, MA1, MA2, N1, N2, P1, P2, SP1, T1, T2
1.30	119.0440 [M + H]$^+$,	117.0195 [M-H]$^-$	$C_4H_6O_4$	*Succinic acid**	ALL
1.43	182.0798 [M + H]$^+$ 165.0536 [M + H-17]$^+$	180.0667 [M-H]$^-$ 163.0399 [M-H-17]$^-$	$C_9H_{11}NO_2$	*Tyrosine*	C1, C2, G1, G2, HT1, HT2, HS1, HS2, J1-1, J1-2, J2-1, J2-2, LH1, LH2, MA1, MA2, HM2, N1, N2, P1, P2, SP2, T1, T2
1.55	245.0754 [M + H]$^+$ 267.0571 [M + Na]$^+$ 113.0342 [M + H-132]$^+$	243.0610 [M-H]$^-$ 200.0558 [M-H-43]$^-$	$C_9H_{12}N_2O_6$	*Uridine*	ALL
1.97	171.0285 [M + H]$^+$ 153.0168 [M + H-18]$^+$	169.0138 [M-H]$^-$	$C_7H_6O_5$	*Gallic acid**	SP2
2.12	268.1032 [M + H]$^+$ 136.0612 [M + H-132]$^+$	266.0888 [M-H]$^-$ 533.1779 [2M-H]$^-$ 312.0848 [M-H + 46]$^-$ 134.0504 [M-H-132]$^-$	$C_{10}H_{13}N_5O_4$	*Adenosine*	ALL

2.69	166.0843 $[M + H]^+$	164.0720 $[M\text{-}H]^-$ 147.0469 $[M\text{-}H\text{-}17]^-$	$C_9H_{11}NO_2$	*Phenylalanine*	C1, C2, G1, G2, HT1, HT2, HS1, HS2, J1-1, J1-2, J2-1, J2-2, LH2, MA1, MA2, HM1, N1, N2, P1, P2, SP1, SP2, T2
4.46	339.0719 $[M + Na]^+$ 355.0597 $[M + K]^+$	315.0740 $[M\text{-}H]^-$	$C_{13}H_{16}O_9$	*Dihydroxybenzoic acid hexose*	ALL
4.94	220.1177 $[M + H]^+$ 242.0981 $[M + Na]^+$ 258.0679 $[M + K]^+$ 202.1063 $[M + H\text{-}18]^+$ 184.0950 $[M + H\text{-}18\text{-}18]^+$ 116.0337 $[M + H\text{-}18\text{-}18\text{-}68]^+$ 90.0557 $[M + H\text{-}18\text{-}18\text{-}94]^+$	218.1035 $[M\text{-}H]^-$ 146.0825$[M\text{-}H\text{-}28\text{-}44]^-$	$C_9H_{17}NO_5$	*Pantothenic acid**	ALL
6.30	205.0969 $[M + H]^+$ 188.0680 $[M + H\text{-}17]^+$ 146.0541 $[M + H\text{-}18\text{-}42]^+$	203.0825 $[M\text{-}H]^-$	$C_{11}H_{12}N_2O_2$	*Tryptophan*	C1, C2, G1, G2, HT2, HS2, J1-1, J1-2, J2-1, J2-2, LH2, MA1, MA2, HM2, N1, N2, P2, SP2, T1, T2
9.68	455.1515 $[M + Na]^+$ 471.1245 $[M + K]^+$	431.1562 $[M\text{-}H]^-$ 477.1599 $[M\text{-}H + 46]^-$	$C_{19}H_{28}O_{11}$	*Benzyl alcohol dihexose I*	C1, C2, G1, G2, HM1, HM2, HT1, HT2, HS1, HS2, J1-1, J1-2, J2-1, J2-2, LH1, LH2, N1, N2, P1, P2, SP1, SP2, T1, T2
9.90	169.0440 $[M + H]^+$ 151.0374 $[M + H\text{-}18]^+$	167.0348 $[M\text{-}H]^-$	$C_8H_8O_4$	*Vanillic acid**	SP2
10.27	629.2398 $[M + Na]^+$ 6452167 $[M + K]^+$	605.2458 $[M\text{-}H]^-$ 651.2495 $[M\text{-}H + 46]^-$ 443.1903 $[M\text{-}H\text{-}162]^-$	$C_{27}H_{42}O_{15}$	*Dehydrophaseic acid dihexose*	C1, C2, G1, G2, HM1, HM2, HT2, HS1, HS2, J1-1, J1-2, J2-1, J2-2, LH1, LH2, MA1, MA2, N1, N2, P1, P2, SP1, SP2, T1, T2
11.11	181.0498 $[M + H]^+$ 163.0376 $[M + H\text{-}18]^+$	179.0345 $[M\text{-}H]^-$	$C_9H_8O_4$	*Caffeic acid**	HT2, LH1, LH2, N1, P2, SP1, SP2, T2

continued

TABLE 8.2 (continued)
Peak Assignment of the Metabolites Present in Different Avocado Samples Using UHPLC–ESI–TOF MS

Retention Time (min)[a]	ESI(+) TOF MS[b]	ESI(-) TOF MS[b]	Molecular Formula [M] Generated	Assignment[c]	Avocado Samples Containing the Compound[d]
11.45	291.0844 $[M+H]^+$ 313.2735 $[M+Na]^+$ 329.0399 $[M+K]^+$ 165.0528 $[M+H-126]^+$	289.0718 $[M-H]^-$	$C_{15}H_{14}O_6$	*Catechin**	G1, C2, HM1, HM2, HT1, HS1, J1-1, J1-2, J2-1, LH1, LH2, MA1, MA2, P1, P2, SP1, SP2, T1, T2
12.32	467.1870 $[M+Na]^+$ 483.1616 $[M+K]^+$ 265.1412 $[M+H-162-18]^-$	443.1914 $[M-H]^-$ 489.1965 $[M-H+46]^-$ 281.1374 $[M-H-162]^-$	$C_{21}H_{32}O_{10}$	*Dehydrophaseic acid hexose I*	C1, C2, G1, G2, HM2, HT2, J1-1, J1-2, LH2, MA1, N1, N2, P2, SP2, T2
12.69	349.0878 $[M+Na]^+$ 365.0542 $[M+K]^+$	325.0930 $[M-H]^-$ 651.1945 $[2M-H]^-$ 163.0407 $[M-H-162]^-$ 145.0305 $[M-H-162-18]^-$ 117.0361 $[M-H-162-18-28]^-$	$C_{15}H_{18}O_8$	*Coumaroyl-hexose/ coumaric acid hexose II*	C1, C2, G2, HM1, HM2, HT2, HS2, J1-1, J1-2, LH2, MA1, MA2, N2, P2, SP1, SP2, T2
12.71	355.1019 $[M+H]^+$ 377.0824 $[M+Na]^+$ 393.0578 $[M+K]^+$ 163.0377 $[M+H-192]^+$	353.0885 $[M-H]^-$ 191.0340 $[M-H-162^e]^-$	$C_{16}H_{18}O_9$	*Chlorogenic acid**	C2, G2, HS1, J1-1, J2-1, LH1, LH2, P1, P2, SP1, SP2, T1
13.06	153.0464 $[M+H]^+$ 125.0575 $[M+H-18]^+$	151.0399 $[M-H]^-$	$C_8H_8O_4$	*Vanillin**	C2, SP2
13.65	467.1863 $[M+Na]^+$ 4831562 $[M+K]^+$ 265.1420 $[M+H-162-18]^-$	443.1915 $[M-H]^-$	$C_{21}H_{32}O_{10}$	*Dehydrophaseic acid hexose II*	ALL

13.87	349.0803 [M + Na]$^+$ 365.0596 [M + K]$^+$	325.0939 [M-H]$^-$ 145.0305[M-H-162-18]$^-$	$C_{15}H_{18}O_8$	*Coumaroyl-hexose/ coumaric acid hexose II*	C1, C2, G2, HM1, HM2, HT2, HS2, J1-1, J1-2, LH2, MA2, N2, P2, SP2, T2
13.88	199.0536 [M + H]$^+$ 221.0442 [M + Na]$^+$ 237.0041 [M + K]$^+$ 181.0480 [M + H-18]$^+$ 155.0663 [M + H-44]$^+$	197.0460 [M-H]$^-$ 182.1967 [M-13]$^-$	$C_9H_{10}O_5$	*Syringic acid**	C2, N1, N2, SP2
14.00	455.1501 [M + Na]$^+$ 471.1248 [M + K]$^+$	431.1555 [M-H]$^-$	$C_{19}H_{28}O_{11}$	*Benzyl alcohol dihexose II*	C1, C2, G1, G2, HM1, HM2, HT1, HT2, HS1, HS2, J1-1, J1-2, J2-1, J2-2, LH1, LH2, N1, N2, P1, P2, SP1, SP2, T1, T2
14.32	451.1878 [M + H]$^+$ 473.1091 [M + Na]$^+$	449.1109 [M-H]$^-$	$C_{21}H_{22}O_{11}$	*Eriodictyol hexose/ Macrocarposide/ Okanin hexose . . .*	C1, C2, G2, HM1, HM2, HT1, HT2, HS1, HS2, J1-1, J1-2, J2-1, J2-2, LH1, LH2, MA1, MA2, N1, N2, P2, SP1, SP2, T1, T2
15.53	165.0478 [M + H]$^+$ 187.1623 [M + Na]$^+$ 203.0073 [M + K]$^+$ 147.0514 [M + H-18]$^+$ 119.0440 [M + H-18-28]$^+$	163.0401 [M-H]$^-$ 119.0529 [M-H-44]$^-$	$C_9H_8O_3$	*p-Coumaric acid**	C1, C2, G2, HM2,HT2, HS2, J1-2, J2-2, LH2, MA2, N2, P2, SP2, T2
16.10	379.1032 [M + Na]$^+$ 395.0710 [M + K]$^+$ 195.0669 [M + H-162]$^+$ 177.0474 [M + H-162-18]$^+$	355.1029 [M-H]$^-$	$C_{16}H_{20}O_9$	*Ferulic acid hexose I*	C1, C2, G2, HM2, HT2, HS2, J1-1, J1-2, J2-2, LH2, MA2, N2, P2, SP1, SP2, T2
16.14	409.1102 [M + Na]$^+$ 425.0838 [M + K]$^+$	385.1128 [M-H]$^-$	$C_{17}H_{22}O_{10}$	*Sinapoyl-hexose/sinapic acid hexose I*	C1, C2, G1, G2, HT1, J1-2, LH2, MA1, MA2, N1, N2, P2, SP1, SP2, T1
16.37	120.0798 [M + H-88]$^+$	206.0821 [M-H]$^-$ 164.0719 [M-H-42]$^-$ 147.0495[M-H-42-17]$^-$	$C_{11}H_{13}NO_3$	*N-acetylphenylalanine*	ALL

continued

TABLE 8.2 (continued)
Peak Assignment of the Metabolites Present in Different Avocado Samples Using UHPLC–ESI–TOF MS

Retention Time (min)[a]	ESI(+) TOF MS[b]	ESI(-) TOF MS[b]	Molecular Formula [M] Generated	Assignment[c]	Avocado Samples Containing the Compound[d]
16.89	425.1405 $[M+Na]^+$ 441.1135 $[M+K]^+$	401.1453 $[M-H]^-$ 447.1534 $[M-H+46]^-$	$C_{18}H_{26}O_{10}$	*Benzyl alcohol hexose-pentose*	C1, C2, G1, G2, HM1, HM2, HT1, HT2, HS1, HS2, J1-1, J1-2, J2-1, J2-2, LH1, LH2, MA2, N1, N2, P1, P2, SP1, SP2, T1, T2
17.21	305.1293 $[M+Na]^+$ 265.1405 $[M+H-18]^+$	281.1392 $[M-H]^-$ 237.1489 $[M-H-44]^-$	$C_{15}H_{22}O_5$	*Dehydrophaseic acid*	C1, C2, G1, G2, HM1, HM2, HT1, HT2, HS2, J1-1, J1-2, J2-1, J2-2, LH1, MA1, N1, N2, P1, P2, SP1, SP2, T1, T2
17.28	379.0988 $[M+Na]^+$ 395.0742 $[M+K]^+$ 195.0622 $[M+H-162]^+$ 177.540 $[M+H-162-18]^+$	355.1048 $[M-H]^-$	$C_{16}H_{20}O_9$	*Ferulic acid hexose II*	C1, C2, G2, HM2, HT2, HS2, J1-1, J1-2, J2-2, LH2, MA2, N2, P2, SP2, T2
17.63	123.0441 $[M+H]^+$	121.0291 $[M-H]^-$	$C_7H_6O_2$	*Benzoic acid**	C1, C2, G1, HM1, HM2, HT1, HS1, HS2, J1-1, J1-2, J2-1, LH1, LH2, N1, P1, P2, SP1, SP2, T1, T2
17.70	291.0850 $[M+H]^+$ 313.0669 $[M+Na]^+$ 329.0386 $[M+K]^+$ 165.0565 $[M+H-126]^+$	289.0714 $[M-H]^-$	$C_{15}H_{14}O_6$	*Epicatechin**	ALL
17.93	495.1401 $[M+Na]^+$ 511.1279 $[M+K]^+$ 147.0677 $[M+H-308-18]^+$	471.1514 $[M-H]^-$ 943.4374 $[2M-H]^-$	$C_{21}H_{28}O_{12}$	*o-Coumaric acid rhamnosyl-hexose/o-coumaric acid coumaroyl-hexose I*	C1, C2, G2, HM1, HM2, HT1, HT2, HS2, J1-1, J1-2, J2-2, LH2, MA1, MA2, N1, N2, P2, SP1, SP2, T1, T2
18.35	409.1677 $[M+Na]^+$ 425.1314 $[M+K]^+$	385.1145 $[M-H]^-$	$C_{17}H_{22}O_{10}$	*Sinapoyl-hexose/sinapic acid hexose II*	C1, C2, G1, G2, HM1, HM2, HT1, HT2, HS1, HS2, J1-1, J1-2, J2-1, J2-2, LH2, MA1, MA2, P2, SP1, SP2, T1, T2

18.40	473.9953 [M + H]$^+$ 495.1450 [M + Na]$^+$ 511.1250 [M + K]$^+$ 165.0505 [M + H-308]$^+$ 147.0431 [M + H-308-18]$^+$	471.1496 [M-H]$^-$ 943.3266 [2M-H]$^-$ 163.0371 [M-H-308]$^-$	$C_{21}H_{28}O_{12}$	*o-Coumaric acid rhamnosyl-hexose/o-coumaric acid coumaroyl-hexose II*	C1, C2, G1, G2, HT1, HT2, HM2, HS2, J1-2, J2-1, J2-2, LH2, MA2, N2, P2, SP2, T2
19.38	409.1114 [M + Na]$^+$ 425.0780 [M + K]$^+$	385.1131 [M-H]$^-$	$C_{17}H_{22}O_{10}$	*Sinapoyl-hexose/sinapic acid hexose III*	ALL
20.61	195.0943 [M-H]$^+$ 233.0067 [M + K]$^+$ 177.0532 [M + H-18]$^+$ 145.0247 [M + H-50]$^+$	193.0499 [M-H]$^-$ 178.0231 [M-H-15]$^-$	$C_{10}H_{10}O_4$	*Ferulic acid**	C1, G2, HM2, HT2, HS2, J1-1, J1-2, J2-2, LH2, MA2, N2, P2, SP2, T2
22.98	413.1155 [M + H]$^+$ 435.0879 [M + Na]$^+$ 451.0594 [M + K]$^+$ 395.0952 [M + H-18]$^+$ 165.0535 [M + H-162-86]$^+$ 147.0445 [M + H-86-162-18]$^+$	411.0922 [M-H]$^-$ 823.1929 [2M-H]$^-$ 367.1029 [M-H-44]$^-$ 205.0470[M-H-44-162]$^-$ 163.0390[M-H-162-86]$^-$	$C_{18}H_{20}O_{11}$	*Coumaric acid malonyl-hexose I*	G1, G2, LH2, N1, N2, P2
23.03	435.0883 [M + H]$^+$	433.0779 [M-H]$^-$	$C_{20}H_{18}O_{11}$	*Qercetin pentose/ hydroxyluteolin pentose/herbacetin pentose...*	G1, G2, HM1, HM2, HT1, HT2, HS1, J1-1, J2-1, MA1, MA2, N1, N2, P1, P2, SP1, SP2, T1, T2
23.31	225.1450 [M + H]$^+$ 207.0645 [M + H-18-32]$^+$ 147.0469 [M + H-18-60]$^+$ 119.0470 [M + H-78-28]$^+$	223.0608 [M-H]$^-$	$C_{11}H_{12}O_5$	*Sinapic acid**	G2, HM2, SP2
24.07	435.0881 [M + Na]$^+$ 451.0622 [M + K]$^+$ 395.0952 [M + H-18]$^+$,	411.0911 [M-H]$^-$ 823.2973 [2M-H]$^-$ 367.1023 [M-H-44]-	$C_{18}H_{20}O_{11}$	*Coumaric acid malonyl-hexose II*	C2, G2, HM2, HT2, HS2, J1-2, J2-2, LH2, MA2, N2, SP2, T2

continued

TABLE 8.2 (continued)
Peak Assignment of the Metabolites Present in Different Avocado Samples Using UHPLC–ESI–TOF MS

Retention Time (min)[a]	ESI(+) TOF MS[b]	ESI(-) TOF MS[b]	Molecular Formula [M] Generated	Assignment[c]	Avocado Samples Containing the Compound[d]
25.87	449.1767 [M + Na]$^+$	425.2253 [M-H]$^-$ 263.1362 [M-H-162]$^-$	$C_{21}H_{30}O_9$	*Abscisic acid hexose ester*	C1, C2, G1, G2, HT1, HT2, HS2, J1-1, J1-2, J2-2, LH1, LH2, MA1, MA2, N2, P1, P2, SP1, SP2, T2
26.39	-	287.0556 [M-H]$^-$	$C_{15}H_{12}O_6$	*Eriodictyol/ dihydrokaempferol/ tetrahydroxyflavanone*	ALL
29.37	201.1263 [M + Na]$^+$	177.0549 [M-H]$^-$ 533.2239 [3M-H]$^-$	$C_{10}H_{10}O_3$	*Methoxycinnamic acid*	C1, C2, G2, HT2, HM1, HM2, HS1, HS2, J1-1, J1-2, J2-1, J2-2, LH2, MA2, N2, SP2, T2
29.75	265.1403 [M + H]$^+$ 287.1235 [M + Na]$^+$ 303.0952 [M + K]$^+$ 247.1306 [M + H-18]$^+$ 229.1199 [M + H-18-18]$^+$ 201.1271 [247-46]$^+$ 187.1048 [247-60]$^+$ 163.0769 [M + H-102]$^+$ 135.0875 [163-28]$^+$	263.1286 [M-H]$^-$ 219.1409 [M-H-44]$^-$ 153.0931[M-H-44-66]$^-$	$C_{15}H_{20}O_4$	*Abscisic acid**	ALL
31.34	333.0587 [M + H]$^+$	331.0452 [M-H]$^-$	$C_{16}H_{12}O_8$	*Laricitrin**	C2, P2, SP2
32.08	273.0745 [M + H]$^+$	271.0602 [M-H]$^-$	$C_{15}H_{12}O_5$	*Naringenin**	C2, P2, SP2
36.25	327.2686 [M + Na]$^+$ 343.1723 [M + K]$^+$	303.2178 [M-H]$^-$	$C_{15}H_{12}O_7$	*Trihydroxypalmitic acid*	C1, C2, G1, G2, HM1, HM2, HS1, HS2, J1-1, J1-2, J2-1, J2-2, LH1, LH2, MA1, MA2, N1, N2, P1, P2, SP1, SP2, T1, T2

36.44	353.2268 $[M + Na]^+$	329.2322 $[M-H]^-$	$C_{18}H_{34}O_5$	*Trihydroxyoctadecenoic acid*	ALL
38.37	255.0634 $[M + H]^+$	253.0499 $[M-H]^-$	$C_{15}H_{10}O_4$	*Chrysin**	C2, P2, SP2
39.78	301.0688 $[M + H]^+$	299.0553 $[M-H]^-$	$C_{16}H_{12}O_6$	*Kaempferide**	C2, P2, SP2
45.39	-	295.2272 $[M-H]^-$	$C_{18}H_{32}O_3$	*Hydroxylinoleic acid/ hydroxyoctadecadienoic acid I...*	ALL
46.49	-	295.2275 $[M-H]^-$	$C_{18}H_{32}O_3$	*Hydroxylinoleic acid/ hydroxyoctadecadienoic acid II...*	C1, C2, G1, G2, HM2, HT2, HS1, HS2, J1-1, J1-2, J2-1,LH1, LH2, MA1, MA2, N2, P1, P2, SP1, SP2, T2
46.94	277.2454 $[M + H]^+$ 299.2779 $[M + Na]^+$ <u>315.2838</u> $[M + K]^+$	275.2373 $[M-H]^-$	$C_{19}H_{32}O$	*2-(Pentadecyl)-furan*	C1, C2, G1, G2, HM1, HM2, HT1, HT2, HS1, HS2, J1-1, J1-2, J2-1, J2-2, LH2, MA1, MA2, N1, N2, P1, P2, SP1, SP2, T1, T2
47.14	-	277.2173 $[M-H]^-$	$C_{18}H_{30}O_2$	*Linolenic acid*	G2, MA2

Source: Adapted from Hurtado-Fernández, E. et al. 2011. *J. Chromatogr. A* 1218:7723–7738. With permission.

[a] Average value. RSD of 7% for the retention time.

[b] For many compounds, different *m/z* values rather than $[M + H]^+/[M-H]^-$ have been detected in the MS spectra; when those ions were more intense than the corresponding $[M + H]^+/[M-H]^-$, they have been underlined. The mentioned different *m/z* values mainly correspond to in-source fragments (typical losses detected have been -17 (NH_3), -18 (H_2O), -28 (CO), -42 (C_2H_2O), -43 (CHNO), -44 (CO_2), -132 (pentose), -162 (hexose)) and to sodium $[M + 23]^+$ and potassium $[M + 49]^+$ or formic acid $[M-H + 46]^-$ adducts, in the positive and negative polarities, respectively.

[c] (I, II) different isomers; (*) identification confirmed by comparison with authentic standards.

[d] C: Colin V33, G: Gem, HM: Hass Motril, HT: Harvest, HS: Hass, J1: Jimenez1, J2: Jimenez 2, LH: Lamb Hass, MA: Marvel, N: Nobel, P: Pinkerton, SP: Sir Prize, T: Tacambaro. Numbers 1 and 2 correspond to the 1st or 2nd ripening degree, respectively.

[e] In this case, loss of 162 does not correspond to a hexose, but to the caffeic acid moiety ($M-H_2O$).

In all the cases, the values obtained for error were lower than 5.3 ppm and the mSigma values ranged from 2.8 to 50.1, considering the *m/z* for ESI(-) TOF.

Prize at the second stage of ripeness, and in Nobel at both. Ripe varieties of Gem, Hass Motril, and Sir Prize were the only ones where sinapic acid was found, with Colin V 33 and Sir Prize being the only varieties that had vanillin. Gallic and vanillic acids were exclusively present in ripe Sir Prize avocado, which turned out to be the richest sample in terms of the number of compounds.

It is important to note that even though there are a wide number of publications that make reference to the determination of some metabolites in avocado fruit using diverse analytical techniques, the only global approach is the one described by Hurtado and coworkers [73]. The other manuscripts, in general, studied a very restricted number of compounds (usually belonging to the same family) or made a global estimation of the content of a chemical group of compounds, and compared their levels with their concentrations in other fruits or vegetables [53,72].

8.2.2 Quantification

Following the typical workflow in metabolomics (Figure 8.2), after the identification of the compounds, the validation of the method is necessary for carrying out the subsequent quantification.

In the manuscript that we are describing in depth in this chapter [73], repeatability, reproducibility, and accuracy were checked, and the calibration curves were established, allowing the calculation of the detection and quantification limits. RSD values for repeatability were lower than 7.01%; accuracy values oscillated between 97.2% and 102.0%; and limits of detection were low, ranging from 1.64 to 730.54 ppb (negative polarity) and from 0.51 to 310.23 ppb (positive polarity). Neither matrix effect nor ion suppression were detected. The authors proved that the method developed for UHPLC–UV–ESI–TOF MS was a very valuable tool, because it was able to determine a wide number of metabolites in a single run and it was reliable from an analytical point of view. For this reason, it was also applied for the quantitative analysis of the 26 samples under study.

The quantitative results obtained in that work for each individual compound are presented in Table 8.3. A total of 20 metabolites were quantified in the 13 avocado varieties. To the best of our knowledge, these are the first quantitative results for abscisic and pantothenic acids.

Quinic, succinic, and pantothenic acids were determined in a wide range of concentrations in the 26 samples under study. Unripe Colin V 33 was the richest variety in quinic acid (18.95 mg/kg); however, the highest amounts of succinic and pantothenic acids were observed for ripe fruits of Colin V 33 (176.88 mg/kg) and Sir Prize (6.14 mg/kg), respectively. It was possible to quantify epicatechin in 24 samples, being Sir Prize at 1st ripening degree (26.72 mg/kg) the richest one. *p*-Coumaric acid was found at the 2nd stage of ripeness in all the samples, ranging from 5.39 mg/kg (for Pinkerton) to 33.78 mg/kg (for Harvest). Among the phenolic acids quantified, caffeic acid (only detected in five samples) is the one with the lowest concentration, not exceeding 0.88 mg/kg.

Table 8.4 shows an estimation of the total content (mg/kg dry sample) of each family of compounds for all the avocado fruit samples, and Figure 8.4 shows a graphical representation of total contents for four selected varieties. Paying attention to these

TABLE 8.3
Quantitative Results (mg of Analyte/Kg of Dry Sample) of the Different Avocado Cultivars Analyzed by UHPLC–ESI(-)–TOF MS Method, Structured in Four Separated Tables: Organic Acids, Flavonoids, Phenolic Acids and Vitamins

Avocado Samples		Organic Acids			
		Abscisic	Benzoic*	Quinic	Succinic
Colin V 33	1st	4.24	n.d	18.95	171.09
	2nd	3.62	3.93	12.92	176.88
Gem	1st	0.96	1.20	1.61	37.06
	2nd	2.66	n.d	3.09	60.83
Harvest	1st	1.67	1.24	11.64	24.31
	2nd	9.94	n.d	10.42	161.21
Hass	1st	n.q	9.85	1.25	25.42
	2nd	2.62	1.14	1.80	47.29
Jiménez 1	1st	1.00	4.11	2.21	16.18
	2nd	7.03	2.46	2.35	57.60
Jiménez 2	1st	1.27	1.75	1.67	45.17
	2nd	2.48	n.d	1.68	77.20
Lamb Hass	1st	5.68	5.09	8.11	29.87
	2nd	10.49	6.01	13.90	97.34
Marvel	1st	0.56	n.d	3.25	38.43
	2nd	4.94	n.d	0.75	117.64
Hass Motril	1st	0.47	1.48	0.70	27.60
	2nd	2.25	0.61	1.35	81.51
Nobel	1st	n.q	0.84	1.25	23.50
	2nd	2.90	n.d	1.10	101.06
Pinkerton	1st	0.92	6.66	13.71	27.34
	2nd	2.00	3.96	17.11	66.70
Sir Prize	1st	2.81	15.78	2.73	33.85
	2nd	6.74	8.31	1.58	166.74
Tacambaro	1st	0.63	8.83	2.44	29.43
	2nd	3.09	4.34	2.00	85.12

continued

TABLE 8.3 (continued)

Quantitative Results (mg of Analyte/Kg of Dry Sample) of the Different Avocado Cultivars Analyzed by UHPLC–ESI(-)–TOF MS Method, Structured in Four Separated Tables: Organic Acids, Flavonoids, Phenolic Acids and Vitamins

Avocado Samples		Flavonoids					
		Catechin	Chrysin	Epicatechin	Kaempferide	Laricitrin	Naringenin
Colin V 33	1st	n.d	n.d	0.86	n.d	n.d	n.d
	2nd	n.q	4.52	n.q	4.69	3.73	4.32
Gem	1st	0.15	n.d	2.60	n.d	n.d	n.d
	2nd	n.d	n.d	0.28	n.d	n.d	n.d
Harvest	1st	0.55	n.d	0.95	n.d	n.d	n.d
	2nd	n.d	n.d	0.60	n.d	n.d	n.d
Hass	1st	0.37	n.d	16.55	n.d	n.d	n.d
	2nd	n.d	n.d	1.85	n.d	n.d	n.d
Jiménez 1	1st	0.34	n.d	8.80	n.d	n.d	n.d
	2nd	0.15	n.d	16.99	n.d	n.d	n.d
Jiménez 2	1st	0.09	n.d	4.05	n.d	n.d	n.d
	2nd	n.d	n.d	0.39	n.d	n.d	n.d
Lamb Hass	1st	n.q	n.d	8.81	n.d	n.d	n.d
	2nd	0.37	n.d	15.35	n.d	n.d	n.d
Marvel	1st	n.q	n.d	0.43	n.d	n.d	n.d
	2nd	n.q	n.d	0.20	n.d	n.d	n.d
Hass Motril	1st	0.12	n.d	3.95	n.d	n.d	n.d
	2nd	0.02	n.d	4.59	n.d	n.d	n.d
Nobel	1st	n.d	n.d	0.23	n.d	n.d	n.d
	2nd	n.d	n.d	n.d	n.d	n.d	n.d
Pinkerton	1st	0.79	n.d	10.90	n.d	n.d	n.d
	2nd	0.43	3.18	7.56	3.32	2.99	3.25
Sir Prize	1st	0.62	n.d	26.72	n.d	n.d	n.d
	2nd	0.21	8.93	1.84	9.15	8.61	8.66
Tacambaro	1st	0.40	n.d	15.34	n.d	n.d	n.d
	2nd	0.11	n.d	9.49	n.d	n.d	n.d

Avocado Samples		Phenolic Acids								
		Caffeic	Chlorogenic	Ferulic	Gallic	p-coumaric	Sinapic	Syringic	Vanillic	Vanillin
Colin V 33	1st	n.d	n.d	n.q	n.d	6.58	n.d	n.d	n.d	n.d
	2nd	n.d	n.q	n.d	n.d	5.42	n.d	6.01	n.d	2.10
Gem	1st	n.d	n.d	n.d	n.d	n.d	n.d	n.d	n.d	n.d
	2nd	n.d	n.q	4.02	n.d	33.58	0.67	n.d	n.d	n.d
Harvest	1st	n.d	n.d	n.d	n.d	n.d	n.d	n.d	n.d	n.d
	2nd	0.46	n.d	3.52	n.d	33.78	n.d	n.d	n.d	n.d
Hass	1st	n.d	1.74	n.d	n.d	n.d	n.d	n.d	n.d	n.d
	2nd	n.d	n.d	1.67	n.d	8.39	n.d	n.d	n.d	n.d
Jiménez 1	1st	n.d	0.91	n.d	n.d	n.d	n.d	n.d	n.d	n.d
	2nd	n.d	n.d	1.37	n.d	7.61	n.d	n.d	n.d	n.d
Jiménez 2	1st	n.d	0.77	n.d	n.d	n.d	n.d	n.d	n.d	n.d
	2nd	n.d	n.d	0.57	n.d	7.86	n.d	n.d	n.d	n.d
Lamb Hass	1st	0.88	n.q	n.d	n.d	n.d	n.d	n.d	n.d	n.d
	2nd	0.54	1.67	2.36	n.d	15.48	n.d	n.d	n.d	n.d
Marvel	1st	n.d	n.d	n.d	n.d	n.d	n.d	n.d	n.d	n.d
	2nd	n.d	n.d	4.29		22.95	n.d	n.d	n.d	n.d
Hass Motril	1st	n.d	n.d	n.d	n.d	n.d	n.d	n.d	n.d	n.d
	2nd	n.d	n.d	1.39	n.d	8.12	0.30	n.d	n.d	n.d
Nobel	1st	n.q	n.d	n.d	n.d	n.d	n.d	3.84	n.d	n.d
	2nd	n.d	n.d	0.53	n.d	8.76	n.d	3.12	n.d	n.d
Pinkerton	1st	n.d	2.20	n.d	n.d	n.d	n.d	n.d	n.d	n.d
	2nd	0.58	1.25	0.85	n.d	5.39	n.d	n.d	n.d	n.d
Sir Prize	1st	0.55	1.17	n.d	n.d	n.d	n.d	n.d	n.d	n.d
	2nd	0.57	0.88	4.85	8.61	27.76	0.75	12.17	6.80	6.28
Tacambaro	1st	n.d	n.q	n.d	n.d	n.d	n.d	n.d	n.d	n.d
	2nd	0.27	n.d	1.01	n.d	6.61	n.d	n.d	n.d	n.d

continued

TABLE 8.3 (continued)

Quantitative Results (mg of Analyte/Kg of Dry Sample) of the Different Avocado Cultivars Analyzed by UHPLC–ESI(-)–TOF MS Method, Structured in Four Separated Tables: Organic Acids, Flavonoids, Phenolic Acids and Vitamins

Avocado Samples		Vitamins
		Pantothenic Acid
Colin V 33	1st	4.00
	2nd	4.87
Gem	1st	3.16
	2nd	3.59
Harvest	1st	2.51
	2nd	3.74
Hass	1st	2.33
	2nd	1.90
Jiménez 1	1st	3.34
	2nd	4.37
Jiménez 2	1st	4.56
	2nd	3.59
Lamb Hass	1st	2.29
	2nd	3.60
Marvel	1st	2.06
	2nd	2.82
Hass Motril	1st	2.35
	2nd	2.95
Nobel	1st	3.62
	2nd	4.59
Pinkerton	1st	1.66
	2nd	2.68
Sir Prize	1st	4.08
	2nd	6.14
Tacambaro	1st	2.61
	2nd	3.45

(Value shown = mean value). RSD in all the cases ≤5%.

nd: non detectable; nq: non quantifiable

* Results achieved using the positive polarity (using the $[M + H]^+$ ion)

TABLE 8.4
Estimation of the Total Content (mg/kg Dry Sample) of Each Chemical Family Determined in Avocado Samples by UHPLC–ESI–TOF MS

		Chemical Family			
Avocado Samples		**Organic Acids**	**Flavonoids**	**Phenolic Acids**	**Vitamins**
Colin V 33	1st	194.28	0.86	6.58	4.00
	2nd	197.35	17.26	13.53	4.87
Gem	1st	40.83	2.75	–	3.16
	2nd	66.58	0.28	38.27	3.59
Harvest	1st	38.86	1.50	–	2.51
	2nd	181.57	0.60	37.76	3.74
Hass	1st	36.52	16.92	1.74	2.33
	2nd	52.85	1.85	10.06	1.90
Jiménez 1	1st	23.50	9.14	0.91	3.34
	2nd	69.44	17.14	8.98	4.37
Jiménez 2	1st	49.86	4.14	0.77	4.56
	2nd	81.36	0.39	8.43	3.59
Lamb Hass	1st	48.75	8.81	0.88	2.29
	2nd	127.74	15.72	20.05	3.60
Marvel	1st	42.24	0.43	–	2.06
	2nd	123.33	0.20	27.24	2.82
Hass Motril	1st	30.25	4.07	–	2.35
	2nd	85.72	4.61	9.81	2.95
Nobel	1st	25.59	0.23	3.84	3.62
	2nd	105.06	–	12.41	4.59
Pinkerton	1st	48.63	11.69	2.20	1.66
	2nd	89.77	20.73	8.07	2.68
Sir Prize	1st	55.17	27.34	1.72	4.08
	2nd	183.37	37.40	68.67	6.14
Tacambaro	1st	41.33	15.74	–	2.61
	2nd	94.55	9.60	7.89	3.45

Note: Total content of each family was the result of a summation of the amount of every analyte that belongs to this family.

results, the concentration of organic and phenolic acids tends to increase as the fruits ripen; a fact also noticeable for vitamins, except for Hass and Jiménez 2 varieties. However, it is difficult to establish a general behavior for flavonoid content, because six samples showed an increase, while the other seven samples experimented a diminution.

Considering total contents just for fruits ready for consumption, there are remarkable differences among the varieties. Colin V 33 is the richest in organic acid content, followed closely by Sir Prize, which contains the highest amount of the rest of the families (vitamins, phenolic acids, and flavonoids).

It is quite complicated to compare the quantitative results achieved by Hurtado et al. [73] with those previously published by other authors. Two main difficulties

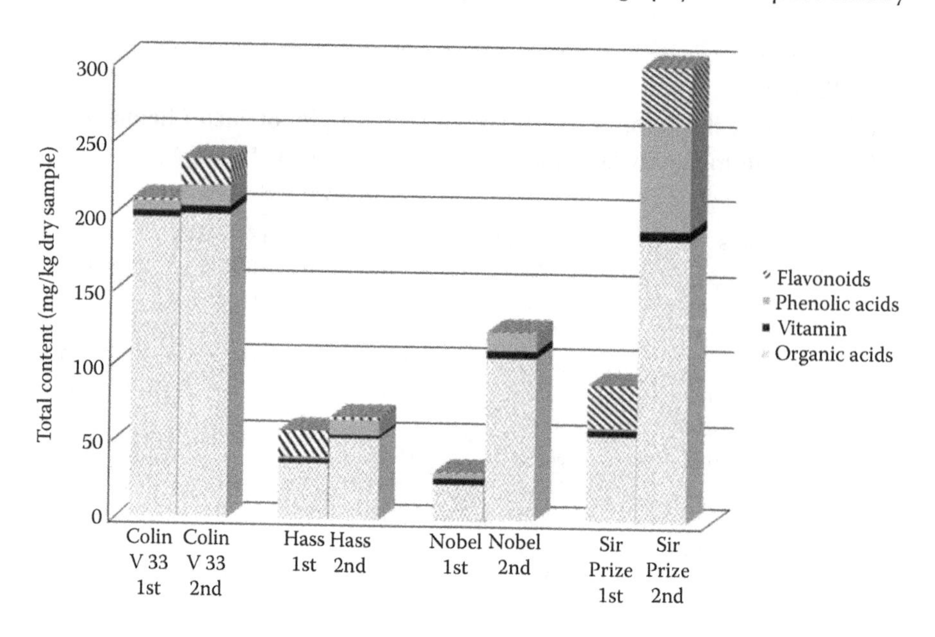

FIGURE 8.4 Estimation of the total content of metabolite families found in some of the avocado cultivars at two different ripening degrees (Hass, Colin V 33, Nobel, and Sir Prize) by UHPLC–ESI–TOF MS.

can be found: the scarce number of publications containing quantitative results (in terms of individual metabolites) and the different approaches used to express the final quantitative results (using the pure standard of a compound to carry out its quantification in the avocado sample or using another analyte). Torres et al. [45] were the first authors studying phenolic acids in avocado mesocarp, identifying 16 and establishing the total content of phenols as gallic acid equivalents. The reason for which is that a direct comparison is not possible. Later on, Golucku and Ozdemir [70] quantified individual phenolic acids in avocado samples. Nevertheless, it is hard to compare both data, since Golucku and Ozdemir used three different avocado varieties and just Hass cultivar is quantified in both studies. Besides, they considered the edible portion weighted to express the results, whereas Hurtado et al. [73] took into account the weight of dry sample. However, some information provided by Golucku and Ozdemir [70] is in strong agreement with that reviewed in this chapter; for instance, the fact that the phenolic composition of avocado fruits shows significant differences depending on the agronomic conditions.

Some other works [36,72] include information about the quantification of phenolic compounds by families and the results are usually expressed as equivalents of catechin or gallic acid, making difficult a proper comparison.

Pantothenic acid, among other vitamins and minerals, was quantified for the Hass variety by Slater et al. [43], but the results were expressed as mg per 100 g of edible pulp, in the same way as for food composition tables.

Apart from the possible discrepancies in the way to express an analytical quantitative result, it is important to bear in mind that the fruit metabolome changes depending on the variety, climatic factors, agricultural practices, and so on; hence, an exhaustive comparison cannot be easily performed unless the conditions are identical or quite similar.

8.3 DATA TREATMENT TO ASSURE AN APPROPRIATE INTERPRETATION OF THE RESULTS

The treatment of multivariate data provided by an instrumental system is included in the field of chemometrics, which is a potent tool that allows grouping or classifying samples that present similar characteristics [94–96]. As was mentioned in the Introduction, chemometrics plays an important role in the *"omics"* technologies in general, and in foodomics in particular, where it has been used to solve different problems within food chemistry.

Among the different chemometric methods, exploratory data analysis and pattern recognition are frequently used in the area of food analysis. Exploratory data analysis is focused on the possible relationships between samples and variables, while pattern recognition studies the behavior between samples and variables [95]. Principal component analysis (PCA) and partial least-squares discriminant analysis (PLS-DA) are the methods most commonly used for exploratory analysis and pattern recognition, respectively. The importance of these statistical tools has been demonstrated by the wide number of works in the field of food science where they have been applied. The majority of the applications are related to the characterization and authentication of olive oil, animal fats, marine and vegetable oils [95], wine [97], fruit juice [98], honey [99], cheese [100,101], and so on, although other important use of statistical tools is the detection of adulterants or frauds [96,102].

In the study developed by Hurtado et al., both PCA and PLS-DA were applied for the classification of avocado samples. The objective went beyond the classification between unripe and ripe avocado fruits. The quantitative results showed that some compounds changed significantly during the ripening process; however, multivariate analysis allowed obtaining a bigger knowledge about the variation of the metabolic profiles during this period, considering all the information of the chromatogram. Hence, the pursued goal was the identification of the metabolites responsible of the discrimination, trying to establish possible ripening markers.

Figure 8.5 shows the score plot of the PCA model and, as it could be expected, the principal source of variability in the data was the ripening degree, whereas the second component is related to the changes among avocado varieties. Samples at second ripening degree were more clustered than unripe fruits. As already commented in Section 8.2, four of the varieties studied were mutants of Hass (Hass Motril, Tacambaro, Jiménez 1, and Jiménez 2); thus, samples from those varieties were closely grouped in the same area of the plot, and also near to the Hass variety sample. It is worth noticing that Colin V 33 and Sir Prize at first ripening degrees were the most separated samples in the first principal component, highlighting the remarkable differences between their metabolic profiles at both stages of ripeness, compared to the rest of the avocado samples.

A cross-validated score plot of PLS-DA model (Q2 [cum] 0.803, R2Y [cum] 0.935) is represented in Figure 8.6. It was built to confirm the metabolic variations

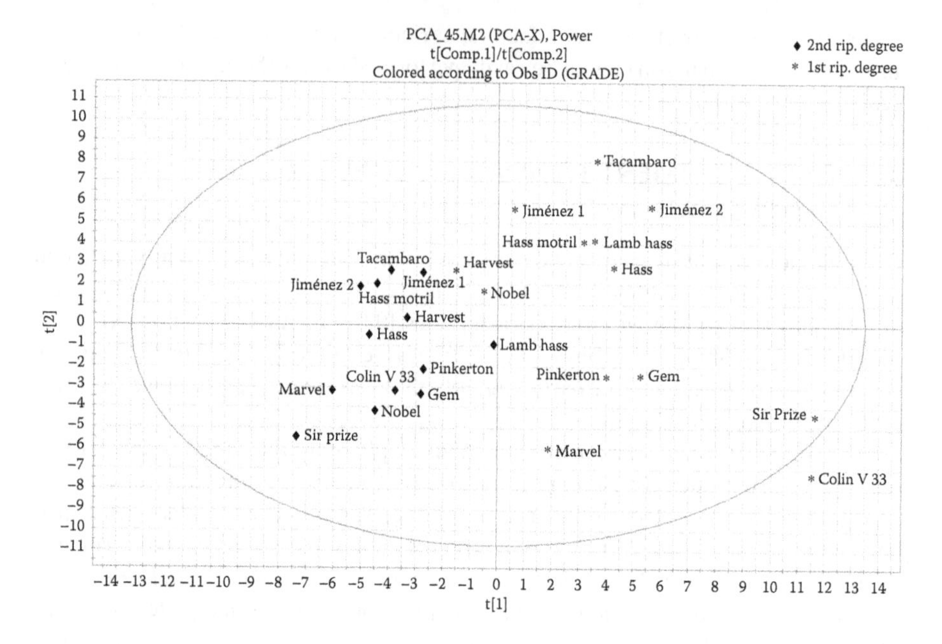

FIGURE 8.5 PCA modeling of UHPLC–ESI–TOF MS data. Samples are colored according to ripening degree. (Adapted from Hurtado-Fernández, E. 2011. *J. Chromatogr. A* 1218:7723–7738. With permission.)

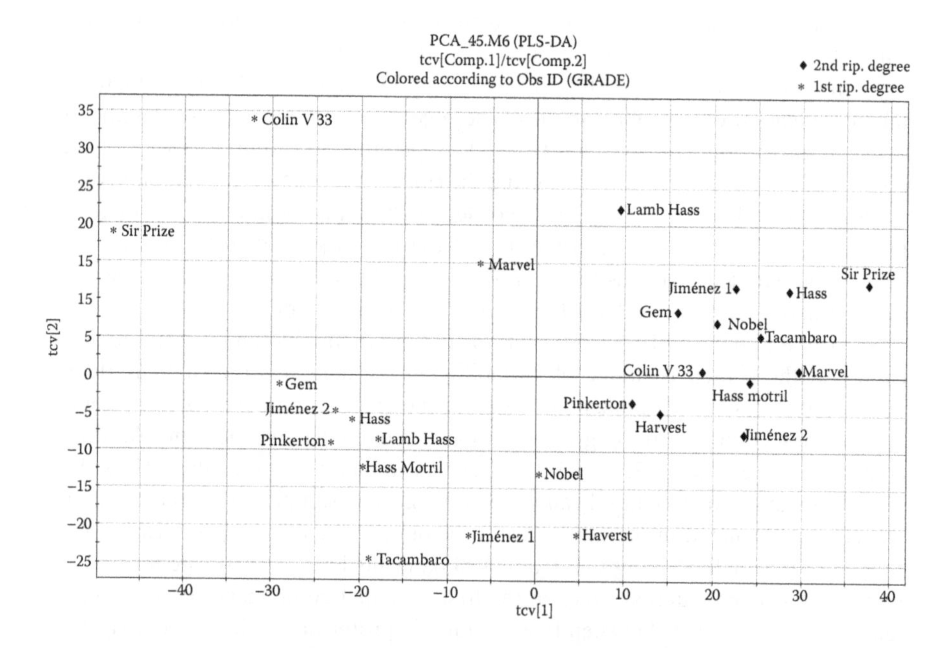

FIGURE 8.6 PLS-DA modeling of UHPLC–ESI–TOF MS data. Samples are colored according to ripening degree. (Adapted from Hurtado-Fernández, E. 2011. *J. Chromatogr. A* 1218:7723–7738. With permission.)

related to ripeness employing the ripening degree as class characteristic. A similar classification to the one obtained with PCA is observed. Samples ready for consumption appeared more clustered than those just harvested, and Sir Prize and Colin V 33 were again the most distant samples from the group. The VIP scores were used to select the most influential features. It was possible to provide a tentative identification for the features that allowed the discrimination (with a VIP score lower than 1.5), taking advantage of the TOF capabilities (high mass accuracy, precise isotopic distribution, etc.), which is included in Table 8.5.

8.4 COMPLEMENTARITY OF TECHNIQUES

The coupling of UHPLC and MS has demonstrated to be a very powerful technique for the identification and quantification of a big number of metabolites in avocado fruits. However, nowadays the use of a single analytical platform does not allow the analysis of each and every one of the metabolites present in a complex matrix, due to the numerous compounds that it could contain which besides present diverse physical and chemical properties. For this reason, the application of different analytical techniques is recommendable to achieve a deep and global knowledge about the whole metabolome of a fruit or vegetable, taking advantage of the strengths of each technique [103]. Vibrational spectroscopies, such as IR (infrared), provide a detailed and reproducible spectrum that will reflect the molecular composition of the analyzed sample, but assignment of IR bands to individual compounds is not normally possible. UV and photodiode array detector (PDA) also provide reproducible and characteristic spectra, but only chromophore-bearing compounds are detectable. The UV spectrum gives useful structural information, but often indicates the metabolite class rather than its exact identity. The requirement of unambiguous annotation essentially limits the metabolic profiling to nuclear magnetic resonance (NMR) spectroscopy and MS-based approaches. High-resolution nuclear magnetic resonance (HR-NMR) spectroscopy allows the detection of a wide range of structurally diverse metabolites simultaneously. It is a nondestructive and inherently quantitative technique that requires relatively simple sample preparation and fast data acquisition per analysis. Other advantages of NMR include robustness and high reproducibility. Its major drawback is its poor sensitivity compared to MS. MS-based techniques have a wide dynamic range and high selectivity and sensitivity but require molecules to be ionized (charged). The largest disadvantage of MS in comparison with NMR is its lower reproducibility. MS-based approaches usually require extensive sample preparation, which can cause metabolite losses, and depending on the ionization technique, metabolites may be differentially ionized. Combination with a separation technique reduces the complexity of the mass spectra and provides additional information on the physico-chemical properties due to the time dimension added [104,105]. Despite good separation, highly complex extracts can still be prone to significant ion suppression/matrix effects. This, together with variable ionization frequencies, makes MS-based quantification more difficult and totally dependent on available reference standards.

In this context, GC–MS can provide a comprehensive analysis of a large variety of metabolites, but requires them to be volatile. This requirement is readily accomplished by chemical derivatization, but at the cost of additional time, processing,

TABLE 8.5
m/z Values and Retention Times of the Possible Classifiers (Decreasing Order of VIP Score) of Avocado Variety and Ripening Degree Obtained by Statistical Analysis (PLS-DA)

Retention Time (min)	Formula	Classifiers Signal[a]		Assignment
		Signal	Other Signals	
29.68	$C_{14}H_{24}O_6$	+ 311.1449 [M + Na]$^+$	211 [M + H-18-60]$^+$, 427 [M + K]$^+$,193 [M + H-78-18]$^+$, 175 [M + H-78-18-18]$^+$, 157 [M + H-78-18-18-18]$^+$, 111 [157-46]$^+$	*Triethyl pentane-1,3,5-tricarboxylate/Bis(2-(hydroxymethyl)butyl) 2-butenedioate/D-Galactono-1,4-lactone, 5,6-O-octylidene*
		− 287.1493	269 [M-H-18]$^-$, 227 [M-H-60 (18 + 42)]$^-$, 209 [M-H-60-18]$^-$, 185 [M-H-60-42]$^-$	
7.47	$C_{16}H_{30}O_{11}$	+ 437.1404 [M + K]$^+$	421 [M + Na]$^+$, 399 [M + H]$^+$	*Butyl 4-O-β-D-galactopyranosyl-β-D-glucopyranoside/tert-butyl 4-O-β-D-galactopyranosyl-β-D-glucopyranoside/3,6,9,12,15,18,21-heptaoxatricosane-1,23-dioic acid*
		− 397.1715	443 [M-H + 46]$^-$	
17.48	$C_{19}H_{30}O_{10}$	+ 457.1471 [M + K]$^+$	419 [M + H]$^+$, 441 [M + Na]$^+$	*9-[3-(3,6,6a-trihydroxy-5-oxo-3,3a-dihydro-2H-furo[3,2-b] furan-6-yl)-5-hydroxyoxolan-2-yl]nonanoic acid*
		− 417.1787	463 [M-H + 46]$^-$	
1.91	$C_{10}H_{13}N_5O_4$	+ 268.1040 [M + H]$^+$	136 [268-132]$^+$	*Adenosine*
		− 266.0888	533[2M-H]$^-$,312[M-H + 46]$^-$,134[M-H-132]$^-$	
13.23	$C_{19}H_{28}O_{11}$	+ 471.1263 [M + K]$^+$	455 [M + Na]$^+$, 427, 441, 450	*Benzyl alcohol dihexose*
		− 431.1555	-	
24.81	$C_{12}H_{22}O_5$	+ 269.1359 [M + Na]$^+$	309, 285 [M + K]$^+$, 211 [M + H-18-18]$^+$, 193 [M + H-18-18-18]$^+$	*Dibutyl malate/Hydroxydodecanedioic acid/dimethyl 3-hydroxydecanedioate/N-ethoxyacetylglycine...*
		− 245.1388	227[M-H-18]$^-$,209[M-H-18-18]$^-$,185[M-H-18-42]$^-$	
15.96	$C_{20}H_{30}O_{12}$	+ 480.2028	485 [M + Na]$^+$, 501 [M + K]$^+$	*2-(3,4-dihydroxyphenyl)ethyl 3-O-(6-deoxy-α-L-mannopyranosyl)-β-D-glucopyranoside/4-[(α-D-mannopyranosyloxy)methyl]benzyl α-D-mannopyranoside*
		− 461.1653	439/393/283/207	

[a] The MS signals which have been underlined were the most intense in the MS spectra.

and variance. Compared to GC, LC does not require sample volatility, simplifying sample preparation, and is able to detect simultaneously thousands of metabolites. While metabolite separation in GC and LC is based on their interaction with the stationary phases, in CE it is based on their mass-to-charge ratio providing different information on the chemical properties. CE provides sensitivity, separation efficiency, and smaller sample and solvent volumes. The main disadvantages of CE are larger variation in migration time and peak shapes and its technical limitations for the analysis of nonpolar metabolites.

To illustrate the complementarity of hyphenated platforms, such as reversed-phase LC, GC, and CE coupled to UV and MS, Figure 8.7 shows the metabolic profiles obtained for different avocado samples and the chemical classes identified with each one.

Combining the information obtained by HPLC–MS [5], UHPLC–MS [73], CE-MS [81], CE-UV [82], and GC–MS [106] methods (using targeted and/or nontargeted approaches), we obtained multiple levels of information useful for structural elucidation, which enabled the identification and quantification of a comprehensive set of metabolites commonly present in avocado pulp. At the same time that allowed us to assess the relative strengths and weaknesses of the hyphenated platforms concerning the analysis of the avocado extracts.

CE-UV/MS were the least informative platforms in terms of the number of metabolites tentatively identified: all of the compounds assigned in the CE profiles could also be identified using LC or GC. However, polar metabolites are not well retained and separated on reversed-phase columns, making CE a powerful and complementary technique for the analysis of such metabolites in short analysis times. Concerning reproducibility, migration times shifted more in CE-MS than the retention times in UHPLC–MS, with RSD values smaller than 11.45% and 7.01%, respectively. The UHPLC–MS showed smaller LODs in the negative mode for the same phenolic compounds quantified by CE-MS. The smaller sensitivity of CE-MS might be explained considering the dilution of the sample by using the coaxial sheath flow liquid and the smaller injection volume.

As can be seen in Figure 8.7, peaks in the UHPLC profile are sharper and better resolved in a shorter analysis time than in the HPLC one; ion suppression was greatly minimized in this way, which improved the ability to identify and quantify a wider number of compounds/classes of compounds. Thus, it appears that UHPLC may be the preferred choice among the liquid separation techniques for metabolic profiling. GC–MS and UHPLC–MS provided the detection of the largest sets of compounds belonging to a bigger number of metabolite classes. The GC–MS method provided the detection of low-molecular-weight metabolites (including amino and organic acids, phenolic compounds, sterols, or fatty acids, among others) with a boiling point low enough after chemical derivatization to allow their elution from the GC column. The UHPLC–MS reversed-phase method provided the complementary detection of higher-molecular-weight compounds of medium-to-high lipophilicity from the same classes mentioned above, with the exception of sterols. On the contrary, glycosides of phenolic compound could not be detected by GC.

It should be noted that, even though platforms such as GC or UHPLC coupled to MS are able to detect thousands of compounds, they cannot provide a comprehensive

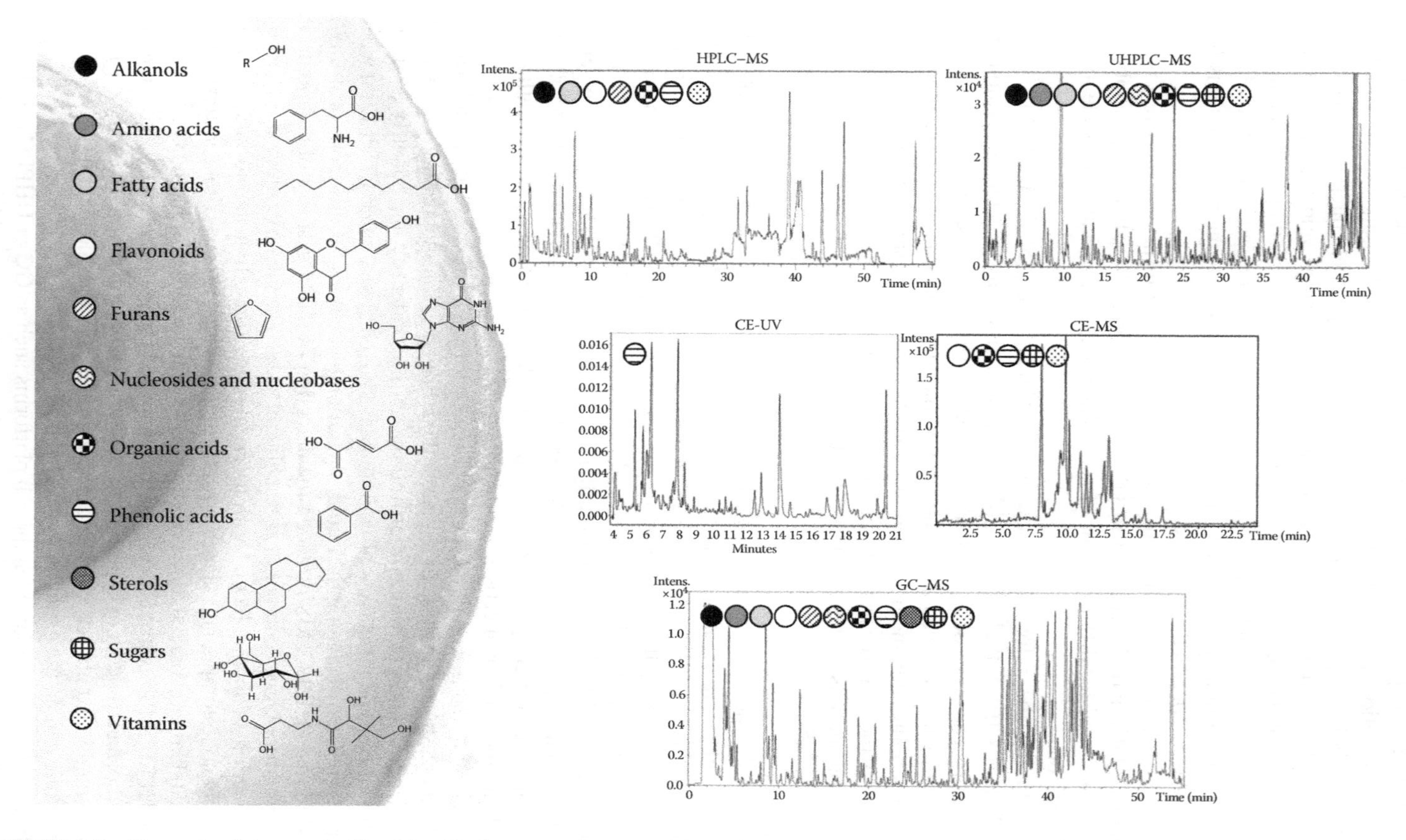

FIGURE 8.7 Example of the metabolites identified in avocado samples by using different analytical techniques.

overview of the whole metabolome when used independently, and, hence, it is becoming common practice to use several analytical platforms to combine the information obtained and meet the demanding requirements of metabonomics.

8.5 CONCLUSIONS

This chapter gives an overview about the composition of *Persea americana* and the analytical techniques that can be used to characterize it, paying particular attention to UHPLC coupled to MS. Since there are very few examples showing a comprehensive characterization of this matrix, the achievements included in the manuscript published by Hurtado et al. in 2011 have been discussed in detail. In that paper, the authors achieved the identification of more than 60 metabolites from different chemical families and the quantification of 20 compounds. In addition, they carried out a comparison by using chemometrics to see how the ripening process influenced the metabolic profile of the fruit.

The last section of the chapter includes a discussion about how the application of different analytical techniques could allow enhancing the coverage of the fruit metabolome, being it impossible for any single platform to obtain information about all the metabolites present in a sample. The mentioned section illustrates the different levels of information obtained for the analysis of avocado fruit by CE-UV, CE-MS, HPLC–MS, UHPLC–MS, and GC–MS (using targeted and/or nontargeted approaches).

Despite its bias against hydrophilic metabolites, overall, and given only one choice, it appears that UHPLC–MS may be the platform of choice for avocado metabolomics studies, due to its ability to determine different kinds of compounds with high sensitivity, reproducibility, and speed.

ABBREVIATIONS

3D-FL	three dimensional fluorescence
AOAC	Association of Official Analytical Chemists
BPC	base peak chromatogram
CE	capillary electrophoresis
DAD	diode array detector
ELSD	evaporative light scattering detector
ESI	electrospray ionization
FID	flame ionization detector
FL	fluorescence detector
FT-IR	Fourier transform infrared spectroscopy
GC	gas chromatography
HA-LAESI	heat-assisted laser ablation electrospray ionization
HR-EI	high-resolution electron impact
HR-FAB	high-resolution fast atom bombardment
HR-NMR	high-resolution nuclear magnetic resonance
ICP	inductively coupled plasma
IR	infrared spectroscopy

LC	liquid chromatography
LSD	light scattering detector
MS	mass spectrometry
NIR	near-infrared spectroscopy
NMR	nuclear magnetic resonance
OES	optical emission spectrometry
PCA	principal component analysis
PDA	photodiode array detector
PLS-DA	partial least-squares discriminant analysis
RI	refractive index
TLC	thin layer chromatography
TOF	time of flight
UHPLC	ultra-high-performance liquid chromatography
USDA	United States Department of Agriculture
UV	ultraviolet detection
UV/Vis	ultraviolet/visible detection
VIP	variable importance in the projection

ACKNOWLEDGMENT

The authors are very grateful to the Andalusia Regional Government (Department of Economy, Innovation and Science, Project P09-FQM-5469), to the University of Granada (pre-doctoral grant) and to the Alfonso Martín Escudero Foundation (post-doctoral contract). They appreciate the support provided by Professor J. I. Hormaza and its research group.

REFERENCES

1. Meyer, M. D., Landahl, S., Donetti, M., and Terry, L. A. 2011. Avocado. In *Health-Promoting Properties of Fruits & Vegetables*, ed. Terry, L. A., CABI International, Wallingford, pp. 27–50, ISBN-13: 978 1 84593 528 3.
2. Alcaraz, M. L. and Hormaza, J. I. 2007. Molecular characterization and genetic diversity in an avocado collection of cultivars and local Spanish genotypes using SSRs. *Hereditas* 144:244–253.
3. Alcaraz, M. L., Montserrat, M., and Hormaza, J. I. 2011. *In vitro* pollen germination in avocado (*Persea americana* Mill.): Optimization of the method and effect of temperature. *Sci. Hortic.* 130:152–156.
4. Statistics Division of the Food and Agriculture Organization for the United Nations 2011. FAOSTAT data. Available at http://faostat.fao.org/site/291/default.aspx
5. Hurtado-Fernández, E., Carrasco-Pancorbo, A., and Fernández-Gutiérrez, A. 2011. Profiling LC-DAD-ESI-TOF MS method for the determination of phenolic metabolites from avocado (*Persea americana*). *J. Agric. Food Chem.* 59:2255–2267.
6. Rodríguez-Fragoso, L., Martínez-Arismendi, J. L., Orozco-Bustos, D., Reyes-Esparza, J., Torres, E., and Burchiel, S. W. 2011. Potential risks resulting from fruit/vegetable–drug interactions: Effects on drug-metabolizing enzymes and drug transporters. *J. Food Sci.* 76:R112–R124.
7. Grant, W. C. 1960. Influence of avocados on serum cholesterol. *California Avocado Society Yearbook* 44:79–88.

8. Alvizouri-Muñoz, M., Carranza-Madrigal, J., Herrera-Abarca, J. E., Chávez-Carbajal, F., and Amezcua-Gastelum, J. L. 1992. Effects of avocado as a source of monounsaturated fatty acids on plasma lipid levels. *Arch. Med. Res.* 23:163–167.
9. Bergh, B. 1992. Nutritious value of avocado. *California Avocado Society Yearbook* 76:123–135.
10. Colquhoun, D. M., Moores, D., Somerset, S. M., and Humphries, J. A. 1992. Comparison of the effects on lipoproteins of a diet high in monounsaturated fatty-acids, enriched with avocado, and a high-carbohydrate diet. *Am. J. Clin. Nutr.* 56:671–677.
11. Lerman-Garber, I., Ichazo-Cerro, S., Zamora-Gónzalez, J., Cardoso-Saldana, G., and Posadas-Romero, C. 1994. Effect of a high-monounsaturated fat diet enriched with avocado in NIDDM patients. *Diabetes Care* 17:311–315.
12. Carranza-Madrigal, J., Alvizouri-Muñoz, M., Alvarado-Jiménez, M. d. R., Chávez-Carbajal, F., Gómez, M., and Herrera-Abarca, J. E. 1995. Effects of avocado on the level of blood lipids in patients with phenotype II and IV dyslipidemias. *Arch. Institut. Cardiol. Mex.* 65:342–348.
13. López-Ledesma, R., Frati-Munari, A. C., Hernández-Domínguez, B. C., Cervantes-Montalvo, S., Hernández-Luna, M. H., Juárez, C., and Morán-Lira, S. 1996. Monounsaturated fatty acid (avocado) rich diet for mild hypercholesterolemia. *Arch. Med. Res.* 27:519–523.
14. Carranza-Madrigal, J., Herrera-Abarca, J. E., Alvizouri-Muñoz, M., Alvarado-Jiménez, M. D., and Chávez-Carbajal, F. 1997. Effects of a vegetarian diet vs. a vegetarian diet enriched with avocado in hypercholesterolemic patients. *Arch. Med. Res.* 28:537–541.
15. Sinyinda, S. and Gramshaw, J. W. 1998. Volatiles of avocado fruit. *Food Chem.* 62:483–487.
16. Pieterse, Z., Jerling, J. C., Oosthuizen, W., Kruger, H. S., Hanekom, S. M., Smuts, C. M., and Schutte, A. E. 2005. Substitution of high monounsaturated fatty acid avocado for mixed dietary fats during an energy-restricted diet: Effects on weight loss, serum lipids, fibrinogen, and vascular function. *Nutrition* 21:67–75.
17. Henrotin, Y., Lambert, C., Couchourel, D., Ripoll, C., and Chiotelli, E. 2011. Nutraceuticals: Do they represent a new era in the management of osteoarthritis?—A narrative review from the lessons taken with five products. *Osteoarthr. Cartilage* 19:1–21.
18. López, H. L. 2012. Nutritional interventions to prevent and treat osteoarthritis. Part II: Focus on micronutrients and supportive nutraceuticals. *Am. Acad. Phys. Med. Rehab.* 4:S155–S168.
19. Lu, Q.-Y., Arteaga, J. R., Zhang, Q., Huerta, S., Go, V. L. W., and Heber, D. 2005. Inhibition of prostate cancer cell growth by an avocado extract: Role of lipid-soluble bioactive substances. *J. Nutr. Biochem.* 16:23–30.
20. Ding, H., Chin, Y. W., Kinghorn, A. D., and D'Ambrosio, S. M. 2007. Chemopreventive characteristics of avocado fruit. *Semin. Cancer Biol.* 17:386–394.
21. D'Ambrosio, S. M., Han, C., Pan, L., Kinghorn, A. D., and Ding, H. 2011. Aliphatic acetogenin constituents of avocado fruits inhibit human oral cancer cell proliferation by targeting the EGFR/RAS/RAF/MEK/ERK1/2 pathway. *Biochem. Bioph. Res. Co.* 409:465–469.
22. Ibiebele, T. I., Nagle, C. M., Bain, C. J., and Webb, P. M. 2012. Intake of omega-3 and omega-6 fatty acids and risk of ovarian cancer. *Cancer Cause Control* 23:1775–1783.
23. Jackson, M. D., Walker, S. P., Simpson-Smith, C. M., Lindsay, C. M., Smith, G., McFarlane-Anderson, N., Bennett, F. I. et al. 2012. Associations of whole-blood fatty acids and dietary intakes with prostate cancer in Jamaica. *Cancer Cause Control* 23:23–33.
24. Oberlies, N. H., Rogers, L. L., Martin, J. M., and McLaughlin, J. L. 1998. Cytotoxic and insecticidal constituents of the unripe fruit of *Persea americana*. *J. Nat. Prod.* 61:781–785.

25. Kawagishi, H., Fukumoto, Y., Hatakeyama, M., He, P., Arimoto, H., Matsuzawa, T., Arimoto, Y., Suganuma, H., Inakuma, T., and Sugiyama, K. 2001. Liver injury suppressing compounds from avocado (*Persea americana*). *J. Agric. Food Chem.* 49:2215–2221.
26. Bredif, S., Baudouin, C., Garnier, S., Buommino, E., Tufano, M. A., and Msika, P. 2010. Natural avocado sugars: A new strategy for skin protection and defense against microorganism aggression. *J. Invest. Dermatol.* 130:2524–2524.
27. Rosenblat, G., Meretski, S., Segal, J., Tarshis, M., Schroeder, A., Zanin-Zhorov, A., Lion, G., Ingber, A., and Hochberg, M. 2011. Polyhydroxylated fatty alcohols derived from avocado suppress inflammatory response and provide non-sunscreen protection against UV-induced damage in skin cells. *Arch. Dermatol. Res.* 303:239–246.
28. Paoletti, I., Buommino, E., Fusco, A., Baudouin, C., Msika, P., Tufano, M. A., Baroni, A., and Donnarumma, G. 2012. Patented natural avocado sugar modulates the HBD-2 and HBD-3 expression in human keratinocytes through Toll-like receptor-2 and ERK/MAPK activation. *Arch. Dermatol. Res.* 304:619–625.
29. Ashton, O. B. O., Wong, M., McGhie, T. K., Vather, R., Wang, Y., Requejo-Jackman, C., Ramankutty, P., and Woolf, A. B. 2006. Pigments in avocado tissue and oil. *J. Agric. Food Chem.* 54:10151–10158.
30. Dembitsky, V. M., Poovarodom, S., Leontowicz, H., Leontowicz, M., Vearasilp, S., Trakhtenberg, S., and Gorinstein, S. 2011. The multiple nutrition properties of some exotic fruits: Biological activity and active metabolites. *Food Res. Int.* 44:1671–1701.
31. Kim, O. K., Murakami, A., Takahashi, D., Nakamura, Y., Torikai, K., Kim, H. W., and Ohigashi, H. 2000. An avocado constituent, persenone A, suppresses expression of inducible forms of nitric oxide synthase and cyclooxygenase in macrophages, and hydrogen peroxide generation in mouse skin. *Biosci. Biotech. Bioch.* 64:2504–2507.
32. Kim, O. K., Murakami, A., Nakamura, Y., Takeda, N., Yoshizumi, H., and Ohigashi, H. 2000. Novel nitric oxide and superoxide generation inhibitors, persenone A and B, from avocado fruit. *J. Agric. Food Chem.* 48:1557–1563.
33. Gorinstein, S., Haruenkit, R., Poovarodom, S., Vearasilp, S., Ruamsuke, P., Namiesnik, J., Leontowicz, M., Leontowicz, H., Suhaj, M., and Sheng, G. P. 2010. Some analytical assays for the determination of bioactivity of exotic fruits. *Phytochem. Anal.* 21:355–362.
34. Villa-Rodríguez, J. A., Molina-Corral, F. J., Ayala-Zavala, J. F., Olivas, G. I., and González-Aguilar, G. A. 2010. Effect of maturity stage on the content of fatty acids and antioxidant activity of 'Hass' avocado. *Food Res. Int.* 44:1231–1237.
35. Wang, W., Bostic, T. R., and Gu, L. W. 2010. Antioxidant capacities, procyanidins and pigments in avocados of different strains and cultivars. *Food Chem.* 122:1193–1198.
36. Rodríguez-Carpena, J. G., Morcuende, D., Andrade, M. J., Kylli, P., and Estevez, M. 2011. Avocado (*Persea americana* Mill.) phenolics, *in vitro* antioxidant and antimicrobial activities, and inhibition of lipid and protein oxidation in porcine patties. *J. Agric. Food Chem.* 59:5625–5635.
37. Wang, M., Zheng, Y., Toan, K., and Lovatt, C. J. 2012. Effect of harvest date on the nutritional quality and antioxidant capacity in 'Hass' avocado during storage. *Food Chem.* 135:694–698.
38. Devalaraja, S., Jain, S., and Yadav, H. 2011. Exotic fruits as therapeutic complements for diabetes, obesity and metabolic syndrome. *Food Res. Int.* 44:1856–1865.
39. Fulgoni, V. L., III, Dreher, M., and Davenport, A. J. 2013. Avocado consumption is associated with better diet quality and nutrient intake, and lower metabolic syndrome risk in US adults: Results from the National Health and Nutrition Examination Survey (NHANES) 2001–2008. *Nutr. J.* 12:1.
40. Bredif, S., Buommino, E., Tufano, M.-A., Baudouin, C., Paoletti, I., Garnier, S., Baroni, A., and Msika, P. 2011. Ability of natural avocado sugars to prevent cutaneous inflammation. *J. Invest. Dermatol.* 131:S9–S9.

41. Kashman, Y., Neeman, I., and Lifshitz, A. 1969. New compounds from avocado pear. *Tetrahedron* 25:4617–4631.
42. Gross, J., Gabai, M., and Lifshitz, A. 1973. Carotenoids in pulp, peel and leaves of *Persea americana. Phytochemistry* 12:2259–2263.
43. Slater, G. G., Shankman, S., Shepherd, J. S., and Alfinslater, R. B. 1975. Seasonal variation in composition of California avocados. *J. Agric. Food Chem.* 23:468–474.
44. Gaydou, E. M., Lozano, Y., and Ratovohery, J. 1987. Triglyceride and fatty-acid compositions in the mesocarp of *Persea americana* during fruit development. *Phytochemistry* 26:1595–1597.
45. Torres, A. M., Maulastovicka, T., and Rezaaiyan, R. 1987. Total phenolics and high performance liquid chromatography of phenolic acids of avocado. *J. Agric. Food Chem.* 35:921–925.
46. Frega, N., Bocci, F., Capozzi, F., Luchinat, C., Capella, P., and Lercker, C. 1991. A new lipid component identified in avocado pear by GC-MS and NMR-spectroscopy. *Chem. Phys. Lipids* 60:133–142.
47. Hierro, M. T. G., Tomás, M. C., Fernández Martín, F., and Santa María, G. 1992. Determination of the triglyceride composition of avocado oil by high performance liquid chromatography using a light scattering detector. *J. Chromatogr. A* 607:329–338.
48. Liu, X., Robinson, P. W., Madore, M. A., Witney, G. W., and Arpaia, M. L. 1999. 'Hass' avocado carbohydrate fluctuations. II. Fruit growth and ripening. *J. Am. Soc. Hortic. Sci.* 124:676–681.
49. de Pascual-Teresa, S., Santos-Buelga, C., and Rivas-Gonzalo, J. C. 2000. Quantitative analysis of flavan-3-ols in Spanish foodstuffs and beverages. *J. Agric. Food Chem.* 48:5331–5337.
50. Domergue, F., Helms, G. L., Prusky, D., and Browse, J. 2000. Antifungal compounds from idioblast cells isolated from avocado fruits. *Phytochemistry* 54:183–189.
51. Dvash, L., Afik, O., Shafir, S., Schaffer, A., Yeselson, Y., Dag, A., and Landau, S. 2002. Determination by near-infrared spectroscopy of perseitol used as a marker for the botanical origin of avocado (*Persea americana* Mill.) honey. *J. Agric. Food Chem.* 50:5283–5287.
52. Liu, X., Sievert, J., Arpaia, M. L., and Madore, M. A. 2002. Postulated physiological roles of the seven-carbon sugars, mannoheptulose, and perseitol in avocado. *J. Am. Soc. Hortic. Sci.* 127:108–114.
53. Piironen, V., Toivo, J., Puupponen-Pimia, R., and Lampi, A. M. 2003. Plant sterols in vegetables, fruits and berries. *J. Sci. Food Agric.* 83:330–337.
54. Ortiz Moreno, A., Dorantes, L., Galindez, J., and Guzmán, R. I. 2003. Effect of different extraction methods on fatty acids, volatile compounds, and physical and chemical properties of avocado (*Persea americana* mill.) oil. *J. Agric. Food Chem.* 51:2216–2221.
55. García-Alonso, M., de Pascual-Teresa, S., Santos-Buelga, C., and Rivas-Gonzalo, J. C. 2004. Evaluation of the antioxidant properties of fruits. *Food Chem.* 84:13–18.
56. Ozdemir, F. and Topuz, A. 2004. Changes in dry matter, oil content and fatty acids composition of avocado during harvesting time and post-harvesting ripening period. *Food Chem.* 86:79–83.
57. Terrab, A. and Heredia, F. J. 2004. Characterisation of avocado (*Persea americana* Mill) honeys by their physicochemical characteristics. *J. Sci. Food Agric.* 84:1801–1805.
58. Terrab, A., Recamales, A. F., González-Miret, M. L., and Heredia, F. J. 2005. Contribution to the study of avocado honeys by their mineral contents using inductively coupled plasma optical emission spectrometry. *Food Chem.* 92:305–309.
59. Chun, J., Lee, J., Ye, L., Exler, J., and Eitenmiller, R. R. 2006. Tocopherol and tocotrienol contents of raw and processed fruits and vegetables in the United States diet. *J. Food Compos. Anal.* 19:196–204.

60. Haiyan, Z., Bedgood, D. R., Bishop, A. G., Prenzler, P. D., and Robards, K. 2007. Endogenous biophenol, fatty acid and volatile profiles of selected oils. *Food Chem.* 100:1544–1551.
61. Pacetti, D., Boselli, E., Lucci, P., and Frega, N. G. 2007. Simultaneous analysis of glycolipids and phospholids molecular species in avocado (*Persea americana* Mill) fruit. *J. Chromatogr. A* 1150:241–251.
62. Meyer, M. D. and Terry, L. A. 2008. Development of a rapid method for the sequential extraction and subsequent quantification of fatty acids and sugars from avocado mesocarp tissue. *J. Agric. Food Chem.* 56:7439–7445.
63. Takenaga, F., Matsuyama, K., Abe, S., Torii, Y., and Itoh, S. 2008. Lipid and fatty acid composition of mesocarp and seed of avocado fruits harvested at Northern Range in Japan. *J. Oleo Sci.* 57:591–597.
64. Blakey, R. J., Bower, J. P., and Bertling, I. 2009. Influence of water and ABA supply on the ripening pattern of avocado (*Persea americana* Mill.) fruit and the prediction of water content using near infrared spectroscopy. *Postharvest Biol. Technol.* 53:72–76.
65. Landahl, S., Meyer, M. D., and Terry, L. A. 2009. Spatial and temporal analysis of textural and biochemical changes of imported avocado cv. Hass during fruit ripening. *J. Agric. Food Chem.* 57:7039–7047.
66. Lu, Q.-Y., Zhang, Y., Wang, Y., Wang, D., Lee, R.-P., Gao, K., Byrns, R., and Heber, D. 2009. California Hass avocado: Profiling of carotenoids, tocopherol, fatty acid, and fat content during maturation and from different growing areas. *J. Agric. Food Chem.* 57:10408–10413.
67. Plaza, L., Sánchez-Moreno, C., de Pascual-Teresa, S., de Ancos, B., and Cano, M. P. 2009. Fatty acids, sterols, and antioxidant activity in minimally processed avocados during refrigerated storage. *J. Agric. Food Chem.* 57:3204–3209.
68. Retief, L., McKenzie, J. M., and Koch, K. R. 2009. A novel approach to the rapid assignment of 13C NMR spectra of major components of vegetable oils such as avocado, mango kernel and macadamia nut oils. *Magn. Reson. Chem.* 47:771–781.
69. Concha-Herrera, V., Lerma-García, M. J., Herrero-Martínez, J. M., and Simo-Alfonso, E. F. 2010. Classification of vegetable oils according to their botanical origin using amino acid profiles established by high performance liquid chromatography with UV–vis detection: A first approach. *Food Chem.* 120:1149–1154.
70. Golukcu, M. and Ozdemir, F. 2010. Changes in phenolic composition of avocado cultivars during harvesting time. *Chem. Nat. Compd.* 46:112–115.
71. Meyer, M. D. and Terry, L. A. 2010. Fatty acid and sugar composition of avocado, cv. Hass, in response to treatment with an ethylene scavenger or 1-methylcyclopropene to extend storage life. *Food Chem.* 121:1203–1210.
72. Poovarodom, S., Haruenkit, R., Vearasilp, S., Namiesnik, J., Cvikrova, M., Martincova, O., Ezra, A., Suhaj, M., Ruamsuke, P., and Gorinstein, S. 2010. Comparative characterisation of durian, mango and avocado. *Int. J. Food Sci. Technol.* 45:921–929.
73. Hurtado-Fernández, E., Pacchiarotta, T., Gómez-Romero, M., Schoenmaker, B., Derks, R., Deelder, A. M., Mayboroda, O. A., Carrasco-Pancorbo, A., and Fernández-Gutiérrez, A. 2011. Ultra high performance liquid chromatography-time of flight mass spectrometry for analysis of avocado fruit metabolites: Method evaluation and applicability to the analysis of ripening degrees. *J. Chromatogr. A* 1218:7723–7738.
74. Vergara-Barberán, M., Escrig-Domenech, A., Lerma-García, M. J., Simo-Alfonso, E. F., and Herrero-Martínez, J. M. 2011. Capillary electrophoresis of free fatty acids by indirect ultraviolet detection: Application to the classification of vegetable oils according to their botanical origin. *J. Agric. Food Chem.* 59:10775–10780.
75. Berasategi, I., Barriuso, B., Ansorena, D., and Astiasaran, I. 2012. Stability of avocado oil during heating: Comparative study to olive oil. *Food Chem.* 132:439–446.

76. Esteve, C., D'Amato, A., Marina, M. L., Concepcion García, M., and Righetti, P. G. 2012. Identification of avocado (*Persea americana*) pulp proteins by nano-LC-MS/MS via combinatorial peptide ligand libraries. *Electrophoresis* 33:2799–2805.
77. Kosivska, A., Karamac, M., Estrella, I., Hernández, T., Bartolomé, B., and Dykes, G. A. 2012. Phenolic compound profiles and antioxidant capacity of *Persea americana* Mill. Peels and seeds of two varieties. *J. Agric. Food Chem.* 60:4613–4619.
78. Lu, Y.-C., Chang, H.-S., Peng, C.-F., Lin, C.-H., and Chen, I.-S. 2012. Secondary metabolites from the unripe pulp of *Persea americana* and their antimycobacterial activities. *Food Chem.* 135:2904–2909.
79. Mooz, E. D., Gaino, N. M., Hayashi Shimano, M. Y., Amancio, R. D., and Fillet Spoto, M. H. 2012. Physical and chemical characterization of the pulp of different varieties of avocado targeting oil extraction potential. *Ciencia Tecnol. Alime.* 32:274–280.
80. Vaikkinen, A., Shrestha, B., Nazarian, J., Kostiainen, R., Vertes, A., and Kauppila, T. J. 2013. Simultaneous detection of nonpolar and polar compounds by heat-assisted laser ablation electrospray ionization mass spectrometry. *Anal. Chem.* 85:177–184.
81. Contreras-Gutiérrez, P. K., Hurtado-Fernández, E., Gómez-Romero, M., Hormaza, J. I., Carrasco-Pancorbo, A., and Fernández-Gutiérrez, A. 2013. Determination of changes in the metabolic profile of avocado fruits (*Persea americana*) by two CE-MS approaches (targeted and non-targeted). *Electrophoresis* 34:2928–2942.
82. Hurtado-Fernández, E., Contreras-Gutiérrez, P. K., Cuadros-Rodríguez, L., Carrasco-Pancorbo, A., and Fernández-Gutiérrez, A. 2013. Merging a sensitive capillary electrophoresis-ultraviolet detection method with chemometric exploratory data analysis for the determination of phenolic acids and subsequent characterization of avocado fruit. *Food Chem.* 141:3492–3503.
83. United States Department of Agriculture (USDA) Nutrient Data Laboratory, Food and Nutrition Information Center (FNIC) and Information Systems Division of the National Agricultural Library 2011. USDA National Nutrient Database for Standard Reference. Available at http://ndb.nal.usda.gov/
84. Capozzi, F. and Bordoni, A. 2013. Foodomics: A new comprehensive approach to food and nutrition. *Genes Nutr.* 8:1–4.
85. Cifuentes, A. 2009. Food analysis and foodomics. *J. Chromatogr. A* 1216:7109.
86. Herrero, M., Simó, C., García-Cañas, V., Ibáñez, E., and Cifuentes, A. 2012. Foodomics: MS-based strategies in modern food science and nutrition. *Mass Spectrom. Rev.* 31:49–69.
87. Ibáñez, C., Valdés, A., García-Cañas, V., Simó, C., Celebier, M., Rocamora-Reverte, L., Gómez-Martínez, A. et al. 2012. Global foodomics strategy to investigate the health benefits of dietary constituents. *J. Chromatogr. A* 1248:139–153.
88. García-Cañas, V., Simó, C., Herrero, M., Ibáñez, E., and Cifuentes, A. 2012. Present and future challenges in food analysis: Foodomics. *Anal. Chem.* 84:10150–10159.
89. Meyer, V. R. 2004. *Practical High-Performance Liquid Chromatography,* Wiley, Weinheim. ISBN: 978 0 470 68218 0
90. Corradini, D. and Phillips, T. M. 2011. *Handbook of HPLC,* ed. Cazes, J., CRC Press. Taylor & Francis Group, Boca Raton. ISBN-13: 978 1 574 44554 1.
91. Allwood, J. W. and Goodacre, R. 2009. An introduction to liquid chromatography-mass spectrometry instrumentation applied in plant metabolomic analyses. *Phytochem. Anal.* 21:33–47.
92. Kanehisa Laboratories and NPO Bioinformatics Japan 1995. Kegg-Ligand Database. Available at http://www.kegg.jp/kegg/ligand.html
93. National Center for Biotechnology Information (NCBI), National Library of Medicine (NLM), National Institutes of Health (NIH) and US Department of Health and Human Services (DHHS) 2004. Pubchem Database. Available at http://pubchem.ncbi.nlm.nih.gov/

94. Cuadros-Rodríguez, L. and Bosque-Sendra, J. M. 2011. Mediterranean chemometrics. *Anal. Bioanal. Chem.* 399:1925–1927.
95. Bosque-Sendra, J. M., Cuadros-Rodríguez, L., Ruiz-Samblás, C., and de la Mata, A. P. 2012. Combining chromatography and chemometrics for the characterization and authentication of fats and oils from triacylglycerol compositional data—A review. *Anal. Chim. Acta* 724:1–11.
96. Santos, P. M., Pereira-Filho, E. R., and Rodriguez-Saona, L. E. 2013. Rapid detection and quantification of milk adulteration using infrared microspectroscopy and chemometrics analysis. *Food Chem.* 138:19–24.
97. Martelo-Vidal, M. J., Domínguez-Agis, F., and Vázquez, M. 2013. Ultraviolet/visible/near-infrared spectral analysis and chemometric tools for the discrimination of wines between subzones inside a controlled designation of origin: A case study of Rias Baixas. *Aust. J. Grape Wine R.* 19:62–67.
98. Abad-García, B., Berrueta, L. A., Garmon-Lobato, S., Urkaregi, A., Gallo, B., and Vicente, F. 2012. Chemometric characterization of fruit juices from Spanish cultivars according to their phenolic compound contents: I. Citrus fruits. *J. Agric. Food Chem.* 60:3635–3644.
99. Cavazza, A., Corradini, C., Musci, M., and Salvadeo, P. 2012. High-performance liquid chromatographic phenolic compound fingerprint for authenticity assessment of honey. *J. Sci. Food Agric.* 93:1169–1175.
100. Bratu, A., Mihalache, M., Hanganu, A., Chira, N.-A., Todasca, M.-C., and Rosca, S. 2012. Gas chromatography coupled with chemometric method for authentication of Romanian cheese. *Rev. Chim.* 63:1099–1102.
101. Guerreiro, J. S., Barros, M., Fernandes, P., Pires, P., and Bardsley, R. 2013. Principal component analysis of proteolytic profiles as markers of authenticity of PDO cheeses. *Food Chem.* 136:1526–1532.
102. Ebrahimi-Najafabadi, H., Leardi, R., Oliveri, P., Casolino, M. C., Jalali-Heravi, M., and Lanteri, S. 2012. Detection of addition of barley to coffee using near infrared spectroscopy and chemometric techniques. *Talanta* 99:175–179.
103. Robertson, D. G., Watkins, P. B., and Reily, M. D. 2011. Metabolomics in toxicology: Preclinical and clinical applications. *Toxicol. Sci.* 120:S146-S170.
104. Xiayan, L. and Legido-Quigley, C. 2008. Advances in separation science applied to metabonomics. *Electrophoresis* 29:3724–3736.
105. Issaq, H. J., Abbott, E., and Veenstra, T. D. 2008. Utility of separation science in metabolomic studies. *J. Sep. Sci.* 31:1936–1947.
106. Hurtado-Fernández, E., Pacchiarotta, T., Longueira-Suárez, E., Mayoboroda, O. A., Fernández-Gutiérrez, A., and Carrasco-Pancorbo, A. 2013. Evaluation of gas chromatography-atmospheric pressure chemical ionization-mass spectrometry as an alternative to gas chromatography-electron ionization-mass spectrometry: Avocado fruit as example. *J. Chromatogr. A* 1313:228–244.

9 UHPLC–MS in Virgin Olive Oil Analysis

An Evolution toward the Rationalization and Speed of Analytical Methods

Aadil Bajoub, Alegría Carrasco-Pancorbo, Noureddine Ouazzani, and Alberto Fernández-Gutiérrez

CONTENTS

9.1 INTRODUCTION

Virgin olive oil (VOO), the emblematic food of the Mediterranean diet, is now one of the foods which are being more studied, since the characterization of its composition and quality has a great interest. Recognized for its various nutritional virtues and the beneficial health effects of several of its compounds, VOO is considered, in many countries, as a basic ingredient for a well-balanced nutrition that promotes vitality, well-being, and protects from many diseases.

For a very long time, the need of characterizing the VOO composition and checking its genuineness has motivated the development of several analytical methods. These methods have allowed, besides getting a more holistic overview about VOO composition, the identification and the characterization of several of the bioactive compounds found in this interesting matrix. The implementation of these modern analytical methods has been possible, in part, due to the spectacular development motivated by the use of chromatographic techniques in this field [1–17].

Indeed, the international norms and standards that regulate the characterization of the VOO composition and genuineness [18–20] consider high-performance liquid chromatography (HPLC) and gas chromatography (GC) very important and valuable techniques to be used to achieve a proper characterization of VOO. Furthermore, the robustness, flexibility, and efficiency of these two analytical techniques, together with a quite sensitive and selective detector such as mass spectrometry (MS), facilitate the implementation of several analytical methods for identification and detailed description of the structure of many compounds with high healthy and nutritional values, such as phenolic compounds, volatiles compound, triglycerides, fatty acids, tocopherols, etc. [21–26].

Apart from the mentioned technological advances, the separation power and selectivity of HPLC have been drastically improved over the last years. In this context, ultra-high-performance liquid chromatography coupled with mass spectrometry (UHPLC–MS) represents a relatively new category of analytical coupled techniques that respects the principles of the classic HPLC–MS while bringing some improvements in chromatographic resolution, speed, and sensitivity.

UHPLC–MS opens up great expectations for the characterization of VOO composition. It stands out as an appropriate tool for the implementation and the development of fast, precise, and reproducible methods for the characterization of various VOO compounds, as well as the authentication, safety, and origin traceability of this product.

This chapter provides an overview of the continuous evolution that HPLC has undergone, including turning into UHPLC, which can be easily coupled to MS to combine the strengths of both techniques. A general description about the importance and composition of VOO will be given, as well as a summary of some of the most relevant applications of UHPLC–MS for analyzing different components of VOO. Future trends and perspectives in the use of UHPLC–MS in this field will be also underlined.

9.2 UHPLC–MS: AN ADVANTAGEOUS ALTERNATIVE TO CLASSICAL CHROMATOGRAPHY TECHNIQUES USED IN VOO ANALYSIS

VOO is a natural product with a very complex composition. Among the large number of substances found in this product, many are found at very low concentrations, even though they play a main role in VOO characteristics, quality, nutritional properties, and health benefits. The chemical components of VOO are commonly distinguished into two fractions: saponifiable and unsaponifiable. The saponifiable fraction represents a high percentage, ranging between 98.5% and 99.5% of the VOO. Triglycerides constitute the most important part of this fraction; the rest is mainly composed of free fatty acids together with other minor components of VOO derived from fatty acids, such as mono- and diacylglycerols, phosphatides, waxes, and esters of sterols. Minor components, which constitute the unsaponifiable fraction, are present in very low amounts (about 2% of VOO weight), and include more than 200 chemical compounds, such as aliphatic and triterpenic alcohols, sterols, hydrocarbons, carotenoids, chlorophylls, volatile compounds, and phenolic compounds [27].

VOO composition and quality are very related to genetics as well as climatic, agronomic, and technological factors [28–38]. Each one of these factors contributes to the wide variability of composition of available VOOs in the market. This diversity has motivated the implementation of diverse analytical techniques, not only able to assure and demonstrate the VOO market value, but also to carry out an appropriate quality control, authenticity tests, or even to guarantee the possible genuineness of a particular oil.

Various analytical techniques and methods for VOO analysis are now available; it is possible to find both official methods and some others developed by research groups trying to develop new analytical methods that could simplify and make more reliable the characterization of VOO composition and quality. Among the analytical tools used in this area, chromatographic techniques are the most common, finding, for instance, more than 1700 published studies about VOO characterization using chromatography according to ISI Web of Knowledge database in recent years. In this regard, LC remains the technique of choice for the qualitative and quantitative determination of several compounds in VOO. Low and high molecular weight VOO compounds can be separated by LC due to the wide number of possible combinations between the mobile and the stationary phases. However, conventional chromatographic methods used for VOO analysis, even if they use HPLC, are time consuming and usually very tedious, being the analysis time often longer than 30 min (Figures 9.1 and 9.2).

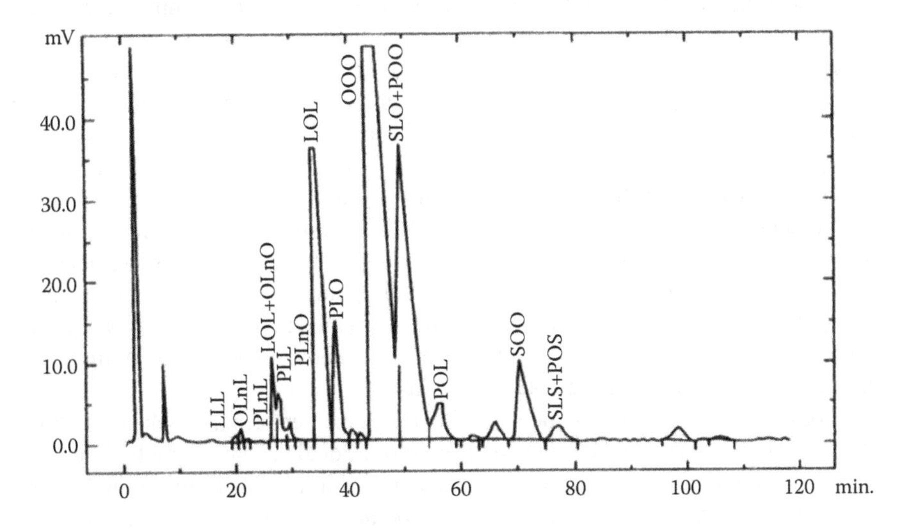

FIGURE 9.1 TAG content with ECN 42 determined by HPLC analysis. (Adapted from International Olive Oil Council. 2001. Determination of the difference between actual and theoretical content of triacyglycerols with ECN 42. Method of analysis COI/T.20/Doc. no. 20/Rev.1.)

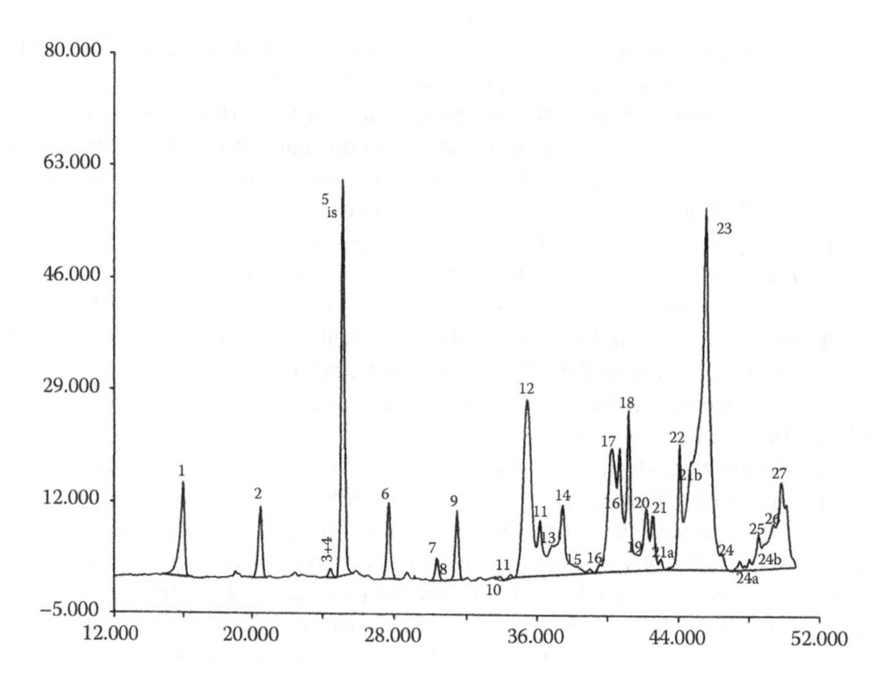

FIGURE 9.2 Determination of biophenols in olive oils by HPLC. Chromatogram recorded at 280 nm for the biophenols profile of an extra-VOO. (Adapted from International Olive Oil Council 2009. Determination of Biophenols in olive oils by HPLC. Method of analysis COI/T.20/Doc no 29.)

The number of samples included in any application is continuously growing, while the time response delivery needs to be greatly reduced. This implies the necessity of a target strategy in order to develop rapid and efficient procedures for performing qualitative and quantitative VOO analysis within reduced analysis times. HPLC started to improve its performance when coupled to MS as an identification and confirmation technique, but shortening the analysis time entails other kinds of strategies, such as using short columns and high flow rates or enlarging the choice of column packages. In this context, we can say that the use of UHPLC is gaining more and more interest over conventional HPLC, because HPLC column chemistry and dimensions affects resolution and sensitivity; whereas sub-2 μm particle stationary phases combined with ultrahigh-pressure pumps (which overcome the high back pressure associated with such columns) have helped to produce narrower peaks, rapid analysis times, and lower detection limits. Moreover, the solvent volume consumed is lower, decreasing the volumes of waste.

In other words, the improvements of the analytical performance when UHPLC–MS is used are beyond the combination of conventional chromatography with MS, and can be found at different levels: resolution as well as sensitivity of analysis by using sub-2 μm particle size, the system is operational at higher pressure, and the mobile phase flows at greater linear velocities as compared to HPLC.

The first application of UHPLC–MS related to VOO analysis was published in 2008; even though the number of manuscripts published since then is not very high, the recent papers cover some of the most important subjects in the field of VOO analysis, a fact that allows us to structure these applications in three different categories:

- Authentication of VOO and detection of adulteration with vegetable oils or any olive oil quality upgrade.
- Characterization of VOO bioactive compounds, particularly phytosterols and phenolic compounds.
- VOO safety: detection of pesticide residue metabolites.

In further sections, we will review in detail some applications belonging to these categories and provide an overview of the advantages offered by UHPLC–MS compared to the classical methods more widely used for these determinations.

9.3 MAIN APPLICATIONS OF UHPLC–MS IN VOO ANALYSIS

9.3.1 UHPLC–MS as a Tool for Checking VOO Genuineness

VOO is defined as naturally edible oil obtained from the fruit of the olive tree solely by mechanical or other physical means. The oil is obtained under conditions (particularly thermal conditions) that do not lead to the deterioration of the oil, and the oil has not undergone any treatment other than washing, decantation, centrifugation, and filtration [20].

Because of its high nutritional value and the beneficial effects that some of its compounds exert, the price of VOO is relatively higher if we compare it with other edible oils. This fact could explain why so many adulterations have been found for VOO; this matrix is susceptible to adulteration with cheaper olive oil categories (olive oil pomace, refined olive oil) and/or other edible oils. Corn, cottonseed, canola, palm, peanut, soybean, and sunflower oils have been detected in adulterated VOOs.

Tremendous efforts have been made in order to detect such fraudulent practices, and different VOO compounds, such as sterols, tocopherols, isoprenoid alcohols, and triglycerides have been monitored as adulteration markers [39–46].

In this regard, various analytical methods have been developed and their potential for the VOO authentication has been evaluated. Thus, spectroscopic tools, such as FT-near-infrared spectroscopy, FT-Raman spectroscopy, fluorescence and ultraviolet–visible detectors, and chromatographic techniques have been widely used in the field of VOO authentication [43–48].

It should be emphasized that the VOO authentication regulatory standards, compiled in the European Commission Regulation (EC Reg No 2568/1991 and its later amendments EC Reg No 1989/2003) [19], the Codex Alimentarius Norm (Codex Alimentarius Commission Draft, 2013) [49], and International Olive Oil Council (IOOC) Trade standards (IOOC/T.15/NC n° 3/Rev.4, 2011) [50], give a particular importance to chromatographic techniques for the analysis of VOO composition in order to detect some fraudulent mixtures. Therefore, for a long time, several HPLC methods have been developed to detect the illegal addition of other oils, including the use of tocopherols, carotenoids, and chlorophylls in various research works to detect adulteration of VOO [51,52].

However, the more widely used compounds to assess the authenticity of VOO are triacylglycerols. As far as TAG are concerned, IOOC has developed a global method for the detection of extraneous oils in VOOs, particularly high linoleic vegetable oils such as sunflower and colza, as well as some high oleic vegetable oils such as hazelnut, high oleic sunflower, and olive pomace oils. The presence of these oils indicates if a typical VOO is genuine or not [53]. This method is based on the absolute difference between the experimental values of triacylglycerols with equivalent carbon number 42 (ECN42) obtained by HPLC with refractive index detection and the theoretical value of TAGs (triacylglycerol) with an equivalent carbon number of 42 (ECN 42 theoretical) calculated from the fatty acid composition. Based on this method an authenticity factor can be calculated as

$$Au = 100 - ECN \times 42(\%)/ECN \times 42(\%)$$

where

Au = Authenticity factor.
ECN = Equivalent carbon number.
CN = Total number of carbon atoms.
X = Number of double bonds.
n = Factor for double-bond contribution.

In practice, even if this method is very reliable for the detection of adulterated VOOs by using other edible oils at a percentage of 2.5%, it remains expensive and time-consuming, as it demands an initial purification of the VOO samples, TAGs separation and grouping according to their equivalent carbon numbers, fatty acid composition determined by GC, and the determination of differences between the analytical results obtained by HPLC and the theoretical content, calculated starting from the fatty acid composition.

Bearing in mind the fact that the described method is quite tedious, different research groups have looked for other alternatives to detect VOO adulterations by using TAG profiles. Indeed, various LC-based improved analytical methodologies with different detectors are currently used as reliable methods able to detect VOO mixtures. In almost all the works developed during the last few years using HPLC–MS (HPLC–APCI–MS [54,56,57], HPLC–APCI–MS/MS [24], and HPLC–ELSD–MS [55]) the step of sample preparation is reduced to a simple dilution of the VOO in a solvent such as hexane, chloroform, or acetone. This helped to reduce the time consumed by the official method of the IOOC.

In addition, the use of these methods, together with the grouping power of chemometric tools, allowed setting up discriminate models that facilitate the detection of the presence of the other vegetable oils in the VOO, or the presence of low-quality olive oil mixed with a VOO (Figures 9.3 and 9.4).

The recent use of the undeniable advantages that UHPLC–MS offers has allowed improving, in a very significant way, the knowledge about TAG VOO composition and its potential to detect blends of vegetable oils with VOOs. Lee and Di Gioia [58], for instance, evaluated the advantages that a UHPLC–DAD–MS method offers in comparison with a conventional HPLC approach for the determination of

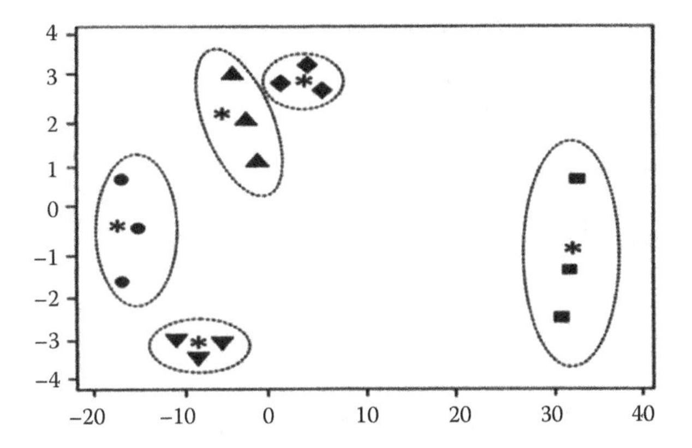

FIGURE 9.3 Discriminant analysis using individual TAG contents as variables. Olive oil–hazelnut oil (50:50) (◆), olive oil–hazelnut oil (70:30) (▲), olive oil–hazelnut oil (90:10) (▼), olive oil (•), hazelnut oil (■), group centroids (*). (From Parcerisa, J. et al. 2000. *J. Chromatogr. A* 881:149–158. With permission.)

VOO TAG profile, and checked the possibility of using this technique as a tool for VOO adulteration control. In this study, the sample preparation consisted, as was the case for the HPLC–MS methods previously described, of a simple dilution of the oils in 2-propanol. The separation was carried out by means of a UHPLC system with a packed column of the following characteristics: 2.1 × 150 mm UPLC 1.7 μm BEH C18.

As highlighted before, one of the most important improvements achieved by UHPLC is the gradual reduction of the analysis time, passing from 80, 50, and 40 min spent, respectively, by the official method of the IOOC and the HPLC–MS methods developed by Parcerisa et al. [54] and Fasciotti et al. [56], to practically 20 min of elution time using a UHPLC–MS method [58]. This represents a great decrease in analysis time, not only if we compare the UHPLC–MS method with the official one, but also with the more advanced HPLC–MS methodologies. Moreover, if we analyze the TAG chromatograms obtained by DAD (diode array detector) for some of the oil analyzed by Lee and Di Gioia [58], we can clearly observe a good separation between the different analytes in about 18 min of analysis, as well as a good reproducibility (Figure 9.5).

For the differentiation of analyzed oil and testing, given the possibility of detecting the presence of vegetable oil blended with VOO, the authors propose the identification of TAG peaks using MS with positive APCI and the calculation of the peak area ratio values, dividing an indicator peak area by a marker peak area. The obtained results clearly show that the peak area ratio values of analyzed oil samples are markedly different and the adulterated VOO samples are detected. Thus, this method is able to detect VOO samples adulterated with 1% of soybean oil and walnut oil, or with 5% of hazelnut oil. The detection threshold for this method is much lower than other LC–MS methods developed previously.

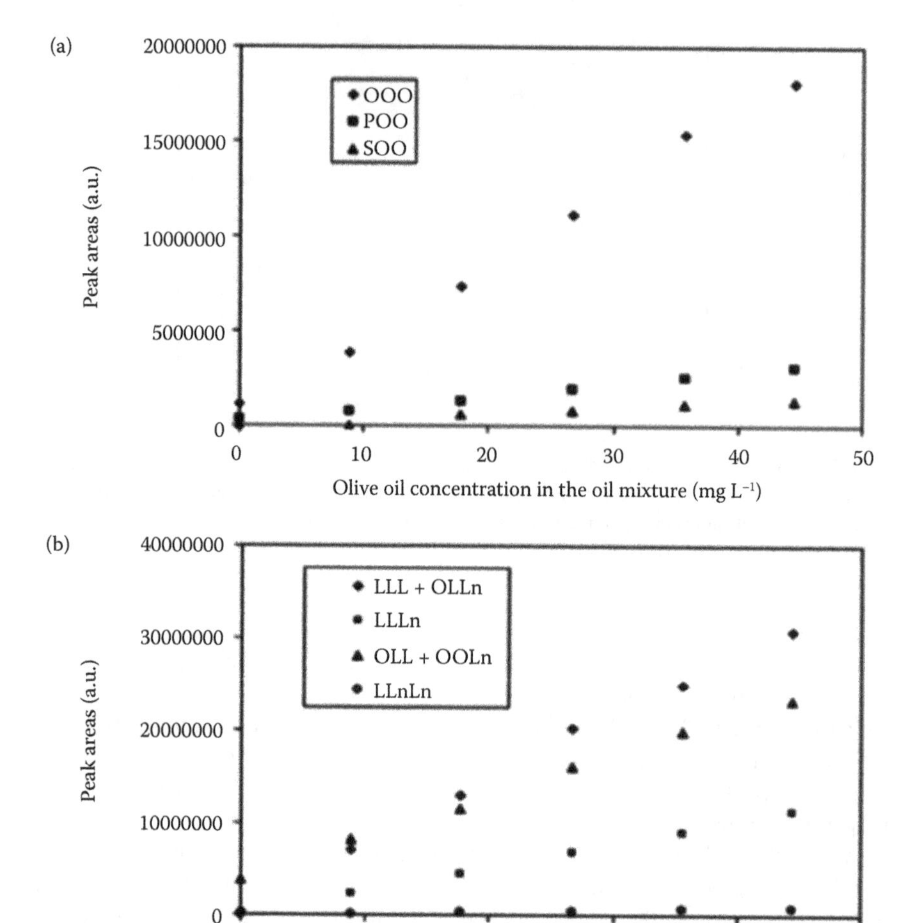

FIGURE 9.4 Lines showing the variation of the areas of oil TAG markers versus oil concentrations (mg/L) in mixtures containing 44.5 mg/L of total oil. (a) Olive oil; (b) soybean oil. O = oleic acid (18:1); P = palmitic acid (16:0); S = stearic acid (18:0); L = linoleic acid (18:2); Ln = linolenic acid (18:3). (From Fasciotti, M. and Pereira Netto A. D. 2010. *Talanta* 81:1116–1125. With permission.)

9.3.2 Characterization of VOO Bioactive Compounds by UHPLC–MS

In the last decade, the relationship between the Mediterranean diet and health has motivated an intense research activity in the characterization bioactive compounds of foods constituting this diet. VOO is the main source of fat in the Mediterranean dietary style. Many studies have been conducted to pinpoint the VOO composition, as well as identify and provide useful information about its bioactive compounds and the roles which they play in preventing a large number of diseases. VOO health-protective potential seems to be mainly attributed to its unsaponifiable fraction. This fraction, which roughly represents 0.5–2.5% of the VOO, has acquired a well-deserved

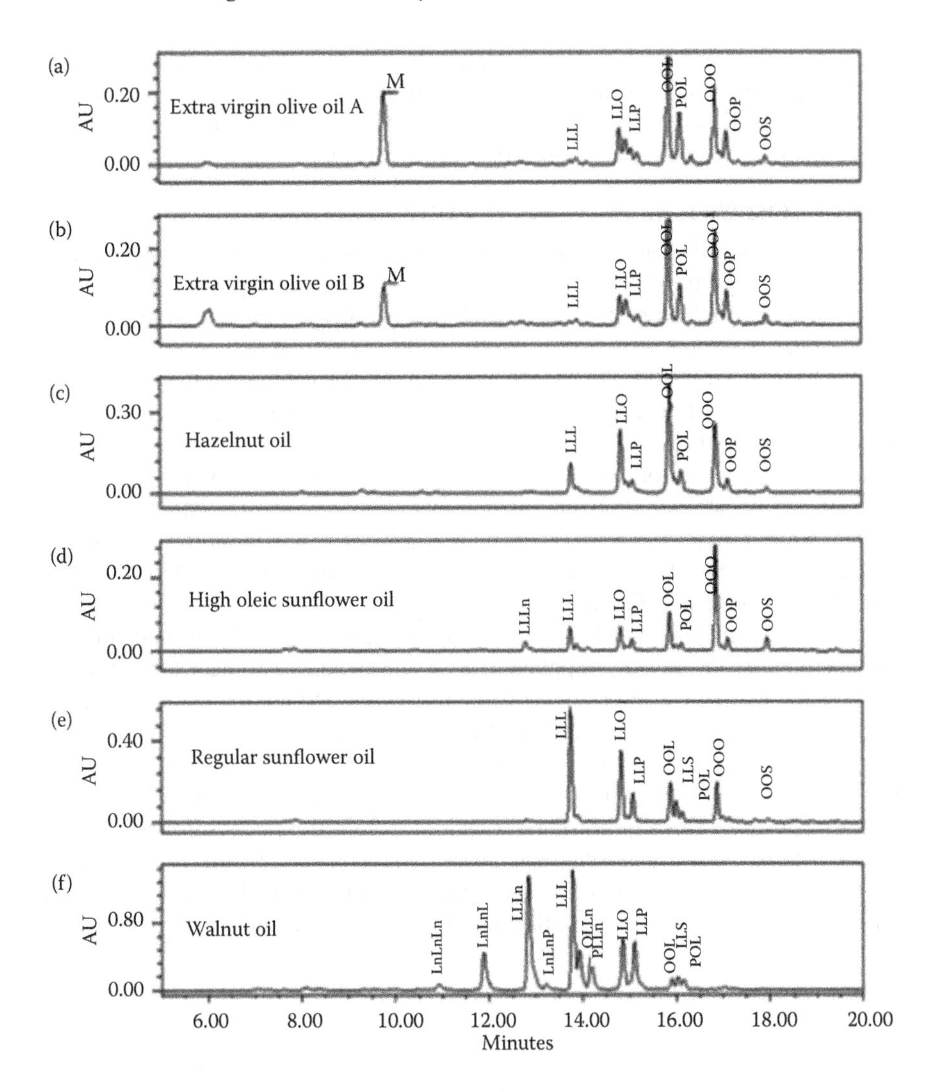

FIGURE 9.5 Comparison of PDA chromatograms (at 210 nm) of six different seed oil samples. TAG peaks are labeled using abbreviations of their fatty acid composition; where, P is palmitic acid, S is stearic acid, O is oleic acid, L is linoleic acid, Ln is linolenic acid. (From Lee, P. J. and Gioia A. J. 2009. *Lipid Techn.* 21:112–115. With permission.)

connotation as the part of VOO joining together nutritional interest and organoleptic quality of VOO [59–66]. Among the constituents of this VOO fraction, phytosterols and phenolic compounds stand out. These two families of bioactive compounds were the central subject of various research works carried out in order to identify them, characterize their structure, as well as study their positive effects on human health.

This vast knowledge of VOO phytosterols and phenolic compounds is widely ascribed to chromatography techniques evolution and, in particular, to the progress in LC–MS instrumentation.

Aware of the importance of these two groups of compounds and the interest in the development of analytical methods for their characterization, the next two sections will be focused on the evolution of the LC–MS methods for these two chemical families and the description of some attractive applications.

9.3.2.1 VOO Phytosterols Analysis by UHPLC–MS

Phytosterols has been found to be beneficial for health due to its antioxidant, anti-inflammatory, and antibacterial activities, as well as for its capacity in protection against cancers, such as breast, colon, and prostate [67–71]. Furthermore, in 2000, the U.S. Food and Drug Administration officially recognized that products containing phytosterols decreased the risk of the cardiovascular diseases if they are associated to low saturated fat and low-cholesterol nutrition [72]. Hence, there is a great interest in determining their content and characterizing their structure.

Numerous phytosterols have been isolated from the unsaponifiable fraction of VOO; the most important are β-sitosterol, Δ5-avenasterol, and campesterol. Several other minor compounds, such as cholesterol, stigmasterol, clerosterol, Δ7-stigmastenol, and Δ7-avenasterol have also been found in VOOs [73,74]. Their structures can be seen in Figure 9.6.

The sterols content in VOO is regulated by the legislation of the European Union [75] and by the trade standards applied to olive oils and olive–pomace oils set by the IOOC [50]. Table 9.1 shows the typical composition of a VOO in terms of sterols.

It should be noted that the content of these compounds varies considerably, but always within the rank of composition fixed by the COI standard, according to several factors, being the most important ones the olive variety [76–78] and the ripening degree of olives [79,80]. The determination of these compounds is also of major interest because their composition can be used to detect adulteration and genuineness or to check geographical origin and varietal authenticity [81]. Thereby, the study of the VOO phytosterols composition usually represents a subject of interest and

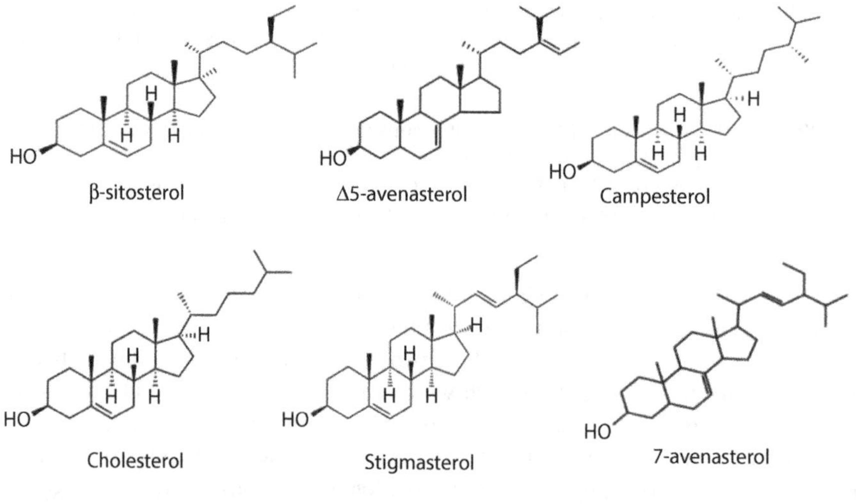

β-sitosterol Δ5-avenasterol Campesterol

Cholesterol Stigmasterol 7-avenasterol

FIGURE 9.6 Chemical structure of the main olive oil phytosterols.

TABLE 9.1
VOO Phytosterols Composition According to International Olive Oil Council Trade Standards

Phytosterols	Concentration
Cholesterol	≤0.5
Brasicasterol	≤0.1
Campesterol	≤0.4
Stigmasterol	≤0.4
β-sitosterol	≤93%
Δ7-stigmastenol	≤0.5
Total phytosterols	≥1000 mg/kg

Source: International Olive Oil Council Trade standards for olive oil and olive pomace oils 2011. COI/T.15/NC n° 3/Rev. 6, November 2011.

implies the need of developing rapid, reliable, and reproducible analytical methods for their characterization.

Official methods for the analysis of phytosterols in olive oil (European Union Commission, 1991; International Olive Oil Council, 2011) involve several steps. First, olive oil saponification is required for the separation of saponifiable and unsaponifiable olive oil fractions with diethyl ether (washing the extract with water), and then separation by thin-layer chromatography (TLC) on silica gel plates and derivatization of the sterols. The sterols as trimethylsilyl derivatives are quantified, afterwards, by means of a capillary GC analysis with FID (flame ionization detector) [50,75].

Even though this is an official method for the determination of phytosterols content in VOO, it requires many steps for the preparation and analysis, making it laborious and time consuming. Several attempts for the improvement of VOO phytosterols separation and subsequent identification steps have been carried out. Segura et al. [82] have tested the possibility of extracting the most important phytosterols from VOO without using TLC. In this study, the authors compared the HPLC–APCI–MS profiles analyzing the extracts obtained after the TLC separation and those obtained by using a much simpler strategy. The latter only required the step of saponification, then the diethyl ether fractions were collected, washed with water, and dried with anhydrous sodium sulfate. Finally, the fractions were filtered and evaporated to dryness using a rotary evaporator. The residue was dissolved in methanol and extracts were filtered through a membrane filter (0.45 mm) before being analyzed. As can be observed in Figure 9.7, the phytosterol profiles of both methods were very similar. This was a very important achievement, as it revealed that this process could remarkably simplify the official method, offering a simpler and less time-consuming alternative.

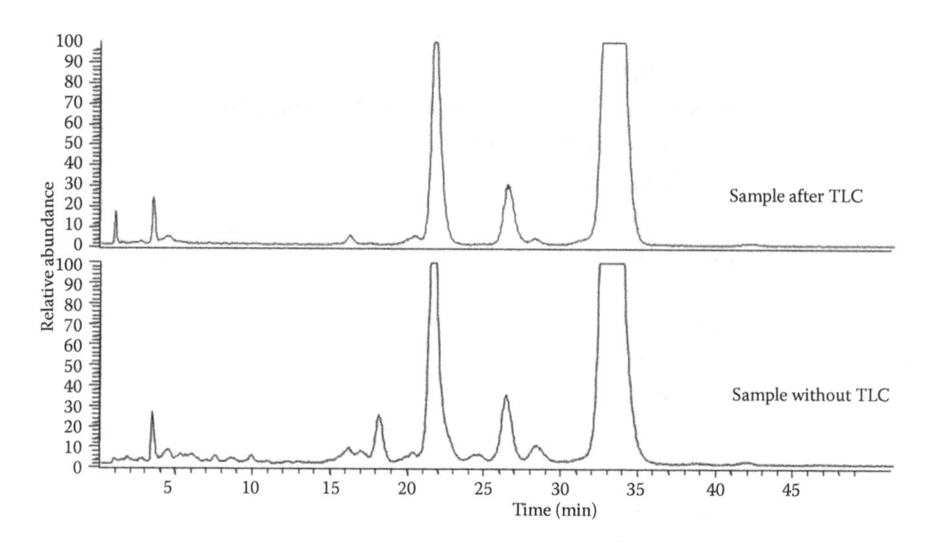

FIGURE 9.7 Chromatogram of two extracts of VOO using isolation of total sterols both with and without TLC. (From Segura-Carretero, A. et al. 2008. *Eur. J. Lipid Sci. Technol.* 110:1142–1149. With permission.)

Recent approaches using new extraction procedures, such as SPE [83] and on-line coupling reversed-phase LC–GC [84] have been proposed, owing to reduce or automate the extraction of phytosterols from VOO.

Likewise, although the quantification of phytosterols in VOO analysis is usually performed by GC with different detectors [80,85,86], recently published research has proposed new analytical methods for the precise determination of phytosterols in olive oil using HPLC–MS [22,87], and more recently, highly sensitive methods based on UHPLC–MS were applied in the determination of VOO phytosterols [88,89].

Lerma-García et al. [89], in their work focused on the evaluation of the potential of UHPLC–MS for the determination of VOO phytosterols to carry out a discrimination according to variety, were able to optimize a method with great analytical characteristics in terms of detection and quantification limits and analysis time. Indeed, comparing the analytical parameters of the UHPLC–MS developed by Lerma-García et al. [89], and the HPLC–MS method used in the study carried out by Cañabate-Díaz et al. [22], it is possible to say that the UHPLC–MS method showed lower detection limit (LOD) and quanitification limit (LOQ). LOD ranged between 0.03 and 0.07 μg/mL, LOQ ranged from 0.10 to 0.25 μg/mL, whereas these values were found between 0.123 and 0.677 μg/mL for LOD and 0.500 and 2.51 μg/mL for LOQ, respectively, in the case of HPLC–MS.

Furthermore, using UHPLC–MS in VOO phytosterols analysis led to a remarkable reduction in the analysis time (both in the sample preparation step and in the phytosterols separation). The separation was achieved in 5 min, time in which 14 phytosterol compounds were identified (Figure 9.8). Thereby, the analysis time using UHPLC–MS is four to six times faster than those of HPLC–MS methods [22,82,90], where the time required is between 25 and 30 min. The comparison is even more

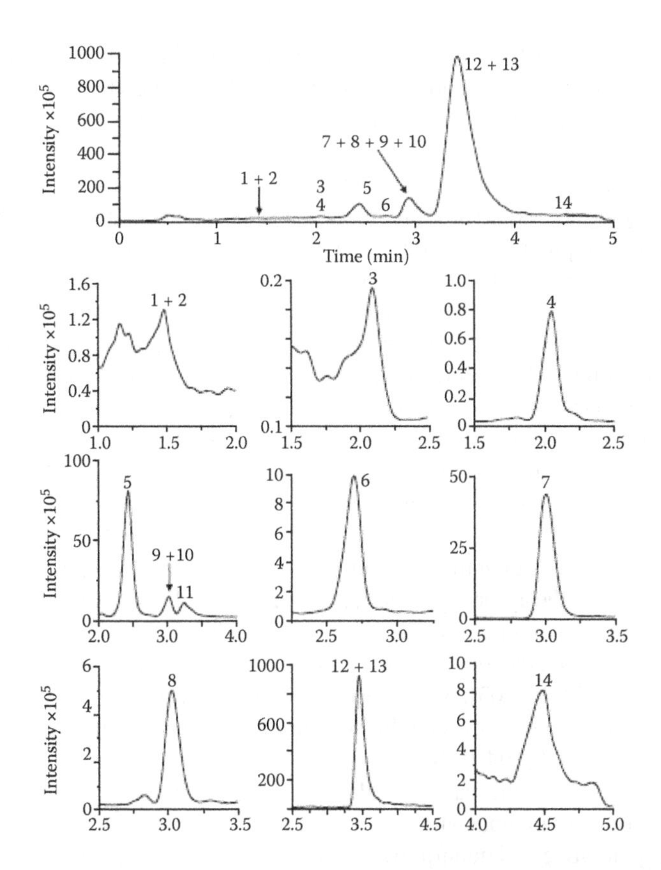

FIGURE 9.8 TIC and SIRs of Borriolenca extra-VOO extracts analyzed by UPLC–MS. 1. Erythrodiol; 2. uvaol; 3. ergosterol; 4. brassicasterol; 5. Δ5-avenasterol; 6. cholesterol; 7. campesterol; 8. campestanol; 9. stigmasterol; 10. clerosterol; 11. Δ5,24-stigmastadienol; 12. β-sitosterol; 13. Δ7-stigmastenol; and 14. sitostanol. (From Lerma-García, M. J. et al. 2011. *Food Res. Int.* 44:103–108. With permission.)

favorable to UHPLC–MS when compared with the official method, where the analysis time is 35 min [91].

However, a resolution decrease (particularly in comparison with phytosterols profiles obtained with the official method) can be observed in Figure 9.8, which includes the total ion chromatogram (TIC) and selected ion recordings (SIRs) of VOO analyzed by UPLC–MS.

We can note that there is an overlapping of some peaks, particularly in the case of β-sitosterol and Δ7stigmastenol, and stigmasterol and clerosterol. However, in spite of this, the improvements achieved by this method are more than enough to consider it as a very powerful alternative, even if the separation and resolution could still be improved.

9.3.2.2 VOO Phenolic Compounds Analysis by UHPLC–MS

As commented before, the protective effects of VOO are attributed to its high content of monounsaturated fatty acids and to the presence of some minor components, which add up to 2% of the weight. Among its several minor constituents, phenolic compounds or polyphenols are attracting considerable attention because of their biological activities, their influence on the organoleptic properties of VOO, and their contribution to its oxidative stability [92].

The phenolic fraction of VOO consists of a heterogeneous mixture of compounds belonging to several families with varying chemical structures. The main components of the phenolic fraction of VOO are hydroxytyrosol, tyrosol, and their derivatives linked to the aldehydic and dialdehydic forms of elenolic acid, which are described as secoiridoids [93,94]. Moreover, significant amounts of the lignans, such as pinoresinol and 1-acetoxypinoresinol, are also present [95,96], as well as flavonoids (luteolin and apigenin) [97,98], phenolic acids (such as caffeic, vanillic, syringic, *p*-coumaric, *o*-coumaric, protocatechuic, sinapic, and *p*-hydroxybenzoic acid) [99], and hydroxy-isochromans [100]. The complexity of this fraction can be seen in Table 9.2, where the most important phenolic compounds of VOO together with their compound name, general chemical structure, and molecular weight are included.

The qualitative and quantitative composition of VOO hydrophilic phenols is strongly affected by the agronomic and technological conditions of production [101]. As far as agronomic parameters are concerned, the cultivar, the fruit ripening degree, the agronomic techniques used and the pedoclimatic conditions are the aspects more extensively studied [102,103]. The influence of variety, extraction system, ripening degree, and storage in the polyphenolic content of a VOO has been widely discussed in the literature as well [104,105].

Bearing in mind all the variables that determine the composition of VOO in terms of phenolic compounds and the heterogeneity of this fraction, it seems obvious that the qualitative and quantitative determination of these phenolic compounds in oils is not an easy task. Various methods have been developed for the extraction of these substances from oils as well as separation methods for their analysis. There is extensive literature concerning the detection and quantification of phenolic compounds in VOO. Starting from the early days, nonspecific analytical methods such as paper, thin-layer, and column chromatography, as well as UV spectroscopy, were applied to polyphenols analysis with limited success [106]. The need to profile and identify individual phenolic compounds meant that traditional methods were replaced and significant progress was achieved when specific analytical methods were used, such as GC, HPLC, and CE. GC has been used with different detectors (mainly FID and MS). In HPLC, it is possible to find research works where UV (photodiode array), fluorescence, electrochemical, biosensors, nuclear magnetic resonance (NMR), and MS detectors are used; while CE has been used with UV as a detection system and, more recently, with MS and fluorescence detectors. It is a quite evident fact that the couplings with MS are the most powerful ones, since the advantages of MS detection include the capability for both determining molecular weight and providing structural information. We will focus on several UHPLC–MS applications published in recent years for characterizing the phenolic compounds from several kinds of oils.

Suárez et al. [107], in 2008, published a very interesting paper in which the authors developed a UHPLC–MS/MS method and compared its analytical characteristics (in

TABLE 9.2
Phenolic Compounds of VOO: Compound Name, Chemical General Structure, and Molecular Weight

Compound Name	Substituent (MW)	Structure
Benzoic and Derivatives Acids		
3-Hydroxybenzoic acid	3-OH (138)	
p-Hydroxybenzoic acid	4-OH (138)	
3,4-Dihydroxybenzoic acid	3,4-OH (154)	
Gentisic acid	2,5-OH (154)	
Vanillic acid	3-OCH_3, 4-OH (168)	
Gallic acid	3,4,5-OH (170)	
Siringic acid	3,5-OCH_3, 4-OH (198)	
Cinnamic and Derivatives Acids		
o-Coumaric acid	2-OH (164)	
p-Coumaric acid	4-OH (164)	
Caffeic acid	3,4-OH (180)	
Ferulic acid	3-OCH_3, 4-OH (194)	
Sinapinic acid	3,5-OCH_3, 4-OH (224)	
Other Phenolic Acids and Derivatives		
p-Hydroxyphenylacetic acid	4-OH (152)	
3,4-Dihydroxyphenylacetic acid	3,4-OH (168)	
4-Hydroxy-3-methoxyphenylacetic acid	3-OCH_3, 4-OH (182)	
3-(3,4-Dihydroxyphenyl) propanoic acid	(182)	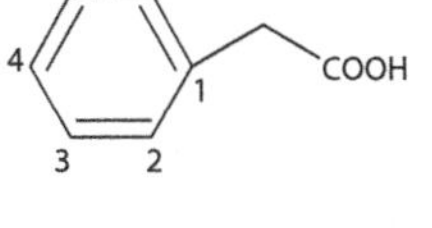
Phenyl Ethyl Alcohols		
Tyrosol [(*p*-hydroxyphenyl) ethanol] or *p*-HPEA	4-OH (138)	
Hydroxytyrosol[(*3,4*-dihydroxyphenyl)ethanol] or 3,4-DHPEA	3,4-OH (154)	
Aglycons Secoiridoids		
Oleuropein aglycon or 3,4-DHPEA-EA	R_1-OH (378)	
Ligstroside aglycon or *p*-HPEA-EA	R_1-H (362)	
Aldehydic form of oleuropein aglycon	R_1-OH (378)	
Aldehydic form of ligstroside aglycon	R_1-H (362)	

continued

TABLE 9.2 (continued)
Phenolic Compounds of VOO: Compound Name, Chemical General Structure, and Molecular Weight

Compound Name	Substituent (MW)	Structure
Dialdehydic Forms of Secoiridoids		
Decarboxymethyl ligstroside aglycon(3,4-DHPEA-EDA)	R_1-OH (304)	R* — O, O, CHO, CHO; dialdehydic form of elenolic acid (EDA)
Decarboxymethyl oleuropein aglycon(*p*-HPEA-EDA)	R_1-H (320)	
Flavonols		
(+)-taxifolin	(304)	OH, OH, HO, O, OH, OH, O
Flavons		
Apigenin	R_1-OH, R_2-H (270)	OH, R_2, HO, O, H, R_1, O
Luteolin	R_1-OH, R_2-OH (286)	
Lignans		
(+)-Pinoresinol	R-H (358)	OCH_3, OH, O, H, R, O, HO, H_3CO
(+)-1-Acetoxypinoresinol	R-$OCOCH_3$ (416)	
(+)-1-Hydroxypinoresinol	R-OH (374)	
Hydroxyisochromans		
1-Phenyl-6,7-dihydroxy-isochroman	R_1,R_2-H (242)	O, R_2, HO, R_1, OH
1-(3′Methoxy-4′hydroxy) phenyl-6,7-dihydroxy-isochroman	R_1-OH,R_2-OCH_3 (288)	

Source: From Bendini, A. et al. 2007. *Molecules* 12:1679–1719.

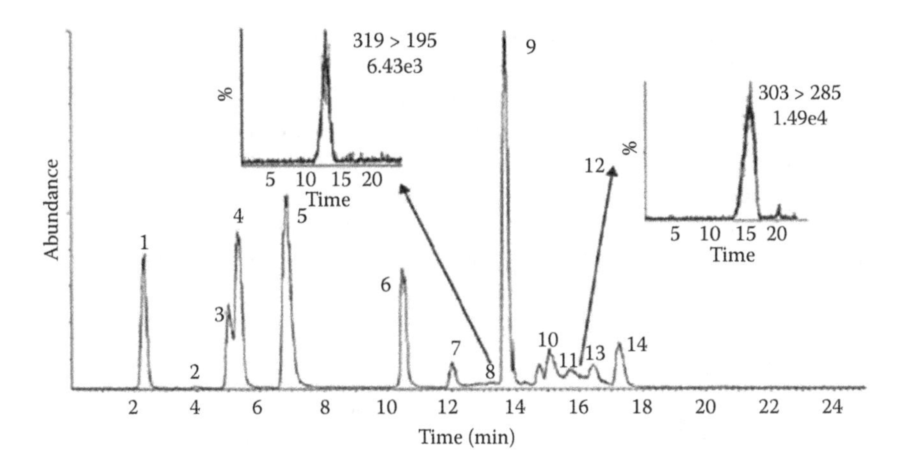

FIGURE 9.9 TIC in SRM acquisition obtained from the analysis of 14 phenolic compounds in a standard solution. Peak designation and its concentration was: (1) hydroxytyrosol, 5 mg/L; (2) tyrosol, 2 mg/L; (3) vanillic acid, 5 mg/L; (4) caffeic acid, 5 mg/L; (5) vanillin, 5 mg/L; (6) luteolin 7-O-G, 1 mg/L; (7) apigenin 7-O-G, 1 mg/L; (8) 3,4-DHPEA-EDA, 5 mg/L; (9) oleuropein, 5 mg/L; (10) luteolin, 1 mg/L; (11) pinoresinol, 5 mg/L; (12) p-HPEA-EDA, 15 mg/L; (13) acetoxypinoresinol, 5 mg/L; and (14) apigenin, 1 mg/L. (From Suárez, M. et al. 2008. *J. Chromatogr. A* 1214:90–99. With permission.)

terms of 14 phenolic compounds) with those of an HPLC-fluorescence method and a UHPLC–DAD approach. When the 1.7 μm column was used, the retention times decreased significantly with respect to conventional HPLC; detection and quantification limits were, in general, lower than for the other two methodologies; and reproducibility was lower than 3.2%. The UHPLC–MS/MS method was applied for the analysis of commercial VOOs from Arbequina cultivar and a wide range of phenolic compounds were determined in the samples under study (Figure 9.9).

The same research group [108], a year later, published an improved method for identifying and quantifying VOO phenolic compounds and their metabolites in human plasma by microelution SPE plate and liquid chromatography–tandem mass spectrometry. The μSPE-UHPLC–ESI–MS/MS method was fully validated and applied to analyze plasma samples collected after the ingestion of 30 mL of VOO in order to identify the studied phenolic compounds and their metabolites. These kinds of approaches are very important to achieve advances in the understanding of VOO polyphenols metabolism; in order to explore and determine the mechanisms of action of VOO polyphenols and their role in disease prevention, to understand the factors that constrain their release from the VOO, it is very important to understand their extent of absorption, and their fate in the organism.

An example that could be classified as a hybrid of traditional HPLC and UHPLC applications is the work carried out by García-Villalba et al. [10], in which the characterization and quantification of extra-VOO phenolic compounds by an RRLC (rapid resolution liquid chromatography) method coupled to diode-array and TOF MS detection systems was developed. The RRLC method, transferred from a conventional HPLC method, achieved better performance with shorter analysis times, separating the phenolic compounds in a column with the following characteristics: C18 column (150 mm × 4.6 mm, 1.8 μm).

Figure 9.10 shows the chromatograms of the same extra-VOO sample (Picual Borges), analyzed using the conventional HPLC and the RRLC method. With the optimum RRLC method the analysis time could be reduced from 60 min (Figure 9.10a) to 20 min (Figure 9.10b and c) and the analyst could achieve even better analytical performance by using that method. This was achieved by using steeper gradients, increased temperature, and higher flow rates.

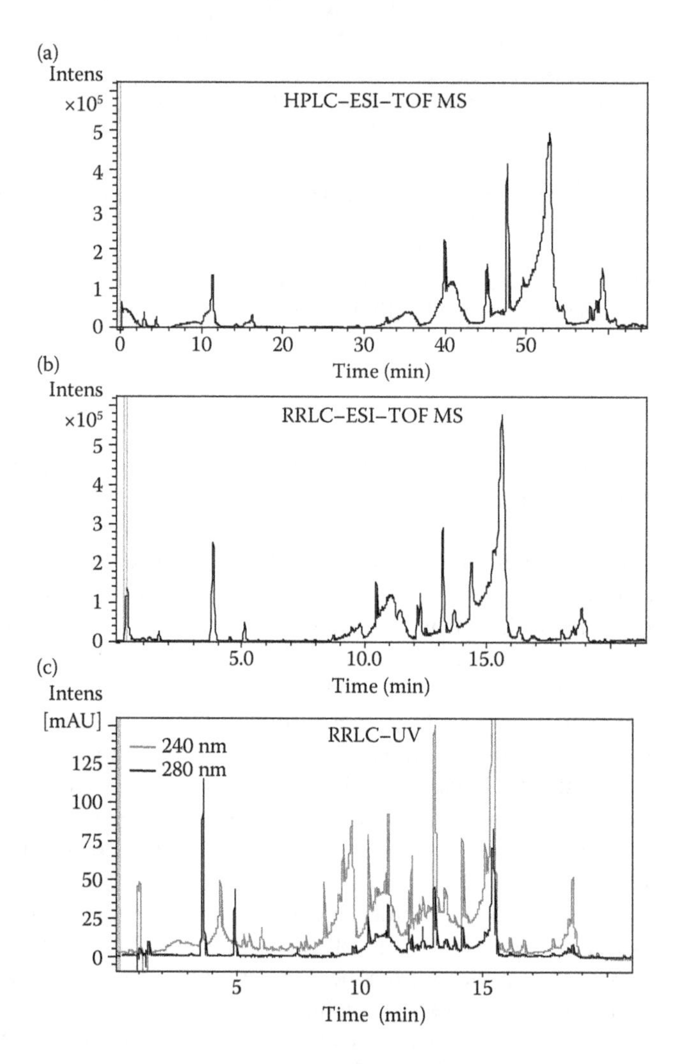

FIGURE 9.10 Comparison of the chromatograms obtained for an extra-VOO sample of Picual variety with an (a) HPLC–ESI–TOF method and a new optimized RRLC method with (b) TOF and (c) UV (280 and 240 nm) as detection systems. (Own data.)

Eight VOOs were analyzed and the quantification of the main phenolic compounds was carried out in three different ways. Taking into account the occurrence of correlations between the phenolic composition of extra-VOO-derived crude phenolic extracts and their antiproliferative abilities toward human breast cancer-derived cell lines, this novel methodological approach could enable a rapid and objective identification of extra-VOO with a potential anticancer value.

In 2012, Alarcón-Flores et al. [15] published a study where a simultaneous determination of several classes of polyphenolic compounds in different kind of oils (olive, sunflower, and soybean oils) was carried out by UHPLC–MS/MS. SPE with Diol and C18 cartridges were compared, the analytical procedure was fully validated, and then applied to analyze the mentioned types of oils. The chromatographic method seemed to be fast (running time 7.5 min), and therefore it could be applied in routine analysis. Moreover, the polyphenols content found in the different studied oils was compared, observing that VOO provides greater polyphenol concentrations than other seed oils.

9.3.3 UHPLC–MS in VOO Safety Control

With the aim of increasing production and upgrading quality, the agricultural practices involved in olive orchard management have undergone significant transformations. Thus, the modernization and intensification of olive orchards are the main reasons for the gradual increase in the use of pesticides, such as organophosphorous insecticides, synthetic pyrethroids, and organochlorine for the chemical protection of olive trees, and herbicides such as diquat, paraquat, and ammonium glufosinate for soil and weed control [109].

Pesticides have usually permitted efficient control of soil weeds and the main olive tree pests and diseases such as the olive fruit fly (*Bactrocera oleae* Gmel), Black scale (*Saissetia oleae* Olivier), olive psyllid (*Euphyllura olivina* Costa), olive leaf spot (*Cycloconium oleaginum* Cast), and olive moth (*Prays oleae* Bern). However, undue use of these products can lead to the accumulation of excessive residues in soil and olive fruits, which may lead to high possibilities of VOO contamination.

Furthermore, to ensure consumer protection by regulating the use of pesticides during the process of the production of olive fruits used in VOO production, both the European Union and the Codex Alimentarius Commission of the Food and Agriculture Organization of the United Nations (FAO) have established maximum residue levels (MRLs) in table olives and olives for oil production [110,111].

The analysis of the residues of pesticides in VOO is a complex activity for three essential reasons:

1. The pesticides used belong to diverse chemical classes, thus, their analysis requires the use of varied and appropriate analytical techniques.
2. The possibility of degradation and interferences among VOO constituents and the active ingredients of these products complicate the analysis.
3. The LODs to reach are very low, for understandable safety reasons, that means that methods have to be very sensitive.

In this context, studies about the development of relevant analytical methods allowing the detection of pesticide residues in VOO are usually focused on an optimization of the various steps of the analysis process, namely extraction, clean-up, identification, and quantitation of pesticide content. The common extraction methods are Soxhlet extraction, microwave-assisted extraction (MAE), supercritical fluid extraction (SFE), and accelerated solvent extraction (ASE). Cleanup methods include SPE, matrix solid-phase dispersion (MSPD), and gel permeation chromatography (GPC).

A large number of analytical methodologies aiming the detection of pesticide residues in VOO at very low concentration levels are being published nowadays. Among them, GC remains the most extensively used technique in routine laboratories for the analysis of these analytes in VOO; it is usually employed in combination with ECD, NPD, and FPD. However, the relatively low sensitivity needed for some pesticides and the difficulties in confirmation of results due to the complexity of the matrix under study force to use GC–MS. With the use of GC–MS/MS techniques, simultaneous determination and confirmation of pesticide residues has been obtained in one analytical run; this improves the analytical accuracy and shortens the analytical time [112,113].

However, some pesticides, such as carbamates and organophosphates, used in olive pest control are more polar, not easily vaporized, and thermally labile, facts that make difficult their determination by GC and cause the use of LC.

Methods using LC for analyzing pesticide residues in VOO have been barely applied in the past, because traditional UV, DAD, and fluorescence detectors have some limitations. However, in the last few years, the use of LC in combination with MS has widely improved the selectivity and sensitivity of this technique for the determination of pesticide analysis in VOO. LC–MS and LC–MS/MS have become widely accepted as reliable tools for identification and quantification of polar and thermally labile pesticides in VOO [114–116].

Moreover, beyond the development of analytical methodologies for the determination of pesticides in VOO, the research is focused on the characterization and detection of degradation and metabolism of pesticides in VOO. The chemical degradation and metabolism are major mechanisms of disappearance of pesticides after application to olive trees or olive grove soil. Studying pesticide metabolites in VOO could allow the characterization of a pesticide's metabolic pathway, providing fundamental information on the residues found in VOO.

Within this context, an interesting methodology using UPLC–QTOF–MS has been developed to achieve this purpose. Hernández et al. [117] were able to evaluate the potential of the UPLC–MS technique for the characterization of phosmet (organophosphate insecticide) metabolites in VOO positives samples.

For investigating the presence of phosmet metabolites, the authors used the MetaboLynx software performance to analyze VOO samples processed from treated and untreated olive fruits with phosmet. The first step was the extraction of phosmet residues with a liquid–liquid extraction (LLE) using acetonitrile saturated in hexane, and filtration of the solution through a nylon syringe filter before injection into the UHPLC–MS system. Then, the MetaboLynx software was used for scanning expected and unexpected metabolites in treated and untreated samples. Thus, comparing TOF MS sample profiles and searching differences in an automated way

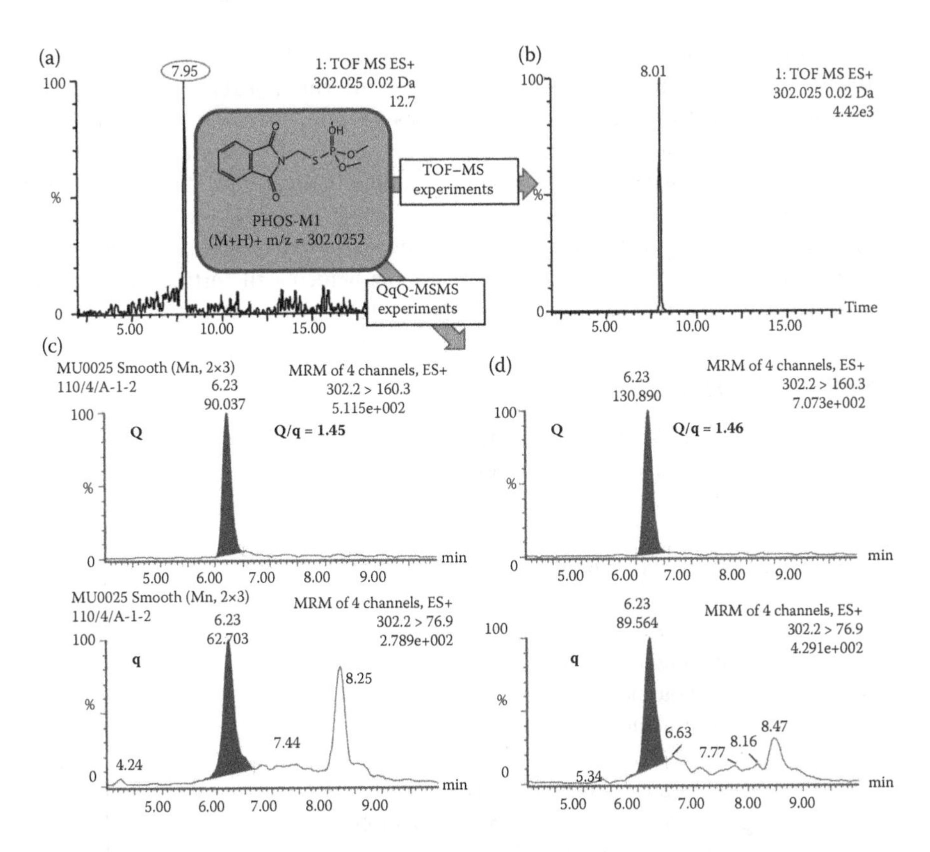

FIGURE 9.11 UHPLC–ESI(+)-TOF EICs at 302.0252 of (a) an olive oil sample where phosmet-oxon was detected (PHOS-M1) and (b) phosmet-oxon standard at 100 ng/mL. HPLC–ESI(+)-QqQ SRM chromatograms of (c) the same olive oil sample than in (a), and (d) phosmet-oxon standard at 1 ng/mL in olive oil matrix. (From Hernández, F. et al. 2009. *J. Sep. Sci.* 32:2245–2261. With permission.)

between TOF MS data of both kinds of samples (treated and untreated), the authors found several phosmet-positive VOO samples in a very short analysis time, as can be seen in Figure 9.11.

9.4 CONCLUSIONS AND FUTURE DIRECTIONS

This chapter gives an overview of the composition of VOO and the analytical techniques that can be used to characterize it, covering three different areas of current interest: authentication of VOO, characterization of the bioactive compounds and detection of pesticide residue metabolites. Particular attention has been paid to UHPLC coupled to MS, trying to show the evolution of the traditional LC methods to the newest trends in UHPLC–MS.

It is important to stand out that even though UHPLC–MS is a very powerful and promising platform for carrying out the kind of applications included in this chapter (mainly due to its rapidity, reproducibility, and proper sensitivity), it is unlikely that a single tool will give to the analyst information about all the relevant compounds present in a sample. This explains why, in some cases, the combination of the information obtained by different analytical platforms is becoming more common to assure the achievement of a holistic overview about the sample under study.

From our point of view, additional improvements are expected, concerning the column chemistry and the MS detection systems, which will further improve the performance of UHPLC–MS.

ABBREVIATIONS

APCI atmospheric pressure chemical ionization
ASE accelerated solvent extraction
CE capillary electrophoresis
DAD diode array detector
ECD electron capture detection
EIC extracted ion chromatogram
ELSD evaporative light scattering detector
FID flame ionization detector
FPD flame photometric detection
FT Fourier transform
GC gas chromatography
GPC gel permeation chromatography
HPLC high-performance liquid chromatography
IOOC International Olive Oil Council
LLE liquid–liquid extraction
LOD detection limit
LOQ quantification limit
MAE microwave-assisted extraction
MRL maximum residue level
MS mass spectrometry
MSPD matrix solid-phase dispersion
NMR nuclear magnetic resonance
NPD nitrogen–phosphorous detection
RP reverse phase
RRLC rapid resolution liquid chromatography
SFE supercritical fluid extraction
SIR selected ion recording
SPE solid-phase extraction
SRM single reaction monitoring
TAG triacylglycerols
TIC total ion chromatogram
TLC thin-layer chromatography
TOF MS time-of-flight mass spectrometry
VOO virgin olive oil

ACKNOWLEDGMENTS

The authors are very grateful to the Andalusia Regional Government (Department of Economy, Innovation and Science, Project P09-FQM-5469), to the Spanish Agency for International Development Cooperation (AECID) (pre-doctoral grant) and to Fruit Tree Productivity Project (PAF)/Millennium Challenge Account Morocco, Olive cultivation register Project/North Moroccan Regions.

REFERENCES

1. Tasioula-Margari, M. and Okogeri, O. 2001. Simultaneous determination of phenolic compounds and tocopherols in virgin olive oil using HPLC and UV detection. *Food Chem.* 74:377–383.
2. Aranda, F., Gómez-Alonso, S., Rivera del Álamo, R. M., Salvador, M. D., and Fregapane, G. 2004. Triglyceride, total and 2-position fatty acid composition of Cornicabra virgin olive oil: Comparison with other Spanish cultivars. *Food Chem.* 86:485–492.
3. Galeano-Díaz, T., Durán-Merás, I., Sánchez-Casas, J., and Alexandre-Franco, M. F. 2005. Characterization of virgin olive oils according to its triglycerides and sterols composition by chemometric methods. *Food Cont.* 16:339–347.
4. Ben Temime, S., Campeol, E., Cioni, P. L., Daoud, D., and Zarrouk, M. 2006. Volatile compounds from Chétoui olive oil and variations induced by growing area. *Food Chem.* 99:315–325.
5. Ollivier, D., Artaud, J., Pinatel, C., Durbec, J. P., and Guérère, M. 2006. Differentiation of French virgin olive oil RDOs by sensory characteristics, fatty acid and triacylglycerol compositions and chemometrics. *Food Chem.* 97:382–393.
6. Baccouri, B., Ben Temime, S., Campeol, E., Cioni, P. L., Daoud D., and Zarrouk, M. 2007. Application of solid-phase micro extraction to the analysis of volatile compounds in virgin olive oils from five new cultivars. *Food Chem.* 102:850–856.
7. Haddada, F. M., Manai, H., Daoud, D., Fernandez, X., Lizzani-Cuvelier, L., and Zarrouk M. 2007. Profiles of volatile compounds from some mono varietal Tunisian virgin olive oils. Comparison with French PDO. *Food Chem.* 103:467–476.
8. Ribeiro, L. H., Costa Freitas, A. M., and Gomes da Silva M. D. R. 2008. The use of head space solid phase micro extraction for the characterization of volatile compounds in olive oil matrices. *Talanta* 77:110–117.
9. Sakouhi, F., Harrabi, S., Absalon, C., Sbei, Kh., Boukhchina, S., and Kallel, H. 2008. α-Tocopherol and fatty acids contents of some Tunisian table olives (*Olea europea* L.): Changes in their composition during ripening and processing. *Food Chem.* 108:833–839.
10. García-Villalba, R., Carrasco-Pancorbo, A., Oliveras-Ferraros, C., Vázquez-Martín, A., Menéndez, J. A., Segura-Carretero, A., and Fernández-Gutiérrez, A. 2010. Characterization and quantification of phenolic compounds of extra-virgin olive oils with anticancer properties by a rapid and resolutive LC-ESI-TOF MS method. *J. Pharm. Biomed. Anal.* 51:416–429.
11. Tsimidou, M. Z. 2010. Squalene and tocopherols in olive oil: Importance and methods of analysis. In *Olives and Olive Oil in Health and Disease Prevention*, ed. Victor, R., and Ronald Ross Watson. Academic Press Life Sciences, Elsevier, Oxford (United kingdom), pp. 561–567.
12. Chen, H., Angiuli, M., Ferrari, C., Tombari, E., Salvetti, G., and Bramanti, E. 2011. Tocopherol speciation as first screening for the assessment of extra virgin olive oil quality by reversed-phase high-performance liquid chromatography/fluorescence detector. *Food Chem.* 125:1423–1429.

13. Ouni, Y., Taamalli, A., Gómez-Caravaca, A. M., Segura-Carretero, A., Fernández-Gutiérrez, A., and Zarrouk M. 2011. Characterization and quantification of phenolic compounds of extra-virgin olive oils according to their geographical origin by a rapid and resolutive LC–ESI-TOFMS method. *Food Chem.* 127:1263–1267.
14. García-Villalba, R., Pacchiarotta, T., Carrasco-Pancorbo, Segura-Carretero, A., Fernández-Gutiérrez, A., Deelder, A. M., and Mayboroda, O. A. 2011. Gas chromatography–atmospheric pressure chemical ionization-time of flight mass spectrometry for profiling of phenolic compounds in extra virgin olive oil. *J. Chromatogr. A* 1218:959–971.
15. Alarcón-Flores, M. I., Romero-González, R., Garrido-Frenich, A., and Martínez-Vidal, J. L. 2012. Analysis of phenolic compounds in olive oil by solid-phase extraction and ultra-high performance liquid chromatography–tandem mass spectrometry. *Food Chem.* 134:2465–2472.
16. Godoy-Caballero, M. P., Acedo-Valenzuela, M. I., and Galeano-Díaz, T. 2012. Simple quantification of phenolic compounds present in the minor fraction of virgin olive oil by LC–DAD–FLD. *Talanta* 101:479–487.
17. Fernández-Arroyo, S., Gómez-Martínez, A., Rocamora-Reverte, L., Quirantes-Piné, R., Segura-Carretero, A., Fernández-Gutiérrez, A., and Ferragut, J. A. 2012. Application of nano LC-ESI-TOF-MS for the metabolomic analysis of phenolic compounds from extra-virgin olive oil in treated colon-cancer cells. *J. Pharm. Biomed. Anal.* 63:128–134.
18. Codex Alimentarius Commission 2003. Draft revised standard for olive oil and olive pomace oils. *Report of the Eighteenth Session of the Codex Committee on Fats and Oils.* 3–7 February of 2003.
19. European Commission 2003. Regulation1989/2003/EC of 6 November 2003 amending regulation (EEC) No 2568/91 on the characteristics of olive oil and olive–pomace oil and the relevant method of analysis. *Off. J. European Union, Brussels* 46: 57–77.
20. International Olive Oil Council 2003.COI/T.15/NC n. 3, 2003, May 25- RIS/6/88/-IV/03 Trade Standards for olive oil and olive pomace oils.
21. Nagy, K., Bongiorno, D., Avellone, G., Agozzino, P., Ceraulo, L., and Vékey, K. 2005. High performance liquid chromatography–mass spectrometry based chemometric characterization of olive oils. *J. Chromatogr. A* 1078:90–97.
22. Cañabate-Díaz, B., Segura-Carretero, A., Fernández-Gutiérrez, A., Belmonte-Vega, A., Garrido-Frenich, A., Martínez-Vidal, J. L., and Marto, J. D. 2007. Separation and determination of sterols in olive oil by HPLC-MS. *Food Chem.* 102:593–598.
23. Oliveras-López, M. J., Innocenti, M., Ieri, F., Giaccherini, C., Romani, A., and Mulinacci, N. 2008. HPLC/DAD/ESI/MS detection of lignans from Spanish and Italian *Olea europaea* L. fruits. *J. Food Chem. Anal* 21:62–70.
24. Fasciotti, M. and Pereira-Netto, A. D. 2010. Optimization and application of methods of triacylglycerol evaluation for characterization of olive oil adulteration by soybean oil with HPLC-APCI-MS-MS. *Talanta* 81:1116–1125.
25. Di Maio, I., Esposto, S., Taticchi, A., Selvaggini, R., Veneziani, G., Urbani, S., and Servili, S. 2011. HPLC–ESI-MS investigation of tyrosol and hydroxytyrosol oxidation products in virgin olive oil. *Food Chem.* 125:21–28.
26. Bakhouche, A., Lozano-Sánchez, J., Beltrán-Debón, R., Joven, J., Segura-Carretero, A., and Fernández-Gutiérrez, A. 2013. Characterization and geographical classification of commercial Arbequina extra-virgin olive oils produced in southern Catalonia. *Food Res. Inter.* 50:401–408.
27. Boskou, D., Blekas, G., and Tsimidou, M. 2006. Olive oil composition. In *Olive Oil: Chemistry and Technology*, ed. Boskou, D., AOCS Press, Champaign, IL, USA, pp. 41–72.
28. Cimato, A., Cantini, C., and Sani, G. 1990. Climate-phenology relationships on olive cv Frantoio. *Acta Hort. (ISHS)* 286:171–174.

29. Lavee, S. and Wodner, M. 1991. Factors affecting the nature of oil accumulation in fruit of olive (*Olea europaea*) cultivars. *J. Hort. Sci.* 66:583–591.
30. Rallo, L. and Martin, G. C. 1991. The role of chilling in releasing olive floral buds from dormancy. *J. Am. Soc. Hortic. Sci.* 116:1058–1062.
31. Fernández-Escobar, R., Benlloch, M., Navarro C., and Martín, G. C. 1992. The time of floral induction in the olive. *J. Am. Soc. Hort. Sci.* 117:304–307.
32. Barranco, D., Milona, G., and Rallo, L. 1994. Épocas de floración de cultivares de olivo en Córdoba. *Invest. Agr. Prod. Prot. Veg.* 9:213–220.
33. Fantozzi, P., Simonetti, M. S., Cossignani, L., and Damiani, P. 1994. An approach to extra virgin olive oil characterization by studying the relationship between place of origin and oil composition. *Acta Hort. (ISHS)* 356:367–371.
34. Alessandri, S., Caselli, S., Agronomi, A., Cimato, A., Modi, G., Tracchi, S., and Crescenzi, A. 1999. The characterization and classification of Tuscan olive oils by zone: Yearly variations of the oil composition and reliability of the classification models. *Acta Hort. (ISHS)* 474:649–652.
35. Inglese, P., Gullo, G., and Pace, L. S. 1999. Summer drought effects on fruit growth, ripening and accumulation and composition of 'Carolea' olive oil. *Acta Hort. (ISHS)* 474:269–274.
36. Ben Rouina, B., Trigui, A., and Boukhris, M. 2002. Effect of the climate and the soil conditions on crops performance of the "Chemlali de Sfax" olive trees. *Acta Hort. (ISHS)* 586:285–289.
37. Rodrigues, M. A. and Arrobas, M. M. 2008. Effect of soil boron application on flower bud and leaf boron concentrations of olives. *Acta Hort. (ISHS)* 791:393–396.
38. Tubeileh, A., Turkelboom, F., Abdeen, M., and Al-Ibrahem, A. 2008. Fruit and oil characteristics of three main syrian olive cultivars grown under different climatic conditions. *Acta Hort. (ISHS)* 791:409–413.
39. Maria, T. and Robert, M. 1987. Authentication of virgin olive oils using principal component analysis of triglyceride and fatty acid profiles: Part 2, detection of adulteration with other vegetable oils. *Food Chem.* 25:251–258.
40. Aparicio, R., Morales, M. T., and Alonso, V. 1997. Authentication of European extra-virgin olive oils by their chemical compounds, sensory attributes and consumers attitudes. *J. Agric. Food Chem.* 45:1076–1083.
41. Bohacenko, I. and Kopicova, Z. 2001. Detection of olive oils authenticity by determination of their sterol content using LC/GC. *Czech J. Food Sci.* 19:97–103.
42. Bowadt, S. and Aparicio, R. 2003. The detection of the adulteration of olive oil with hazelnut oil: A challenge for the chemist. *Inform.* 14:342–344.
43. Cercaci, L., Rodriguez-Estrada, M. T., and Lercker, G. 2003. Solid-phase extraction-thin layer chromatography-gas chromatography method for the detection of hazelnut oil in olive oils by determination of esterified sterols. *J. Chromatogr. A* 985:211–220.
44. Al-Ismail, Kh. M., Alsaed, A. K., Ahmad, R., and Al-Dabbas, M. 2010. Detection of olive oil adulteration with some plant oils by GLC analysis of sterols using polar column. *Food Chem.* 121:1255–1259.
45. Lerma-García, M. J., Ramis-Ramos, G., Herrero-Martínez, J. M., and Simó- Alfonso, E. F. 2010. Authentication of extra virgin olive oils by Fourier-transform infrared spectroscopy. *Food Chem.* 118:78–83.
46. Lerma-García, M. J. 2012. Characterization and authentication of olive and other vegetable oils. *Springer Theses.* pp. 228.
47. Rohman, A. and Che Man, Y. B. 2010. Fourier transform infrared (FTIR) spectroscopy for analysis of extra virgin olive oil adulterated with palm oil. *Food Res. Int.* 43:886–892.
48. Bertran, E., Blanco, M., Coello, J., Iturriaga, H., Maspoch, S., and Montolin, I. 2000. Near infrared spectrometry and pattern recognition as screening methods for the authentication of virgin olive oils of very close geographical origins. *J. Near Infrared Spectrosc.* 8:45–52.

49. Codex Alimentarius Commission 2013. Amended draft revised standard for olive oil and olive pomace oils. *Report of the Twenty-third Session of the Codex Committee on Fats and Oils*. February 2013.
50. International Olive Oil Council Trade standards for olive oil and olive pomace oils 2011. COI/T.15/NC n° 3/Rev. 6, November 2011.
51. Dionisi, F., Prodolliet, J., and Tagliaferri, E. 1995. Assessment of olive oil adulteration by reversed-phase high-performance liquid chromatography/amperometric detection of tocopherols and tocotrienols. *J. AOAC* 72:1505–1511.
52. Bonvehi, J. S., Ventura Coll, F., and Rius, I. A. 2000. Liquid chromatographic determination of tocopherols and tocotrienols in vegetable oils, formulated preparations, and biscuits. *J. AOAC Inter.* 83:627–634.
53. International Olive Oil Council. 2001. Determination of the difference between actual and theoretical content of triacyglycerols with ECN 42. Method of analysis COI/T.20/ Doc. no. 20/Rev.1.
54. Parcerisa, J., Casals, I., Boatella, J.,Codony, R., and Rafecas, M. 2000. Analysis of olive and hazelnut oil mixtures by high performance liquid chromatography-atmospheric pressure chemical ionisation mass spectrometry of triacylglycerols and gas-liquid chromatography of non-saponifiable compounds (tocopherols and sterols). *J. Chromatogr. A* 881:149–158.
55. Cunha, S. C. and Oliveira M. B. P. P. 2006. Discrimination of vegetable oils by triacylglycerols evaluation of profile using HPLC/ELSD. *Food Chem.* 95:518–524.
56. Miroslav, L., Holcapek, M., and Bohac, M. 2009. Statistical evaluation of triacylglycerol composition in plant oils based on high-performance liquid chromatography atmospheric pressure chemical ionization mass spectrometry data. *J. Agric. Food Chem.* 57:6888–6898.
57. Holcapek, M. and Miroslav, L. 2009. Statistical evaluation of triacylglycerol composition by HPLC/APCI-MS. *Lipid Techn.* 21:261–265.
58. Lee, P. J. and Gioia A. J. 2009. Rapid seed oil analysis using UPLC for quality control and authentication. *Lipid Techn.* 21:112–115.
59. Owen, R. W., Haubner, R., Wurtele, G., Hull, W. E., Spiegelhalder, B., and Bartsch, H. 2004. Olives and olive oil in cancer prevention. *Eur. J. Cancer Prev.* 13:319–326.
60. Owen, R. W., Giocoso, A., Hull, W. E., Spiegelhalder, B., and Bartsch, H. 2000. The antioxidant/anticancer potential of phenolic compounds isolated from olive oil. *Eur J. Cancer* 36:1235–1247.
61. Andrews, P., Busch, J. L. H. C., Joode, T. D., Groenewegen, A., and Alexandre, H. 2003. Sensory properties of virgin olive oil polyphenols: Identification of deacetoxy-ligstroside agglycon as a key contributor to pungency. *J. Agric. Food Chem.* 51:1415–1420.
62. Soler-Rivas, C., Espin, J. C., and Wichers, H. J. 2000. Oleuropein and related compounds. *J. Sci. Food Agric.* 80:1013–1023.
63. Gutiérrez-Rosales, F., J., Ríos, M., and Gómez-Rey, L. 2003. Main polyphenols in the bitter taste of virgin olive oil. Structural confirmation by on line HPLC electrospray ionization mass spectrometry. *J. Agric. Food Chem.* 51:6021–6025.
64. Covas, M. I., Nyyssönen, K., Poulsen, H. E., Kaikkonen, J., Zunft, H. J., Kiesewetter, H., Gaddi, A. et al. 2006a. The effect of polyphenols in olive oil on heart disease risk factors: A randomized trial. *Ann. Int. Med.* 145:333–341.
65. Covas, M. I., Ruiz-Gutiérrez, V., DelaTorre, R., Kafatos, A., Lamuela-Raventós, R. M., Osada, J., Owen, R. W., and Visioli, F. 2006b. Olive oil minor components: Evidence to date of health benefits in humans. *Nutr. Rev.* 64:20–30.
66. Cerretani, L., Salvador, M. D., Bendini, A., and Fregapane, G. 2008. Relationship between sensory evaluation performed by italian and Spanish official panels and volatile and phenolic profiles of virgin olive oils. *Chem. Percept.* 1:258–267.

67. Awad, A. B. and Fink, C. S. 2000. Phytosterols as anticancer dietary components: Evidence and mechanism of action. *J. Nutr.* 130:2127–2130.
68. Orzechowski, A., Ostaszewski, P., Jank, M., and Berwid, S. J. 2002. Bioactive substances of plant origin in food, impact on genomics. *Rep. Nutr. Dev.*42:461–477.
69. St-Onge, M. P. and Jones, P. J. H. 2003. Phytosterols and human lipid metabolism: Efficacy, safety, and novel foods. *Lipids* 38:367–375.
70. Trautwein, E. A., Duchateau, G., Lin, Y. G., Melnikov, S. M., Molhuizen, H. O. F., and Ntanios, F. Y. 2003. Proposed mechanisms of cholesterol-lowering action of plant sterols. *Eur. J. Lipid. Sci. Techn.* 105:171–185.
71. Brufau, G., Canela, M. A., and Rafecas, M. 208. Phytosterols: Physiologic and metabolic aspects related to cholesterol-lowering properties. *Nutr. Res.* 28:217–225.
72. Food and Drug Administration 2007. Food and drugs. Food labeling. Specific requirements for health claims. Health claims: Plant sterol/stanol esters and risk of coronary heart disease. Retrieved 14 August 2007.
73. Matos, L. C., Cunha, S. C., Amaral, J. S., Pereira, J., Andrade, P., Seabra, R. M., and Oliveira, B. P. P. 2007. Chemometric characterization of three varietal olive oils (Cvs. Cobrançosa, Madural and VerdealTransmontana) extracted from olives with different maturation indices. *Food Chem.* 102:406–414.
74. Lazzez, A., Perri, E., Caravita, M. A., Khlif, M., and Cossentini, M. 2008. Influence of olive maturity stage and geographical origin on some minor components in virgin olive oil of the Chemlali variety. *J. Agric. Food Chem.* 56:982–988.
75. EEC 1991. On the characteristics of olive oil and olive-residue oil and on the relevant methods of analysis. EEC Regulation 2568. EEC Official Report L248, pp. 1–48.
76. Issaoui, M., Mechri, B., Echbili, A., Dabbou, S., Yangui, A., Belguith, H., Trigui, A., and Hammami, M. 2008. Chemometric characterization of five Tunisian varietals of *Olea europaea* L. olive fruit according to different maturation indices. *J. Food Lipids* 15:277–296.
77. Mailer, R. J., Ayton, J., and Graham, K. 2010. The influence of growing region, cultivar, and harvest timing on the diversity of Australian olive oil. *J. AOCS.* 87:877–884.
78. Vekiari, S. A., Oreopoulou, V., Kourkoutas, Y., Kamoun, N., Msallem, M., Psimouli, V., and Arapoglou, D. 2010. Characterization and seasonal variation of the quality of virgin olive oil of the Throumbolia and Koroneiki varieties from Southern Greece. *Grasas y Aceites.* 61:221–231.
79. Sánchez-Casas, J., Bueno, E. O., García, A. M. M., and Cano, M. M. 2004. Sterol and erythrodiol + uvaol content of virgin olive oils from cultivars of Extremadura (Spain). *Food Chem.* 87:225–230.
80. Lukic, M., Lukic, I., Krapac, M., Sladonja, B., and Pilizota, P. 2013. Sterols and triterpenediols in olive oil as indicators of variety and degree of ripening. *Food Chem.* 136:251–258.
81. Lerma-García, M. J., Concha-Herrera, V., Herrero-Martínez, J. M., Simó-Alfonso, E. F. 2009. Classification of extra virgin olive oils produced at la Comunitat Valenciana according to their genetic variety using sterol profiles established by high-performance liquid chromatography with mass spectrometry detection. *J. Agric. Food Chem.* 57:10512–10517.
82. Segura-Carretero, A., Carrasco-Pancorbo, A., Cortacero, S., Gori, A., Cerretani, L., and Fernández-Gutiérrez, A. 2008. A simplified method for HPLC-MS analysis of sterols in vegetable oil. *Eur. J. Lipid Sci. Technol.* 110:1142–1149.
83. Azadmard-Damirchi, S., and Dutta, P. C. 2006. Novel solid-phase extraction method to separate 4-desmethyl-,4-monomethyl-, and 4,4′dimethylsterols in vegetable oils. *J. Chromatogr. A* 1108:183–187.
84. Toledano, R. M., Cortés, J. M., Rubio-Moraga, A., Villén, J., and Vázquez, A. 2012. Analysis of free and esterified sterols in edible oils by online reversed phase liquid

chromatography-gas chromatography (RPLC-GC) using the through oven transfer adsorption desorption (TOTAD) interface. *Food Chem.* 135:610–615.

85. Alves, R. M., Cunha, S. C., Amaral, Pereira, J. A., and Oliveira, M. B. P. P. 2005. Classification of PDO olive oils on the basis of their sterol composition by multivariate analysis. *Anal. Chim. Acta* 549:166–178.
86. Sivakumar, G., Briccoli, B. C., Perri, E., and Uccella, N. 2006. Gas chromatography screening of bioactive phytosterols from mono-cultivar olive oils. *Food Chem.* 95:525–528.
87. Romero-González, R., Garrido-Frenich, A., and Martínez-Vidal, J. L. 2010. Liquid chromatography-mass spectrometry determination of sterols in olive oil. In *Olives and Olive Oil in Health and Disease Prevention,* ed. Victor, R., and Ronald Ross Watson Academic Press Life Sciences, Elsevier, Oxford (United Kingdom), pp. 591–601.
88. Lerma-García, M. J., Simó-Alfonso, E. F., Méndez, A., Lliberia, J. L., and Herrero-Martínez, J. M. 2010. Fast separation and determination of sterols in vegetable oils by ultra-performance liquid chromatography with atmospheric pressure chemical ionization mass spectrometry detection. *J. Agric. Food Chem.* 58:2771–2776.
89. Lerma-García, M. J., Simó-Alfonso, E. F., Méndez, A., Lliberia J. L., and Herrero-Martínez, J. M. 2011. Classification of extra virgin olive oils according to their genetic variety using linear discriminant analysis of sterol profiles established by ultra-performance liquid chromatography with mass spectrometry detection. *Food Res. Inter.* 44:103–108.
90. Martínez-Vidal, J. L., Garrido-Frenich, A., Escobar-García, M. A., and Romero-González, R. 2007. LC–MS Determination of sterols in olive oil. *Chromatographia* 65:695–699.
91. International Olive Oil Council 2001. Determination of the composition and content of sterols by capillary-column gas chromatography. Method of analysis COI/T.20/Doc. no. 10/Rev.1.
92. Bendini, A., Cerretani, L., Carrasco-Pancorbo, A., Gómez-Caravaca, A. M., Segura-Carretero, A., Fernández-Gutiérrez, A., and Lercker, G. 2007. Phenolic molecules in virgin olive oils: A survey of their sensory properties, health effects, antioxidant activity and analytical methods. An overview of the last decade. *Molecules* 12:1679–1719.
93. Montedoro, G. F., Servili, M., Baldioli, M., Selvaggini, R., Miniati, E., and Macchioni, A. 1993. Simple and hydrolyzable compounds in virgin olive oil. Spectroscopic characterizations of the secoiridoid derivatives. J. *Agric. Food Chem.* 41:2228–2234.
94. Morelló, J. R., Vuorela, S., Romero M. P., Motilva, M. J., and Heinonen, M. 2002. Antioxidant activity of olive pulp and olive oil phenolic compounds of the arbequina cultivar. *J. Agric. Food Chem.* 53:2002–2008.
95. Bonoli, M., Bendini, A., Cerretani, L., Lercker, G., and Gallina Toschi, T. 2004. Qualitative and semiquantitative analysis of phenolic compounds in extra virgin olive oft as a function of the ripening degree of olive fruits by different analytical techniques. *J. Agric. Food Chem.* 52:7026–7032.
96. Brenes, M., García, A., Ríos, J. J., García, P., and Garrido, A. 2002. Use of 1-acetoxypinoresinol to authenticate Picual olive oils. *Int. J. Food Sci. Technol.* 37:615–625.
97. Pinelli, P., Galardi, C., Mulinacci, N., Vincieri, F. F., Cimato, A., and Romani, A. 2003. Minor polar compound and fatty acid analyses in monocultivar virgin olive oils from Tuscany. *Food Chem.* 80:331–336.
98. Brenes, M., García, A., García, P., Ríos, J. J., and Garrido, A. 1999. Phenolic compounds in Spanish olive oils. *J. Agric. Food Chem.* 47: 3535–3540.
99. Carrasco-Pancorbo, A., Cruces-Blanco, C., Segura-Carretero, A., and Fernández-Gutiérrez, A. 2004. Sensitive determination of phenolic acids in extra-virgin olive oil by capillary zone electrophoresis. *J. Agric. Food Chem.* 52:6687–6693.
100. Bianco, A., Coccioli, F., Guiso, M., and Marra, C. 2001. Analysis by HPLC-MS/MS of biophenolic components in olives and oils. *Food Chem.* 77:405–411.

101. Servili, M., Selvaggini, R., Esposto, S., Taticchi, A., Montedoro, G., and Morozzi. G. 2004. Health and sensory properties of virgin olive oil hydrophilic phenols: Agronomic and technological aspects of production that affect their occurrence in the oil. *J. Chromatogr. A* 1054:113–127.
102. Tovar, M. J., Motilva, M. J., and Romero, M. P. 2001. Changes in the phenolic composition of virgin olive oil from young trees (*Olea europaea* L. cv Arbequina) grown under linear irrigation strategies. *J. Agric. Food Chem.* 49:5502–5508.
103. Uceda, M., Hermoso, M., García-Ortiz, A., Jiménez, A., and Beltrán, G. 1999. Intraspecific variation of oil contents and the characteristics of oils in olive cultivars. *Acta Hort. (ISHS)* 474:659–652.
104. Aparicio, M. and Luna, G. 2002. Characterization of monovarietal virgin olive oils. *Eur. J. Lipid Sci. Techn.* 104:614–627.
105. Boskou, D. 2009. Phenolic compounds in olives and olive oil. In *Olive Oil. Minor Constituents and Health*, ed. Boskou, D. CRC press, NY, USA, pp. 11–44, ISBN 978-1-4200-5993-9.
106. Carrasco-Pancorbo, A., Cerretani, L., Bendini, A., Segura-Carretero, A., Gallina-Toschi, T., and Fernández-Gutiérrez, A. 2005. Analytical determination of polyphenols in olive oils. *J. Sep. Sci.* 28:837–858.
107. Suárez, M., Macia, A., Romero, M. P., and Motilva, M. J. 2008. Improved liquid chromatography tandem mass spectrometry method for the determination of phenolic compounds in virgin olive oil. *J. Chromatogr. A* 1214:90–99.
108. Suárez, M., Romero, M. P., Macia, A., Valls, R. M., Fernandez, S., Sola, R., and Motilva, M. J. 2009. Improved method for identifying and quantifying olive oil phenolic compounds and their metabolites in human plasma by micro-elution solid-phase extraction plate and liquid chromatography-tandem mass spectrometry. *J. Chromatogr. B Analyt. Technol. Biomed. Life Sci.* 877:4097–4106.
109. International Olive Oil Council 2007. *Production Techniques in Olive Growing*, ed Artegraf, S. A., Madrid (Spain), pp. 348.
110. Pesticide EU-MRLs Data base. 2005. Regulation (EC) n. 396/2005. http://ec.europa.eu/sanco_pesticides/public/index.cfm.
111. Codex Alimentarius. Pesticide MRL Database. http://www.mrldatabase.com.
112. Marinas, A., Lafont, F., Aramendia, M. A., García, I. M., Marinas, J. M., and Urbano, F. J. 2010. Multiresidue analysis of low- and medium-polarity pesticides in olive oil by GC-MS/MS. In *Olives and Olive Oil in Health and Disease Prevention*, ed. Victor, R., and Ronald Ross Watson Academic Press Life Sciences, Elsevier, Oxford (United Kingdom), pp. 667–683.
113. Anagnostopoulos, C. and Miliadis, G. E. 2013. Development and validation of an easy multiresidue method for the determination of multiclass pesticide residues using GC–MS/MS and LC–MS/MS in olive oil and olives. *Talanta* 112:1–10.
114. García-Reyes, J. F., Ferrer, C., Gómez-Ramos, M. J., Fernández-Alba, A. R., and Molina-Díaz, A. 2007. Determination of pesticide residues in olive oil and olives. *TrAC Trends in Anal. Chem.* 26:239–251.
115. Gilbert-López, B., García-Reyes, J. F., Fernández-Alba, A. R., and Molina-Díaz, A. 2010. Evaluation of two sample treatment methodologies for large-scale pesticide residue analysis in olive oil by fast liquid chromatography–electrospray mass spectrometry. *J. Chromatogr. A* 1217:3736–3747.
116. Benincasa, C., Perri, E., Iannotta, N., and Scalercio, S. 2011. LC/ESI–MS/MS method for the identification and quantification of spinosad residues in olive oils. 2011. *Food Chem.* 125:1116–1120.
117. Hernández, F., Grimalt, S., Pozo, O. J., and Sancho, J. V. 2009. Use of ultra-high-pressure liquid chromatography–quadrupole time-of-flight MS to discover the presence of pesticide metabolites in food samples. *J. Sep. Sci.* 32:2245–2261.

10 Vitamin Analysis in Food by UPLC–MS

Ahmad Aqel, Kareem Yusuf, Asma'a Al-Rifai, and Zeid Abdullah Alothman

CONTENTS

10.1 INTRODUCTION

10.1.1 Vitamin Structure and Function

Vitamins are defined as a biologically active group of organic compounds that have a relatively low molecular weight. They are minor, but essential for an organism's normal health and growth. Humans need to obtain them from food or supplements [1,2]. These nutrients facilitate the metabolism of proteins, carbohydrates, and fats. They are reported to reduce damage from free radicals [3], and insufficient levels may result in deficiency diseases [4].

Vitamins are relatively labile compounds that are affected by factors such as heat, light, air, pH, other food components, and food processing conditions [5–7]. Most vitamins and their related compounds are now synthetically produced and widely used as food or feed additives, medical or therapeutic agents, health aids, or cosmetic and technical aids [8].

10.1.2 Classification of Vitamins

Vitamins are classified by their biological and chemical activity, not their structure. Thus, there are 13 vitamins identified that are classified according to their solubility into fat-soluble vitamins (A, E, D, and K) and water-soluble vitamins (B-group vitamins and vitamin C) [9]. Table 10.1 shows the vitamin classes and their chemical name, structure, sources, functions, and deficiency diseases.

10.2 METHODS OF VITAMIN ANALYSIS

10.2.1 Importance of Analysis

Vitamin analysis in food has a variety of purposes; it is used to provide quality assurance for supplemented products; to study changes in vitamin content attributable to food processing, packing, and storage; to provide data for food composition tables; and to check compliance with contract specifications and nutrient labeling regulation [16]. This section gives a short overview of techniques for the analysis of the vitamin content in food and some of the problems associated with these techniques.

10.2.2 Methods of Analysis

Many vitamins have sensitivity to different conditions, regardless of the type of assay applicable. Therefore, certain precautions need to be taken to prevent any deterioration throughout the analytical process. For example, in bioassays, some steps need to be followed with the test material throughout the feeding period, while in microbiological and physicochemical methods, they are required during extraction and analytical procedure [4,9].

Various vitamins belong to different classes of organic substances, so analysis of vitamins is performed by different chemical, physical, and biological methods [17]. Vitamin analysis methods can be classified as follows [18]:

- Bioassays involving humans and animals
- Microbiological assays using organisms, bacteria, and yeast
- Physicochemical assays, including spectrophotometric, fluorometric, chromatographic, enzymatic, immunological, and radiometric methods

- *Bioassay methods* Bioassays are procedures that can determine the concentration of purity or biological activity of a substance such as vitamin, hormone, and plant growth factor [19]. They measure the enzyme under the influence of a vitamin and the phenotypic effects of their deficiency [18]. Bioassay methods are rarely used clinically for vitamin analysis; they are most commonly used for the analysis of vitamins B12 and D [20]. There are two types of bioassays:
  - *In vivo* bioassays Although this method is close to reality, it has many disadvantages, including that it is not applicable for all vitamins, is time consuming, is an indirect method, is expensive, and has no clinical application [21].
  - *In vitro* bioassays Compared with *in vivo* bioassays, this method is easier, is less time consuming, is inexpensive, can be used to analyze several vitamins, and is not close to reality, and the cell cultures are at risk of contamination [22].

TABLE 10.1
List of Vitamins with Their Chemical Name, Structure, Sources, Functions, and Deficiency Diseases

Vitamin Generic Descriptor Name	Vitamer Chemical Name	Chemical Structure	Solubility	Food Sources	Function	Deficiency Diseases
Vitamin A	Retinol	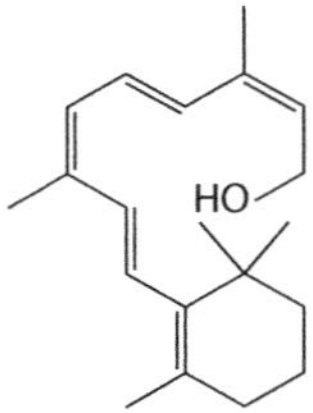	Fat	Orange, ripe yellow fruits, leafy vegetables, carrots, milk, cheese, ice cream, liver	Necessary for normal vision, immune function, reproduction	Night blindness, hyperkeratosis, keratomalacia
Vitamin B1	Thiamine	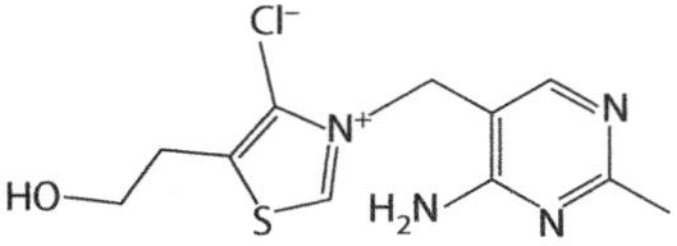	Water	Oatmeal, brown rice, vegetables, potatoes, liver, eggs	Allows the body to process carbohydrates and some protein	Beriberi, Wernicke–Korsakoff syndrome
Vitamin B2	Riboflavin	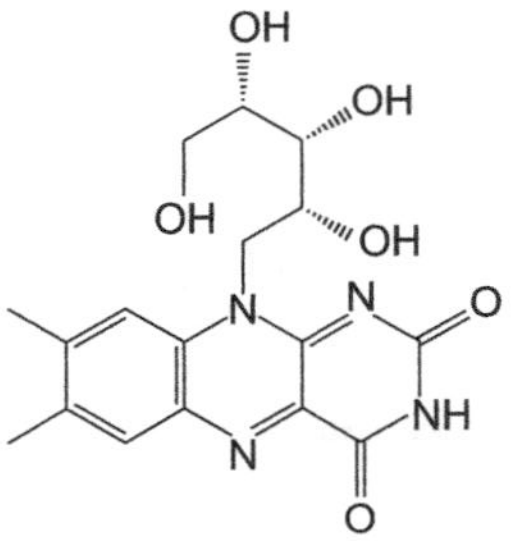	Water	Dairy products, bananas, popcorn, green beans, asparagus	Key in the metabolism and the conversion of food into energy, helps produce red blood cells	Ariboflavinosis

continued

TABLE 10.1 (continued)
List of Vitamins with Their Chemical Name, Structure, Sources, Functions, and Deficiency Diseases

Vitamin Generic Descriptor Name	Vitamer Chemical Name	Chemical Structure	Solubility	Food Sources	Function	Deficiency Diseases
Vitamin B3	Niacin	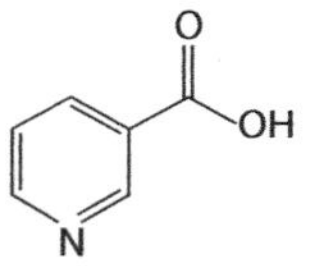	Water	Meat, fish, eggs, many vegetables, mushrooms, tree nuts	Assists in digestion and the conversion of food into energy, important in the production of cholesterol	Pellagra
Vitamin B5	Pantothenic acid	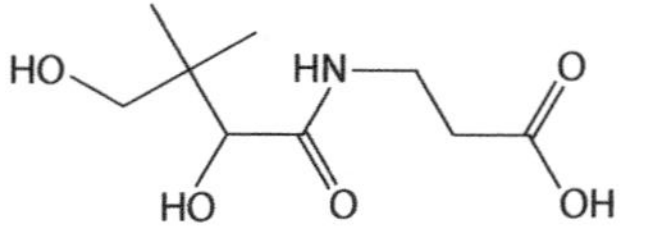	Water	Meat, broccoli, avocados	Important in fatty acid metabolism	Paresthesia
Vitamin B6	Pyridoxine	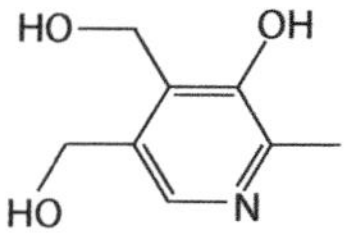	Water	Meat, vegetables, tree nuts, bananas	Important for the nervous system, helps the body metabolize proteins and sugar	Anemia peripheral neuropathy
Vitamin B7	Biotin	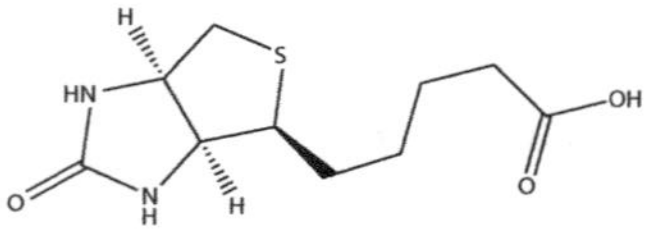	Water	Raw egg yolk, liver, peanuts, certain vegetables	Helps with the synthesis of fats, glycogen, and amino acids	Dermatitis, enteritis

Vitamin B9	Folic acid	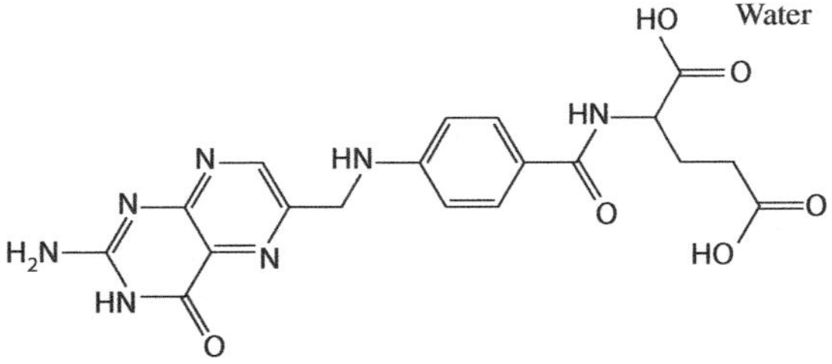	Water	Leafy vegetables, pasta, bread, cereal, liver	Key for the development of cells, protein metabolism, and heart health, and in pregnant women, helps prevent birth defects	Megaloblast and deficiency during pregnancy is associated with birth defects, such as neural tube defects
Vitamin B12	Cyanocobalamin	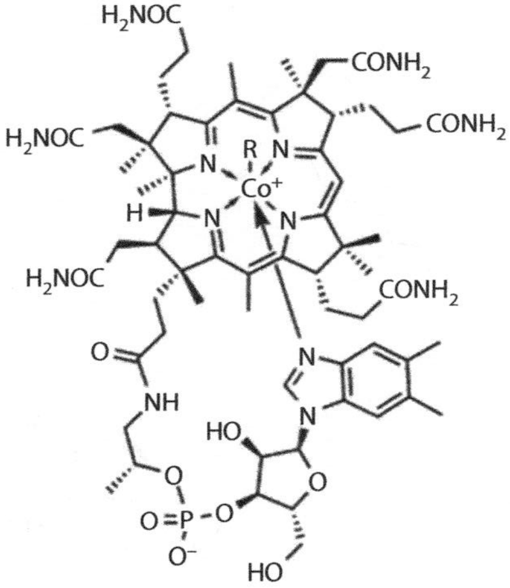 R = 5′-deoxyadenosyl, me, OH, CN	Water	Meat and other animal products	Important in the production of red blood cells	Megaloblastic anemia

continued

TABLE 10.1 (continued)
List of Vitamins with Their Chemical Name, Structure, Sources, Functions, and Deficiency Diseases

Vitamin Generic Descriptor Name	Vitamer Chemical Name	Chemical Structure	Solubility	Food Sources	Function	Deficiency Diseases
Vitamin C	Ascorbic acid	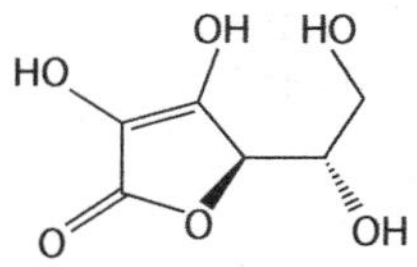	Water	Many fruits and vegetables, liver	Antioxidant that protects against cell damage, boosts the immune system, and forms collagen in the body	Scurvy
Vitamin D	Cholecalciferol	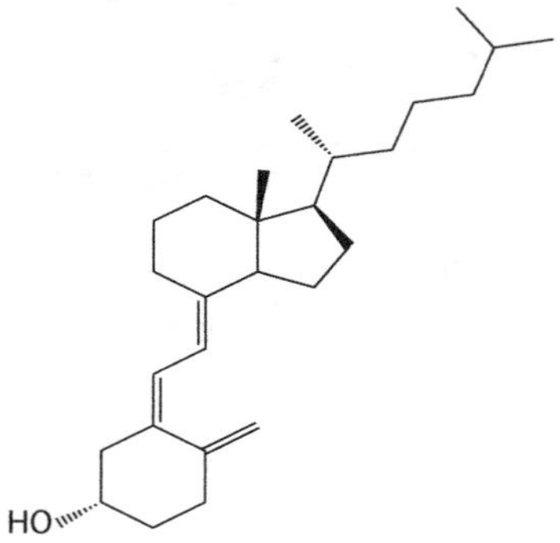	Fat	Fish, eggs, liver, mushrooms, milk	Crucial in metabolizing calcium for healthy bones	Rickets, osteomalacia

Vitamin E	Tocopherols	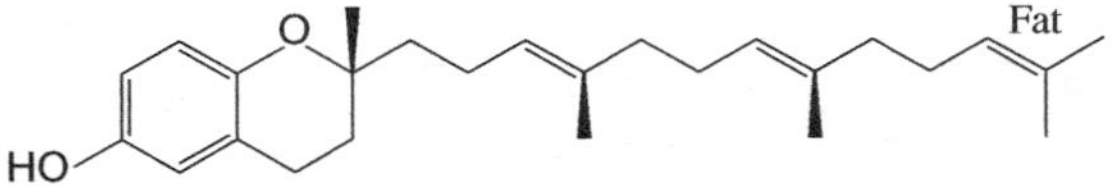 HO, O	Fat	Many fruits and vegetables, nuts and seeds, apples	Antioxidant that protects cells against damage	Deficiency is very rare, mild hemolytic anemia in newborn infants
Vitamin K	Phylloquinone	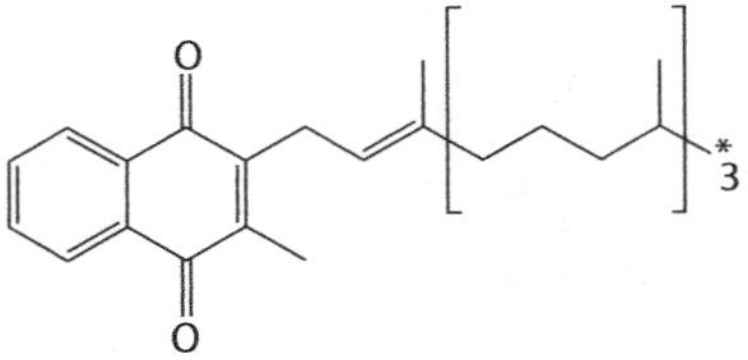 O, O, *, 3	Fat	Leafy green vegetables, egg, liver, kiwi	Important in blood clotting and bone health	Bleeding diathesis

Source: Adapted from AIN (American Institute of Nutrition). 1990. *J. Nutr.* 120:643–644; Booher, L., Hartzler, E, and Hewston, E. 1942. *U.S. Dept. Agric. Cir.* 638:1–244; Booth, S. et al. 1993. *J. Food Comp. Anal.* 6:109–120; Kamal-Eldin, A. and Appelqvist, L. 1996. *Lipids.* 31:671–701; Sheppard, A. and Pennington, J. 1993. *Analysis and Distribution of Vitamin E in Vegetable Oils and Foods.* Marcel Dekker, New York; Rucker, R. et al. 2001. *Handbook of Vitamins.* 3rd Ed. Marcel Dekker, New York.

- *Microbiological assay methods* This is a method of measuring compounds, such as vitamins and amino acids, using microorganisms. It determines or estimates the concentration or potency of an antibiotic by means of measuring and comparing the area of zone of inhibition or turbidity produced by the test substance with that of standard over a suitable microbe under standard conditions. The applicability of microbiological assays is limited to water-soluble vitamins; they are most commonly applied to niacin, cyanocobalamin, and pantothenic acid [4]. These methods are highly sensitive and specific for each vitamin. Although they are somewhat time consuming, they can generally be used for the analysis of a relatively wide array of biological matrices without major modifications [4]. An example of microbiological assays mentioned in the literature is for the determination of vitamins in tarhana [23].
- *Physicochemical assay methods* Because of their relative simplicity, accuracy, and precision, the physicochemical methods, in particular the chromatographic methods using high-performance liquid chromatography (HPLC), are the preferred method for vitamin analysis.

  HPLC has been used increasingly in the analysis of food samples to separate and detect additives and contaminants. HPLC can separate a large number of compounds both rapidly and at high sensitivity, reduce separation times, and reduce the volume of sample needed. HPLC is ideally suited for compounds of limited thermal stability, but requires sample pretreatment such as extraction and filtration. In addition, HPLC requires careful selection of mobile phase and sample pumping rate [24].

  Today, HPLC is used as a reference technique to analyze any type of vitamin [25]. This technique possesses simultaneous detection and determination of vitamins in one sample [17]. HPLC is often used for the simultaneous qualitative and quantitative analysis of water-soluble vitamin and fat-soluble vitamin in biological matrices such as plasma and urine [26–28].

A reversed-phase HPLC chromatographic procedure has been developed for the determination of water-soluble vitamins. As fat-soluble vitamins are highly lipophilic molecules, they are analyzed by normal-phase or even reversed-phase HPLC with a methanol/acetonitrile mobile phase [29–33].

Nowadays, coupling mass spectrometer (MS) to HPLC has added a new dimension to vitamin analysis since it enables the analysis of nonvolatile compounds. In general, LC–MS methods are now available for each fat- and water-soluble vitamin. The LC–MS assays are characterized by high sensitivity and high specificity [34,35]. In the last few years, the trend within the industry has been the adoption of multianalyte methods to improve lab efficiency and productivity. Modern HPLC systems have been improved to work at much higher pressures, and therefore be able to use much smaller particle sizes in the columns (<2 μm). The use of these systems allows for faster analyses, with significant improvements in sensitivity as well as in resolution compared to traditional HPLC methods [36]. These are ultraperformance liquid chromatography systems (UPLCs) [37,38]. The term UPLC is a trademark technology of the Waters Corporation, but people sometimes use it as a general term for the overall technique.

In addition to chromatographic techniques, several methods have been used for the determination and analysis of vitamins in different samples, such as immunoassays [34] and electrochemical [6] and capillary electrophoresis [39].

10.3 IMPETUS BEHIND DEVELOPMENT OF UPLC

HPLC is a well-known technique that has proven to be the predominant analytical technology used in laboratories worldwide during recent years. HPLC is well established as a routine separation technique, this is in large part due to the many features offered by state-of-the-art HPLC such as robustness, ease of operation, well-understood separation principles, sensitivity, and tunable selectivity. However, although many chromatographers feel that HPLC is a mature analytical technology, the evolutions of HPLC are still ongoing. The main limitation of HPLC has long been known to be the lack of high efficiency, especially when compared to gas chromatography (GC) and capillary electrophoresis (CE) [40–42].

The primary driver factor responsible for the growth and development of this technique has been the evolution of packing materials used to effect the separation. The underlying principles of this evolution are governed by the van Deemter equation, which describes the relationship between linear velocity and plate height.

Since particle size is one of the variables, a van Deemter curve can be used to investigate chromatographic performance. According to the van Deemter equation, when the particle size of the chromatographic sorbent is decreased, the efficiency of the separation process increases and the efficiency does not diminish at higher flow rates or linear velocities [43–45]. By using smaller particles, speed and peak capacity can be extended to new limits, termed UPLC. As shown in Figure 10.1, smaller

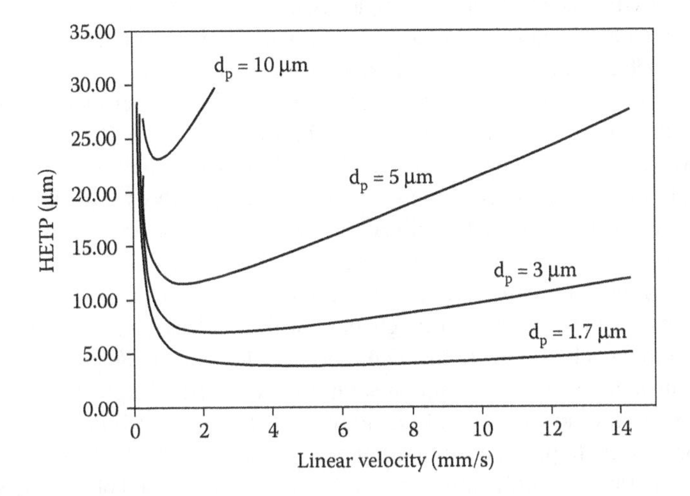

FIGURE 10.1 Van Deemter curves for different particle sizes (10, 5, 3, 1.7 μm). (Adapted from Novakova, L., Matysova, L., and Solich, P. 2006. *Talanta*. 68:908–918. With permission.)

particles provide increased efficiency as well as the ability to work at increased linear velocity without a loss of efficiency, providing both resolution and speed [46].

UPLC technology takes full advantage of chromatographic principles to run separations using columns packed with smaller particles and higher flow rates for increased speed, with superior resolution and sensitivity [45,47]. Working with smaller particles (<2.5 μm) provide both resolution and speed. Efficiency is the primary separation parameter behind UPLC since it relies on the same selectivity and retentivity as HPLC.

The resolution is proportional to the square root of plate numbers. However, since plate numbers are inversely proportional to particle size (as the particle size is lowered by a factor of three, from, e.g., 5 μm [HPLC scale] to 1.7 μm [UPLC scale]), plate numbers are increased by three and the resolution is increased by the square root of three, or 1.7. Plate numbers are also inversely proportional to the square of the peak width. This illustrates that the narrower the peaks, the easier they are to separate from each other.

Also, peak height is inversely proportional to peak width, so, as the particle size decreases to increase plate numbers and subsequently the resolution, an increase in sensitivity is obtained, since narrower peaks are taller peaks. Narrower peaks also mean more peak capacity per unit time. Efficiency is proportional to column length and inversely proportional to the particle size [45,48]. Therefore, the column can be shortened by the same factor as the particle size without loss of resolution.

In practice, the design and development of sub-2 μm particles is a significant challenge. As particle size decreases, the optimum flow to reach maximum plate numbers increases and the use of smaller particles is then substantially limited by a rapid increase in pressure drop [49]. Standard HPLC instruments have a maximum operating pressure of about 400 bar (6000 psi), and these systems simply do not have the capability to take full advantage of sub-2 μm particles.

Other factors and instrument parts should also be taken into account to take full advantage of UPLC separations. A completely new system design with advanced technology in the solvent and sample manager, packing method, introduction method, injection valves, detector sampling rate, data system, and service diagnostics is required [50].

In order to overcome the pressure limitations and the other challenges, Waters Corporation introduced Acquity UPLC™, the first commercially available system that addresses the challenge of using elevated pressure and sub-2 μm particles, which makes it a particularly attractive and promising analysis tool [51]. The UPLC instrumentation is designed to deliver mobile phase at pressures up to 1034 bar (15,000 psi). The researchers were very active in this area for some time to manufacture columns that can withstand these rigorous pressures [52,53].

Using sub-2 μm particles, half-height peak widths of less than 1 s can be obtained, posing significant challenges for the detector. In order to accurately and reproducibly integrate an analyte peak, the detector sampling rate must be high enough to capture enough data points across the peak. In addition, the detector cell must have minimal dispersion volume to preserve separation efficiency.

Mass detection is significantly enhanced by UPLC; increased peak concentrations with reduced chromatographic dispersion at lower flow rates promotes increased source ionization efficiencies (reduced ion suppression) for improved sensitivity.

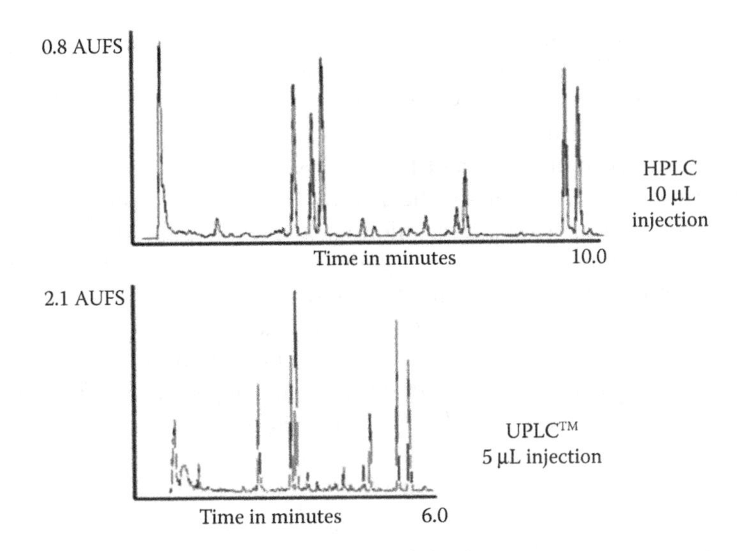

FIGURE 10.2 Comparison of HPLC and UPLC for the separation of a ginger root extract. (Adapted from Swartz, E. 2005. *J. Liq. Chromatogr. Related Technol.* 28:1253–1263. With permission.)

Enhanced selectivity and sensitivity, and rapid generic gradients made LC–MS the predominate technology for both quantitative and qualitative analyses [54–56].

UPLC can be regarded as a new invention for liquid chromatography. Compared with conventional HPLC, UPLC can offer significant improvements in speed, sensitivity, and resolution, which also mean reduced solvent consumption and analysis cost. Figure 10.2 shows an HPLC versus UPLC separation comparison of a ginger root extract sample, where both speed and resolution are improved, and sensitivity is increased [57].

With the increasing number, diversity, and complexity of compounds being analyzed, UPLC presents the possibility to extend and expand the utility of separation science. Today, UPLC is widely used for metabolite identification; analysis of natural products and herbal medicines; pharmacokinetic, toxicity, degradation, bioanalysis, and bioequivalence studies; quality control and in drug discovery; determination of pesticides; and separation of various pharmaceutical-related small organic molecules, proteins, and peptides. UPLC is also used for impurity profiling, method development, and validation performed in quality assurance and quality control laboratories [46,47,56–69].

10.4 UPLC–MS FOR VITAMINS ANALYSIS IN FOOD

The determination of vitamins in food represents a complex analytical problem. Vitamins naturally present in foods at very low levels and in very complex matrixes. Furthermore, vitamins are easily destroyed by strong acids or alkali; affected by many factors such as pH, heat, light, air, extraction solvent, and time; and also affected by other food components.

Despite the development of several HPLC and LC–MS methods for vitamin analysis in food [6,70,71], there is still a need for better resolution and sensitivity when analyzing complex food matrices containing different vitamin derivatives. Owing to its high resolution and sensitivity, as well as its high sample throughput, UPLC is particularly attractive in the area of food analysis where sample matrices are very complex. However, research in this field is still limited, and much efforts and time are needed in order to provide more information and to improve the existing methods. Table 10.2 summarizes the most significant recently published UPLC methods for the determination of vitamins in different food samples.

The content of vitamin C in food commodities (the sum of the contents of L-ascorbic acid and dehydroascorbic acid) is used as an index of health-related quality of products, since, as compared with other beneficial compounds, it is more sensitive for degradation by processing and storage. Therefore, interest in the simultaneous analysis of these molecules has increased greatly in food analysis [72–74]. Spinola et al. [75] determined the total vitamin C content in several fruits and vegetables (lemons, passion fruits, papayas, strawberries, broccoli, green and red peppers) using UPLC-photodiode array (PDA) system.

The system was equipped with a trifunctional high-strength silica (100% silica particle) analytical column (Acquity HSS T3; 100 × 2.1 mm i.d., 1.8 μm particle size), using an isocratic mobile phase composed of aqueous 0.1% (v/v) formic acid at a flow rate of 250 μL/min. The injection volume was 20 μL, while the detection wavelength for the PDA was set at 245 nm, and the analytical column was kept at room temperature [75].

The flow rate and the mobile phase composition significantly influenced L-ascorbic acid retention time. Careful selection of the eluent systems demonstrated that the chromatographic separation of the L-ascorbic acid could be achieved within 2 min. The best flow value (250 μL/min) was compromised between the backpressure (approximately 5800 psi), retention times, and peak resolution. The combination of the shorter running time with a smaller flow rate also reduced drastically the solvent consumption and thus is more environmental friendly than conventional HPLC [75].

Vitamin C refers to all compounds exhibiting equivalent biological activity of L-ascorbic acid, including its oxidation products, esters, and synthetic forms [73,76]. L-Ascorbic acid is the main biologically active form of vitamin C and is reversibly oxidized to dehydroascorbic acid, due to the presence of two hydroxyl groups in its structure [77–79], or diketogulonic acid, which has no biological function, and the reaction is no longer reversible [72,76]. The type of extraction medium is therefore very important in order to prevent the irreversible oxidation of L-ascorbic acid.

The use of a high ionic strength acidic extraction solvent is required to suppress metabolic activity upon disruption of the cell and to precipitate proteins. A metal chelator such as ethylenediaminetetraacetic acid (EDTA) is also usually required. The use of metaphosphoric acid is the best way to extract and stabilize L-ascorbic acid, as suggested by various authors [72,78,80,81]. For this reason, the samples were extracted with 3% metaphosphoric acid–8% acetic acid–1 mM EDTA solution, which is known to limit L-ascorbic acid degradation to less than 5% [82].

The L-ascorbic acid content ranged between 27.8 ± 2.3 and 223.7 ± 16.5 mg/100 g of edible portion in the different horticultural products. Peppers (red and green) and

TABLE 10.2
UPLC Methods for Determination of Vitamins in Different Food Samples

Analytes	Food Sample	Sample Preparation	Column	Mobile Phase	LOD	Flow Rate (mL/min)	Run Time (min)	Ref.
L-Ascorbic acid and dehydroascorbic acid	Lemons, passion fruits, papayas, strawberries, broccoli, green and red peppers	Extraction with 3% metaphosphoric acid–8% acetic acid–1 mM EDTA	HSS T3 (100 × 2.1 mm, 1.8 µm)	Aqueous 0.1% (v/v) FA	22 ng/mL	0.25	2	[75]
Cis- and *trans*-phylloquinone	Infant formula (milk-, rice-, and soy-based)	Digest samples with lipase, extract with methanol, sodium hydroxide solution, and *n*-hexane mixture	BEH Shield RP_{18}, BEH C_{18}, T3 (100 × 2.1, 1.7 µm), YMC30 (150 × 4.6 mm, 3 µm)	0.025% FA and 2.5 mmol/L NH_4HCO_2 in CH_3OH	0.011 µg/100 g for *trans*-form and 0.01 µg/100 g for *cis*-form	0.6	13.5	[83]
Vitamins B5, B8, B9, and B12	Fortified infant foods (milk and rice)	Proteins and lipids removal with chloroform, extraction with 10 mmol/L ammonium acetate	BEH C_{18} column (100 × 2.1 mm, 1.7 µm)	Gradient elution of CH_3CN and H_2O	5, 30, 6, and 6 ng/L for vitamin B5, B8, B9, and B12	0.2	6	[85]
Thiamine, riboflavin, nicotinamide and pyridoxal	Human milk	Proteins removal by acetonitrile and trichloroacetic acid and Amicon Ultra-0.5 centrifugal filters, extraction with water and methanol, extraction of the nonpolar constituents with diethyl ether	HSS T3 column (50 × 2.1 mm, 1.8 µm)	Gradient of 10 mM NH_4HCO_2 and CH_3CN	0.01, 0.1, 0.5, and 4 µg/L for thiamine, riboflavin, nicotinamide, and pyridoxal	0.3	2	[86]

continued

TABLE 10.2 (continued)
UPLC Methods for Determination of Vitamins in Different Food Samples

Analytes	Food Sample	Sample Preparation	Column	Mobile Phase	LOD	Flow Rate (mL/min)	Run Time (min)	Ref.
Thiamine, riboflavin, pantothenic acid, pyridoxine, pyridoxal, biotin, folic acid, cyanocobalamin, ascorbic acid, L-carnitine, choline, and taurine	Fortified infant formula (rice flour and wheat powder)	Extraction with 10 mM aqueous ammonium acetate	BEH Shield RP_{18} (100 × 2.1 mm, 1.7 μm)	CH_3OH and 10 mM aqueous $NH_4C_2H_3O_2$	Ranged between 0.0042 ng/mL for riboflavin and 28.872 ng/mL for ascorbic acid	0.2	10	[91]
Vitamins D2 and D3	NIST 1849, ready-to-eat cereal, bread, mushrooms, eggs, yogurt, cheese, fish, butter, milk (RTF infant formula), and premix	Saponification (hot), extraction of unsaponifiables with hexane	HSS C_{18} (100 × 2.1 mm, 1.8 μm)	Gradient elution of CH_3OH and NH_4HCO_2	0.20 and 0.47 μg/100 g for vitamins D2 and D3	0.4	6	[92]

Vitamins D_2 and D_3	Infant formula and adult/pediatric nutritional supplements	Saponification (ambient), extraction of unsaponifiables with hexane	Hypersil GOLD and GOLD aQ (100 × 2.1 mm, 1.9 μm)	Gradient elution of 0.1% FA CH_3OH and H_2O	0.65 and 0.83 ng/g for vitamins D_3 and D_2	0.25–0.5	10	[94]
5-CH_3-H_4folate, H_4folate, 5-HCO-H_4folate, 10-HCO-folic acid, and folic acid	Egg yolk, pickled beetroots, strawberries, orange juice, soft drink, and dry baker's yeast	Extraction with 0.1 M phosphate buffer pH 6.1, 2% sodium ascorbate and 0.1% 2,3-dimercaptopropanol. Extract purification by SPE on SAX Isolute cartridges. Elution with 0.1 M sodium acetate, 10% sodium chloride, 1% ascorbic acid, and 0.1% 2,3-dimercaptopropanol	BEH C_{18} (100 × 2.1 mm, 1.7 μm)	Gradient elution of 30 mM KH_2PO_4 buffer at pH 2.3 and CH_3CN	Ranged from 0.02 to 2.5 ng/mL	0.5–0.7	3.5	[95]
Folic acid and 5-CH_3-H_4folate	Fortified breads	Hydrolysis with α-amylase in a water bath at 75°C, extraction with phosphate buffer (pH 6.1), concentration using SPE on styrene divinylbenzene cartridges	HSS T3 (100 × 0.1 mm, 1.8 μm)	Gradient elution of 0.1% FA in H_2O and CH_3CN	9.0 ng/g for folic acid and 4.3 ng/g for 5-CH_3-H_4folate	0.2	6	[101]

continued

TABLE 10.2 (continued)
UPLC Methods for Determination of Vitamins in Different Food Samples

Analytes	Food Sample	Sample Preparation	Column	Mobile Phase	LOD	Flow Rate (mL/min)	Run Time (min)	Ref.
H_4folate, 5-CH_3-H_4folate, 10- CH_3-H_4folate, 10-HCO-folic acid, 5-HCO-H_4folate, and folic acid	Rice	Hydrolysis with α-amylase, protease, and deconjugase, extraction with phosphate buffer at pH 7.5	HSS T3 (150 × 2.1 mm, 1.8 μm)	Gradient elution of 0.1% FA in H_2O and CH_3CN	Ranged between 0.06 and 0.45 μg/100 g	0.6	8	[108]
Retinol, α-, δ-, γ- tocopherols	Milk	Lipophilic components extraction by *n*-hexane/ethyl acetate, extraction of xanthophylls and vitamin E by ethanol/water, saponification hexanic phase (hot), purification by *n*-hexane/ethyl acetate	HSS T3 (150 × 2.1 mm, 1.8 μm)	Gradient elution of CH_3CN–CH_2Cl_2–CH_3OH (75–10–15) and $NH_4C_2H_3O_2$ 0.05 M in H_2O	0.8 and 2.6 ng/injection for retinol and α-tocopherol	0.4	46	[110]

papayas were the species with the highest initial vitamin C contents. The food commodities with the lowest content of total vitamin C were passion fruits [75].

Good linearity was achieved in both ranges of concentrations tested, with correlation coefficients higher than 0.999 in all cases. The proposed method showed a good sensitivity, with limit of detection (LOD) and limit of quantification (LOQ) of 22 and 67 ng/mL, respectively. The method also showed satisfactory precision. All relative standard deviation (RSD) values achieved for peak areas were lower than 4%. The results obtained showed that the applied method has a good reproducibility and that is stable and reliable. The recovery rate of L-ascorbic acid and total vitamin C was evaluated to assess the extraction efficiency of the proposed method and matrix effects. Satisfactory results were obtained for L-ascorbic acid (96.6 ± 4.4%) and total vitamin C (103.1 ± 4.8%) [75].

The results demonstrated by Spinola and his coworkers revealed an attractive and very promising approach for the analysis of total vitamin C and L-ascorbic acid and for routine use in laboratory. The major improvement of this methodology makes use of a very simple but very effective mobile phase that promotes ion suppression at very high pressures with a trifunctional high-strength silica column specially designed for polar compounds, overcomes the problems normally encountered in HPLC systems, and is suitable for the analysis of large batches of samples without L-ascorbic acid degradation.

Infant food such as human milk and fortified formula provide an essential source of nutrition for baby feed. This nutrition is essential for normal growth and functioning of the human body; therefore, particular attention should be paid to ensure an adequate and balanced intake of vitamins. Several methods have been developed for the analysis of vitamins in infant foods. Huang and coworkers determined vitamin K1 isomers (*cis*- and *trans*-forms) in infant formulas using UPLC–MS/MS [83].

Vitamin K1 (phylloquinone) exists naturally only in the *trans*-form. *Trans*- and *cis*-isomers are formed during UV light exposure or synthetic production of vitamin K1. This *cis*-isomer is considered to have low bioactivity [84]. Therefore, inactive *cis*-vitamin K1 is necessary to measure individually to evaluate the true nutritional status of fortified foods. The developed method proved to be rapid and sensitive, which allows *cis*- and *trans*-vitamin K1 extracted from milk-based, rice-based, and soy-based infant formula to be simultaneously determined by UPLC–MS/MS [83].

Sample preparation procedure mainly involves the lipase hydrolysis and the liquid-to-liquid extraction. After hydrolysis at 37°C for 2 h, the sample was extracted with a mixture solution of methanol, sodium hydroxide solution, and *n*-hexane. The upper phase was then washed with distilled water. The upper phase, containing *n*-hexane with nitrogen, is evaporated to near dryness and reconstituted with methanol. After the filtration, 5 μL of the vitamin K1-enriched extract was directly injected into the UPLC–MS/MS for analysis.

Four columns and several mobile phases were compared in this study. Acquity UPLC BEH Shield RP_{18} (100 × 2.1 mm i.d., 1.7 μm particle size), Acquity UPLC BEH C_{18} (100 × 2.1 mm i.d., 1.7 μm particle size), Acquity UPLC T_3 (100 × 2.1 mm i.d., 1.7 μm particle size), and YMC30 (150 × 4.6 mm i.d., 3 μm particle size) were used for comparisons to the separation capability in vitamin K1 isomers. The type of column plays an important role in the separation of vitamin K1 isomers. As a result,

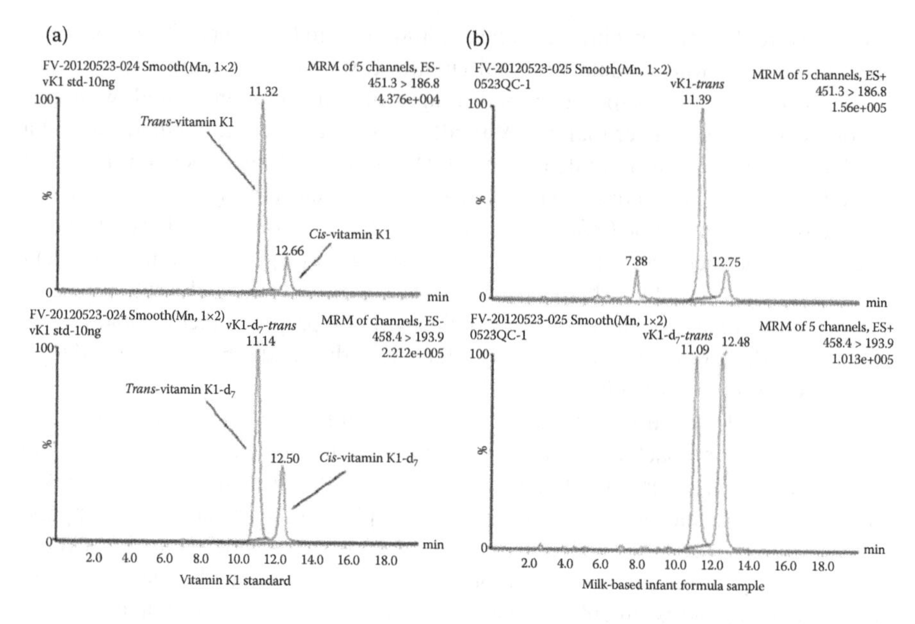

FIGURE 10.3 MRM diagrams of standards and a sample: (a) vitamin K1 and vitamin K1-d_7 standards and (b) a milk-based infant formula sample. (Adapted from Huang, B. et al. 2012. *Eur. Food Res. and Technol.* 235:873–879. With permission.)

the biggest peak areas of vitamin K1 isomers were obtained when using 0.025% formic acid and 2.5 mmol/L ammonium formate in methanol as a mobile phase [83].

The study showed that the level of lipase use and enzymatic hydrolysis time play key roles in vitamin K1 quantification in infant formulas. Chromatograms of vitamin K1 standards and a sample were shown in Figure 10.3.

Trans-vitamin K1 was separated from *cis*-isomer through YMC30 column (150 × 4.6 mm, 3 μm particle size), and vitamin K1-d_7 was used as an internal standard. The components were detected using electrospray ionization (ESI) in a positive ion and quantified by multiple reaction monitoring (MRM) mode. The transition m/z 451.2 → 186.9 was selected for vitamin K1, and the transition m/z 458.2 → 194.0 was selected for vitamin K1-d_7. Calibration curves were linear over the range of 9.3–464.75 ng/mL for *trans*-vitamin K1 ($r^2 > 0.999$) and 1.71–85.25 ng/mL for *cis*-vitamin K1 ($r^2 > 0.999$). The limit of detection for *trans*- and *cis*-vitamin K1 was 0.011 μg/100 g and 0.01 μg/100 g, respectively, while the limit of quantification was 0.037 μg/100 g and 0.031 μg/100 g, respectively. The intra- and interbatch variations in terms of RSD were less than 5%. The *cis*-vitamin K1 isomer contributes to 7.05–17.21% of the total vitamin K1 in certain infant formulas [83].

Reversed-phase UPLC separation system was also used for the determination of four water-soluble B vitamins, including B5, B8, B9, and B12 in fortified infant foods using Waters Acquity UPLC BEH C_{18} column (100 × 2.1 mm i.d., 1.7 μm particle size), and a binary gradient, acetonitrile–water mobile phase [85].

Prior to analysis, the coextracted proteins and lipids should be removed from the prepared solution. Out of the tested solvents (chloroform, acetonitrile, and ethyl acetate), chloroform precipitated proteins and lipids more effectively than acetonitrile and ethyl acetate did. Thus, chloroform was used as a solvent for eliminating proteins and lipids.

In order to obtain complete extraction of the vitamins from food samples, three extraction media including water, 0.1% formic acid aqueous solution, and 10 mmol/L ammonium acetate aqueous solution were tested. Methotrexate was selected as an internal standard because it is usually absent in fortified infant foods, its molecular structure is similar to vitamin B9 (MW = 454), and it has a suitable retention time (4.78 min) according to the retention times of the four vitamins in this LC system. Recovery tests showed that recoveries of methotrexate were similar to those of the concerned vitamins [85].

Formic acid is spiked into the mobile phase to enhance the ionization efficiency. Tandem MS–MS analysis is performed in MRM. Product ion traces at m/z 220.1 → 89.9 for vitamin B5, 245.1 → 227.1 for vitamin B8, 442.3 → 295.2 for vitamin B9, and 678.9 → 147.0 for vitamin B12 are used for quantitation of the corresponding vitamins, and traces at m/z 455.5 → 308.0 are used for methotrexate (internal standard). LOQs are 0.016, 0.090, 0.020, and 0.019 μg/L for vitamin B5, B8, B9, and B12, respectively. LODs are 0.005, 0.030, 0.006, and 0.006 μg/L for vitamin B5, B8, B9, and B12, respectively. Intra- and interday precisions for the determination of the four vitamins are better than 6.84% and 12.26% in RSD, and recoveries for the four vitamins are in the range of 86.0 ~ 101.5% [85].

From the three tested extraction media, it was found that 10 mmol/L ammonium acetate aqueous solutions gave the best recoveries, corresponding to 93.6%, 99.1%, 85.3%, and 95.0% for vitamins B5, B8, B9, and B12, respectively.

In preliminary tests, four ODS C_{18} columns with different column size and/or different particle size were compared regarding the separation of the four vitamins and methotrexate internal standard. Column A is a Waters Atlantis C_{18} column (150 × 2.1 mm i.d., 5 μm particle size), column B is a Waters Sunfire C_{18} column (150 × 2.1 mm i.d., 5 μm particle size), column C is a Waters Acquity UPLC BEH C_{18} column (50 × 2.1 mm i.d., 1.7 μm particle size), and column D is a Waters Acquity UPLC BEH C18 column (100 × 2.1 mm i.d., 1.7 μm particle size).

It was found (as shown in Figure 10.4) that columns A and B, two frequently used columns for LC–MS, could neither separate the four vitamins and methotrexate effectively, nor generate reproducible peak areas in parallel runs, while column C could not separate methotrexate and vitamin B12. Only column D could separate the vitamins and methotrexate and support constant retention times and peak areas in parallel experiments. Therefore, column D was selected for the real sample studies.

The developed method was applied to analyze 10 brands of fortified milk powders and 9 brands of fortified rice powders, which were purchased from a local food market (Hangzhou, China). The obtained results are consistent with those claimed by the producers. So, the method can be applied successfully to analyze vitamins B5, B8, B9, and B12 in the fortified rice and milk powders [85].

Hampel et al. described simultaneous determination of several B vitamins (thiamine, riboflavin, nicotinamide, and pyridoxal) in human milk samples using the UPLC–MS/MS method [86].

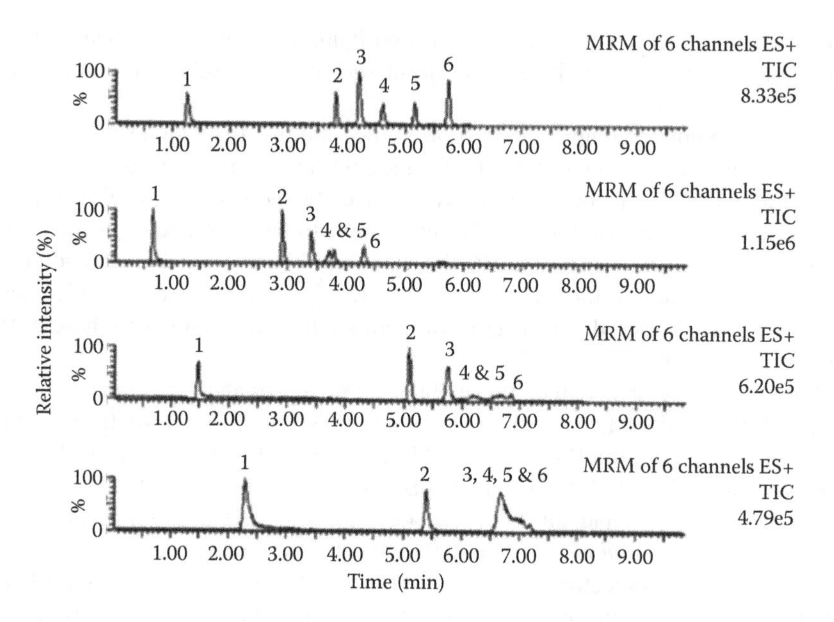

FIGURE 10.4 The LC–MS/MS chromatograms obtained with four different LC columns. (Adapted from Lu, B. et al. 2008. *J. Chromatogr. Sci.* 46:225–232. With permission.)

Sample preparation of human milk depends strongly on the vitamins in question and limits the number of different vitamins that can be analyzed simultaneously [87–89]. Sample preparation was carried out under subdued light and on ice to protect the analytes from degradation. Targeted vitamins were extracted by water and methanol after several dilution, centrifugation, dryness, and filtration steps. The nonpolar constituents of the matrix were then extracted with diethyl ether [90].

Additional sample preparation steps included protein removal by acetonitrile and trichloroacetic acid and Amicon Ultra-0.5 centrifugal filters, while hexane was used to remove the nonpolar constituents. Amicon Ultra-0.5 centrifugal filters were used for the efficiency of protein removal. Choosing an appropriate molecular weight cut-off allows the vitamins as small molecules (MW < 1000) to elute through the filter unit and to be recovered in the flow through, while the protein fraction with a high molecular weight (MW > 10,000) is retained in the filter unit.

Using Acquity UPLC HSS T3 column (50 × 2.1 mm i.d., 1.8 μm particle size) reversed-phase chromatography, water, 10 mM ammonium formate, 0.1% ammonium formate, 10 mM ammonium acetate, and 10 mM formic acid were tested for the aqueous buffer component of the mobile phase, while methanol and acetonitrile were examined as the organic constituent. Best peak shape, resolution, and signal-to-noise ratio was achieved using a shallow gradient of 10 mM ammonium formate and acetonitrile with an initial ratio of 95:5. Optimal chromatographic separations occurred at a flow rate of 0.3 mL/min, and a column temperature of 40°C. The mass spectrometer was operated via positive ion mode ESI, and the vitamins were

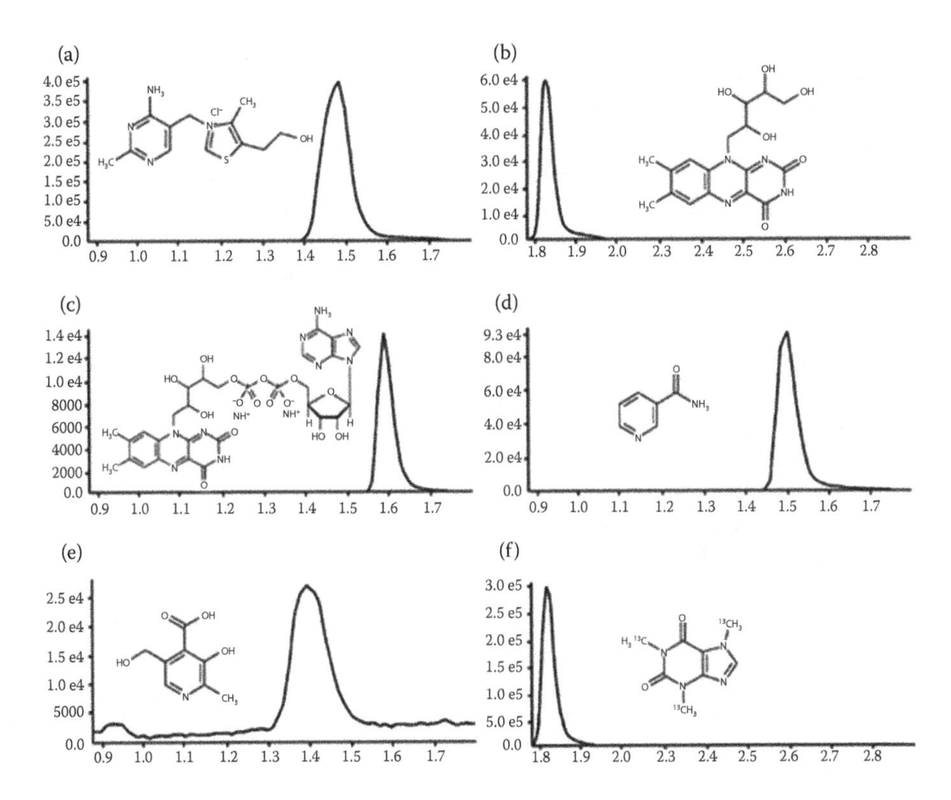

FIGURE 10.5 Extracted ion chromatogram of the analytes and system control standard (A: thiamine, B: riboflavin, C: flavin adenine dinucleotide, D: nicotinamide, E: pyridoxal, F: $^{13}C_3$-caffeine). (Adapted from Hampel, D., York, E., and Allen, L. 2012. *J. Chromatogr. B.* 903:7–13. With permission.)

detected in MRM. Resolution by retention time or MRM for the vitamins has been optimized within 2 min. Figure 10.5 shows the extracted ion chromatogram of the analytes and system control standard [86].

Intra- and interassay variability for all analytes ranged from 0.4% to 7.9% and from 2.2% to 5.2%, respectively. Quantification was done by ratio response to the stable isotope-labeled internal standards. $^{13}C_4$-Thiamine hydrochloride, $^{13}C_4$, $^{15}N_2$-riboflavin, and $^{2}H_3$-pyridoxal hydrochloride were added to the samples as internal standards for quantification. The standard addition method determined recovery rates for each vitamin (73.0–100.2%). The limit of quantitation for all vitamins was between 0.05 and 5 μg/L depending on the vitamin.

The method was applied to 80 human milk samples from Cameroon, China, India, Malawi, and the United States [86]. The results show that the vitamin content of human milk varies depending on the geographic origin, most likely due to differences in diet and fortification that remains to be explored; moreover, the influence of the lactation stage during sample collection needs to be investigated and its contribution to the variation of concentration determined.

The UPLC–MS/MS method has also been developed for fast simultaneous separation and determination of 14 different water-soluble vitamins and vitamin-like compounds in infant formula (thiamine, riboflavin, pantothenic acid, nicotinic acid, nicotinamide, pyridoxine, pyridoxal, biotin, folic acid, cyanocobalamin, ascorbic acid, L-carnitine, choline, and taurine) [91]. Methotrexate was also used as an internal standard for riboflavin, cyanocobalamin, biotin, and folic acid, while nicotinamide was used as an internal standard for the other compounds.

Choosing a chromatographic column gives the possibility to perform various water-soluble vitamin and vitamin-like compound separations in one run and with better resolution. Chromatographic separation was performed using Waters Acquity UPLC BEH Shield RP_{18} column (100 × 2.1 mm i.d., 1.7 μm particle size) at a flow rate of 0.2 mL/min, and the injection volume was 10 μL. The mobile phase also has a significant effect on peak shape, sensitivity, and resolution. Methanol with 10 mM aqueous ammonium acetate gave the best overall performance and was deemed to be most suitable when all 14 compounds were considered.

Identification and quantification were performed by using MRM with one parent ion, one daughter ion for each analyte, and ESI in positive ion scan mode. Under these conditions, a chromatographic separation within 10 min was achieved. Recoveries at three concentration levels (low, medium, and high) for all analytes ranged between 81.8% and 106.1%. The limit of detection ranged between 0.0042 ng/mL for riboflavin and 28.872 ng/mL for ascorbic acid, with an RSD of 1.17% to 5.09% for intraday and 2.61% to 7.43% for interday analysis [91].

Different infant formula, rice flour and wheat powder commercial samples were tested using the developed method. The results demonstrated that the proposed method is both sensitive and selective enough to be applied for the fast determination of these compounds at a wide range of concentration levels in real samples. Vitamin C was not detected from all the samples. The values for riboflavin and nicotinamide were 10^3 times higher than the value for cyanocobalamin. The vitamin compounds were also found at different low concentration-level ng/mL.

Stevens and Dowell reported a method for the determination of vitamin D (total vitamins D_2 and D_3) in infant formula and adult nutritionals by UPLC–MS/MS [92]. Vitamin D_2, ergocalciferol, and vitamin D_3, cholecalciferol, are two forms of vitamin D found in the body. Because vitamin D is not readily available in many foods, the main source remains fortified foods [93]. Because of the importance of vitamin D levels in foods, methods that produce accurate results in a timely manner are needed.

Vitamin D has been analyzed in different food samples, including NIST 1849 (infant/adult nutritional formula), ready-to-eat cereal, bread, mushrooms, eggs, yogurt, cheese, fish, butter, milk (RTF infant formula), and premix (mixture of vitamins). After saponification with ethanol and potassium hydroxide, and sequential liquid–liquid extraction steps using ethanol, water, and *n*-heptane, the food samples were eluted through a disposable SPE silica gel column. The extracted samples were assayed using a methanol and ammonium formate gradient on a UPLC–MS/MS system equipped with an HSS C_{18} UPLC (100 × 2.1 mm i.d., 1.8 μm particle size) column at 0.4 mL/min flow rate. The injection volume was 20 μL [92].

Under these conditions, the chromatographic separation for the two vitamins was achieved within 6 min. The method was established in the range of 0.005–50 μg/mL. The recovery range was 93.4–100.9% for vitamin D_2 and 102.4–106.2% for vitamin D_3. The LOD and LOQ for vitamin D_2 were reported as 0.20 and 0.61 μg/100 g, respectively; for vitamin D_3, the reported values were 0.47 and 1.44 μg/100 g, respectively [92].

Vitamin D has also been measured in infant formula and adult/pediatric nutritional supplements using UPLC–MS/MS and compared with conventional HPLC–MS/MS [94]. The accuracy of this method was evaluated by the analysis of NIST selected reaction monitoring (SRM) 1849 infant formula reference material. Both saponification and liquid–liquid extraction procedures have been applied for the extraction of vitamins prior to analysis.

The mass spectrometer was operated via APCI at 21 V collision energy, 50 psi ion gas pressure, and 320°C ionization temperature. Under gradient elution of methanol and water with both including 0.1% formic acid at a flow rate 0.25–0.5 mL/min with 5 μL injection volume, and using Hypersil GOLD and GOLD aQ columns (100 × 2.1 mm i.d., 1.9 μm particle size), all the previtamins of vitamin D_3, D_2, and isotope-labeled vitamin D_3 were baseline separated from their corresponding vitamins.

The average concentration of vitamin D_3 was 0.251 ± 0.012 μg/g, which was in an excellent agreement with the certified value of 0.251 ± 0.027 μg/g. In addition, the spike recovery from a commercial infant formula matrix was in the range 100–108% for both vitamins D_3 and D_2. The LOQ values determined were 0.0022 and 0.0028 μg/g for vitamins D_3 and D_2, respectively, while the LOD values were 0.65 and 0.83 ng/g for vitamins D_3 and D_2, respectively. Precision ranged from 3.7% to 8.2% for vitamin D_2 and 3.7% to 7.0% for vitamin D_3 [94].

The analytical range was determined to be 0.0015–1.350 μg/g. This range made the method flexible and useful to deal with the wide concentration range of vitamin D in various samples. The method was robust; measured values generally fell below 5% based on the results of changing the parameters of LC separation and MS measurement. This accurate and reliable vitamin D method significantly increased instrument efficiency and analysis productivity [94].

Compared to the HPLC method, the run time of the chromatographic separation was shorted by 50% (from 20 min in HPLC–MS/MS to 10 min in UPLC–MS/MS) [94], while equal or better separation efficiency was achieved to deal with complex food matrixes.

Jastrebova and his coworkers compared both UPLC and HPLC for the separation and determination of the most common dietary folates (a group of derivatives of folic acid); 5-methyltetrahydrofolate (5-CH_3-H_4folate), tetrahydrofolate (H_4folate), 5-formyltetrahydrofolate (5-HCO-H_4folate), 10-formylfolic acid (10-HCO-folic acid), and folic acid [95]. Four alkyl-bonded columns, two UPLC columns—Acquity BEH C_{18} and Acquity HSS T3, 100 × 2.1 mm i.d., 1.7 μm particle size—and two HPLC columns—Xbridge C_{18} and Atlantis d18, 150 × 4.6 mm i.d., 3.5 μm particle size—were tested for this purpose. In respect to the surface chemistry, the Acquity BEH C_{18} column is similar to the XBridge C_{18} column, while the Acquity HSS T3 column is similar to the Atlantis d18 column.

The separation of folates by HPLC was conducted at 23°C column temperature, 0.4 mL/min flow rate, and 20 μL injection volume. On the other hand, UPLC analysis was conducted at 30°C for the first column and 60°C for the second one, and the flow rate was 0.5 or 0.7 mL/min and 10 μL sample injection volume [95].

In both systems, the mobile phase was a binary gradient mixture of 30 mM potassium phosphate buffer at pH 2.3 and acetonitrile. In the UPLC system, a binary mixture of 10 mM formic acid and acetonitrile was also tested at the mobile phase. Folic acid was quantified using UV detection at 290 nm, whereas other forms were quantified by using fluorescence detection: excitation 290 nm and emission 360 nm for H_4folate, 5-CH_3-H_4folate, and 5-HCO-H_4folate, and excitation 360 nm and emission 460 nm for 10-HCO-folic acid.

Folate samples included egg yolk, pickled beetroots, strawberries, orange juice, a soft drink, and dry baker's yeast. Samples were prepared as described by Patring and Jastrebova [96] and purified as described by Jastrebova et al. [97]. An extraction buffer containing 0.1 M phosphate buffer pH 6.1 with 2% sodium ascorbate (w/v) and 0.1% 2,3-dimercaptopropanol (v/v) was added for each sample. Purification of the obtained extracts was performed by SPE on strong anion exchange (SAX) Isolute cartridges. The retained folates were eluted at low flow rate (not exceeding one drop/s), with 0.1 M sodium acetate containing 10% (w/v) sodium chloride, 1% (w/v) ascorbic acid, and 0.1% (v/v) 2,3-dimercaptopropanol. The purified extracts were finally stored at 0°C overnight prior to analysis.

Figure 10.6 shows the comparison between both UPLC and HPLC for folate analysis using different columns [95]. Both UPLC and HPLC provided high-quality chromatograms. Using UPLC showed a great advantage in terms of the signal-to-noise ratio, which could be improved by a factor of 2–50 for different folate derivatives, and the run time that could be reduced fourfold without sacrificing separation efficiency.

The BEH C_{18} column at 30°C provided the best LOD values ranging from 0.02 to 2.5 ng/mL followed by HSS T3 column at 30°C, while the LOQ values were the lowest (0.2–30 ng/mL) with the BEH C_{18} column and HSS T3 column at 30°C. Linearity tests revealed marked differences between all columns tested at low concentrations. The BEH C_{18} column provided the best linearity, followed by the HSS T3 column, whereas deviations from linearity were considerably greater for the HSS T3 column at 60°C and HPLC columns XBridge and Atlantis d18 at 23°C. At higher concentrations, the linearity was good for all columns and gradients tested. In conclusion, the obtained results showed that the application of a UPLC for determining folates in foods is of great interest, because it can provide much higher resolution, lower detection limits, and shorter analysis time compared with traditional HPLC.

The best performance column (BEH C_{18} column) was used for the determination of folates in real samples at optimum chromatographic conditions. It was found that the fortified drink contained only synthetic folic acid in a concentration of 48 μg/100 mL, whereas two folate derivatives, 5-CH_3-H_4folate and 10-HCO-folic acid, were found in egg yolk. The concentrations of 5-CH_3-H_4folate and 10-HCO-folic acid in the egg yolk were 226 and 46 μg/100 g, respectively. Three folate derivatives, H_4folate, 5-CH_3-H_4folate, and 10-HCO-folic acid, were found in dry baker's yeast in concentrations of 697, 2.520, and 212 μg/100 g, respectively. Pickled

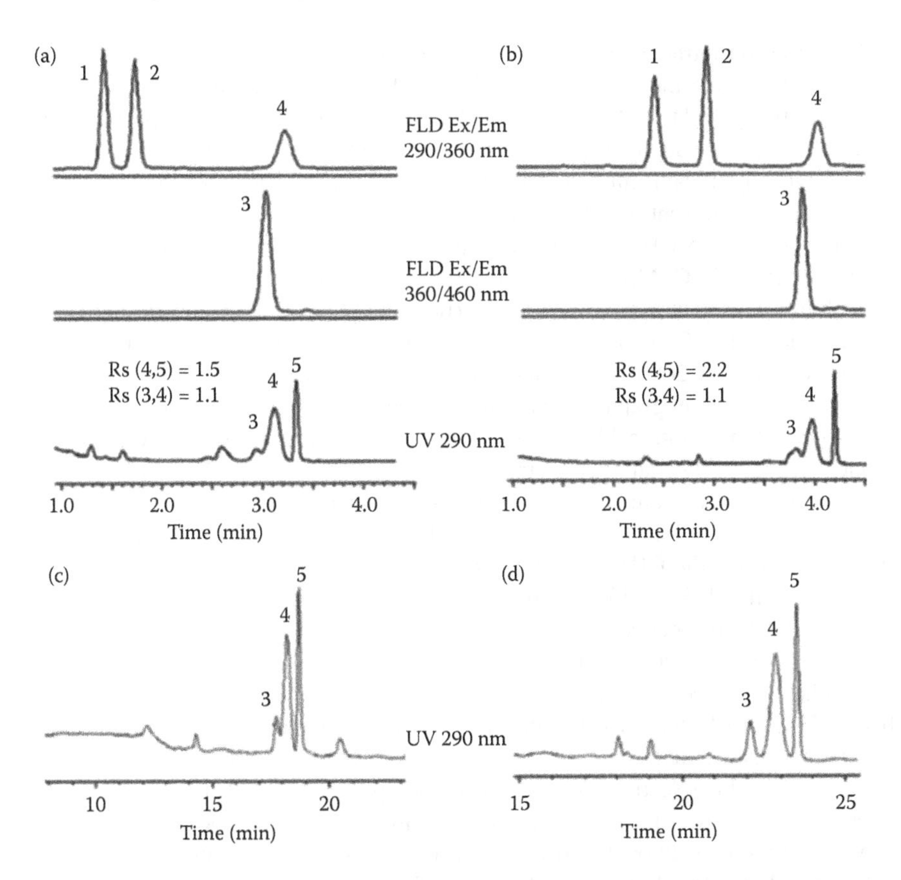

FIGURE 10.6 Separation of folates on Acquity BEH C_{18} (a) and Acquity HSS T3 (b) using gradient program B at 30°C and XBridge C_{18} (c) and Atlantis d18 (d) using gradient program A at 23°C. Peaks: 1—H_4folate; 2—5-CH_3-H_4folate; 3—10-HCO-folic acid; 4—5-HCO-H_4folate; and 5—folic acid. (Adapted from Jastrebova, J. et al. 2011. *Chromatograph.* 73:219–225. With permission.)

beetroots, strawberries, and orange juice contained 5-CH_3-H_4folate, which is known to be the predominant folate form in vegetables, fruits, and berries [97–99]. The concentration of 5-CH_3-H_4folate in pickled beetroots, strawberries, and orange juice was 15, 78, and 22 μg/100 g, respectively. The obtained values showed that the UPLC method provided good potential for separation of folates from matrix compounds with high sensitivity and fast sample throughput.

Rapid UPLC method for the determination of retinol and α-, γ-, and δ-tocopherols in foods was accomplished by Shim and his coworkers [100]. The method was validated in terms of precision, accuracy, and linearity. The separation was performed on a reversed-phase C_{18} column with 2 μm particle size, 2 mm i.d., and 75 mm length, followed by fluorescence detection. The recovery of retinol was more than 84.5% in all cases, while the detection and quantitation limits of the UPLC analysis were 0.015 and 0.045 mg/kg, respectively. The precision values were less than

9.12% based on intraday and interday measurements. The recoveries of α-, γ-, and δ-tocopherols were more than 81.3%; the LODs were 0.014, 0.002, and 0.001 mg/kg, respectively, and the LOQs were 0.042, 0.005, and 0.004 mg/kg, respectively. All calibration curves had good linearity ($r^2 = 0.99$) within the test ranges. This method can provide significant improvements in terms of speed, sensitivity, and resolution compared with conventional HPLC methods [100].

Folic acid and 5-CH_3-H_4folate have also been accurately measured in fortified breads by the UPLC–MS/MS method [101]. Reversed-phase Waters Acquity HSS T3 column (100 × 0.1 mm i.d., 1.8 μm particle size) has been eluted with a gradient mobile phase of 0.1% (v/v) formic acid in Milli-Q water and acetonitrile, the mobile phase delivered at 200 μL/min, and the sample injection volume fixed at 20 μL. Under these chromatographic conditions, the total run time was 6 min.

Using isotope dilution MS method, folates were detected in positive ESI SRM mode and accurately quantified by internal standards [102–104]. $^{13}C_5$ formic acid and $^{13}C_5$ 5-CH_3-H_4folate were added into various food matrices as internal standards for the quantification of folic acid and 5-CH_3-H_4folate, respectively [102–107]. For quantitative purposes, the CID transitions monitored were m/z 442 → 295, 447 → 295, 460 → 313, and 465 → 313 for folic acid, $^{13}C_5$ folic acid, 5-CH_3-H_4folate, and $^{13}C_5$ 5-CH_3-H_4folate, respectively.

Samples of boiled rice, corn starch, and tapioca starch were analyzed in the search for a folate-free food matrix. These samples were treated with α-amylase in a water bath at 75°C for 1 h and then extracted with phosphate buffer (pH 6.1). Folates were then selectively concentrated using SPE on styrene divinylbenzene cartridges, which have a broad hydrophobic selectivity for both polar and apolar organics but exhibit enhanced interaction with aromatic compounds when compared with traditional C_{18} reversed-phase sorbents. In addition, the assay has been used to measure the two folate levels in commercially produced bread samples from different brands in Australia [101].

Standard calibration curves for the two analytes were linear over the range of 0.018–14 μg folic acid/g of fresh bread ($r^2 = 0.997$) and 9.3–900 ng 5-CH_3-H_4folate/g of fresh bread ($r^2 = 0.999$). The absolute recoveries were 90% and 76% for folic acid and 5-CH_3-H_4folate, respectively. Inter- and intraday coefficients of variation were 2.6% and 3% for folic acid and 17.1% and 18% for 5-CH_3-H_4folate, respectively [101].

The LOD determined using preextracted tapioca starch as the blank matrix was 9.0 ng/g for folic acid and 4.3 ng/g for 5-CH_3-H_4folate. The assay is rugged, fast, accurate, and sensitive; applicable to a variety of food matrices; and is capable of the detection and quantification of the naturally occurring low levels of 5-CH_3-H_4folate in wheat breads. The findings of this study revealed that the folic acid range in Australian fortified breads was 79–110 μg/100 g of fresh bread and suggest that the flour may not have the mandated folic acid fortification level (200–300 μg/100 g of flour), though this cannot be determined conclusively from experimental bread data alone, as variable baking losses have been documented by other authors [101].

Brouwer and his coworkers applied a UPLC method for rapid and sensitive quantitative determination of folates in rice. Six monoglutamate folates in rice have been examined, including tetrahydrofolate, 5-methyltetrahydrofolate, 10-methenyltetrahydrofolate, 10-formylfolic acid, 5-formyltetrahydrofolate, and folic acid [108].

UPLC was performed under gradient conditions consisting of 0.1% of formic acid in water and acetonitrile. Ten microliter of samples were injected on a Waters Acquity HSS T3 column (150 × 2.1 mm i.d., 1.8 μm particle size), followed by tandem mass spectrometry detection. The instrument was operated in the ESI positive mode and the data were acquired in MRM mode.

Sample preparation was carried out under subdued light. The folate samples were treated with α-amylase, protease, and deconjugase and then extracted with phosphate buffer at pH 7.5. Matrix effects were compensated by use of isotopically labelled internal standards.

The developed UPLC–MS/MS method has been fully validated and was found to be suitable for determining the concentrations of the six folates in a total time of 8 min. The applicability of this analytical method was demonstrated by analyzing two types of rice; wild-type and genetically modified rice samples (biofortified with folates).

The method was found to have good accuracy and reproducibility for all analytes. Accuracy varied between 90.3% and 104.3%, while the intra- and interday precision had variation coefficients lower than 15%, except for folic acid (17.6%). LOD and LOQ varied between 0.06, 0.45, 0.12, and 0.91 μg/100 g, respectively. With a total run time of 8 min, this method is more sensitive for every single folate (up to 10 times) and offers faster separation than other developed HPLC methods (LOQ ranging from 0.6 to 4.0 μg/100 g, with a total run time of 20 min) [109].

The validated method has been successfully applied to determine folates in real rice samples. 5-Methyltetrahydrofolate was the predominant natural folate form in both wild-type (60%) and transgenic (90%) rice. The analysis of the distribution and levels of folates in wild-type and folate-biofortified rice showed up to 50-fold enrichment in biofortified rice, with total folate levels of up to 900 μg/100 g rice. Figure 10.7 shows the UPLC–MS/MS chromatogram for the folates in the two different samples. Some biofortified lines yielded folate levels of up to 900 μg/100 g, whereas levels in wild-type rice were around 20 μg/100 g [108].

One more study was developed by Chauveau-Duriot and his coworkers for the determination of vitamins A and E in milk by UPLC [110]. Fifty-six milk samples were collected from cows fed rations known to contain a broad variation of micronutrient contents [111]. Milk samples were stored in the dark at −20°C until extraction.

Each sample was deproteinized by adding the same volume of ethyl alcohol (containing internal standards). *n*-Hexane/ethyl acetate (9/1, v/v) was added to extract lipophilic components, and the resulting organic phases were collected. In the resulting organic phase, ethanol/water (90/10, v/v) was added to extract xanthophylls and vitamin E and preserve them from saponification. The lower ethanolic phase was collected in a new tube. The remaining upper hexanic phase was saponified in a solution of 10% KOH in ethyl alcohol (w/v) for 1 h at 60°C under darkness. Then, *n*-hexane/ethyl acetate (9/1, v/v) was added for vitamin purification. The final dry residue was dissolved in tetrahydrofuran (THF) and diluted with acetonitrile–dichloromethane–methanol (75–10–15) prior to automatic sampling, using 40 μL for HPLC and 10 μL for UPLC.

In all samples, internal standards were added as calibrants to compensate for losses in handling. Retinyl acetate and tocopheryl acetate were not used as internal standards for both vitamins because they are hydrolyzed by the saponification step

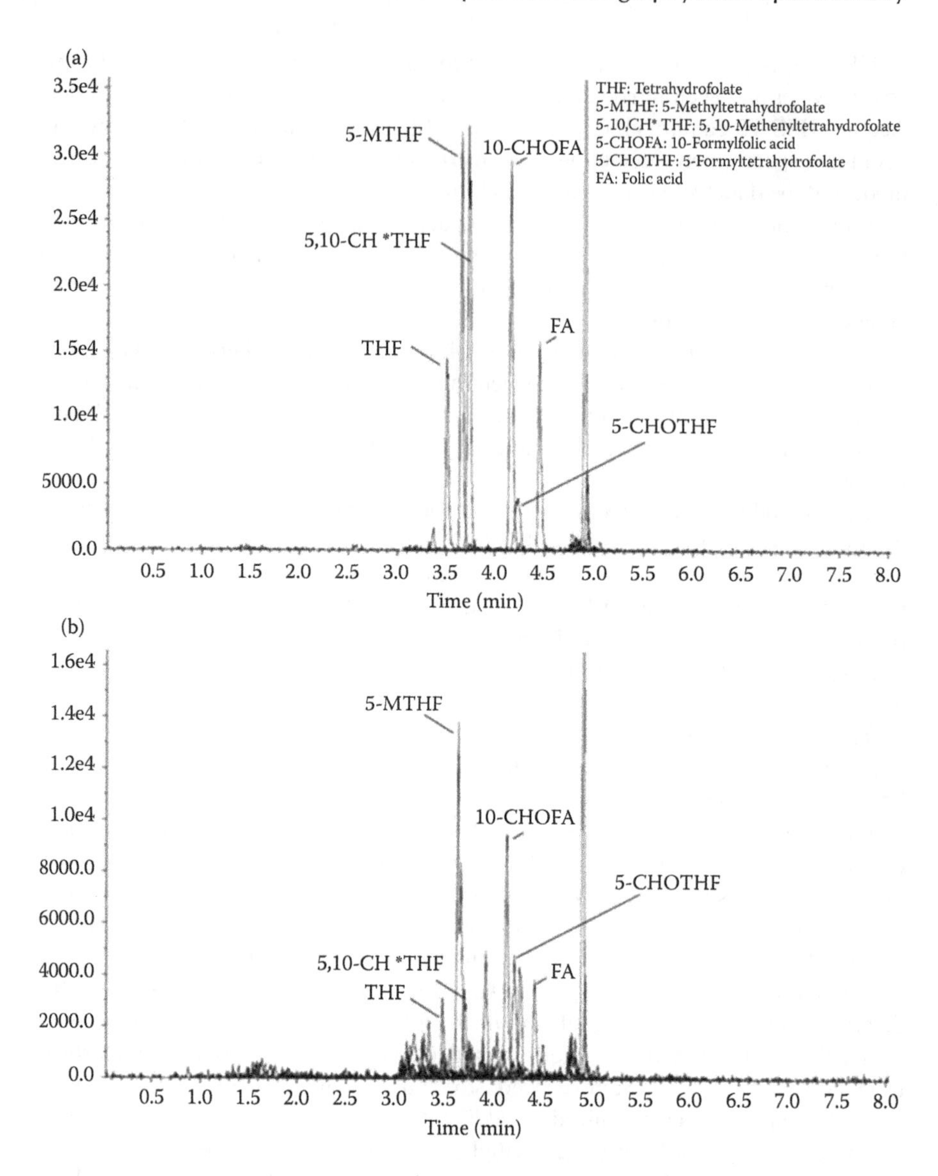

FIGURE 10.7 UPLC–MS/MS chromatogram (a) of a standard mixture of the six monoglutamate folates spiked to a 10 times diluted rice matrix (concentration of the individual folates varying between 4 and 5 µg/100 g) and (b) of a wild-type rice sample (total folate content is 19 µg/100 g) for the chromatographic conditions on the HSS T3 column. (Adapted from Brouwer, V. et al. 2010. *J. Chromatogr. B.* 878:509–513. With permission.)

performed to disrupt dairy fat during the extraction procedures. Hence, echinenone and δ-tocopherol were used as internal standard for vitamin A and E, respectively.

HPLC separation used two columns in series: 150 × 4.6 mm i.d., RP C_{18}, 3 µm particle size Nucleosil column coupled with a 250 × 4.6 mm i.d. RP C_{18}, 5 µm particle size 201 Vydac TP54 column. The isocratic mobile phase was a mix of acetonitrile–dichloromethane–ammonium acetate 0.05 M in methanol–water (70–10–15–5). The flow rate applied was 2 mL/min and all analysis was performed at room temperature. A 150 × 2.1 mm i.d. Acquity UPLC HSS T3, 1.8 μm particle size column was used with a gradient of acetonitrile–dichloromethane–methanol (75–10–15) and ammonium acetate 0.05 M in water. The flow rate was 0.4 mL/min. Column temperature was maintained at 35°C using a column oven. Vitamins A and E were detected at 325 and 292 nm, respectively [110].

The UPLC allowed a better resolution, an equal or better sensitivity according to gradient, and a better reproducibility of peak areas and retention times, but did not reduce the time required for analysis. Figures 10.8 and 10.9 show chromatograms of pure standards of vitamins A and E, respectively, obtained with HPLC and UPLC.

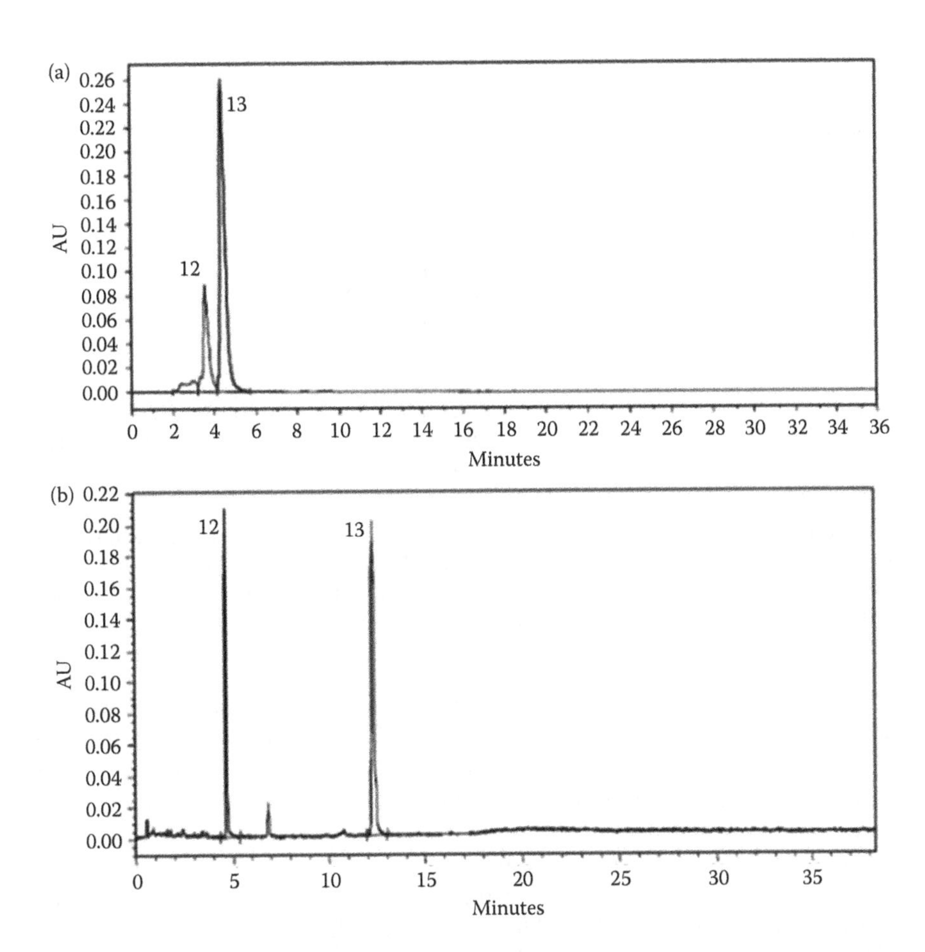

FIGURE 10.8 Chromatograms of a mix of standard vitamin A obtained with either HPLC (a) or UPLC (b). Detection was performed at 325 nm: (12) retinol and (13) retinyl acetate. AU: absorbance units. (Adapted from Chauveau-Duriot, B. et al. 2010. *Anal. Bioanal. Chem.* 397:777–790. With permission.)

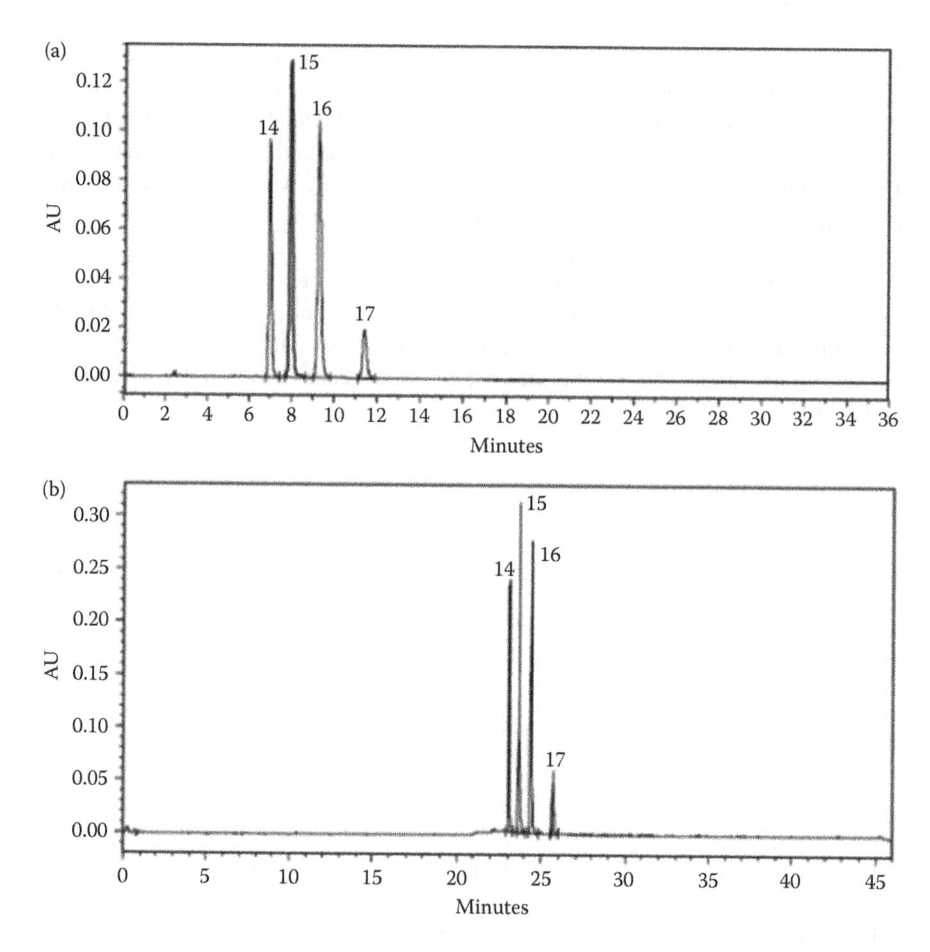

FIGURE 10.9 Chromatograms of a mix of standard vitamin E obtained with either (a) HPLC or (b) UPLC. Detection was performed at 292 nm: (14) δ-tocopherol, (15) γ-tocopherol, (16) α-tocopherol, and (17) tocopheryl acetate. AU: absorbance units. (Adapted from Chauveau-Duriot, B. et al. 2010. *Anal. Bioanal. Chem.* 397:777–790. With permission.)

Retinol was detected in all milks. α-Tocopherol was the most commonly observed form of vitamin E in these samples, but γ-tocopherol was sometimes detected with the UPLC but not by HPLC, due to the better (10-fold higher) sensitivity of UPLC for this molecule. Retinol concentrations ranged from 0.03 to 0.6 μg/mL, and the mean values for α-tocopherol ranged from 0 to 0.9 μg/mL. Extraction recoveries for all analytes were up to 70%. Vitamin resolution was similar and higher to 1 with both methods for vitamin E (α, δ, γ tocopherols and tocopheryl acetate), and was much higher with UPLC than with HPLC for vitamin A [110].

The UPLC system did not offer shorter analysis times. This could be explained by the difference in the column's chemistry, as well as the proportion of water in the mobile phase. However, it gave the best reproducibility of retention time due to temperature control, and the best peak area due to better chromatographic resolution.

In conclusion, the arrival of the UPLC in laboratories has permitted the improvement of the quality and the productivity of analyses, with an increased sensitivity, a better selectivity, a high throughput, and simplification of sample preparation. To increase the productivity of mass spectrometers, UPLC can also be a solution, with a reduction of run time, a better resolution, and better sensitivity, in comparison to classical HPLC systems.

ACKNOWLEDGMENT

This work was supported by King Saud University, Deanship of Scientific Research, College of Science Research Center.

REFERENCES

1. Moreno, P. and Salvado, V. 2000. Determination of eight water- and fat-soluble vitamins in multi-vitamin pharmaceutical formulations by high-performance liquid chromatography. *J. Chromatogr. A*. 870:207–215.
2. Finglas, P., Faure, U., and Wagstaffe, P. 1993. Improvements in the determination of vitamins in food through intercomparisons and preparation of RMs for vitamin analysis within the BCR[1] programme. *Fresenius J. Anal. Chem.* 345:180–184.
3. Jacab, R. and Sotoudeh, G. 2002. Vitamin C function and status in chronic disease. *Nutri. Clin. Care*. 5:66–74.
4. Blake, C. 2007. Analytical procedures for water-soluble vitamins in foods and dietary supplements: A review. *Anal. Bioanal. Chem.* 389:63–76.
5. Ottaway, P. 1993. *The Technology of Vitamins in Food*. Chapman & Hall, New York.
6. Nollet, L. 1992. *Food Analysis by HPLC*. 2nd Ed. Marcel Dekker, New York.
7. Machlin, L. 1991. *Handbook of Vitamins*. 2nd Ed. Marcel Dekker, New York.
8. Anthea, M., Hopkins, J., McLaughlin, C., Johnson, S., Warner, M., LaHart, D., and Wright, J. 1993. *Human Biology and Health*. Prentice Hall, Englewood Cliffs, New Jersey, USA.
9. Ball, G. 2006. *Vitamins in Foods. Analysis, Bioavailability, and Stability*. CRC Press, Boca Raton, Florida.
10. AIN (American Institute of Nutrition). 1990. Nomenclature policy: Generic descriptors and trivial names for vitamins and related compounds. *J. Nutr.* 120:643–644.
11. Booher, L., Hartzler, E., and Hewston, E. 1942. A compilation of the vitamin values of foods in relation to processing and other variants, *U.S. Dept. Agric. Cir.* 638:1–244.
12. Booth, S., Sadowski, J., Weihrauch, J., and Ferland, G. 1993. Vitamin K1 (phylloquinone) content of foods: A provisional table. *J. Food Comp. Anal.* 6:109–120.
13. Kamal-Eldin, A. and Appelqvist, L. 1996. The chemistry and antioxidant properties of tocopherols and tocotrienols. *Lipids*. 31:671–701.
14. Sheppard, A. and Pennington, J. 1993. *Analysis and Distribution of Vitamin E in Vegetable Oils and Foods*. Marcel Dekker, New York.
15. Rucker, R., Suttie, J., McCormick, D., and Machlin, L. 2001. *Handbook of Vitamins*. 3rd Ed. Marcel Dekker, New York.
16. Nollet, L. 2000. *Food Analysis by HPLC*. 2nd Ed. Marcel Dekker, New York.
17. Kozlov, E., Solunina, I., Lyubareva, M., and Nadtochii, M. 2003. HPLC determination of Vitamins A, D, and E in multivitamin composition. *Pharm. Chem. J.* 37:560–562.

18. Pegg, R., Landen, W., and Eitenmiller, R. 2010. *Vitamin Analysis*. Springer Science + Business Media, New York.
19. United States Environmental Protection Agency (EPA). 2002. Methods for Measuring the Acute Toxicity of Effluents and Receiving Waters to Fresh water and Marine Organisms. Washington, DC. Document No. EPA-821-R-02-012.
20. Ball, G. 1998. *Bioavailability & Analyzsis of Vitamins in Foods*. Chapman & Hall, London.
21. Carlucci, A. and Bowes, P. 1972. Determination of vitamin B12, thiamine, and biotin in Lake Tahoe waters using modified marine bioassay techniques. *National Agricultural Library*. 17:774–777.
22. Gregory, J. and Litherland, S. 1986. Efficacy of the rat bioassay for the determination of biologically available vitamin B-612. *J. Nutr.* 116:87–97.
23. Ibanoglu, S., Ainsworth, P., Wilson, G., and Hayes, G. 1995. Effect of formulation on protein breakdown, *in vitro* digestibility, rheological properties and acceptability of tarhana, a traditional Turkish cereal food. *Int. J. Food Sci.* 30:579–585.
24. Lee, H. 2000. *HPLC Analysis of Phenolic Compounds*. 1st Ed. Marcel Dekker, New York.
25. De Leenheer, A., Lambert, W., and De Ruyter, M. 1985. *Modern Chromatographic Analysis of the Vitamins, Chromatographic Science Series*. 2nd Ed. Marcel Dekker, New York.
26. Gentili, A., Caretti, F., D'Ascenzo, G., Marchese, S., Perret, D., Corcia, D., and Rocca, L. 2008. Simultaneous determination of water-soluble vitamins in selected food matrices by liquid chromatography/electrospray ionization tandem mass spectrometry. *Rapid Commun. Mass Spectrom.* 22:2029–2043.
27. Luttseva, A. and Maslov, L. 1999. Methods of control and standardization of drugs containing water-soluble vitamins (a review). *J. Pharma. Chem.* 33:490–498.
28. Luttseva, A., Maslov, L., and Seredenko, V. 2001. Methods of control and standardization of drugs containing fat-soluble vitamins (a review). *J. Pharma. Chem.* 35: 567–572.
29. Ekinci, R. and Kadakal, C. 2005. Determination of seven water-soluble vitamins in tarhana, a traditional Turkish cereal food, by high-performance liquid chromatography. *Acta Chromatogr. A*. 15:289–297.
30. Rongjie, F., Jianzhong, L., and Wang, Y. 2010. Fat-Soluble Vitamins Analysis on an Agilent ZORBAX Eclipse PAH Polymeric C18 Bonded Column. Application Note. Agilent Technologies, USA.
31. Giorgi, M., Howland, K., Martin, C., and Bonner, A. 2012. A novel HPLC method for the concurrent analysis and quantitation of seven water-soluble vitamins in biological fluids (plasma and urine): A validation study and application. *Scientific World Journal*. 2012:1–8.
32. Doughty, E., Herwehe, K., and Yearick, V. 1996. Fat soluble vitamin analyses by HPLC. *The Reporter*. 15:4–5.
33. Laura, A., Armas, G., Hollis, B., and Heaney, R. 2004. Vitamin D2 is much less effective than vitamin D3 in humans. *J. Clin. Endocrinol. Metab.* 89:5387–5391.
34. Strathmann, F., Laha, T., and Hoofnagle, A. 2011. Quantification of 1α,25 dihydroxy vitamin D by immunoextraction and liquid chromatography-tandem mass spectrometry. *Clin. Chem.* 57:1279–1285.
35. Leporati, A., Catellani, D., Suman, M., Andreoli, R., Manini, P., and Niessen, W. 2005. Application of a liquid chromatography tandem mass spectrometry method to the analysis of water-soluble vitamins in Italian pasta. *Anal. Chim. Acta* 531:87–95.
36. *2nd International Vitamin Conference*, 23–25 May 2012, Copenhagen.
37. Swartz, M. 2005. Principal Scientist, Waters Corporation, Milford, Massachusetts, Separation Science Redefined. on-line. Available: www.chromatographyonline.com.
38. Villiers, A., Lestremau, F., Szucs, R., Gelebart, S., David, F., and Sandra, P. 2006. Evaluation of ultra-performance liquid chromatography. Part I: Possibilities and limitations. *J. Chromatogr. A*. 1127:60–69.
39. Heiger, D. 2000. *High Performance Capillary Electrophoresis—An Introduction*. Agilent Technologies, Germany.

40. Welsch, T. and Michalke, D. 2003. (Micellar) electrokinetic chromatography: An interesting solution for the liquid phase separation dilemma. *J. Chromatogr. A.* 1000:935–951.
41. Zhang, Y., Gong, X., Zhang, H., Larock, R., and Yeung, E. 2000. Combinatorial screening of homogeneous catalysis and reaction optimization based on multiplexed capillary electrophoresis. *J. Comb. Chem.* 2:450–452.
42. Zhou, C., Jin, Y., Kenseth, J., Stella, M., Wehmeyer, K., and Heineman, W. 2005. Rapid pk_a estimation using vacuum-assisted multiplexed capillary electrophoresis (VAMCE) with ultraviolet detection. *J. Pharmac. Sci.* 94:576–589.
43. Van Deemter, J., Zuiderweg, E., and Klinkenberg, A. 1956. Longitudinal diffusion and resistance to mass transfer as causes of non-ideality in chromatography. *Chem. Eng. Sci.* 5:271–289.
44. Jerkovich, A., Mellors, J., and Jorgenson, J. 2003. The use of micron-sized particles in ultrahigh-pressure liquid chromatography. *LCGC.* 21:600–610.
45. Swartz, M. 2005. Ultra Performance Liquid Chromatography (UPLC): An introduction, separation science re-defined. *LCGC.* 23:8–14.
46. Novakova, L., Matysova, L., and Solich, P. 2006. Advantages of application of UPLC in pharmaceutical analysis. *Talanta.* 68:908–918.
47. Swartz, M. 2004. Ultra performance liquid chromatography: Tomorrow's HPLC technology today. *LabPlus Int.* 18:6–8.
48. Lars, Y. and Honore, H. 2003. On-line turbulent-flow chromatography–high-performance liquid chromatography–mass spectrometry for fast sample preparation and quantitation. *J. Chromatogr. A.* 1020:59–67.
49. Swartz, M., Murphy, B., and Sievers, D. 2004. UPLC: Expanding the limits of HPLC. *GIT Lab J.* 8:43–45.
50. Wu, N., Lippert, J., and Lee, M. 2001. Practical aspects of ultrahigh pressure capillary liquid Chromatography. *J. Chromatogr. A.* 911:1–12.
51. MacNair, J., Patel, K., and Jorgenson, J. 1999. Ultrahigh-pressure reversed-phase capillary liquid chromatography: Isocratic and gradient elution using columns packed with 1.0 mm particles. *Anal. Chem.* 71:700–708.
52. Patel, K., Jerkovich, A., Link, J., and Jorgenson, J. 2004. In-depth characterization of slurry packed capillary columns with 1.0-microm nonporous particles using reversed-phase isocratic ultrahigh-pressure liquid chromatography. *Anal. Chem.* 76:5777–5786.
53. De Villiers, A., Lestremau, F., Szucs, R., Gelebart, S., David, F., and Sandra, P. 2006. Evaluation of ultra-performance liquid chromatography. Part I. Possibilities and limitations. *J. Chromatogr. A.* 1127:60–69.
54. Plumb, R., Dear, G., Mallett, D., and Ayrton, J. 2001. Direct analysis of pharmaceutical compounds in human plasma with chromatographic resolution using an alkyl-bonded silica rod column. *Rapid Commun. Mass Spectrom.* 15:986–993.
55. Bayliss, M., Little, D., Mallett, D., and Plumb, R. 2000. Parallel ultra-high flow rate liquid chromatography with mass spectrometric detection using a multiplex electrospray source for direct, sensitive determination of pharmaceuticals in plasma at extremely high throughput. *Rapid Commun. Mass Spectrom.* 14:2039–2045.
56. Castro-Perez, J., Plumb, R., Granger, J., Beattie, I., Joncour, K., and Wright, A. 2005. Increasing throughput and information content for *in vitro* drug metabolism experiments using ultra-performance liquid chromatography coupled to a quadrupole time-of-flight mass spectrometer. *Rapid Commun. Mass Spectrom.* 19:843–848.
57. Swartz, E. 2005. UPLC™: An introduction and review. *J. Liq. Chromatogr. Related Technol.* 28:1253–1263.
58. Guoyu, D. and Xiangru, Z. 2009. A picture of polar iodinated disinfection byproducts in drinking water by (UPLC/) ESI-tqMS, SAR. *Environ. Sci. Technol.* 43:9287–9293.

59. Srivastava, B., Sharma, B., Baghel, U., Want, Y., and Sethi, N. 2010. Ultra performance liquid chromatography (UPLC): A chromatography technique. *Int. J. Pharm. Quality Assur.* 2:19–25.
60. Broske, A. 2004. *Agilent Technologies Application Note*. 5988EN.
61. Goodwin, L., White, S., and Spooner, N. 2007. Evaluation of ultra-performance liquid chromatography in the bioanalysis of small molecule drug candidates in plasma. *J. Chromatogr. Sci.* 45:298–304.
62. Plumb, R., Castro-Perez, J., Granger, J., Beattie, I., Joncour, K., and Wright, A. 2004. Ultra-performance liquid chromatography coupled to quadrupole-orthogonal time-of-flight mass spectrometry. *Rapid Commun. Mass Spectrom.* 18:2331–2337.
63. Wang, L., Yuan, K., and Yu, W. 2010. Studies of UPLC fingerprint for the identification of *Magnolia officinalis* cortex processed. *Pharmacognosy Magazine*. 6:83–88.
64. Mazzeo, J., Wheat, T., Gillece-Castro, B., and Lu, Z. 2006. Next generation peptide mapping with ultra performance liquid chromatography. *BioPharm. Int.* 19:22–25.
65. Novakova, L., Solichova, D., and Solich, P. 2006. Advantages of ultra-performance liquid chromatography over high-performance liquid chromatography: Comparison of different analytical approaches during analysis of diclofenac gel. *J. Sep. Sci.* 29:2433–2443.
66. Wilson, I., Plumb, R., Granger, J., Major, H., Williams, R., and Lenz, E. 2005. HPLC-MS-based methods for the study of metabonomics. *J. Chromatogr. B*. 817:67–76.
67. Plumb, R. and Wilson, I. 2004. High throughput and high sensitivity LC/MS-OA-TOF and UPLC/TOF-MS for the identification of biomarkers of toxicity and disease using a metabonomics approach. *Abstr. Pap. Am. Chem. Soc.* 228:189.
68. Castro-Perez, J., Plumb, R., Granger, J., Beattie, I., Joncour, K., and Wright, A. 2005. Increasing throughput and information content for *in vitro* drug metabolism experiments using ultra-performance liquid chromatography coupled to a quadrupole time-of-flight mass spectrometer. *Rapid Commun. Mass Spectrom.* 19:843–848.
69. Wilson, I., Nicholson, J., Castro-Perez, J., Granger, J., Johnson, K., Smith, B., and Plumb, R. 2005. High resolution "ultra-performance" liquid chromatography coupled to oa-TOF mass spectrometry as tool for differential metabolic pathway profiling in functional genomic studies. *J. Proteome. Res.*, 4:591–598.
70. Wilfried, M. 2006. *Liquid Chromatography–Mass Spectrometry*. 3rd Ed. CRC Press, Taylor & Francis Group, LLC, London.
71. De Leenheer, A., Lambert, W., and Van Bocxlaer, J. 2000. *Modern Chromatographic Analysis of Vitamins*. Marcel Dekker, New York.
72. Novakova, L. and Solich, P. 2008. HPLC methods for simultaneous determination of ascorbic and dehydroascorbic acids. *TraC. Trends Anal. Chem.* 27:942–958.
73. Fenoll, J. and Martínez, A. 2011. Simultaneous determination of ascorbic and dehydroascorbic acids in vegetables and fruits by liquid chromatography with tandem-mass spectrometry. *Food Chem.* 127:340–344.
74. Odriozola-Serrano, I. and Hernández-Jover, T. 2007. Comparative evaluation of UV-HPLC methods and reducing agents to determine vitamin C in fruits. *Food Chem.* 105:1151–1158.
75. Spinola, V., Mendes, B., Camara, J., and Castilho, P. 2012. An improved and fast UHPLC-PDA methodology for determination of L-ascorbic and dehydroascorbic acids in fruits and vegetables. Evaluation of degradation rate during storage. *Anal. Bional. Chem.* 403:1049–1058.
76. Lee, S. and Kader, A. 2000. Preharvest and postharvest factors influencing vitamin C content of horticultural crops. *Postharvest Biol. Technol.* 20:207–220.
77. Johnston, C. and Bowling, D. 2007. Ascorbic acid. In: *Handbook of Vitamins*. 4th Ed. Zempleni, J., Rucker, R.B., McCormick, D.B., and Suttie, J.W. (Eds.) CRC Press, Boca Raton, Florida.

78. Phillips, K. and Tarrago-Trani, M. 2010. Stability of vitamin C in frozen raw fruit and vegetable homogenates. *J. Food Compos. Anal.* 23:253–259.
79. Valente, A. and Albuquerque, T. 2011. Ascorbic acid content in exotic fruits: A contribution to produce quality data for food composition databases. *Food Res. Int.* 44:2237–2242.
80. Hernández, Y. and Lobo, M. 2006. Determination of vitamin C in tropical fruits: A comparative evaluation of methods. *Food Chem.* 96:654–664.
81. Frenich, A. and Torres, M. 2005. Determination of ascorbic acid and carotenoids in food commodities by liquid chromatography with mass spectrometry detection. *J. Agric. Food Chem.* 53:7371–7376.
82. Eitenmiller, R., Ye, L., and Landen, W. 2008. Ascorbic acid: Vitamin C. In: *Vitamin Analysis for the Health and Food Sciences*. 2nd Ed. CRC Press, Boca Raton, Florida.
83. Huang, B., Zheng, F., Fu, S., Yao, J., Tao, B., and Ren, Y. 2012. UPLC-ESI-MS/MS for determining *trans*- and *cis*-vitamin K1 in infant formulas: Method and applications. *Eur. Food Res. and Technol.* 235:873–879.
84. Report of the 30th Session of the Codex Committee on Nutrition and Foods for Special Dietary Uses Cape Town. 2008. pp 65–66. www.codexalimentarius.org.
85. Lu, B., Ren, Y., Huang, B., Liao, W., Cai, Z., and Tie, X. 2008. Simultaneous determination of four water-soluble vitamins in fortified infant foods by ultra-performance liquid chromatography coupled with triple quadrupole mass spectrometry. *J. Chromatogr. Sci.* 46:225–232.
86. Hampel, D., York, E., and Allen, L. 2012. Ultra-performance liquid chromatography tandem mass-spectrometry (UPLC–MS/MS) for the rapid, simultaneous analysis of thiamin, riboflavin, flavin adenine dinucleotide, nicotinamide and pyridoxal in human milk. *J. Chromatogr. B.* 903:7–13.
87. Sakurai, T., Furukawa, M., Asoh, M., Kanno, T., Kojima, T., and Yonekubo, A. 2005. Fat-soluble and water-soluble vitamin contents of breast milk from Japanese women. *J. Nutr. Sci. Vitaminol.* 51:239–247.
88. Stuetz, W., Carrara, V., McGready, R., Lee, S., Erhardt, J., Breuer, J., Biesalski, H., and Nosten, F. 2011. Micronutrient status in lactating mothers before and after introduction of fortified flour: Cross-sectional surveys in Maela refugee camp. *Eur. J. Nutr.* 51:425–434.
89. Greer, F. 2001. Do breastfed infants need supplemental vitamins? *Pediatr. Clin. North Am.* 48:415–423.
90. Van Herwaarden, A., Wagenaar, E., Merino, G., Jonker, J., Rosing, H., Beijnen, J., and Schinkel, A. 2007. Multidrug transporter ABCG2/breast cancer resistance protein secretes riboflavin (vitamin B2) into milk. *Mol. Cell. Biol.* 27:1247–1253.
91. Zhang, H., Chen, S., Liao, W., and Ren, Y. 2009. Fast simultaneous determination of multiple water-soluble vitamins and vitamin like compounds in infant formula by UPLC-MS/MS. *J. Food Agric. Environ.* 7:88–93.
92. Stevens, J. and Dowell, D. 2012. Determination of vitamins D_2 and D_3 in infant formula and adult nutritionals by ultra-pressure liquid chromatography with tandem mass spectrometry detection (UPLC-MS/MS): First action 2011.12. *J. AOAC Int.* 95:577–582.
93. Centers for Disease Control and Prevention. July 2008. http://www.cdc.gov/nutritionreport/part_2b.html.
94. Huang, M. and Winters, D. 2011. Application of ultra-performance liquid chromatography/tandem mass spectrometry for the measurement of vitamin D in foods and nutritional supplements. *J. AOAC. Int.* 94:211–223.
95. Jastrebova, J., Strandler, H., Patring, J., and Wiklund, T. 2011. Comparison of UPLC and HPLC for analysis of dietary folates. *Chromatograph.* 73:219–225.
96. Patring, J. and Jastrebova, J. 2007. Application of liquid chromatography-electrospray ionisation mass spectrometry for determination of dietary folates: Effects of buffer nature and mobile phase composition on sensitivity and selectivity. *J. Chromatogr. A.* 1143:72–82.

97. Jastrebova, J., Witthoft, C., Grahn, A., Svensson, U., and Jagerstad, M. 2003. HPLC determination of folates in raw and processed beetroots. *Food Chem.* 80:579–588.
98. Vahteristo, L., Lehikoinen, K., Ollilainen, V., and Varo, P. 1997. Application of an HPLC assay for the determination of folate derivatives in some vegetables, fruits and berries consumed in Finland. *Food Chem.* 59:589–597.
99. Stralsjo, L., Ahlin, H., Witthoft, C., and Jastrebova, J. 2003. Folate determination in berries by radioprotein-binding assay (RPBA) and high performance liquid chromatography (HPLC). *Eur. Food Res. Technol.* 216:264–269.
100. Shim, Y., Kim, K., Seo, D., Ito, M., Nakagawa, H., and Ha, J. 2012. Rapid method for the determination of vitamins A and E in foods using ultra-high-performance liquid chromatography. *J. AOAC. Int.* 95:517–522.
101. Chandra-Hioe, M., Bucknall, M., and Arcot, J. 2011. Folate analysis in foods by UPLC-MS/MS: Development and validation of a novel, high throughput quantitative assay; folate levels determined in Australian fortified breads. *Anal. Bioanal. Chem.* 401:1035–1042.
102. Freisleben, A., Schieberle, P., and Rychlik, M. 2002. Syntheses of labeled vitamers of folic acid to be used as internal standards in stable isotope dilution assays. *J. Agric. Food Chem.* 50:4760–4768.
103. Freisleben, A., Schieberle, P., and Rychlik, M. 2003. Specific and sensitive quantification of folate in foods by stable isotope dilution assays using high-performance liquid chromatography-tandem mass spectrometry. *Anal. Bioanal. Chem.* 376:149–156.
104. Freisleben, A., Schieberle, P., and Rychlik, M. 2003. Comparison of folate quantification in foods by high-performance liquid chromatography-fluorescence detection to that by stable isotope dilution assays using high-performance liquid chromatography-tandem mass spectrometry. *Anal. Biochem.* 315:247–255.
105. Pawlosky, R., Flanagan, V., and Pfeiffer, C. 2001. Determination of 5-methyltetrahydrofolic acid in human serum by stable-isotope dilution high-performance liquid chromatography-mass spectrometry. *Anal. Biochem.* 298:299–305.
106. Pawlosky, R., Flanagan, V., and Doherty, R. 2003. A mass spectrometric validated high-performance liquid chromatography procedure for the determination of folates in foods. *J. Agric. Food Chem.* 51:3726–3730.
107. Pawlosky, R., Hertrampf, E., Flanagan, V., and Thomas, P. 2003. Mass spectral determinations of the folic acid content of fortified breads from Chile. *J. Food Compos. Anal.* 16:281–286.
108. Brouwer, V., Storozhenko, S., Stove, C., Daele, J., Straeten, D., and Lambert, W. 2010. Ultra-performance liquid chromatography–tandem mass spectrometry (UPLC–MS/MS) for the sensitive determination of folates in rice. *J. Chromatogr. B.* 878:509–513.
109. De Brouwer, V., Storozhenko, S., Van De Steene, J., Wille, S., Stove, C., Van Der Straeten, D., and Lambert, W. 2008. Optimisation and validation of a liquid chromatography-tandem mass spectrometry method for folates in rice. *J. Chromatogr. A.* 1215:125–132.
110. Chauveau-Duriot, B., Doreau, M., Nozière, P., and Graulet, B. 2010. Simultaneous quantification of carotenoids, retinol, and tocopherols in forages, bovine plasma, and milk: Validation of a novel UPLC method. *Anal. Bioanal. Chem.* 397:777–790.
111. Citova, I., Havlikova, L., Urbanek, L., Solichova, D., Novakova, L., and Solich, P. 2007. Comparison of a novel ultra-performance liquid chromatographic method for determination of retinol and alpha-tocopherol in human serum with conventional HPLC using monolithic and particulate columns. *Anal. Bioanal. Chem.* 388:675–681.

11 UPLC–MS as an Analytical Tool for the Determination of Aflatoxins in Food

Kareem Yusuf, Ahmad Aqel, Ayman Abdel Ghfar, and Zeid Abdullah Alothman

CONTENTS

11.1 INTRODUCTION

Aflatoxins, on a worldwide scale, are important mycotoxins in human foods and animal feedstuffs [1].

Aflatoxin contamination causes economic losses of corn, cottonseed, peanuts, sorghum, wheat, rice, and other commodities, as well as economic losses of processed food and feedstuffs. As commodities considered unsafe for human consumption can be incorporated into animal feedstuffs [2,3], there exists opinion that aflatoxicosis in domestic animals is considerably more prevalent than it is diagnosed. Health effects occur in companion animals, livestock, poultry, and humans because aflatoxins are potent hepatotoxins, immunosuppressants, mutagens, and carcinogens [4–6]. Aflatoxins are teratogenic [7].

Aflatoxicosis in the human population, especially in areas stricken by poverty and drought and other adverse growing conditions, is an important public health problem [1].

Most aflatoxins are chemically and structurally diverse. Since the majority of secondary metabolites are synthesized by simple biosynthetic reactions from small molecules (acetates, pyruvates, etc.), this is surprising, however, this leads to the compounds having such a diverse range of toxic effects, both acute and chronic [8].

11.2 AFLATOXINS AND FOOD CONTAMINATION

11.2.1 Mycotoxins

The contamination of food by the intentional use of chemicals, such as pesticides or veterinary drugs, is a worldwide public health concern. However, food contamination due to natural toxicants, such as mycotoxins, can also compromise the safety of food and feed supplies and adversely affect health (WHO/FAO, 2001) in humans and animals [9].

Mycotoxins are secondary metabolites of fungi. Many foods and feeds can become contaminated with mycotoxins since they can form in commodities before harvest, during the time between harvesting and drying, and in storage. Commodities and products frequently contaminated with mycotoxins include corn, wheat, barley, rice, oats, nuts, milk, cheese, peanuts, and cottonseed. The major fungal genera producing mycotoxins include *Aspergillus*, *Fusarium*, and *Penicillium*. The most common mycotoxins are aflatoxins, ochratoxin A, fumonisins, deoxynivalenol, T-2 toxin, and zearalenone (Table 11.1) [10]. Mycotoxins produce a wide range of adverse and toxic effects in animals in addition to being foodborne hazards to humans [11].

Among all mycotoxins, aflatoxins are the most toxic, widespread, and the strongest natural carcinogens (Table 11.2). The International Agency for Research on Cancer (IARC) has defined aflatoxin B1 (AFB1) as a carcinogen [11].

11.2.2 Aflatoxins

Aflatoxins, in the late 1950s and the early 1960s, were identified as the cause of the mysterious turkey "X" disease in Great Britain [4,5]. They have also been identified as carcinogens found in rainbow trout [11].

TABLE 11.1
Major Chemical Types of Mycotoxins

Mycotoxins	Main Producing Fungi
Aflatoxins B1, B2, G1, G2	*Aspergillus flavus, A. parasiticus, A. nomius*
Ochratoxin A	*Penicillium verrucosum, A. alutaceus, A. carbonarius*
Patulin	*P. expansum, A. clavatus, Byssochlamys nivea*
Fumonisins	*Fusarium moniliforme, F. proliferatum*
Deoxynivalenol (trichothecenes)	*F. graminearum, F. culmorum, F. crookwellense*
Zearalenone	*F. graminearum, F. culmorum, F. crookwellense*

In the United States, studies on aflatoxins incriminated aflatoxins as the cause of epizootic hepatitis in dogs and as the cause of moldy corn poisoning in pigs [12].

Aflatoxins (AFs) are a family of structure-related mycotoxins produced as secondary metabolites by the spoilage of fungi *Aspergillus*, particularly *A. flavus* and *A. parasiticus* [13–15].

The most important members are AFB1, AFB2, AFG1, AFG2, AFM1, and AFM2. Among the major AFs of concern, AFB1 is the most frequent metabolite in contaminated samples and is clarified in group I as a human carcinogen, the carcinogenic mechanism of which is achieved by affecting the pericellular membrane, interfering with the inductive style of specific enzymes and inhibiting the synthesis of RNA [16–18].

AFB2, AFG1, and AFG2 are also clarified in group I as carcinogens to humans. Although the toxicity of AFM1 is lower than AFB1, it is known for its hepatotoxic and carcinogenic effects [19].

Aflatoxin in food is one of the most widely spread food contaminations. It can be found in over a hundred kinds of agro-products and foods, such as peanuts, corn, rice, soy sauce, vinegar, plant oil, pistachios, tea, Chinese medicinal herb, eggs, milk, feed, and so on. Some have also been detected in animal organisms. Besides

TABLE 11.2
Mycotoxins Notification

Hazard	2002	2003	2004	2005	2006	2007	2008	2009	2010	2011
Aflatoxins	288	762	839	946	801	705	902	638	649	585
Deoxynivalenol (DON)						10	4	3	2	11
Fumonisins		15	14	2	15	9	2	1	3	4
Ochratoxin A	14	26	27	42	54	30	20	27	34	35
Patulin				6	7		3			
Zearalenone					1	6	3			
Total mycotoxins	**302**	**803**	**880**	**996**	**878**	**760**	**933**	**669**	**688**	**635**

Source: Adapted from European Commission. 2011. The Rapid Alert System for Food and Feed, Annual Report. http://ec.europa.eu/food/food/rapidalert/docs/rasff_annual_report_2011_en.pdf. With permission.

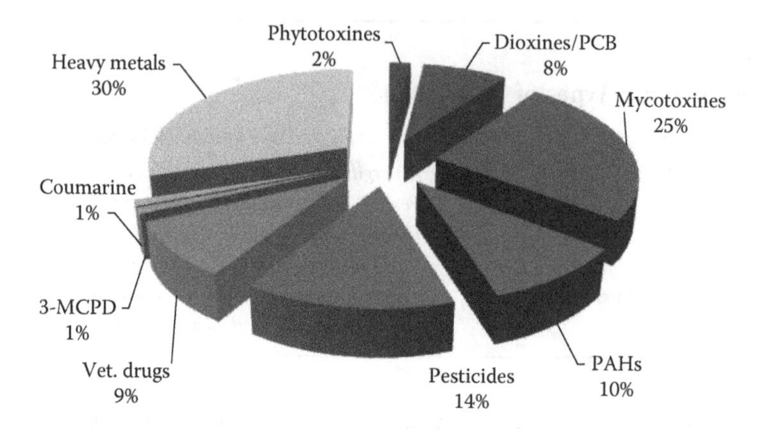

FIGURE 11.1 Illustration of the most common chemical hazards in food and feedstuff in European Union chemical alerts 2007.

these, aflatoxin can spread and be accumulated in the environment, for example, in rivers or agricultural fields [20].

Aflatoxin contributes around 25% of the total toxins in foods affecting human health (Figure 11.1).

11.2.3 Chemistry of Aflatoxins

Aflatoxins have a difuranocoumarin chemical structure. Approximately 18 aflatoxins have been chemically characterized. Aflatoxins are in two chemical groups, the difurocoumarocyclopentenone series (includes AFB_1, AFB_2, AFB_{2A}, AFM_1, AFM_2, AFM_{2A}, and aflatoxicol) and the difurocoumarolactone series (includes AFG_1 and AFG_2) (Figure 11.2).

The "B" group is fluorescent blue in long-wavelength ultraviolet light and the "G" group is fluorescent green. The primary aflatoxins of concern in feedstuffs are AFB_1, AFB_2, AFG_1, and AFG_2 (Figure 11.2). Analytical results for aflatoxins generally are the sum of the concentrations of these four toxins. AFB1 is the most potent aflatoxin and this chemical form is generally the most abundant in feedstuffs and foods. The order of toxicity is $AFB_1 > AFG_1 > AFB_2 > AFG_2$. Hydroxylated aflatoxin metabolites are excreted in milk and the important metabolites are AFM_1 and AFM_2 [21]. AFM_1 is the toxic metabolite of AFB_1 and AFM_2 is the hydroxylated form of AFB_2. Although AFM_1 and AFM_2 are commonly associated with milk and other edible animal products, these compounds can also be produced by aflatoxigenic fungi. The chemical methods are the most reliable for testing a wide variety of substances for aflatoxins.

11.2.4 Carcinogenicity

Aflatoxins are highly toxic, mutagenic, teratogenic, and carcinogenic compounds. AFB1, for example, has a toxicity 10 times that of potassium cyanide, 68 times that of arsenic, and 416 times that of melamine. Furthermore, their carcinogenicity is over 70 times that of dimethylnitrosamine and 10,000 times of benzene hexachloride

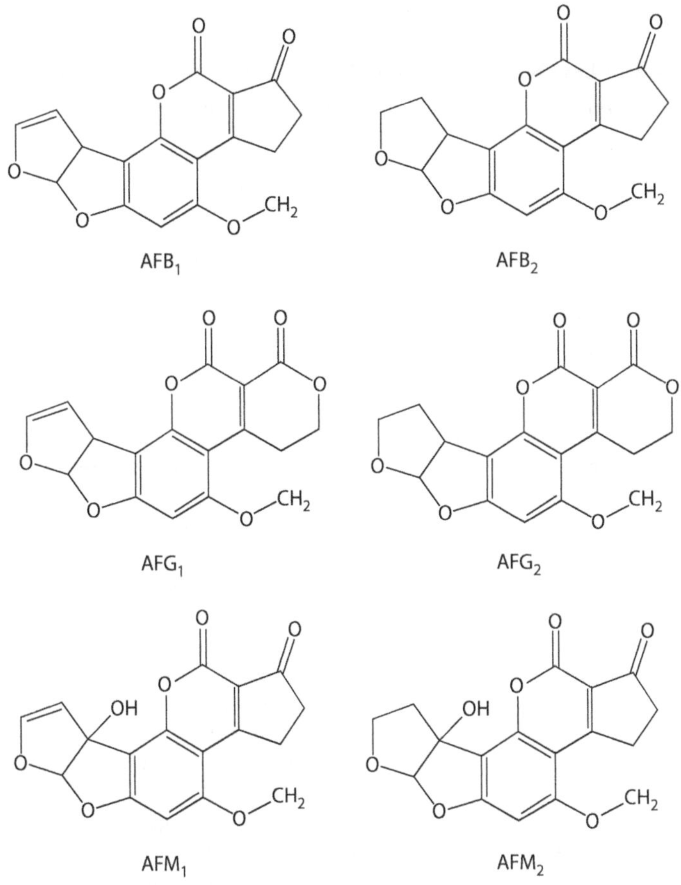

FIGURE 11.2 Structural formula of aflatoxins.

(BHC). The IARC of the World Health Organization (WHO) accepted that aflatoxin should be classified as a Group 1 carcinogen in 1987, and then AFB1 is classified as Group 1 (carcinogenic to humans) by the WHO–IARC in 1993 [20].

According to the most recent research conducted at the University of Pittsburgh, aflatoxin may play a causative role in 4.6–28.2% of all global hepatocellular carcinoma (HCC) cases [22].

11.2.5 Regulations

The toxicity of the aflatoxins has led many countries to set up regulations for their control in foods of plant origin that are intended for human or animal consumption (Table 11.3) (Commission Regulation, 2006; FAO, 2003; National Standard of PR China, 2005). In addition, in order to minimize the levels of mycotoxins in cereals, the European Union has also promoted several good agricultural practices from the cultivation to the distribution of cereals, such as crop rotation or dry storage. Regulations for major mycotoxins in commodities and food exist in at least 100

TABLE 11.3

U.S. FDA Action Levels and European Union Regulations on Maximum Levels for Aflatoxins in Foodstuffs and Animal Feedstuffs

United States		European Union[a]	
Product	**Level (ppb)**	**Product**	**Level (ppb)**
All foods	20	Groundnuts, nuts, and dried fruits, and processed products (direct human consumption)	4 (2)
Cottonseed meal intended for beef cattle/swine/poultry feedstuffs (regardless of age or breeding status)	300	Groundnuts (to undergo physical processing before human consumption)	15 (8)
Maize and peanut products intended for breeding beef cattle/swine or mature poultry	100	Nuts and dried fruit (to undergo physical processing before human consumption)	10 (5)
Maize and peanut products intended for finishing swine of 100 pounds or greater	200	Cereals (for direct human consumption or to undergo physical processing before human consumption)	4 (2)
Maize and peanut products intended for finishing beef cattle	300	Spices (*Capsicum* spp., *Piper* spp., *Myristica fragans*, *Zingiber officinale*, *Curcuma longa*)	10 (5)
		Feed materials with the exception of	(50)
		– Groundnut, copra, palm-kernel, cotton seed, babassu, maize, and products derived from the processing thereof	(20)
		Complete feedingstuffs for cattle, sheep, and goats with the exception of	(50)
		– Dairy cattle	(5)
		– Calves and lambs	(10)
		Complete feedingstuffs for pigs and poultry (except young animals)	(20)
		Other complete feedingstuffs	(10)
		Complementary feedingstuffs for cattle, sheep, and goats (except for diary animals, calves, and lambs)	(50)
		Complementary feedingstuffs for pigs and poultry (except young animals)	(30)
		Other complementary feedingstuffs	(5)

Source: Adapted from Zheng, M. Z., Richard, J. L., and Binder, J. 2006. *USA Mycopathol.* 161:261–273. With permission.

[a] Numbers in parentheses refer to a separate standard for aflatoxin B1 alone.

countries, most of which are for aflatoxins; maximum tolerated levels differ greatly among countries [23].

11.3 ANALYTICAL METHODS FOR AFLATOXIN DETERMINATION IN FOOD

Most aflatoxins are chemically stable, so they tend to survive storage and processing, even when cooked at quite high temperatures such as those reached during baking bread or breakfast cereal production. This makes it important to avoid the conditions that lead to aflatoxin formation, which is not always possible and not always achieved in practice. Aflatoxins are notoriously difficult to remove and the best method of control is prevention [24].

The presence of a recognized toxin-producing fungus does not, in fact, necessarily mean that the associated toxin will also be present, as many factors are involved in its formation. Equally, the absence of any visible mold will not guarantee freedom from toxins, as the mold may have already died out while leaving the toxin intact.

Fungi generally tend to develop in isolated pockets and are not evenly distributed in stored commodities. Therefore, it is important to develop a protocol to ensure that if a sample is taken for analysis, it is representative of the whole consignment. Grab samples have been reported to generally give very low estimates of mycotoxin content. In fact, nearly 90% of the error associated with aflatoxins assays could be attributed to how the original sample was collected. Since aflatoxins are not evenly distributed in grain or in mixed feeds, taking a feed or grain sample that will give a meaningful result in aflatoxins analyses is reported to be difficult [25,26].

The fact that most aflatoxins are toxic in very low concentrations requires sensitive and reliable methods for their detection. Sampling and analysis is of critical importance since failure to achieve a satisfactory verified analysis can lead to unacceptable consignments being accepted or satisfactory loads being unnecessarily rejected. Owing to the varied structures of these compounds, it is not possible to use one standard technique to detect all aflatoxins, as each will require a different method. What works well for some molecules could be inappropriate for others of similar properties, or for the same molecule in a different environment/matrix. Likewise, practical requirements for high-sensitivity detection and the need for a specialist laboratory setting create challenges for routine analysis. Therefore, depending on the physical and chemical properties, procedures have been developed around existing analytical techniques, which offer flexible and broad-based methods of detecting compounds. It would be desirable to have simple detection methods to be used by nonscientific personnel that are both fast and inexpensive. The application of simpler, cheaper, and effective solutions for the detection of aflatoxins is increasingly being required, due to their perceived importance, based around their toxicity and requirements of legislation for limits on amounts in foods. A successful detection method should be robust, be sensitive, and have a high degree of flexibility, over a wide range of compounds, but can also be very specific when required. All techniques should be reproducible to a high level, and the results gained must be relevant and easy to analyze. For fieldwork, the system should also be rapid and portable. There are many methods used, of which many are lab-based, but there is no single technique that stands out above the rest, although analytical liquid chromatography, commonly linked with mass spectroscopy, is gaining popularity. Many of the techniques described below have been combined to form protocols, which are used in laboratories today [8].

11.3.1 Sample Pretreatment Methods

Most methods used for determination of aflatoxins must rely on the correct extraction and clean-up methods (with the exception of enzyme-linked immunosorbent assay (ELISA), which may not require clean-up) [27]. These steps are vital for a successful protocol, as they are time consuming (sample preparation is the main time factor in an analysis and takes approximately two-thirds of the total) and will affect the final choice of detection procedure. The extraction method used to remove the aflatoxins from the biological matrix is dependent on the structure of the toxin. Hydrophobic toxins such as AFT rely on use of organic solvents [28,29]. These can be direct extractions, or may be partitioned with other solvents, such as *n*-hexane for partial clean-up, to remove excess components of the biological matrix. The choice of extraction solvent is also dependent on the matrix from which the extraction is required, as the differing chemical mixtures can affect it [30]. The use of chlorinated chemicals for extraction is being gradually reduced, as they are proven to be ecological hazards [31]. The clean-up procedure used in a protocol is the most important step, as the purity of the sample affects the sensitivity of the results. Trace amounts of a target molecule may be masked by interfering compounds, found not only in the matrix but in the chemicals, materials, and solvents used in the technique. Glassware should also be free of contamination, such as alkaline detergents, which can form salts with the compounds and result in lower detection rates [29].

Several methods exist and have all been recorded for use with cleaning up aflatoxins samples [32]. Some of those that are widely used have been described below in this section.

11.3.1.1 Liquid–Liquid Extraction

Liquid–liquid extraction (LLE) involves exploiting the different solubilities of the toxin in aqueous phase and in immiscible organic phase, to extract the compound into one solvent, leaving the rest of the matrix in the other. Thus, solvents such as hexane and cyclohexane are used to remove nonpolar contaminants, for example, lipids and cholesterol. The procedure is effective for several toxins and works well in small-scale preparations [33].

However, it is time consuming, and is dependent on which matrix is being used and which compounds are been determined. Disadvantages lie with possible loss of sample by adsorption onto the glassware.

11.3.1.2 Supercritical Fluid Extraction

Supercritical fluid extraction (SFE) uses a supercritical fluid, such as CO_2, to extract the required compound from the matrix. This works well due to the high solvating power and density of the solvating liquid. Supercritical fluid chromatography on fused silica capillary columns has been applied previously for separating toxins [34] but it is not a successful technique owing to the problems related to SFE [35].

Further, this technique is not suitable for routine analysis due to high costs and the need for specialized equipment [36].

11.3.1.3 Solid-Phase Extraction

The basic principle of SPE technology is a variation of chromatographic techniques based around small disposable cartridges packed with silica gel, or bonded phases that are in the stationary phase. The sample is loaded in one solvent, generally under reduced pressure; rinsed, where most of the contaminants are removed; and eluted in another solvent [37].

This system can be used "on" and "off" line. These cartridges have a high capacity for binding of small molecules and contain different bonding phases, ranging from silica gel, C-18 (octadecylsilane), florisil, phenyl, aminopropyl, ion exchange materials (both anionic and cationic), to affinity materials such as immunoadsorbents and molecular imprinted polymers (MIPs) [38–49].

In addition to cleaning the sample, they can also be used to preconcentrate the sample, providing better detection results. SPE has found widespread use and is an integral part of many extraction and detection protocols [50].

11.3.2 Chromatographic Technique

11.3.2.1 TLC Technique

Aflatoxins possess significant UV absorption and fluorescence properties, so techniques based on chromatographic methods with UV or fluorescence detection have always predominated. Originally, the chromatographic separation was performed by thin layer chromatography (TLC): since aflatoxins were first identified as chemical agents, it has been the most widely used separation technique in aflatoxin analysis in various matrices, like corn, raw peanuts [51], and cotton seed [52–54], and it has been considered the Association of Official Agricultural Chemists (AOAC) official method for a long period. This technique is simple and rapid and the identification of aflatoxins is based on the evaluation of fluorescent spots observed under a UV light. AFB1 and AFB2 show a blue fluorescent color, while it is green for AFG1 and AFG2. TLC allows qualitative and semiquantitative determinations by comparison of sample and standard analyzed in the same conditions. Many TLC methods for aflatoxins were validated more than 20 years ago and again more recently, though the performance of the methods has often been established at contamination levels too high to be of relevance to current regulatory limits. The combination of TLC methods with the much-improved modern clean-up stage offers the possibility to be a simple, robust, and relatively inexpensive technique [55] that after validation can be used as a viable screening method. Moreover, given the significant advantages of the low cost of operation, the potential to test many samples simultaneously, and the advances in instrumentation that allow quantification by image analysis or densitometry, TLC can also be used in laboratories of developing countries as an alternative to other chromatographic methods that are more expensive and require skilled and experienced staff to operate. Improvements in TLC techniques have led to the development of high-performance thin-layer chromatography (HPTLC), successfully applied to aflatoxins analysis [56].

Over pressured-layer chromatographic technique (OPLC), developed in the 1970s, has been used for quantitative evaluation of aflatoxins in foods [57] , as well as in fish, corn, wheat samples that can occur in different feedstuffs [58].

11.3.2.2 Capillary Electrophoresis

For a short time, capillary electrophoresis has been a technique of interest in aflatoxin separation, in particular its application as micellar electrokinetic capillary chromatography with laser-induced fluorescence detection [59], but it has not found application in routine analysis.

11.3.2.3 High-Performance Liquid Chromatographic Technique

Because of its higher separation power, higher sensitivity, and accuracy, and the possibility of automating the instrumental analysis, HPLC is now the most commonly used technique in analytical laboratories. HPLC using fluorescence detection has already become the most accepted chromatographic method for the determination of aflatoxins. For its specificity in the case of molecules that exhibit fluorescence, Commission Decision 2002/657/EC, concerning the performance of analytical methods, considers the HPLC technique coupled with fluorescence detector a suitable confirmatory method for aflatoxin identification.

However, HPTLC and HPLC techniques complement each other: the HPTLC for preliminary work to optimize LC separation conditions during the development of a method or it may also use as screening for the analysis of a large number of samples to limit the HPLC analysis only to positive samples. Liquid chromatographic methods for aflatoxin determination include both normal and reverse-phase separations, although current methods for aflatoxin analysis typically rely upon reverse-phase HPLC, with mixtures of methanol, water, and acetonitrile for mobile phases.

Aflatoxins are naturally strongly fluorescent compounds, so the HPLC identification of these molecules is most often achieved by fluorescent detection. Reverse-phase eluents quench the fluorescence of AFB1 and AFG1 [60]; for this reason, to enhance the response of these two analytes, chemical derivatization is commonly required, using pre- or postcolumn derivatization with suitable fluorophore, improving detectability.

The precolumn approach uses trifluoroacetic acid (TFA) with the formation of the corresponding hemiacetals [61–63] that are relatively unstable derivatives. The postcolumn derivatization is based on the reaction of the 8,9-double bond with halogens. Initially, the postcolumn reaction used iodination [64], but it has several disadvantages, such as peak broadening and the risk of crystallization of iodine. An alternative method is represented by bromination by an electrochemical cell (Kobra cell) with potassium bromide dissolved in an acidified mobile phase or by the addition of bromide or pyridinium hydrobromide perbromide (PBPB) to a mobile phase and using a short reaction coil at ambient [65–69]. The bromination methods offer the advantage to be rapid, simple, and easy to automate, improving reproducibility and ruggedness and reducing analysis time. A postcolumn derivatization method that seems analytically equivalent to iodination and bromination is the photochemical one: it is based on the formation of hemiacetals of AFB1 and AFG1 as the effect of the irradiation of the HPLC column eluate by a UV light [70,71].

A method based on the formation of an inclusion complex between aflatoxins and cyclodextrins (CDs) has been recently developed [72], and specific CDs are added to mobile phase (water–methanol), including aflatoxins in their cyclic structure, enhancing AFB1 and AFG1 fluorescence [73].

11.3.3 Bioassay Technique

11.3.3.1 Enzyme-Linked Immunosorbent Assay

ELISA methods for aflatoxins assay have been available for more than a decade. The technology is based on the ability of a specific antibody to distinguish the three-dimensional structure of specific aflatoxins. The direct competitive ELISA is commonly used in aflatoxin analysis [74].

A conventional microtiter plate, ELISA requires equilibrium of the antibody–antigen reaction that would require an incubation time of approximately 1–2 h. Currently, most of the commercially available ELISA test kits for aflatoxins are working in the kinetics phase of antibody–antigen binding, which reduces the incubation time to minutes. Although reduction of incubation time may lead to some loss of assay sensitivity, the test kit can provide accurate and reproducible results [75].

A typical principle of direct competitive ELISA is shown in Figure 11.3. After an aflatoxin is extracted from a ground sample with solvent, a portion of the sample

○ Aflatoxin-enzyme conjugate
● Aflatoxin
Y Anti-aflatoxin antibody
S Substrate

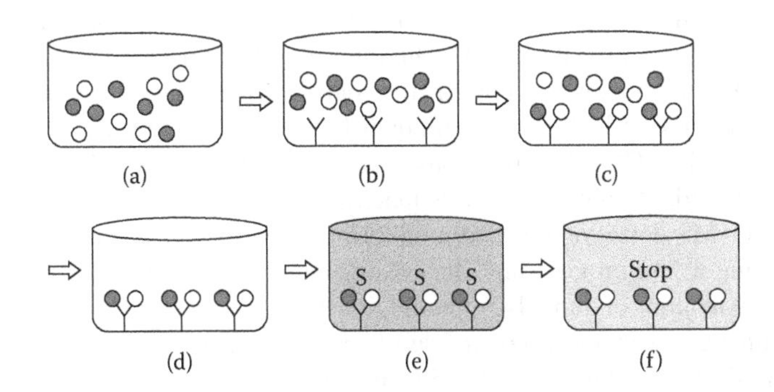

FIGURE 11.3 Principle of competitive ELISA for aflatoxin analysis. (a) Sample mixed with conjugate; (b) mixed content added to antibody-coated well; (c) aflatoxin binds to antibody in the first incubation; (d) unbound materials are rinsed away in the washing step; (e) substrate is added to develop color; (f) stop solution is added to stop the reaction. (Adapted from Zheng, M. Z., Richard, J. L., and Binder, J. 2006. *Mycopathologia* 161:261–273. With permission.)

extract and a conjugate of an enzyme-coupled aflatoxin are mixed and then added to the antibody-coated microtiter wells. Any aflatoxin in the sample extract or control standards is allowed to compete with the enzyme-conjugated aflatoxin for the antibody binding sites. After washing, an enzyme substrate is added and a blue color develops. The intensity of the color is inversely proportional to the concentration of aflatoxin in the sample or standard. A solution is then added to stop the enzyme reaction. The intensity of the solution color in the microtiter wells is measured, optically using an ELISA reader with an absorbance filter of 450 nm. The optical densities (ODs) of the samples are compared to the ODs of the standards and an interpretative result is determined [75,76].

ELISA test kits are favored as high-throughput assays with low sample volume requirements and often less sample extract clean-up procedures compared to conventional methods such as TLC and HPLC. The methods can be fully quantitative. They are rapid, simple, specific, sensitive, and portable for use in the field for the detection of aflatoxins in foods and feeds [77]. Although the antibodies have the advantage of high specificity and sensitivity, because the target compounds are aflatoxins but not the antigens, compounds with similar chemical groups can also interact with the antibodies. This so-called matrix effect or matrix interference commonly occurs in ELISA methods, resulting in underestimates or overestimates in aflatoxin concentrations in commodity samples [78]. Additionally, insufficient validation of ELISA methods causes the methods to be limited to those matrices for which they were validated [79]. Therefore, an extensive study on the accuracy and precision of an ELISA method over a wide range of commodities is needed and a full validation for an ELISA method is essential and critical [80].

11.3.4 Chromatography and Mass Spectrum Combination Technique

11.3.4.1 GC and GC–MS Techniques

GC is regularly used to identify and quantify the presence of aflatoxins in food samples, and many protocols have been developed for these materials. Normally, the system is linked to MS, flame ionization detector (FID), or Fourier transform infrared spectroscopy (FTIR) detection techniques in order to detect the volatile products [81–83]. Most aflatoxins are not volatile and therefore have to be derivatized for analysis using GC [83]. Several techniques have been developed for the derivatization of aflatoxins. Chemical reactions such as silylation or polyfluoroacylation are employed in order to obtain a volatile material [32].

The GC–MS detection allowed monitoring of up to four compounds simultaneously during a 23-min GC run. The volatile fungal metabolites were measured in grain as indicators of fungal contamination [84]. The GC–MS system was compared with electronic nose, showing superior performance of the first one, since the GC–MS misclassified only three of 37 samples and the electronic nose, seven of 37 samples.

While as shown above, a number of examples do exist on the successful application of GC for analysis of aflatoxins, there are several disadvantages. First, the samples that need to be analyzed are those that are volatile or those that can be converted into volatile samples. Further, thermal stability is a problem because heating sometimes degrades the samples. In some cases, injection of a sample has been shown to be a problem that needs addressing. This is mainly because the sample gets lost when it comes into contact with the heated areas of the injector, leading to loss in vaporization. However, the use of GC detection is not expected for commercial protocols due to the existence of cheaper and faster alternatives such as HPLC.

11.3.4.2 LC–MS Technique

The introduction of mass spectrometry and the subsequent coupling of liquid chromatography to this very efficient system of detection has resulted in the development of many LC–MS or LC–MS/MS methods for aflatoxin analysis. Because of the advantages of specificity and selectivity, chromatographic methods coupled to mass spectrometry continue to be developed: they improve detection limits and are able to identify molecules by means of mass spectral fragmentation patterns.

Some of them comprise a single-liquid extraction and direct instrumental determination without a clean-up step [85–87]. This assumption relies on the ability of the mass analyzer to filter out by mass any coeluting impurities. However, many authors assert that further sample preparation prior to LC–MS analysis would benefit analysis [88–90] because ionization suppression can occur by matrix effects. A number of instrument types have been used: single quadrupole, triple quadrupole, and linear ion trap [15,88–90].

Atmospheric pressure chemical ionization (APCI) is the ionization source that provides lower chemical noise and, subsequently, lower quantification limit than electrospray ionization (ESI) which is more robust. The use of mass spectrometric methods can be expected to increase, particularly as they become easier to use and the costs of instrumentation continue to fall. Despite the enormous progress in analytical technologies, methods based on HPLC with fluorescence detection are the most used today for aflatoxins instrumental analysis, because of the large diffusion of this configuration in routine laboratories.

In Table 11.4, some analytical methods for aflatoxin determination have been included with their performance characteristics.

11.4 UPLC–MS ANALYSIS OF AFLATOXINS IN FOOD

The recent availability of analytical columns with reduced size of the packing material has improved chromatographic performance. Today, numerous manufacturers commercialize columns packed with sub-2 μm particles to use devices that are able to handle pressure higher than 400 bar, such as Ultra-Performance Liquid Chromatography® (UPLC). This strategy allows a significant decrease in analysis time: aflatoxin runs are completed in 3–4 min, with a decrease of over 60% compared to traditional HPLC. In addition, solvent usage has been reduced by 85%, resulting in greater sample throughput and significant reduction of costs of analysis. UPLC system can be coupled to a traditional detector or, using a mobile phase of water/methanol with 0.1% formic acid, to a mass spectrometry detector. The aim of this chapter is to summarize a number of the most important recent research on using UPLC coupled with MS for aflatoxin analysis in different food matrices. Table 11.5 shows some examples of recent studies on aflatoxin analysis in food using UPLC/MS [91].

TABLE 11.4
Some Analytical Methods for Aflatoxins Determination and Their Performance Characteristics

Aflatoxin	Matrix	Method	Sample Preparation	LOD (μg/kg)	LOQ (μg/kg)	R%	RSD_R (%)	Reference
B1	Corn	HPLC/fluorescent (fluor). precolumn der. trifluoroacetic acid (TFA), postcolumn pyridinium hydrobromide perbromide (PBPB)	IAC	–	–	82–84	19–37	[68]
B1, B2 G1, G2	Corn, raw peanut, peanut butter	Thin layer chromatography (TLC)/densit.	SPE	–	–	95–139	26–84 (B1)	[51]
B1, B2 G1, G2	Corn, raw peanut, peanut butter	HPLC/fluor. postcolumn der. (iodine)	IAC	–	–	97–131	11–108	[92]
B1, B2 G1, G2 M1	Mold cheese	LC–MS/MS triple quadrupole (electrospray ionization (ESI) source)	Only extraction	0.3 (M1) 0.8 (B-G)	0.6 (M1) 1.6 (B-G)	96–143	2–12	[86]
B1, B2 G1, G2	Fish, corn, wheat	Over pressure layer chromatography (OPLC)	Extraction and L–L partition	2	–	73–104	7–13 (RSDr)	[58]
B1	Corn	Capillary electrophoresis/ laser-induced fluor.	SPE or IAC	0.5	–	85	–	[59]
B1, B2 G1, G2	Peanuts	HPLC/fluor.	MSPD	–	0.125–2.5	78–86	4–7 (RSDr)	[15]
M1	Milk	HPLC/fluor. precolumn der. (TFA)	SPE or IAC	0.027–0.031	–	82–92 (RSDr)	15–19	[15]
M1	Milk	Colorimetric ELISA	None	0.006	–	100 (RSDr)	11	[62]
M1	Milk, soft cheese	HPLC/fluor. postcolumn der. (PBPB)	SPE	0.001– 0.005	–	76–90	3–9 (RSDr)	[66]

M1	Hard cheese	HPLC/fluor. postcolumn der. (PBPB)	SPE	0.008	0.025	67	4–7 (RSDr)	[69]
M1	Milk	HPLC/fluor.	IAC	–	0.005	74	21–31	[93]
M1	Milk	HPLC/fluor.	IAC	0.006	0.015	91	8–15	[94]
M1	Milk	Chemiluminescent enzyme-linked immunosorbent assay (ELISA)	None	0.00025	0.001	96–122	2–8	[95]
M1	Milk	LC–MS/MS linear ion trap (ESI and APCI source)	Carbograph-4 cartridge	–	0.006–0.012	92–96	3–8	[89]
M1	Milk	Membrane-based flow through enzyme immunossay	IAC	0.05	–	97	–	[96]
M1	Milk	Electrochemical biosensor	None	0.01	–	–	–	[97]
M1	Milk, milk powder	LC–MS/MS triple quadrupole (ESI source)	IAC	0.59–0.66	–	78–87	–	[88]
M1	Milk, milk powder	LC–MS/MS triple quadrupole (ESI source)	Multifunction column	9–14	–	7–16	–	[88]

Source: Adapted from Manetta, A. C. 2011. Aflatoxins: Their measure and analysis. In: *Aflatoxins—Detection, Measurement and Control,* ed. Dr Irineo Torres-Pacheco, InTech Publisher, http://www.intechopen.com/books/aflatoxins-detection-measurement-and-control/aflatoxins-their-measureand-analysis. With permission.

TABLE 11.5
Examples for Aflatoxins Analysis in Food Using UPLC/MS

Aflatoxin	Matrix	UPLC	Mass Spectrometry	Column	Sample Preparation	LOD (μg/kg)	LOQ (μg/kg)	R%	RSD%	Reference
B_1, B_2, G_1, G_2, M_1	Corn, peanut butter	Acquity (Waters, USA)	Micromass Quattro Ultima triple-quadrupole (Micromass, UK)	BEH C18 (Waters, USA) (100 mm–2.1 mm–1.7 μm)	SPE	0.003, 0.006(G_2)	0.01, 0.02(G_2)	91–104	3.56–5.18	[131]
B_1, B_2, G_1, G_2, M_1	Maize, walnut, biscuit, breakfast cereals	Acquity (Waters, USA)	Acquity TQD tandem quadrupole (Waters, UK).	BEH C18 (Waters, USA) (100 mm–2.1 mm–1.7 μm)	SLE	0.02(B_1), 0.2(G_1, G_2), 0.01(M_1), 0.1(B_2)	–	71–108	5.8– 21.9	[132]
B_1, B_2, G_1, G_2, M_1, M_2	Chinese medicines	Acquity (Waters, USA)	Micromass Quattro Ultima triple-quadrupole (Micromass, UK)	HSS T3 (Waters, USA) (100 mm–2.1 mm–1.8 μm)	SPE (Homemade cartridge)	0.13(B_1), 0.16(B_2), 0.17(G_1), 0.14(G_2), 0.13(M_1), 0.15(M_2)	0.16(B_1), 0.33(B_2), 0.25(G_1), 0.18(G_2), 0.18(M_1), 0.24(M_2)	85– 113	1.2–15.9	[133]
B_1, B_2, G_1, G_2, M_1, M_2	Fresh peanuts, musty peanuts, peanut butters	Acquity (Waters, USA)	Micromass Quattro Ultima triple-quadrupole (Micromass, UK)	HSS T3 (Waters, USA) (100 mm–2.1 mm–1.8 μm)	SPE (Homemade cartridge)	0.009(B_1), 0.056(B_2), 0.085(G_1), 0.212(G_2), 0.017(M_1), 0.106(M_2)	0.012(B_1), 0.084(B_2), 0.182(G_1), 0.273(G_2), 0.021(M_1), 0.138(M_2)	80–88	1.9–9.4	[134]

B_1, B_2, G_1, G_2	Wheat, cucumber, red wine	Acquity (Waters, USA)	Acquity TQD tandem quadrupole (Waters, UK).	BEH C18 (Waters, USA) (100 mm–2.1 mm–1.7 μm)	QuEChERS	–	5.4(B_1), 4.4(B_2), 3.8(G_1), 4(G_2)	71–110	3–20	[125]
B_1, B_2, G_1, G_2, M_1	Baby food, milk	Acquity (Waters, USA)	Acquity TQD tandem quadrupole (Waters, UK).	BEH C18 (Waters, USA) (50 mm–2.1 mm–1.7 μm)	IAC	4(B_1), 5(B_2), 3.5(G_1), 3(G_2), 2(M_1)	12(B_1), 25(B_2), 11(G_1), 10.5(G_2), 6.5(M_1)	79–112	3–10	[126]
B_1, B_2, G_1, G_2	Cereals	Finnegan TSQ quantum ultra mass (Thermo, USA)	Finnegan TSQ quantum ultra mass (Thermo, USA)	C18 (Thermo, USA) (50 mm–2.1 mm–1.9 μm)	SLE	0.3(B_1), 0.5 (B_2), 0.08(G_1), 0.7(G_2)	0.55(B_1), 0.9(B_2), 0.15(G_1), 1.25(G_2)	83–107	6.8–9.9	[127]
B_1, B_2, G_1, G_2	Nonalcoholic beverages (Waters, USA)	Acquity (Waters, USA) (Micromass, UK)	Micromass Quattro Ultima triple-quadrupole (Micromass, UK) (Waters, USA)	BEH C18 (Waters, USA) (100 mm–2.1 mm–1.7 μm)	LLE 0.003(B_2),	0.003(B_1), 0.003(B_2), 0.01(B_2), 0.001(G_1), 0.002(G_2)	0.01(B_1), 0.01(B_2), 0.004(G_1), 0.007(G_2)	86–95	0.01–2.3	[128]
B_1, B_2, G_1, G_2	Barley	Accela (Thermo Fisher, USA)	Orbitrap (Thermo Fisher, USA)	HSS T3 (Waters, USA) (100 mm–2.1 mm–1.8 μm)	MSPD Modified QuEChERS SLE	–	–	73–81 75–82 80–85	14–18 9–12 12–17	[129]
B_1, B_2, G_1, G_2	Cereals	Acquity (Waters, USA)	Acquity TQD tandem quadrupole (Waters, UK)	BEH C18 (Waters, USA) (50 mm–2.1 mm–1.7 μm)	SLE	–	–	–	–	[130]

11.4.1 Instruments

HPLC combined with fluorescence detection is proven to be more accurate and has been studied extensively in different materials [98–101]. However, conventional HPLC methods often cost a lot of time to separate the target analytes and, additionally, in order to improve detection limits of AFB1 and AFG1, a tedious pre- or post-column derivatization must be done [98,102]. These problems have been successfully solved in the present study by introducing UPLC–MS/MS method. Reduction of the particle diameter from 5 μm (HPLC) to 1.7 μm (UPLC) results in greatly increased speed, while the introduction of the MS/MS detection avoids the tedious derivatization process. Despite the high sensitivity and selectivity of the LC–MS/MS method, the variable matrix effects limit its application. As a result, the previously established LC–MS/MS method could not be applied to determine AFs in different medicinal materials [103,104].

The term "UPLC" is a trademark of the Waters Corporation, but is often used to refer to the more general technique. The Waters Acquity Ultra-High-Performance LC system (Waters, Milford, MA, USA) seems to be the most popular UPLC system in aflatoxins analysis in food [105–111]. Other UPLC systems have been used successfully, such as the Finnegan TSQ quantum ultra mass (Thermo Scientific, CA, USA) system [112], and the An Accela U-HPLC system (Thermo Fisher Scientific, San Jose, CA, USA). The hyphenated MS/MS detectors were Acquity TQD tandem quadrupole mass spectrometer (Waters, Manchester, UK) [23,108,109,111], tandem quadrupole mass spectrometer (Micromass, Manchester, UK) [105–107,110], Finnegan TSQ quantum ultra mass (Thermo Scientific, CA, USA) system [113] and single-stage Orbitraps mass spectrometer (Exactive; Thermo Fisher Scientific, Bremen, Germany) [112].

Chromatographic separations were achieved on an Acquity UPLC HSS T3 column (1.8 μm, 100 × 2.1 mm I.D., Waters, Milford, MA, USA) [106,107,112], UPLC BEH C18 column (1.7 μm, 100 × 2.1 mm I.D., Waters, Milford, MA, USA) [23,105,108,110], UPLC BEH C18 column (1.7 μm, 50 × 2.1 mm I.D., Waters, Milford, MA, USA), [109,111], and C18 column (1.9 μm, 50 × 2.1 mm I.D., Thermo Scientific, CA, USA) [113]. Zheng Han et al. [106] compared four candidate columns with different lengths and particle sizes, that is, (1) Agilent SB-C18 column (2.1 × 150 mm, 1.8 μm particle size), (2) Acquity UPLC HSS T3 column (2.1 × 100 mm, 1.8 μm particle size), (3) Atlantics RC18 column (2.1 × 150 mm, 1.8 μm particle size), and (4) UPLC BEH Shield RP18 column (2.1 × 100 mm, 1.7 μm particle size), in the pilot test to get a complete separation of the AFs. The separation efficiency of columns 1, 2, and 3 was obviously better than that of column 4. The sensitivity was greatly improved when choosing column 2 compared to other candidate columns.

11.4.2 Sample Pretreatment

Pretreatment of the sample (protein precipitation, defatting, extraction, and filtration) is an important phase for removing many interferences and for having, in this way, extracts without impurities to allow accuracy and reproducibility in the subsequent instrumental step. The first phase is the extraction of the toxins from the matrices: it generally involves chloroform, dichloromethane, or aqueous mixtures of polar organic solvents such as methanol, acetone, or acetonitrile. The aqueous mixture should be the one most recently used because it will be more compatible not only with the environment but also with the antibodies involved in the subsequent step of clean-up with immunoaffinity columns that are increasingly utilized.

Clean-up is another very critical step. It is necessary to remove many of the coextracted impurities and obtaining cleaner extracts for the subsequent instrumental determination, to have the most accurate and reproducible results. The traditional techniques, such as liquid–liquid partition or purification of conventional glass columns packed with silica, are time and solvent consuming. Nowadays, new sample preparation technologies, based on extraction by adsorbent materials, are available.

After extraction using acetonitrile aqueous solution, homogenization and filtration take place [105], then cleaned-up an aliquot of 15 mL of filtrate by passing through the Mycosep 226 Aflazon+ Multifunctional cartridges (PN. COCMY 2226, Romer Labs, Tulln, Austria). The sample was then dried, redissolved by a mixture of methanol and ammonium acetate, and shaken briefly for about 30 s by vortex to mix the content of the tube. Finally, the solution was passed through a 0.22 μm nylon filter and ready for injection. A Chinese group [106,107] has prepared a homemade cartridge and used it for the clean-up of the food samples for aflatoxins analysis in two different research. They prepared their cartridge simply as two layers of silica gel and alumina in a 6 mL hollow SPE cartridge, then covered by a cribriform plate to ensure the supine surface is smooth and flat.

R. Romero-Gonzalez et al. [108] compared three different pretreatment methods:

Method A: The well-known QuEChERS methodology [114] (quick, easy, cheap, effective, rugged, and safe). QuEChERS-based methodologies have been applied for the extraction of compounds with a wide range of physicochemical properties from different samples [115] using an acetate buffer [116]. For cucumber and red wine samples, 10 g of sample was weighed in a 50 mL polypropylene centrifuge tube. For wheat, 5 g of homogenized sample was weighed and 5 mL of water was added, soaking for 1 h. Subsequently, 10 mL of 1% acetic acid in acetonitrile (v/v) was added, and the tubes were shaken for 1 min with a vortex. Then, 4 g of anhydrous magnesium sulfate and 1.5 g of sodium acetate were added and the tubes were shaken immediately for 1 min. After centrifugation at 5000 rpm ($4136 \times g$) for 5 min, the supernatant was taken and filtered through a Millex-GN nylon filter (0.20 μm, Millipore, Carrightwohill, Ireland) prior to UHPLC–MS/MS analysis.

Method B: Sonication extraction. A sample of 5 g was weighed into a 50 mL polypropylene centrifuge tube and 10 mL of a mixture of acetonitrile/water 80:20 (v/v) was added. The mixture was vortexed for 2 min and then the tube was kept in an ultrasonic bath for 30 min. Then, the mixture was centrifuged for 10 min at 5000 rpm ($4136 \times g$), and the supernatant was filtered through a Millex-GN nylon filter and transferred into an autosampler vial prior to UHPLC–MS/MS analysis.

Method C: Generic extraction procedure, developed by Mol et al. [117]. Analytes were extracted using a method based on the procedure previously described by Mol et al. [112], where 2.5 g of sample was weighed into a 50 mL polypropylene centrifuge tube and 5 mL of water was added. The mixture was shaken with a vortex for 1 min. If wheat matrix was studied, the mixture was allowed to soak for 1 h, then 15 mL of acetonitrile (1% formic acid, v/v) was added, and the sample was extracted by end-over-end shaking for 1 h at 50 rpm. After that, the mixture was centrifuged for 10 min at 5000 rpm (4136 × *g*) and the supernatant was filtered through a Millex-GN nylon filter and transferred into an autosampler vial prior to UHPLC–MS/MS analysis.

In order to evaluate the performance of the three selected methods, wheat blank samples spiked at 50 μg/kg were treated, applying the three procedures showing the obtained results. It can be observed that the best results were obtained when QuEChERS procedure was used, allowing the extraction of more than 80 compounds with suitable recoveries (70–120%) and relative standard deviation (RSD) lower than 20%. When the ultrasound method was applied, more than 80 compounds were extracted, but only 36 compounds, including all the mycotoxins and biopesticides assayed in this study, were quantitatively extracted, whereas this approach was not suitable for most of the selected pesticides. Finally, an intermediate situation was obtained when the procedure described by Mol et al. was applied. More than 50 compounds were extracted with recoveries ranging from 70% to 120% and RSD values lower than 20%.

Rubert et al. [118] also compared different procedures for Barley sample pretreatment for aflatoxins analysis via UPLC–MS/MS.

Matrix solid-phase dispersion (MSPD): Barley samples were homogenized by mixing them thoroughly. Homogenized and representative 1 g portions were weighed and placed into a glass mortar (50 mL) and gently blended with 1 g of C18 for 5 min using a pestle, to obtain a homogeneous mixture. This mixture was introduced into a 100 × 9 mm i.d. glass column, and eluted dropwise with 1 Mm ammonium formate in 10 mL of acetonitrile/methanol (50/50, v/v) by applying a light vacuum. Then, an aliquot (1 mL) of extract was filtered through a 22 μm nylon filter prior to injection into the UPLC–Orbitrap MS.

Modified QuEChERS: This procedure was employed to extract aflatoxins from the examined matrix [118,119]. Homogenized and representative portions of 2 g were weighed into a 50 mL PTFE centrifuge tube (conical-bottom centrifuge tube), and then 10 mL of 0.1% formic acid in deionizer water was added. The mixture was mixed for 3 min and waited for the next step for 10 min. Afterward, 10 mL acetonitrile were added, and consecutively, the mixture was vigorously shaken (3 min). The following step, 4 g MgSO4 and 1 g NaCl were added, and then the mixture was shaken for 3 min again. Once the extraction was completed, the sample was centrifuged (5 min, 11,000 rpm, 20°C). Then, an aliquot (1 mL) filtered through a 22 μm nylon filter before their injection into the UPLC–Orbitrap MS.

Solid–liquid extraction (SLE): Representative portions of 2 g samples were accurately weighed and transferred to a PTFE centrifuge tube (50 mL). Samples were extracted by shaking with 10 mL acetonitrile/water/acetic acid (79:20:1, v/v/v) on an automatic shaker for 90 min, and then centrifuged (5 min, 11,000 rpm, 20°C). Afterward, the supernatant extract was twofold diluted with HPLC-grade water, taking an aliquot of 0.5 mL and diluting to 1 mL. After that the sample was filtered through a 0.22 μm filter, consecutively the sample was injected into the UPLC–MS/MS.

SPE clean-up method: The previous SLE extract was used for clean-up. The extraction procedure was used according to Vendl et al. [120]. C18-SPE clean-up procedure was performed with Oasis HLB cartridges (150 mg) from Waters. The 2 mL of SLE extract was diluted with 30 mL of water in order to obtain a required maximum concentration of 5% organic solvent. The columns were prewashed with 10 mL of acetonitrile, and further conditioned with 10 mL of 5% acetonitrile in deionized water. Consequently, the diluted sample was loaded onto the C18 cartridge. After that, SPE columns were washed with 10 mL of 5% acetonitrile in water. The cartridges were then dried for 30 min. In the last step, the aflatoxins were eluted by adding 5 mL of acetonitrile. Then, the extract was transferred into a 15 mL conical tube and evaporated to dryness at 35°C with Buchi Rotavapor. The residue was reconstituted to a final volume of 1 mL with methanol/water (50:50, v/v) and filtered through a 0.22 μm Millex-G nylon filter, before the injection. To sum up, modified QuEChERS was selected for further studies in order to take advantage of its potential for simultaneous extraction of selected compounds. The data comparison showed that QuEChERS offered an acceptable range of recoveries and low RSDs. Furthermore, QuEChERS took very little time during the extraction procedure, and it was also easier and cheaper than MSPD, SLE, and SPE clean-up. For these reasons, QuEChERS was the most efficient and effective extraction procedure evaluated.

Immunoaffinity has been used by Eduardo Beltrán et al. [109] as a clean-up procedure to analyze aflatoxins in baby food and milk. In order to prepare the extracts for the immunoaffinity clean-up, acetonitrile of the extract was removed by using a turbo evaporator system (water bath at 50°C under gentle nitrogen stream). Then, the extracts were diluted with water up to 20 mL final volume. The 20 mL aqueous extracts were passed through an AflaOchra HPLC™ column at 1–2 drops per second. Then, the column was washed with 5 mL of HPLC water. Aflatoxins were eluted from the column with 4 mL methanol. To ensure complete elution of the bound toxin from the antibody, the solvent remained in contact with the column at least 1 min before starting the elution. The methanolic elutes were dried under gentle nitrogen stream at 50°C and reconstituted with 1 mL of HPLC-grade water. Finally, 20 μL extracts were injected into the UHPLC–ESI–MS/MS system. The use of a mixed-mode antibodies column has made possible the determination of all targeted aflatoxins in one single analysis. Their results showed that immunoaffinity columns allowed the simultaneous clean-up and analyte preconcentration, obtaining satisfactory chromatograms, with recoveries in the range of 79–112%.

11.5 CONCLUSION

In conclusion, a broad range of detection techniques used for practical analysis and detection of aflatoxins are available. This chapter highlighted some recent developments and new techniques about aflatoxins analysis in food via UPLC/MS. As shown, though there have been several recent successes in detection of aflatoxins, new methods are still required to achieve higher sensitivity and address other challenges that are posed by these toxins.

UPLC provides an efficient, fast, and high-resolution separation, and the application of MS in conjunction with other tools for decreasing limits of detection has been of increased interest in recent times. Future trends will focus on rapid assays and tools that would measure multiple toxins from a single matrix. Since matrix interferences were detected during the UHPLC–MS/MS analysis of the sample extracts, additional analyte identification suitable for extensive multianalyte methods needs to be investigated.

ACKNOWLEDGMENTS

The authors would like to extend their sincere appreciation to King Saud University, Deanship of Scientific Research, College of Science Research Center for its supporting of this book chapter.

REFERENCES

1. Williams, J. H., Phillips, T. D., Jolly, P. E., Stile, J. K., Jolly, C. M., and Aggarwal, D. 2004. Human aflatoxicosis in developing countries: A review of toxicology, exposure, potential health consequences, and interventions. *Am. J. Clin. Nutr.* 80:1106–1122.
2. Coppock, R. W. and Swanson, S. P. 1986. Aflatoxins. In: *Current Veterinary Therapy: Food Animal Practice*, 2nd Ed., Howard JL, Saunders, Philadelphia , PA.
3. Gourami, H. and Bullerman, L. B. 1995. *Aspergillus flavus* and *Aspergillus parasiticus*: Aflatoxigenic fungi of concern in foods and feeds: A review. *J. Food Protect.* 58:1395–1404.
4. Clegg, F. G. and Bryson, H. 1962. An outbreak of poisoning in stored cattle attributed to Brazilian groundnut meal. *Vet. Rec.* 74:992–994.
5. Allcroft, R. and Lewis, G. 1963. Groundnut toxicity in cattle: Experimental poisoning of calves and a report on clinical effects in older cattle. *Vet. Rec.* 75:487–494.
6. Eaton, D. L. and Gallagher, E. P. 1994. Mechanisms of aflatoxin carcinogenesis. *Annu. Rev. Pharmacol. Toxicol.* 34:135–172.
7. Robens, J. F. and Richard J. L. 1992. Aflatoxins in animal and human health. *Rev. Environ. Contam. Toxicol.* 127:69–94.
8. Turner, N. W., Subrahmanyam, S., and Piletsky, S. A. 2009. Analytical methods for determination of mycotoxins: A review. *Anal. Chim. Acta* 632:168–180.
9. Frenich, A. G., Martínez Vidal, J. L., Romero-González, R., and Aguilera-Luiz, M. M. 2009. Simple and high-throughput method for the multimycotoxin analysis in cereals and related foods by ultra-high performance liquid chromatography/tandem mass spectrometry. *Food Chem.* 117:705–712.
10. Council for Agricultural Science and Technology (CAST). 2003. Mycotoxins: Risks in Plant, Animal and Human Systems. Task Force Report, No. 139, CAST, Ames, Iowa.
11. European Commission. 2011. The Rapid Alert System for Food and Feed, Annual Report, http://ec.europa.eu/food/food/rapidalert/docs/rasff_annual_report_2011_en.pdf
12. Newberne, P. M., Russo, R., and Wogan, G. N. 1966. Acute toxicity of aflatoxin B1 in the dog. *Pathol. Vet.* 3:331–340.
13. Sweeney, M. J. and Dobson, A. D. W. 1998. Mycotoxins production by *Aspergillus, Fusarium* and *Penicillium* species. *Int. J. Food Micro.* 43:141–158.
14. Creppy, E. E. 2002. Update of survey, regulation and toxic effects of mycotoxins in Europe. *Toxicol. Lett.* 127:19–28.
15. Blesa, J., Soriano, J. M., Molto, J. C., Marin, R., and Manes, J. 2003. Determination of aflatoxins in peanuts by matrix solid-phase dispersion and liquid chromatography. *J. Chromatogr. A.* 1011:49–54.

16. Simon, P., Delsaut, P., Lafontaine, M., Morele, Y., and Nicot, T. 1998. Automated column-switching high-performance liquid chromatography for the determination of aflatoxin M1. *J. Chromatogr. B.* 712:95–104.
17. Georggiett, O. C., Muino, J. C., Montrull, H., Brizuela, N., Avalos, S., and Gomez, R. M. 2000. Relationship between lung cancer and aflatoxin B1. *Rev. Fac. Cien. Med. Univ. Nac. Cordoba.* 57:95–107.
18. Van Vleet, T. R., Watterson, T. L., Klein, P. J., and Coulombe, R. A. 2006. Aflatoxin B1 alters the expression of p53 in cytochrome P450-expressing human lung cells. *Toxicol. Sci.* 89:399–407.
19. Sadeghi, N., Oveisi, M. R., Jannat, B., Hajimahmoodi, M., Bonyani, H., and Jannat, F. 2009. Incidence of aflatoxin M1 in human breast milk in Tehran, Iran. *Food Control.* 20:75–78.
20. Li, P., Zhang, Q., Zhang, D., Guan, D., Liu, D. X., Fang, S., Wang, X., and Zhang, W. 2011. Aflatoxin measurement and analysis. In: *Aflatoxins—Detection, Measurement and Control.* ed. Dr Irineo Torres-Pacheco, InTech Publisher, Available from: http://www.intechopen.com/books/aflatoxins-detection-measurement-and-control/aflatoxin-measurement-and-analysis.
21. Garrido, N. S., Iha, M. H., Santos Ortolani, M. R., and Duarte Favaro, R. M. 2003. Occurrence of aflatoxins M(1) and M(2) in milk commercialized in Ribeirao Preto-SP, Brazil. *Food Addit. Contam.* 20:70–73.
22. Liu, Y. and Wu, F. 2010. Global burden of aflatoxin-induced hepatocellular carcinoma: A risk assessment. *Environ. Health Perspect.* 118(6): 818–824.
23. Frenich, A. G., Vidal, J. M., Romero-González, R., and Aguilera-Luiz, M. M. 2009. Simple and high-throughput method for the multimycotoxin analysis in cereals and related foods by ultra-high performance liquid chromatography/tandem mass spectrometry. *Food Chem.* 117:705–712.
24. Bullerman, L. B., Schroeder, L. L., and Park, K. Y. 1984. Formation and control of mycotoxins in food. *J. Food Prot.* 47:637–646.
25. Mycotoxin Sampling, Testing, and Test Kits, http://www.ces.ncsu.edu/gaston/Agriculture/mycotoxins/mycotest.html.
26. Lauren, D. R., Jensen, D. J., and Smith, W. A. 2006. Mycotoxin contamination in graded fractions of maize (*Zea mays*) in New Zealand. *J. Crop Horticul. Sci.* 34:63–72.
27. Chu, F. S. 1992. Development and use of immunoassays in the detection of ecologically important mycotoxins. In: *Handbook of Applied Mycology; Mycotoxins in Ecological Systems,* ed. D. Bhatnagar, E. B. Lillehoj and D. K. Arora, Marcel Dekker, New York, NY.
28. Holcomb, M., Wilson, D. M., Trucksess, M. W., and Thompson, H. C. 1992. Determination of aflatoxins in food products by chromatography. *J. Chromatogr.* 624:341–352.
29. Association of Analytical Communities. 1997. AOAC Int. Gaithersburg. Maryland, 35.
30. Wilkes, J. G. and Sutherland, J. B. 1998. Sample preparation and high-resolution separation of mycotoxins possessing carboxyl groups. *J. Chromatogr. B* 717:135–156.
31. Montreal Protocol. 1998. http://www.uneptie.org/ozonaction/compliance/protocol/main.html.
32. Scott, P. M. 1995. Mycotoxin methodology. *Food Addit. Contam.* 12:395–403.
33. Bauer, J. and Gareis, M. 1987. Ochratoxin A in the food chain. *J. Vet. Med. Ser. B* 34:613–627.
34. Young, J. C. and Games, D. E. 1992. Supercritical fluid chromatography of *Fusarium* mycotoxins. *J. Chromatogr.* 627:247–254.
35. Engelhardt, H. and Hass, P. 1993. Possibilities and limitations of SFE in the extraction of aflatoxin B1 from food matrices. *J. Chromatogr. Sci.* 31:13–19.
36. Holcomb, M., Thompson, H. C., Cooper, W. M., and Hopper, M. L. 1996. SFE extraction of aflatoxins (Bt, Bz, Gt, and Gs) from corn and analysis by HPLC. *J. Supercrit. Fluids.* 9:118–126.

37. European Mycotoxin Awareness Network (EMAN) co-ordinated by Leatherhead Food Research Association (UK). 2003. http://www.lfra.co.uk/eman/index.htm.
38. Zambonin, C. G., Monachi, L., and Aresta, A. 2001. Determination of cyclopiazonic acid in cheese samples using solid-phase microextraction and high performance liquid chromatography. *Food Chem.* 75:249–254.
39. Jornet, D., Busto, D. O., and Guasch, J. 2000. Solid-phase extraction applied to the determination of ochratoxin A in wines by reversed-phase high-performance liquid chromatography. *J. Chromatogr. A.* 882:29–35.
40. Visconti, A., Pascale, M., and Centonze, G. 2000. Determination of ochratoxin A in domestic and imported beers in Italy by immunoaffinity clean-up and liquid chromatography. *J. Chromatogr. A* 888:321–326.
41. Sharma, M. and Marquez, C. 2001. Determination of aflatoxins in domestic pet foods (dog and cat) using immunoaffinity column and HPLC. *Anim. Feed Sci. Technol.* 93:109–114.
42. Visconti, A. and Pascale, M. 1998. Determination of zearalenone in corn by means of immunoaffinity clean-up and high-performance liquid chromatography with fluorescence detection. *J. Chromatogr. A* 818:133–139.
43. Supelco. 1998. Guide to solid phase extraction. *Supelco Bull.* 910:1–6.
44. Mateo, J. J., Mateo, R., Hinojo, M. J., Lorens, A., and Jimenez, M. 2002. Liquid chromatographic determination of toxigenic secondary metabolites produced by *Fusarium* strains. *J. Chromatogr. A.* 955:245–256.
45. Vatinno, R., Vuckovic, D., Zambonin, C. G., and Pawliszyn, J. 2008. Automated high-throughput method using solid-phase microextraction–liquid chromatography–tandem mass spectrometry for the determination of ochratoxin A in human urine. *J. Chromatogr. A.* 1201:215–221.
46. Katerere, D. R., Stockenström, S., Thembo, K. M., Rheeder, J. P., Shephard, G. S., and Vismer, H. F. 2008. A preliminary survey of mycological and fumonisin and aflatoxin contamination of African traditional herbal medicines sold in South Africa. *Hum. Exp. Toxicol.* 27:793–798.
47. Giraudi, G., Anfossi, L., Baggiani, C., Giovannoli, C., and Tozzi, C. 2007. Solid-phase extraction of ochratoxin A from wine based on a binding hexapeptide prepared by combinatorial synthesis. *J. Chromatogr. A* 1175:174–178.
48. Muñoz, K., Vega, M., Rios, G., Muñoz, S., and Madariaga, R. 2006. Preliminary study of Ochratoxin A in human plasma in agricultural zones of Chile and its relation to food consumption. *Food Chem. Toxicol.* 14:1884–1890.
49. Hernandez, M. J., Moreno, M. G., Duran, E., Guillen, D., and Barroso, C. G. 2006. Validation of two analytical methods for the determination of ochratoxin A by reversed-phased high-performance liquid chromatography coupled to fluorescence detection in musts and sweet wines from Andalusia. *Anal. Chim. Acta* 566:117–121.
50. Piletsky, S. A., Turnera, N. W., and Subrahmanyamb, S. 2009. Analytical methods for determination of mycotoxins: A review. *Anal. Chim. Acta* 632:168–180.
51. Park, D. L., Trucksess, M. W., Nesheim, S., Stack, M. E., and Newell, R. F. 1994. Solvent efficient thin-layer chromatographic method for the determination of aflatoxins B1, B2, G1 and G2 in corn and peanut products: Collaborative study. *J. AOAC Int.* 77:637–646.
52. Pons, W. A., Lee, L. S., and Stoloff, L. 1980. Revised method for aflatoxin in cottonseed products and comparison of thin layer chromatography and high pressure liquid chromatography determinative steps: Collaborative study. *J. Assoc. Offi. Anal. Chem.* 63:899–906.
53. Trucksess, M. W., Stoloff, L., Pons, W. A., Cucullu, A. F., Lee, L. S., and Franz, A. O. 1977. Thin layer chromatographic determination of aflatoxin B1 in eggs. *J. Assoc. Offi. Anal. Chem.* 60, 4:795–798.
54. Van Egmond, H. P., Paulsch W. E., and Schuller, P. L. 1978. Confirmatory test for aflatoxin M1 on thin layer plate. *J. Assoc. Offi. Anal. Chem.* 61:809–812.

55. Vargas, E. A., Preis, R. A., Castro, L., and Silva, C. G. 2001. Co-occurrence of aflatoxins B1, B2, G1, G2, zearalenone and fumonisin B 1 in Brazilian corn. *Food Addit. Contam.* 18:981–986.
56. Nawaz, S., Coker, R. D., and Haswell, S. J. 1992. Development and evaluation of analytical methodology for the determination of aflatoxins in palm kernels. *Analyst* 117:67–71.
57. Otta, K. H., Papp, E., Mincsovics, E., and Záray, G. 1998. Determination of aflatoxins in corn by use of the personal OPLC basic system. *J. Plan. Chromatogr.* 11:370–378.
58. Otta, K. H., Papp, E., and Bagócsi, B. 2000. Determination of aflatoxins in food by overpressured-layer chromatography. *J. Chromatogr. A* 882:11–16.
59. Maragos, C. M. and Greer, J. 1997. Analysis of aflatoxin B1 in corn using capillary electrophoresis with laser-induced fluorescence detection. *J. Agric. Food Chem.* 45:4337–4341.
60. Kok, W. T. 1994. Derivatization reactions for the determination of aflatoxins by liquid chromatography with fluorescence detection. *J. Chromatogr. B* 659:127–137.
61. Stubblefield, R. D. 1987. Optimum conditions for formation of aflatoxin M1-trifluoroacetic acid derivative. *J. Assoc. Offi. Anal. Chem.* 70:1047–1049.
62. Simonella, A., Scortichini, G., Manetta, A. C., Campana, G., Di Giuseppe, L., Annunziata, L., and Migliorati, G. 1998. Aflatossina M1 nel latte vaccino: Ottimizzazione di un protocollo analitico di determinazione quali-quantitativa basato su tecniche cromatografiche, di immunoaffinità e immunoenzimatiche. *Vet. Italiana, Anno XXXIV* 27:25–32.
63. Akiyama, H., Goda, Y., Tanaka, T., and Toyoda, M. 2001. Determination of aflatoxins B1, B2, G1 and G2 in spices using a multifunctional column clean-up. *J. Chromatogr. A* 932:153–157.
64. Shepherd, M. J. and Gilbert, J. 1984. An investigation of HPLC post-column iodination conditions for the enhancement of aflatoxin B1 fluorescence. *Food Addit. Contam.* 1:325–335.
65. Stroka, J., von Host, C., Anklam E., and Reutter, M. 2003. Immunoaffinity column cleanup with liquid chromatography using post-column bromination for determination of aflatoxin B1 in cattle feed: Collaborative study. *J. AOAC Int.* 86:1179–1186.
66. Manetta, A. C., Di Giuseppe, L., Giammarco, M., Fusaro, I., Simonella, A., Gramenzi, A., and Formigoni, A. 2005. High-performance liquid chromatography with post-column derivatisation and fluorescence detection for sensitive determination of aflatoxin M1 in milk and cheese. *J. Chromatogr. A* 1083:219–222.
67. Senyuva, H. Z. and Gilbert, J. 2005. Immunoaffinity column cleanup with liquid chromatography using post-column bromination for determination of aflatoxins in hazelnut paste: Interlaboratory study. *J. AOAC Int.* 88:526–535.
68. Brera, C., Debegnach, F., Minardi, F., Pannunzi, E., De Santis, B., and Miraglia, M. 2007. Immunoaffinity column cleanup with liquid chromatography for determination of aflatoxin B1 in corn samples: Interlaboratory study. *J. AOAC Int.* 90:765–772.
69. Manetta, A. C., Giammarco, M., Di Giuseppe, L., Fusaro, I., Gramenzi, A., Formigoni, A., Vignola, G., and Lambertini, L. 2009. Distribution of aflatoxin M1 during Grana Padano cheese production from naturally contaminated milk. *Food Chem.* 113:595–599.
70. Joshua, H. 1993. Determination of aflatoxins by reversed-phase high performance liquid chromatography with post-column in-line photochemical derivatization and fluorescence detection. *J. Chromatogr.* 654:247–254.
71. Waltking, A. and Wilson, D. 2006. Liquid chromatographic analysis of aflatoxin using postcolumn photochemical derivatization: Collaborative study. *J. AOAC Int.* 89:678–692.
72. Chiavaro, E., Dall'Asta, C., Galaverna, G., Biancardi, A., Gambarelli, E., Dossena, A., and Marchelli, R. 2001. New reversed-phase liquid chromatographic method to detect aflatoxins in food and feed with cyclodextrins as fluorescence enhancers added to the eluent. *J. Chromatogr. A* 937:31–40.

73. Agha Mohammadi, M. and Alizadeh, N. 2007. Fluorescence enhancement of the aflatoxin B1 forming inclusion complexes with some cyclodestrins and molecular modeling study. *J. Luminesc.* 127:575–584.
74. Chu, F. S. 1996. Recent studies on immunoassays for mycotoxin. In: *Immunoassays for Residue Analysis*, ed. R. C. Beier and L. H. Stanker, American Chemical Society, Washington, DC.
75. Zheng, M. Z., Richard, J. L., and Binder, J. 2006. A review of rapid methods for the analysis of mycotoxins. *Mycopathologia* 161:261–273.
77. Trucksess, M. W. 2001. Rapid analysis (thin layer chromatographic and immunochemical methods) for mycotoxins in foods and feeds). In: *Mycotoxins and Phycotoxins in Perspective at the Turn of the Millennium,* ed. de Koe, W. J., Samson, R. A., van Egmond, H. P., Gilbert, J., and Sabino, M. IUPAC, The Netherlands, Wageningen.
78. Trucksess, M. W. and Koeltzow, D. E. 1995. Evaluation and application of immunochemical methods for mycotoxins in food. In: *Immunoanalysis of Agrochemicals in Emerging Technologies,* ed. J. O., Nelson, A. E. Karu, and R. B., Wong, American Chemical Society, Washington, DC.
79. Gilbert, J. and Anklam, E. 2002. Validation of analytical methods for determining mycotoxins in foodstuffs. *Trends Anal. Chem.* 21:468–486.
80. Zheng, Z., Humphrey, C. W., King, R. S., and Richard, J. L. 2005. Validation of an ELISA test kit for the detection of total Aflatoxins in grain and grain products. *Mycopathologia* 159:255–263.
81. Young, J. C. and Games, D. E. 1994. Analysis of *Fusarium* mycotoxins by gas chromatography—Fourier transform infrared spectroscopy. *J. Chromatogr. A* 663:211–218.
82. Suzuki, T., Kurisu, M., Hoshimo, Y., Ichinoe, M., Nose, N., Tokumaru, Y., and Watanabe, A. 1981. Production of trichothecene mycotoxins of *Fusarium* species in wheat and barley harvested in Saitama prefecture. *J. Food Hyg. Soc.* 22:197–205.
83. Onji, Y., Aoki, Y., Tani, N., Umebayashi, K., Kitada, Y., and Dohi, Y. 1998. Direct analysis of several *Fusarium* mycotoxins in cereals by capillary gas chromatography–mass spectrometry. *J. Chromatogr. A* 815:59–65.
84. Olsson, J., Borjesson, T., Lundstedt, T., and Schnurer, J. 2002. Detection and quantification of ochratoxin A and deoxynivalenol in barley grains by GC-MS and electronic nose. *Int. J. Food Microbiol.* 72:203–214.
85. Cappiello, A., Famiglini, G., and Tirillini, B. 1995. Determination of aflatoxins in peanut meal by LC/MS with a particle beam interface. *Chromatographia* 40:411–416.
86. Kokkonen, M., Jestoi, M., and Rizzo, A. 2005. Determination of selected mycotoxins in mould cheeses with liquid chromatography coupled to tandem with mass spectrometry. *Food Addi. Contam.* 22:449–456.
87. Júnior, J., Mendonça, X., and Scussel, V. M. 2008. Development of an LC-MS/MS method for the determination of aflatoxins B1, B2, G1, and G2 in Brazil nut. *Int. J. Environ. Anal. Chem.* 88:425–433.
88. Chen, C. Y., Li, W. J., and Peng, K. Y. 2005. Determination of aflatoxin M1 in milk and milk powder using high-flow solid-phase extraction and liquid chromatography-tandem mass spectrometry. *J. Agric. Food Chem.* 53:8474–8480.
89. Cavaliere, C., Foglia, P., Pastorini, E., Samperi, R., and Laganà, A. 2006. Liquid chromatography/tandem mass spectrometric confirmatory method for determining aflatoxin M1 in cow milk: Comparison between electrospray and atmospheric pressure photoionization sources. *J. Chromatogr. A* 1101:69–78.
90. Lattanzio, V. T., Solfrizzo, M., Powers, S., and Visconti, A. 2007. Simultaneous determination of aflatoxins, ochratoxin A and *Fusarium* toxins in maize by liquid chromatography/tandem mass spectrometry after multitoxin immunoaffinity cleanup. *Rap. Commun. Mass Spec.* 21:3253–3261.

91. Manetta, A. C. 2011. Aflatoxins: Their measure and analysis. In: *Aflatoxins—Detection, Measurement and Control,* ed. Dr Irineo Torres-Pacheco, InTech Publisher, http://www.intechopen.com/books/aflatoxins-detection-measurement-and-control/aflatoxins-their-measureand-analysis.
92. Trucksess, M. W., Stack, M. E., Nesheim, S., Page, S. W., Albert, R., Hansen, T. J., and Donahue, K. F. 1991. Immunoaffinity column coupled with solution fluorometry or liquid chromatography postcolumn derivatization for determination of aflatoxins in corn, peanuts and peanut butter: Collaborative study. *J. Assoc. Offi. Anal. Chem.* 74:81–88.
93. Dragacci, S., Grosso, F., and Gilbert, J. 2001. Immunoaffinity column cleanup with liquid chromatography for determination of aflatoxin M1 in liquid milk: Collaborative study. *J. AOAC Int.* 84, 2:437–443.
94. Muscarella, M., Lo Magro, S., Palermo, C., and Centonze, D. 2007. Validation according to European Commission Decision 2002/657/EC of a confirmatory method for aflatoxin M1 in milk based on immunoaffinity columns and high performance liquid chromatography with fluorescence detection. *Anal. Chim. Acta* 594:257–264.
95. Magliulo, M., Mirasoli, M., Simoni, P., Lelli, R., Portanti, O., and Roda, A. 2005. Development and validation of an ultrasensitive chemiluminescent enzyme immunoassay for aflatoxin M1 in milk. *J. Agri. Food Chem.* 53:3300–3305.
96. Sibanda, L., De Saeger, S., and Van Peteghem, C. 1999. Development of a portable field immunoassay for the detection of aflatoxin M1 in milk. *Int. J. Food Micro.* 48:203–209.
97. Paniel, N., Radoi, A., and Marty, J. L. 2010. Development of an electrochemical biosensor for the detection of aflatoxin M1 in milk. *Sensors* 10:9439–9448.
98. Ip, S. P. and Che, C. T. 2006. Determination of aflatoxins in Chinese medicinal herbs by high-performance liquid chromatography using immunoaffinity column cleanup: Improvement of recovery. *J. Chromatogr. A* 1135:241–244.
99. Tassaneeyakul, W., Razzazi-Fazeli, E., Porasuphatana, S., and Bohm, J. 2004. Contamination of aflatoxins in herbal medicinal products in Thailand. *Mycopathologia* 158:239–244.
100. Zhang, X. H. and Chen, J. M. 2005. HPLC analysis of alfatoxins in medicinal herb extracts by immunoaffinity column cleanup and post-column bromination. *J. Chin. Mat. Med.* 30:182–186.
101. Reif, K. and Metzger, W. 1995. Determination of aflatoxins in medicinal herbs and plant extracts. *J. Chromatogr. A* 692:131–136.
102. Ali, N., Hashim, N. H., Saad, B., Safan, K., Nakajima, M., and Yoshizawa, T. 2005. Evaluation of a method to determine the natural occurrence of aflatoxins in commercial traditional herbal medicines from Malaysia and Indonesia. *Food Chem. Toxicol.* 43:1763–1772.
103. Ventura, M., Gomez, A., Anaya, I., Diaz, J., Broto, F., Agut, M., and Comellas, L. 2004. Determination of aflatoxins B1, G1, B2 and G2 in medicinal herbs by liquid chromatography-tandem mass spectrometry. *J. Chromatogr. A* 1048:25–29.
104. Alcaide-Molina, M., Ruiz-Jimenez, J., Mata-Granados, J. M., and Luque de Castro, M.D. 2009. High through-put aflatoxin determination in plant material by automated solid-phase extraction on-line coupled to laser-induced fluorescence screening and determination by liquid chromatography–triple quadrupole mass spectrometry. *J. Chromatogr. A* 1216:1115–1125.
105. Rena, Y., Zhang, Y., Shao, S., Cai, Z., Feng, L., Pan, H., and Wang, Z. 2007. Simultaneous determination of multi-component mycotoxin contaminants in foods and feeds by ultra-performance liquid chromatography tandem mass spectrometry. *J. Chromatogr. A* 1143:48–64.
106. Hana, Z., Zhenga, Y., Luana, L., Caib, Z., Renb, Y., and Wua, Y. 2010. An ultra-high-performance liquid chromatography-tandem mass spectrometry method for

simultaneous determination of aflatoxins B1, B2, G1, G2, M1 and M2 in traditional Chinese medicines. *Anal. Chim. Acta* 664:165–171.

107. Huanga, B., Hanb, Z., Caia, Z., Wub, Y., and Rena, Y. 2010. Simultaneous determination of aflatoxins B1, B2, G1, G2, M1 and M2 in peanuts and their derivative products by ultra-high-performance liquid chromatography–tandem mass spectrometry. *Anal. Chim. Acta* 662:62–68.
108. Romero-Gonzaleza, R., Garrido Frenicha, A., Martinez Vidala, J. L., Prestesb, O. D., and Grioc, S. L. 2011. Simultaneous determination of pesticides, biopesticides and mycotoxins in organic products applying a quick, easy, cheap, effective, rugged and safe extraction procedure and ultra-high performance liquid chromatography–tandem mass spectrometry. *J. Chromatogr. A* 1218:1477–1785.
109. Beltrán, E., Ibáñez, M., Sancho, J. V., Cortés, M. A., Yusà, V., and Hernández, F. 2011. UHPLC–MS/MS highly sensitive determination of aflatoxins, the aflatoxin metabolite M1 and ochratoxin A in baby food and milk. *Food Chem.* 126:737–744.
110. Khan, M. R., Alothman, Z. A., Ghfar, A. A., and Wabaidur, S. M. 2013. Analysis of aflatoxins in nonalcoholic beer, using liquid–liquid extraction and ultraperformance LC-MS/MS. *J. Sep. Sci.* 36:572–577.
111. Oueslati, S., Romero-González, R., Lasram, S., Frenich, A. G., and Vidal, J. M. 2012. Multi-mycotoxin determination in cereals and derived products marketed in Tunisia using ultra-high performance liquid chromatography coupled to triple quadrupole mass spectrometry. *Food Chem. Toxicol.* 50:2376–2381.
113. Soleimany, F., Jinap, S., Faridah, A., and Khatib, A. 2012. A UPLC MS/MS for simultaneous determination of aflatoxins, ochratoxin A, zearalenone, DON, fumonisins, T-2 toxin and HT-2 toxin, in cereals. *Food Control* 25:647–653.
112. Rubert, J., Dzuman, Z., Vaclavikova, M., Zachariasova, M., Soler, C., and Hajslova, J. 2012. Analysis of mycotoxins in barley using ultra high liquid chromatography high resolution mass spectrometry: Comparison of efficiency and efficacy of different extraction procedures. *Talanta* 99:712–719.
114. Anastassiades, M., Lehotay, S. J., Stajnbaher, D., and Schenck, F. J. 2003. Fast and easy multiresidue method employing acetonitrile extraction/partitioning and "dispersive solid-phase extraction" for the determination of pesticide residues in produce. *J. AOAC Int.* 86:412–431.
115. Wilkowska, A. and Biziuk, M. 2011. Determination of pesticide residues in food matrices using the QuEChERS methodology. *Food Chem.* 125:803–812.
116. Lehotay, S. J., Mastovska, K., and Lightfield, A. R. 2005. Use of buffering and other means to improve results of problematic pesticides in a fast and easy method for residue analysis of fruits and vegetables. *J. AOAC Int.* 22:615–523.
117. Mol, H. J., Plaza-Bolanos, P., Zomer, P., de Rijk, T. C., Stolker, A. M., Mulder, P. J., and Mulder, P. P. 2008. Toward a generic extraction method for simultaneous determination of pesticides, mycotoxins, plant toxins, and veterinary drugs in feed and food matrixes. *Anal. Chem.* 80:9450–9459.
118. Zachariasova, M., Lacina, O., Malachova, A., Kostelanska, M., Poustka, J., Godula, M., and Hajslova, J. 2010. Novel approaches in analysis of *Fusarium* mycotoxins in cereals employing ultra performance liquid chromatography coupled with high resolution mass spectrometry. *Anal. Chim. Acta* 662:51–61.
119. Vaclavik, L., Zachariasova, M., Hrbek, V., and Hajslova, J. 2010. Analysis of multiple mycotoxins in cereals under ambient conditions using direct analysis in real time (DART) ionization coupled to high resolution mass spectrometry. *Talanta* 82:1950–1957.
120. Vendl, O., Berthiller, F., Crews, C., and Krska, R. 2009. Simultaneous determination of deoxynivalenol, zearalenone, and their major masked metabolites in cereal-based food by LC–MS–MS. *Anal. Bioanal. Chem.* 395:1347–1354.

12 Determination of Perfluorochemicals in Food and Drinking Water Samples Using UHPLC–MS Technique

Fabio Gosetti, Eleonora Mazzucco, Maria Carla Gennaro, and Emilio Marengo

CONTENTS

12.1 INTRODUCTION

Recent alarms concern the potential toxicity of polyfluorinated compounds, in consideration of their ubiquitous presence in the environment, their persistence, and potential bioaccumulation and biomagnification properties.

Polyfluorinated compounds comprise hundreds of chemicals characterized by hydrophobic linear alkyl chains partially or fully fluorinated (as the perfluorinated compounds [PFCs]) and containing different functional groups. Polyfluorinated compounds include perfluoroalkyl sulfonamides (PFASAs), fluorotelomer alcohols (FTOHs), polyfluorinated alkyl phosphates (PFAPs), fluorotelomer unsaturated carboxylic acids (FTUCAs), perfluoroalkyl acids (PFAAs), and their salts. The most common PFAAs are perfluoroalkyl carboxylic acids (PFCAs) and perfluoroalkyl sulfonic acids (PFASs). In particular, PFASs contain one or more fluorinated alkyl chains bonded to a polar head, which at neutral pH can be charged (anionic, cationic, and amphiphilic surfactants) or noncharged (nonionic surfactants).

While PFC persistence and volatility depend on the functional groups bonded to the alkyl chain, the strong carbon–fluorine bonds and perfluorinated moieties confer to PFCs characteristics of rigidity, low chemical reactivity, and good stability to thermal and biological degradation, often associated with both oleophobic and hydrophobic properties. Of late, these properties have made PFCs excellent products for a great variety of industrial applications and uses. In 2000, the global production of the two most diffused PFCs, namely, perfluorooctanoic acid (PFOA) and perfluorooctylsulfonate (PFOS), was estimated around 500 and 3500 metric tons, respectively [1]. PFOA is primarily used in the production of fluoroelastomers and fluoropolymers as polytetrafluoroethylene (PTFE or Teflon®), which finds many applications, ranging from coating of cookware to waterproof breathable membranes of clothing, to material used for labware and analytical instrumentation. PFOS is the precursor of several products. Over the past 50 years PFCs have been used in the production of surfactants; lubricants; inks; paints; polishes; adhesives; cleaning agents; food packaging; fire-retarding foams; refrigerants; components of pharmaceuticals; nanomaterials; cosmetics and personal care products; insecticides; papers; and textile coatings resistant to oil, grease, water, and stain. PFCs also find large use in automotive, mechanical, aerospace, chemical, electrical, and medical fields, as well as in building and construction industries. In particular, PFAPs and FTOHs have been used as surface active agents in domestic products as carpet treatment, paints, cleaning agents, and in surface protection products for food contact coatings.

As mentioned, the chemical structures make poly- and perfluorinated alkyl chain resistant to biological, chemical, and thermal degradation, as well as to hydrolysis, photolysis, and metabolic processes. The half-lives for PFOS and PFOA are estimated to be longer than 41 and 92 years in water (at 25°C), respectively [2].

An unexpected consequence of the widespread use of PFC and stability is its wide release to the whole environment: PFCs have been found in surface water, aquatic environments, sediments, soils, sludges, aerosol [13,14], as well as in fish, herring gull eggs, seal liver [15–18], and in human blood, milk, and many human tissues [11,19–25].

Even if the major sources of release are the discharges from fluorochemical industries, PFCs also enter the environment during the use and disposal of PFC-based products [26]. From contaminated water, air, and feed, PFCs enter plants and animals at the bottom of the food chain and, through bioaccumulation and biomagnification, they pass into food-producing animals and then to food. In particular, the contamination of water cycle has been identified as one of the major causes of the presence of PFCs in food, also because the processes of drinking water and sewage treatment are generally scarcely efficacious in PFC removing [2,7,27–29]. PFOA and other PFCs have been found at levels around nanograms per liter in tap water [29], used not only for drinking but also for washing and cooking food, to which PFCs can easily be transferred. PFOA, perfluorohexanoic acid (PFHxA), perfluoroheptanoic acids (PFHpA), and perfluoroundecanoic acid (PFUDA) have already been found in food, PFOS mostly being present in meat, fish, and shellfish at levels around low nanograms per gram [17,29]. PFASAs in particular can pass to food from packaging materials [29].

Human exposure to PFCs can take place already through breast-feeding and then via intake of contaminated food and drinking water, but also via involuntary ingestion of settled indoor dust and soil particles as well as inhalation of contaminated air. Dermal intake of PFOA has also been reported but, in general, exposure through direct contact with PFC-containing products, even if not excluded, seems to be negligible [30]. Food preparation or cooking was indicated as another source of contamination, but the results available up to now indicated little or no effect on the influence of domestic cookware on PFC levels [29].

All of these continuous and widespread sources of contamination can easily give rise to the so-called chronic exposure [25,31]. Studies were performed to possibly correlate PFC concentration levels present in the major environmental sources of intake (dust, wastewaters, soil, air) with PFC content in the food produced in the territory as well as in blood and human tissues of people living in the area [31], to evaluate both intake and accumulation levels. Other environmental monitoring studies were devoted to understanding distribution, source, transportation, and behavior of PFC in the water environment [26].

With regard to intake via food, only chemicals with a molecular weight lower than 1000 g/mol (corresponding to a ca. C70 linear hydrogenated alkyl chain) are considered able to pass through the human intestinal barrier [32]. PFOS and PFCAs have long half-lives in humans and they have been proven to exhibit toxicity in laboratory animals, causing liver cancer, affecting the lipid metabolism, and disturbing immune [33,34] and reproductive systems [13]. This leads to hepatotoxicity, immunotoxicity, and potential carcinogenicity [26,35].

Ionic PFCs are absorbed when orally ingested [36], whereas the amphiphilic species do not preferentially accumulate in adipose tissues but bind to blood proteins influencing hormone feedback systems [8]. PFOSs and PFOAs that are also absorbed by inhalation and dermal contact can cause hepatic diseases and hepatocarcinogenesis in rats and mice [37].

PFASs and PFCAs adhere to serum albumin and accumulate in blood, kidney, and mainly in the liver [2,25,35]. Several PFASs show adverse environmental and toxicological effects as endocrine disruption, which causes cancer cell proliferation

[38]. The longer (>C5) the PFCA perfluorinated chains, the stronger has been shown the persistence and the strength of binding to proteins, which in turn favors bioaccumulation processes [32].

Notwithstanding these evidences, hundreds of PFC-related chemicals (including homologous, PFOA, and telomeres, which can potentially degrade to PFCAs) are not regulated yet [31]. To date, toxicological information is available only for PFOS and PFOA. Based on laboratory animal feeding studies, no observed adverse effect level (NOAEL) value was estimated for PFOS as 0.1 mg/kg/day, and the lowest observed adverse effect level (LOAEL) value as 0.4 mg/kg/day [28]. As it regards PFC production, already in 2000, the 3M Company, the major producer of PFOS, voluntarily phased out production, but PFOS and many related PFCs are still produced by other manufacturers [39]. Since April 2003, the U.S. Environmental Protection Agency (US EPA) released two risk assessments about potential human exposure to PFOA, and in 2006, launched a voluntary stewardship program to reduce PFOA and related chemicals in the environment by 95% by 2010, and to work toward their elimination by 2015 [40]. In October 2006, the European Union issued a directive that prohibited the general use of PFOS and derivatives after June 2008 [41], and in 2010, the US EPA initiative for the industry confirmed the decision to voluntarily eliminate the U.S. production of long-chain PFCA precursors by 2015 [32].

Even if the US EPA suggested PFOA possible carcinogenicity, a final risk assessment is not yet completed. The U.K. Health Protection Agency fixed PFOS and PFOA maximum acceptable concentrations in drinking water at 0.3 and 10 μg/L, respectively [42]. In 2008, the European Food Safety Authority (EFSA) set a tolerable daily intake (TDI) of 150 ng/kg bw/day for PFOS and 1500 ng/kg bw/day for PFOA, based on the contamination in the food chain [29]. In the same year, the *EFSA Journal* reported that the data available allow only indicative values, and the US EPA included PFOA in the Third Drinking Water Contaminant Candidate List (CCL 3) [43]. In January 2009, the US EPA established the Provisional Health Advisories (PHA) limits for PFOS (0.2 μg/L) and for PFOA (0.4 μg/L) in drinking water, [40,44] underlining any way that PHA values were still subjected to possible changes as a function of new information [40]. In the 2009 Stockholm Convention, the United States categorized PFOS and their salts as persistent organic pollutants (POPs) [2,44]. The March 17, 2010 European Commission suggested that the member states monitor the presence of PFCs in food over 2010 and 2011 in order to define exposure levels [45]. The EFSA scientific report, based on 54,195 observations obtained on 7560 food samples collected in the period 2006–2012 from 13 European countries, showed that the highest PFC concentrations were found more frequently in fish and seafood, in meat and edible offal (liver in particular), and to a lesser extent in fruit, vegetables, and drinking water [46]. In 2012, PFOS was added to the EU list of priority substances in the field of water policy according to the Water Framework Directive [47]. The chronic dietary exposure to PFOS and PFOA was observed in all classes of age, but for both average and high consumers, it fortunately resulted well below the TDI. For the other 25 PFCs studied, no TDI value is available and it was, therefore, not possible to evaluate their relevance for human health [46].

It is, therefore, necessary to have available reliable and sensitive methods for PFC identification and determination. This review is devoted to presenting and discussing the most recent analytical methods for PFC analysis in food and drinking water.

A complete and detailed review published in 2011 presents the methods developed between 2004 and the beginning of 2010. Exposure, assessment, and food occurrence of PFCs, as well as the state of the art in extraction, clean-up, detection, confirmation, and quantification, highlighting the advantages and limitations of each technique are here considered [29]. The analytical methods are mostly based on liquid chromatography/mass spectrometry (LC–MS) and gas chromatography/mass spectrometry (for semivolatile PFASAs and FTOHs). High-performance liquid chromatography coupled with mass spectrometry (HPLC–MS) or tandem mass spectrometry (HPLC–MS/MS) generally makes use of electrospray ionization (ESI) in negative ion (NI) mode.

In 2013, a review concerning the determination of PFCs in aquatic organisms mainly through HPLC–MS technique has been published [18].

In this chapter, we report the results of the most recent manuscripts devoted to the determination of PFCs in food based on ultra-high-performance liquid chromatography (UHPLC) coupled with mass spectrometry (MS) technique. Owing to the column packing particles lower than 2 μm and pressures up to 1200 bar, UHPLC technique allows for rapid and efficient separation of a number of substances. By reducing the particle diameter, resolution and sensitivity increase; however, since in these conditions column backpressure also increases, specially designed chromatographic instruments are required. The extra resolution provided by UHPLC systems gives greater information and reduces the risk to not detect potentially important coeluting analytes. With respect to HPLC, UHPLC provides fast and high-resolution separation, which increases LC–MS sensitivity and minimizes matrix interference. The minimization of the component coelution increases mass spectra purity and improves the screening process. In addition, UHPLC technique allows for shorter analysis times, which is very advantageous in routine analysis.

12.2 UHPLC–MS/MS ANALYSIS

12.2.1 Evaluation and Overcoming of External PFC Contamination

Owing to the widespread diffusion of PFC-based products, the first step in PFC determination is to evaluate and possibly eliminate any external PFC contamination that can occur in food samples during the steps of sampling, pretreatment, and analysis. The potential sources of PFC contamination are, besides labware and instrumentation, all the items of common lab use that may contain fluorinated materials, such as polypropylene centrifugation tubes, polyvinylidene fluoride filters, glassware, and rotary evaporators as well as ultrapure water, solvents, chemicals, and sorbent cartridges. Also, the possibility of PFC losses due to adsorption onto containers and glass surfaces must be considered. To control contamination or loss of PFCs is very important, due to the very low concentrations of PFCs (typically low parts per trillion or pg/g) present in most of the food samples. Strategies to minimize contamination include the replacement of PTFE tubing and fittings by ones made from materials that do not contain PFCs, such as polyether ether ketone (PEEK) or stainless steel [38,48].

In some cases, a short C18 HPLC column was inserted between the mixer and the sample loop, with the aim to delay the elution of possible contaminants coming from the LC system and to permit the injected analytes to elute earlier [48]. In many methods, a selective extra guard column (PFC isolator column) was inserted between the pump and the injector to retain PFCs originating from the mobile phase and the UHPLC system [11,17,25,35,38,49]. In other cases, UHPLC columns were used to trap PFCs possibly present as contaminants in the mobile phase [14,50]. The amount of PFOA released from Teflon junctures, capillaries, and connection (total length of about 1 m) of the chromatographic system and collected on the solid-phase extraction (SPE) cartridge during a chromatographic run has been evaluated through the standard addition method as 418 ± 17 ng/L: this PFOA contribution was subtracted in each PFOA quantification [11].

Other authors evaluated the "blank of the instrument" (instrumentation potential PFC contamination) through direct injection of pure methanol into the UHPLC–MS/MS system [2]. Laboratory disposable equipment was solicited in methanol and analyzed for PFCs [50]. In other cases, all the materials were decontaminated by rinsing with ultrapure water and methanol [31]. MilliQ water and MilliQ apparatus were tested for contamination [48]. To choose the purest ultrapure water and solvents from various suppliers, 50 mL of solvent were evaporated to dryness and reconstituted in 200 μL of methanol. In many samples, the presence of detectable amounts of PFCs, in particular PFHxA, PFOA, and PFDA at concentrations ranging from 0.1 to 0.4 pg/mL was evidenced [49].

12.2.2 Matrices Analyzed and Their Pretreatment

Sample pretreatment is a step of particular importance in the analysis, since it helps in increasing recovery and lowering matrix effect. The choice of pretreatment and extraction conditions mainly depends on the kind of the matrix. Besides some evidence of liquid–liquid extraction (LLE), SPE with different kinds of cartridges is generally used in PFC determination in food, since it offers advantages of simplicity, speed, low consumption of organic solvent, and high reproducibility. Recently, online SPE UHPLC–MS methodologies that comprise the steps of extraction and analysis have been optimized [11]. Alkaline digestion, which can often help the release of the analytes bound to matrix components, did not improve recovery [48].

The different pretreatment methods used in the UHPLC–MS/MS determination of PFCs in drinking water and food are summarized here for the different matrices investigated, while details are reported in Table 12.1.

12.2.2.1 Drinking Water

For PFC extraction, drinking water is generally filtered and undergoes SPE by using different cartridges [2,17,51]. Recently, a new material has been synthesized that works as a selective sorbent to extract from water and concentrate PFCs [12]. Sorption is based on the fluorous affinity taking place through the interior walls of the decyl-perfluorinated magnetic mesoporous

TABLE 12.1
List of Analytical Methodologies for the Determination of PFCs

Analytes	Analytical Method	Experimental Conditions	LOD, LOQ, MDL, MQL	Matrix	Sample Pretreatment	Ref
FOSA, N-MeFOSA, PFDA, PFDoA, PFHxA, PFHxS, PFNA, PFOA, PFOS, and PFUDA	UHPLC–MS/MS (QQQ)	Stationary phase: Kinetex C18 Mobile phase: mixture of methanol and 10 mM of *N*-methylmorpholine aqueous solution Gradient elution Flow rate: 0.9 mL/min	LODs and LOQs of the order of ng/L and ng/g according to the matrix	Food and drinking water	For food samples: alkaline digestion and automated SPE (HLB disks) For water samples: pH correction to 3.5 and automated SPE (HLB disks)	[2]
FOSA, PFBS, PFHpA, PFHxS, PFOA, PFODA, PFOS, PFPeA, and PFTeDA	UHPLC–MS/MS (QQQ)	Stationary phase: Zorbax Eclipse XDB-C18 Mobile phase: mixture of 0.01% NH_4OH solution in 5 mM ammonium acetate and 0.01% NH_4OH solution in acetonitrile Gradient elution Flow rate: 1.0 mL/min	LODs between 3 and 15 ng/L LOQs between 10 and 50 ng/L	Food, river, and biological samples	Homogenization in methanol and on-line SPE (Poros® HQ column)	[11]
PFBA, PFBS, PFDA, PFDoA, PFDS, PFHpA, PFHxA, PFHxS, PFNA, PFOA, PFOS, PFPeA, PFTeDA, PFTrDA, and PFUDA	UHPLC–MS/MS (QQQ)	Stationary phase: Acquity BEH C18 Mobile phase: 2 mM ammonium acetate in water and 2 mM ammonium acetate in methanol Gradient elution Flow rate: 0.4 mL/min	/	Drinking water, fish, and shellfish	Water samples: SPE (WAX sorbent) Food samples: homogenization, sonication, and hexane extraction	[17]

continued

TABLE 12.1 (continued)
List of Analytical Methodologies for the Determination of PFCs

Analytes	Analytical Method	Experimental Conditions	LOD, LOQ, MDL, MQL	Matrix	Sample Pretreatment	Ref
PFDA, PFDoA, PFHpA, PFHxA, PFHxS, PFOS, PFNA, PFTeDA, and PFUDA	UHPLC–MS/MS (QQQ)	Stationary phase: Acquity BEH C18 Mobile phase: 2 mM ammonium acetate in water and 2 mM ammonium acetate in methanol Gradient elution Flow rate: 0.4 mL/min	LOQs ranging from 0.02 ng/mL and 0.50 ng/mL	Breast milk	SPE (WAX sorbent)	[25]
PFDA, PFDoA, PFHpA, PFHxS, PFNA, PFOA, PFOS, and PFUDA	UHPLC–MS/MS (QQQ)	Stationary phase: Acquity BEH C18 Mobile phase: 2 mM ammonium acetate aqueous solution and acetonitrile Gradient elution Flow rate: 0.4 mL/min	LODs from 0.5 to 1.0 µg/L LOQs from 1.2 to 3.5 µg/L	Tap water	SPE (HLB cartridge)	[26]
PFOA and PFOS	UHPLC–MS/MS (QQQ)	Stationary phase: Acquity BEH C18 Mobile phase: 0.1% formic acid in acetonitrile and 0.1% formic acid in water Gradient elution Flow rate: 0.45 mL/min	LOQs from 0.02 to 1.50 ng/mL	Blood, drinking water, and food	Extraction with acetonitrile, sonication, centrifugation and SPE (ENVI-carb sorbent)	[30]
PFBS, PFDA, PFDoA, PFHpA, PFHxA, PFHxS, PFNA, PFOS, and PFUDA	UHPLC–MS/MS (QQQ)	Stationary phase: Acquity BEH C18 Mobile phase: 2 mM ammonium acetate in water and 2 mM ammonium acetate in methanol Gradient elution Flow rate: 0.4 mL/min	LODs from 0.001 to 0.200 ng/g	Raw, cooked, and packaged foodstuffs	SPE (WAX sorbent)	[31]

PFOA, PFOS, and anionic and nonionic polyfluorinated surfactants	UHPLC–MS (QTOF)	Stationary phase: Acquity BEH C18 Mobile phase: methanol and water/methanol 95/5 (v/v), both at pH 9.7 for NH_4OH Gradient elution Flow rate: 0.28 mL/min	LODs from 0.0076 to 1.81 mg/L	Food packaging	Solid liquid extraction with ethanol/water 95/5 (v/v) at 60°C	[32]
PFDA, PFDoA, PFHpA, PFNA, PFOA, PFOS, and PFUDA	UHPLC–MS/MS (QTOF)	Stationary phase: Acquity BEH C18 Mobile phase: 0.1% formic acid in acetonitrile and 0.1% formic acid in water Gradient elution Flow rate: 0.45 mL/min	LODs between 0.1 and 0.8 ng/mL LOQs between 0.3 and 2.3 ng/mL	Popcorn packaging	PLE both with methanol and saliva simulant	[34]
PFBA, PFBS, PFDA, PFDoA, PFDS, PFHpA, PFHxA, PFHxDA, PFHxS, PFNA, PFOA, PFOcDA, PFOS, PFPeA, PFTeDA, PFTrDA, PFUdA, and THPFOS	UHPLC–MS/MS (QQQ)	Stationary phase: Acquity BEH C18 Mobile phase: 2 mM ammonium acetate in methanol and in water Gradient elution Flow rate: 0.3 mL/min	/	Foodstuffs	Liquid extraction (with sodium hydroxide and methanol) and then SPE (WAX and ENVI-carb sorbents)	[35]
Screening of 20 polyfluorinated alkyl surfactants	UHPLC–MS (QTOF)	Tested different stationary phases, mobile phases, compositions, and flow rates gradient elution	/	Popcorn packaging	/	[38]
FOSA, N-EtFOSA, N-EtFOSE, N-MeFOSA, and N-MeFOSE PFBA, PFBS, PFDA, PFDoA, PFDPA, PFDS, PFHpA, PFHxA, PFHxPA, PFHxS, PFNA, PFOA, PFOPA, PFOS, PFPeA, PFTeDA, PFTrDA, and PFUDA	UHPLC–MS/MS (QQQ)	Stationary phase: Acquity HSS T3 Mobile phase: 5 mM ammonium acetate in water and methanol Gradient elution Flow rate: 0.3 mL/min	LOQs between 0.001 and 0.006 μg/kg in milk LOQs between 0.001 and 0.013 μg/kg in fish	Fish and Milk	QuEChERS extraction Clean-up with dispersive SPE (MgSO4, C18, and ENVI-carb sorbents)	[48]

continued

TABLE 12.1 (continued)
List of Analytical Methodologies for the Determination of PFCs

Analytes	Analytical Method	Experimental Conditions	LOD, LOQ, MDL, MQL	Matrix	Sample Pretreatment	Ref
PFDA, PFDoA, PFHpA, PFHxA, PFHxS, PFOA, PFOS, PFNA, and PFUDA	UHPLC–MS/MS (QQQ)	Stationary phase: Acquity BEH C18 Mobile phase: 2 mM ammonium acetate both in methanol and in methanol/water 10/90 (v/v) Gradient elution Flow rate: 0.4 mL/min	MDLs from 0.3 to 6.6 pg/g MQLs from 0.6 to 13 pg/g	Dietary food composites	Ion-pair extraction with tetrabutyl ammonium hydrogen sulfate in ultrasonic bath. SPE (Florisil and ENVI-carb sorbents)	[49]
Five mono-PFAPs and eight di-PFAPs	UHPLC–MS/MS (QQQ)	Stationary phase: Acquity BEH C8 Mobile phase: methanol and 0.1% NH_4OH in water Gradient elution Flow rate: 0.2 mL/min	LODs from 0.05 to 12.00 ng/L LOQs from 0.1 to 40.0 ng/L	Drinking water	SPE (WAX sorbent)	[51]
PFOA and PFOS	UHPLC–MS/MS (QQQ)	Stationary phase: Acquity BEH C18 Mobile phase: 20 mM ammonium acetate in water and methanol Gradient elution Flow rate: 0.5 mL/min	LOQs: 24 ng/L for PFOA and 15 ng/L for PFOS	Human milk	Protein precipitation with acetone and SPE (Oasis HLB and ENVI-carb sorbents)	[52]
PFBS, PFDA, PFDoA, PFHpA, PFHxA, PFHxS, PFNA, PFOA, PFOS, PFPeA, PFTeDA, PFTrDA, and PFUDA	UHPLC–MS/MS (QQQ)	Stationary phase: Acquity BEH C18 Mobile phase: methanol and 2 mM ammonium acetate in water Gradient elution Flow rate: 0.4 mL/min	LODs from 0.6 to 30.0 pg/g	Seafood	Sonication and SPE (WAX cartridge)	[54]
PFOA and PFOS	UHPLC–MS/MS (QQQ)	Stationary phase: Eclipse XDB C18 Mobile phase: acetonitrile and 10 mM ammonium acetate in water Gradient elution Flow rate: 0.25 mL/min	/	Food packaging	PLE with methanol	[55]

microspheres of the sorbent. In addition, silanol groups on the exterior surface of the microspheres contribute to the good water dispersibility in water of the material, which results in the particularly efficient extraction of PFOA, perfluorononanoic acid (PFNA), perfluorododecanoic acid (PFDoA), and PFOS. Figure 12.1 shows a typical UHPLC–MS/MS chromatogram of five mono-PFAPs and eight di-PFAPs in a drinking water sample.

12.2.2.2 Food Samples

A great variety of food has been analyzed for PFC content. The food matrices were chosen in order to be representative of dietary exposure studies and with the aim to possibly correlate PFC content in food samples with their water, protein, and fat content. Often, composite samples of food were prepared and analyzed according, in particular, to the "duplicate diet," which considers duplicate portions of all the food and drinks generally consumed by one male adult during one day. The individual food samples included in the "duplicate diet composite" were prepared or cooked as consumed [49].

The food matrices analyzed are rice, flour, meats, pork, beef, chicken, pork liver, eggs, whole milk, seafood, oysters, clams, shrimps, squid [2], salmon, grass carps, and other fish [11]. The total daily intake homogenate (duplicate diet) comprises four composites: baby food, fish, meat, and vegetable [49]. Forty food composite samples contained fish (anchovy, red mullet, sole, clam, mussel, seafood, shellfish, canned sardine, and tuna), vegetables (lettuce, carrot, potato, tuber), fruit (apple, orange), milk and yogurt, meat (boiled ham, Frankfurt-type sausage, veal), oils, and bakery products [35]. Other authors analyzed six groups of food: vegetables (potato, chicory, onion, tomato, carrot, lettuce, and leek), fruit (apple and strawberry), meat (beef, pork, and chicken), fish (eel, cod), eggs, milk, tap water, and beer [30]. Other composite samples were prepared and analyzed: veal steak (raw, grilled, and fried), pork loin (raw, grilled, and fried), chicken breast (raw, grilled, and fried), black pudding uncooked, lamb liver (raw), marinated salmon (homemade and packaged), lettuce (fresh and packaged), paté of pork liver, foie gras of duck, Frankfurt sausages, and fried chicken nuggets [31].

As evidenced, a very extensive and representative panorama of different food samples has been analyzed.

As it concerns the pretreatment and the extraction methods applied by the different authors on these matrices, we provide a panoramic view, with the details for each method reported in Table 12.1.

For solid food samples, when high water content can affect extraction and clean-up performance, prior to the sample work-up, freeze-drying at −80°C for 24 h was employed [17,49]. Then the food samples or the food composites were homogenized, divided into aliquots, and stored at −18°C [35,49] or −20°C [17] in polypropylene containers until analysis. In other cases, food samples were digested in alkaline methanol [2], homogenized in methanol, centrifuged [11], and then subjected to SPE [2], on-line SPE purification [11], graphitized carbon clean-up [48], ion-pair extraction into methyl *tert*-butyl ether (MTBE) with SPE clean-up [49], methanol extraction and SPE in weak anion exchange [35], or to grounding, extraction, and ultrasonication [17].

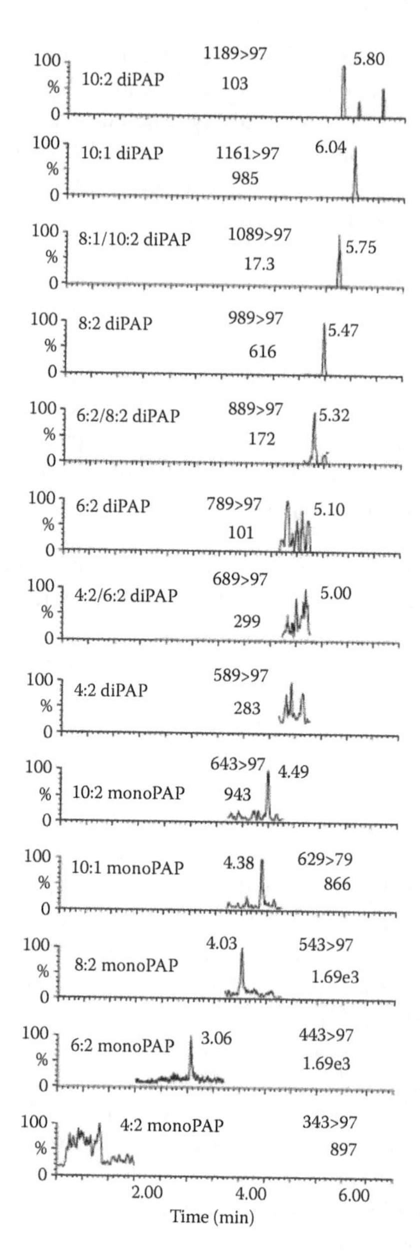

FIGURE 12.1 UHPLC–MS/MS chromatogram of five mono-PFAPs and eight di-PFAPs in a drinking water sample. (Adapted from Ding, H. et al. 2012. *J. Chromatogr. A* 1227:245–252. With permission.)

Milk was subjected to digestion in potassium hydroxide water solution and automated SPE [2,25]. Milk and fish were homogenized, LLE and SPE [31,52], or QuEChERS (quick, easy, cheap, rugged, and safe) extracted in acetonitrile [48]. The recently proposed QuEChERS procedure is based on PFC extraction in water/acetonitrile mixture and their subsequent transfer, induced by the addition of inorganic salts and formic acid into the acetonitrile layer.

12.2.2.3 Packaging Matrices

As mentioned, the presence of PFCs in food is mainly due to bioaccumulation processes that favor the transfer of PFCs, through the food chain, from environmental sources to food-producing animals and plants. However, PFCs can reach food also for direct migration from their packaging. *N*-EthylPFOSAs, incorporated in materials used for food paper production, has often been found in food. Alkaline digestion, followed by sulfuric acid washing and SPE, was also applied [31]. PFAPs, anionic and nonionic polyfluorinated alkyl surfactants (PFSs), were extracted from popcorn and a popcorn bag with 95% ethanol [53]. Pressurized liquid extraction (PLE) was applied to extract PFCs from popcorn prepared for microwave cooking and from its packaging. PLE variables (extraction, temperature, time, and extraction step) were chemometrically optimized through a central composite design consisting of a 2^2 factorial design [34]. Figure 12.2 shows the chromatogram of six PFCAs and PFOS determined in popcorn packaging.

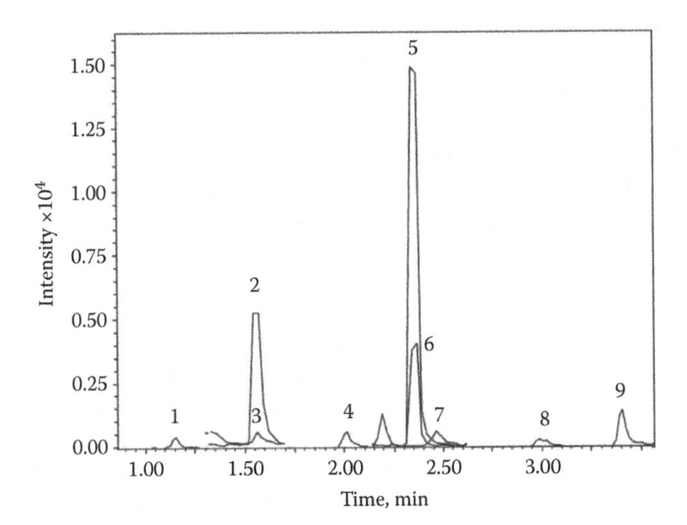

FIGURE 12.2 UHPLC–MS/MS chromatogram of a popcorn packaging sample extract. Peak identification: (1) PFHpA; (2) $^{13}C_8$-PFOA; (3) PFOA; (4) PFNA; (5) $^{13}C_4$-PFOS; (6) PFOS; (7) PFDA; (8) PFUDA; and (9) PFDoA. (Adapted from Martínez-Moral, M. P. and Tena, M. T. 2012. *Talanta* 101:104–109. With permission.)

12.3 ANALYTICAL UHPLC–MS/MS METHODS

12.3.1 Chromatographic and Mass Spectrometry Conditions

The details of the chromatographic and MS detection conditions adopted in the UHPLC–MS/MS methods devoted to PFC determination in food are reported in Table 12.1. The methods make use of C18, C8, ion-exchange, and perfluorinated stationary phases. The extracts to be analyzed are generally brought to pH values around 9.7, at which point both nonionic and anionic PFSs are ionized [32,38]. An alkaline pH 10.47 (0.1% NH_4OH aqueous solution) of the mobile phase was shown to reduce tailing of mono-PFAP chromatographic peaks and to increase the signal intensity in MS detection [51].

When UHPLC is coupled with MS detection, the most frequently used analyzer for PFC determination is the triple quadrupole (QQQ) [2,11,17,25,26,30,31,35,48,49, 51,52,54,55]. This choice is due to its rapid acquisition scan rate and its best sensitivity when working in a targeted scan mode. Only three methods out of all considered here are based on the UHPLC technique coupled with a high-resolution QTOF MS analyzer [32,34,38], which combines the information provided by the high-resolution MS with that obtained by using MS/MS.

MS detection generally operates in ESI NI, that is, the most suitable ionization source for PFCs, because of the strong electronegative character of the perfluorinated alkyl chain [11]. For example, in the UHPLC–QTOF–MS and UHPLC–QTOF–MS/MS methods used to screen PFASs, fluoroacrylates and sulfonamidoalcohols in food contact materials, ESI NI was very suitable because surfactants are easily ionized and the negative charges are stabilized by the high electronegativity of fluorine atoms. The result combines high PFAS sensitivity with low chemical background noise [38].

In addition, ESI NI of fluorinated compounds has the advantages that specific fluorinated product ions and neutral losses are formed, as, for example, neutral loss of $\Delta m = 20$ Da (HF), which is highly diagnostic [32,38].

12.3.2 Matrix Effect

When analyzing complex matrices in MS detection, the matrix can affect the determination through the occurrence of many possible concurrent mechanisms. One of them is the presence of isobaric mass interference among the endogenous substances. For example, the presence of the bile salt taurodeoxycholate in liver can interfere with the most sensitive transition of PFOS (m/z 499◊80) [53,56]. The matrix effects on the ionization efficiency are very difficult to control: coelution of matrix components is easy to occur and to present the phenomena of ion suppression or enhancement. The first operation to reduce or overcome matrix effects consists of the optimization of the steps of pretreatment and extraction of the analytes. To evaluate presence and sign of matrix effect, a *t-test* at 95% confidence level can be usefully employed, which compares the slopes of the external calibration plot and of the standard addition plot built for the real samples investigated [11,34]. Signal suppression/enhancement can also be evaluated through the behavior of the signal observed when spiking, prior injection, the matrix extracts with analyte [51] or internal standard solutions [49].

The results obtained in the UHPLC–MS/MS analysis of different food matrices show that matrix effect depends on the kind of matrix. For instance, signal suppression for most of the PFCs considered was generally lower in drinking water, higher in fish, and the highest in liver. In agreement, method detection limit (MDL) values are much lower in drinking water than in solid matrices.

Ionization signal suppression around –14% was observed for PFUDA in a vegetable extract and a signal enhancement of 22% for PFUDA and PFDoA in fish and duplicate diet extract [49].

No significant matrix effect was instead observed in popcorn packaging methanol extract containing PFOA, PFOS, and some PFCAs, when comparing the slopes of the plots signal/concentration built for methanol solutions of the standard analytes and for the methanol extract [34].

Matrix effect can also depend on the analytes and can be different for the different analytes present in the same matrix. For instance, in the determination of PFACs in beef and liver, the shorter-chain analytes such as PFHxA, PFOA, and PFNA generally showed higher ion suppression than the longer-chain ones, and PFASAs showed higher ion suppression than other PFCs [2]. In a fish sample, the matrix effect was found to range from a signal suppression of around –29% (for perfluorohexane sulfonate, PFHxS) to a signal enhancement around +29% (for perfluorotetradecanoic acid, PFTeDA), being absent for perfluorooctane sulfonamide (FOSA) and PFOA [11]. When the matrix effect observed only depends on the analytes, small deviations are observed in signal area responses measured in food extracts and in solvent standard solutions of the analytes [49].

12.3.3 PFC Quantitation

In the quantitation process, isotopic-labeled standards when available and/or certified standards of PFCs in methanol were used [13,17,25,34,35,48,49]. Matrix-matched calibration was also employed [48].

Even if a number of labeled stable isotopes (for instance, for PFCAs, PFASs, and FOSA) are commercially available, most PFCs of commercial standards are unavailable (including many PFAPs, fluorinated oxethanes, perfluoropolyethers, and cationic surfactants) and an MS approach is, therefore, needed to screen for target and nontarget PFCs [48].

The development of quantitative MS/MS methods is hard, not only due to the lack of commercial analytical standards but also because other difficulties must be considered, such as the low general solubility of PFASs in common LC solvents and their tendency to form micelles, the adhesion of the PFASs to labware and instruments, the difficulty of separating polymeric PFASs in conventional LC columns, and the ability of PFASs to form aggregate and adducts that influence ESI responses [38].

Quantitative analysis in MS/MS systems was often conducted in multiple reaction monitoring (MRM) mode. $[MSO_3]^-$ and $[MCOOH]^-$ were the precursor

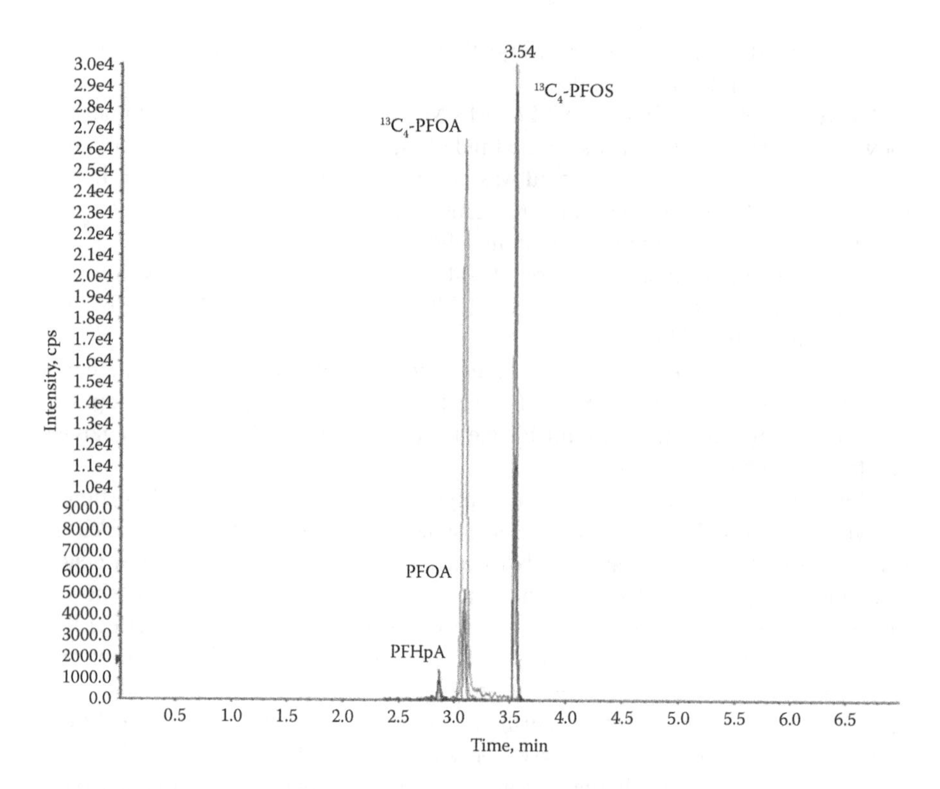

FIGURE 12.3 On-line SPE UHPLC–MS/MS chromatogram of a fish sample extract. (Adapted from Gosetti, F. et al. 2010. *J. Chromatogr. A* 1217:7864–7872. With permission.)

ion for sulfonates and carboxylates, respectively [31]. The product ions of the destroyed PFCAs and PFASs molecules were anion carboxylic groups [M–COO]$^-$ or [M–$(CF_2)_x$–COO]$^-$ for PFCA and sulfonate group [FSO_3]$^-$ or [SO_3]$^-$ for PFASs [11,26,35]. Figure 12.3 shows a typical UHPLC–MS/MS chromatogram of a fish sample, in which each chromatographic peak is monitored by two MRM transitions.

Since PFCs generally present many transitions, the most intense can be used for quantitative analysis (quantifier transition) and the second one (qualifier transition) to confirm the identification. Taking into account the performance criteria of the EU Commission, the MRM ratio between the abundances of the two selected precursor and product ion transitions (qualifier transition to quantifier transition) was used by many authors to confirm analyte identification [11,17,31]. However, it has been reported that the use of only two transitions could result in false-positive or false-negative results when an interfering matrix compound coelutes with the analyte of interest [57–59]. The alternative would be to monitor more than two transitions, but this is not always possible, since it depends on the analyte and on the generation of stable and characteristic MS–MS spectra [59].

12.4 RESULTS AND DISCUSSION

Altogether, the results confirm the widespread diffusion of PFCs all over the world.

It is indeed impossible to directly extend the results obtained for the PFC content in a given matrix to the average content in matrices sampled in different parts of the world, because, for instance, the PFC content in fish not only depends on the total PFC amount in its native water but also on the characteristic of the specific kind of fish, its habits, and bioaccumulation properties. Still, general results can be drawn out.

For instance, levels of PFOA, PFOS, PFNA, and PFUDA resulted always the highest in shellfish and in fatty fish [2,17,35,48,54].

A study that compared PFC content in human milk in different countries evidenced comparable contents of PFOS and PFOA [25,52].

The examples reported here underline the high diffusion of PFCs in food.

In 16 canned (vegetal oil) and fresh fish samples, the presence of PFCAs, PFOS, and FOSA was found, being that PFCA content is greater in canned than in fresh samples [48]. In some milk samples, PFOS and perfluorodecane sulfonate (PFDS) were present and no correlation was found between PFC and milk fat content, as confirmation that PFCs, due to their chemical structures, do not easily accumulate in fat [48]. PFCAs and PFASs have been evidenced even in baby food samples, as shown in Figure 12.4.

In a study in which 18 PFASs were searched for in different kinds of food, PFOS resulted the compound present in the highest number of samples (33 out of 80),

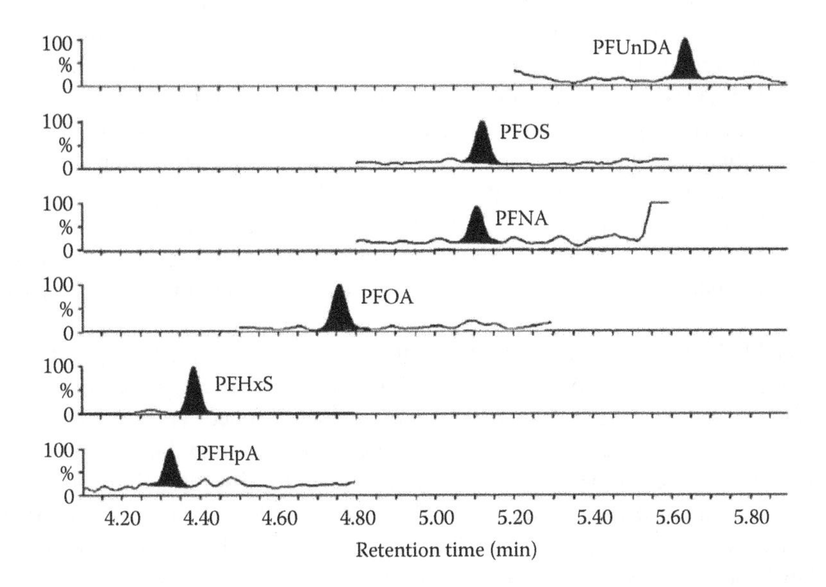

FIGURE 12.4 UHPLC–MS/MS chromatogram of PFCAs and PFSAs in a baby food sample extract. (Adapted from Vestergren, R. et al. 2012. *J. Chromatogr. A* 1237:64–71. With permission.)

followed by PFOA, PFHpA, PFHxS, PFDA, and PFDS. In particular, the greatest PFAS concentrations were found in fresh and canned fish and in shellfish, which also contain PFOA, PFDS, PFOS, and PFDA [35]. PFOA was also found in vegetables, milk, yogurt, ham, and sausages; PFDA in potatoes and fruit; and PFOS in meat, fish, vegetables and tubers, pulses, oils, and bakery products. In the 80 individual food samples analyzed, only perfluoropentanoic acid (PFPeA), perfluorohexadecanoic acid (PFHxDA), and perfluorooctadecanoic acid (PFOcDA) could not be detected [35].

In general, it can be said that PFC concentration and presence tends to increase with chain length. While C4–C6 PFC chains were present in only a few samples, C7–C14 PFCs were detected in most of the samples. In addition, long C9–C14 chain PFCAs with an odd number of carbons (C9, C11, C13) were found at higher concentrations [48].

Given the complete detailed information reported in Table 12.1, some results of particular interest are reported here. They concern food PFC contamination from packaging, the possible influence of the cooking process on qualitative and quantitative PFC food composition, and the possible correlation that can be made between PFC environment (water, soil, atmosphere) contamination and amounts found in food locally produced and in tissues of humans living in the area.

12.4.1 Packaging

For paper and board used in food packaging, specific legislation does not yet exist, and PFCs have no specific migration limit. Nevertheless, according to the Framework Regulations that apply to all types of food contact materials (EU Commission, 2004), all parts of the food packaging chain must ensure that migration of chemical from food contact materials to food should not occur in levels harmful to human health. In lack of harmonized EU legislation, national legislations can be used. In particular, the German national legislation for paper and board lists 10 PFCs allowed [60], and in the United States an explicit regulation includes lists of coatings for paper and board [40]. The council of the EU issued a list of coatings and inks for paper and board, which is not legally binding but informative on their uses [32,61].

When PFC-coated paper and board is in direct contact with food, PFOA and other PFCs can migrate to the food, and packaging becomes a direct source of human exposure. Migration of PFOA from packaging paper into popcorn resulted in mg/kg levels in microwave-produced popcorn, and PFCA was also found in the gas phase of microwaved popcorn [62–64]. Also the presence of di-substituted thioether phosphate surfactants (S-diPAPs) was observed in several microwave popcorn migrates but their quantitation was impossible, due to the lack of di-PAPS standard (manufactured by Ciba under the name Lodyne P208E) [32].

More than 115 molecular structures were found in international industrial blends of food packaging (paper and board). They belong to the groups of mono-PFAPs, di-PFAPs, S-diPAPs, polyfluoroalkyl-ethoxylates, polyfluoroalkyl-acrylates, polyfluoroalkyl-amino acids, polyfluoroalkyl-sulfonamide phosphates, and thioacids, some of which are prohibited in Europe and North America. Also, residual and synthesis by-products (such as PFOS, FOSA, and tri-PFAPs) were found to be present [32].

Higher PFOS concentrations were found in packaged lettuce than in the unpackaged lettuce, and PFOA and PFOS levels in packaged marinated salmon were about twice those found in home-made marinated salmon [31].

12.4.2 Effect of Cooking on PFC Content in Food

To simulate possible effects of the cooking process on PFC content, samples of a fatty fish, in which PFHpA and PFOA were identified and quantified, were cooked at 180°C for 30 min in two nonstick pans, one characterized by a new and integral coating and the other by a damaged one. Comparison analyses, performed for the uncooked and the cooked samples, showed similar results to indicate that cooking process did not modify PFC concentration [11].

According to other results, uncooked meat samples (veal, chicken nuggets, Frankfurt sausages) showed greater PFHxA content than those cooked in nonstick cookware, thus excluding hypothetical migration of PFCs into food [31].

12.4.3 Possible Correlation between Environment PFC Contamination, Food, and Humans

An interesting application tends to evaluate a possible correlation between environment sources of PFCs and their transfer to food locally produced and tissues of humans living in the same area [25,35]. Other studies trend to trace PFC diffusion in the environment [17,26,30,32,65,66].

A study performed in a highly industrial zone in Japan compared the analysis performed for Yangtze river water and local food. Even if estimated daily intake (EDI) of PFOA and PFOS were always lower than TDI value established by the EFSA, the relatively high concentration of PFCs in river water resulted to be correlated with the high levels of PFCs in local tap water and fish [25,54,67].

Also, PFAP content in drinking water from six water supply plants in China varied in a large concentration range (between 1.4 and 76.7 ng/L) that depends on sampling location [51]. The high correlation between the concentration of PFCs in tap and surface water observed by other authors [26] confirms the failure of water purification processes in removing PFCs. Removal rates for PFOA and PFOS were estimated as only about 18% and 27%, respectively [26]. According to other authors, PFCs are neither removed nor incorporated during the treatment process or during transport through the distribution network, and the initial concentration in raw water influences the residue levels in tap water [17].

A correlation was observed between the concentration levels of PFCs found in blood, milk, and liver of humans belonging to populations for which dietary exposure, dust, and indoor air exposition has been previously assessed [35]. Other results indicate that exposure is dominated by intake from food (potatoes, fish and seafood, fruit, and vegetables) in which PFC concentration is much higher than in soil, dust, and air [30].

12.5 CONCLUSIONS

The results indicate that the amount of a single PFC in each sample is always lower than the threshold concentration given for PFOA and PFOS, and that TDI does not reach the threshold concentration. However, the results of the analyses also indicate that in practically all food samples considered, some or more PFCs have been found (even if each at low concentration). Also, considering the PFC ubiquitous presence in water, soil, atmosphere, indoor dust, and air, and their presence in so many products of our quotidian life (our clothes), we can say that we are surrounded by PFCs. Also taking into account their stability and their accumulation properties, the total PFC amount that can pass to humans can become enormous.

Serious proposals about stopping production are now in progress, but it is really very difficult at present to substitute them with other products. When PFCs were first synthetized, their very high stability and inertness seemed to guarantee their safety. Once again, we have to consider how difficult it is for scientists to foresee side effects.

ABBREVIATIONS

EDI	Estimated daily intake
EFSA	European Food Safety Authority
ESI	Electrospray ionization
FOSA	Perfluorooctane sulfonamide
FTOHs	Fluorotelomer alcohols
FTUCAs	Fluorotelomer unsaturated carboxylic acids
HPLC–MS	High-performance liquid chromatography coupled with mass spectrometry
HPLC–MS/MS	High-performance liquid chromatography coupled with tandem mass spectrometry
LC–MS	Liquid chromatography–mass spectrometry
LLE	Liquid–liquid extraction
LOAEL	Lowest observed adverse effect level
LOD	Limit of detection
LOQ	Limit of quantification
MDL	Method detection limit
MQL	Method quantification limit
MRM	Multiple reaction monitoring
MS	Mass spectrometry
MTBE	Methyl *tert*-butyl ether
N-EtFOSA	*N*-ethylperfluorooctanesulfonamide
N-EtFOSE	*N*-ethylperfluorooctanesulfonamidoethanol
NI	Negative ion
N-MeFOSA	*N*-methylperfluorooctanesulfonamide
N-MeFOSE	*N*-methylperfluorooctanesulfonamidoethanol
NOAEL	No observed adverse effect level
PEEK	Polyether ether ketone
PFAAs	Perfluoroalkyl acids

PFAPs	Polyfluorinated alkyl phosphates
PFASAs	Perfluoroalkyl sulfonamides
PFASs	Perfluoroalkyl sulfonic acids
PFBA	Perfluorobutanoic acid
PFBS	Perfluorobutane sulfonate
PFCAs	Perfluoroalkyl carboxylic acids
PFCs	Perfluorinated compounds
PFDA	Perfluorodecanoic acid
PFDoA	Perfluorododecanoic acid
PFDPA	Perfluorodecylphosphonic acid
PFDS	Perfluorodecane sulfonate
PFHpA	Perfluoroheptanoic acid
PFHxA	Perfluorohexanoic acid
PFHxDA	Perfluorohexadecanoic acid
PFHxPA	Perfluorohexylphosphonic acid
PFHxS	Perfluorohexane sulfonate
PFNA	Perfluorononanoic acid
PFOA	Perfluorooctanoic acid
PFOcDA	Perfluorooctanoicdecanoic acid
PFODA	Perfluorooctadecanoic acid
PFOPA	Perfluorooctylphosphonic acid
PFOS	Perfluorooctylsulfonate
PFPeA	Perfluoropentanoic acid
PFSs	Polyfluorinated surfactants
PFTeDA	Perfluorotetradecanoic acid
PFTrDA	Perfluorotridecanoic acid
PFUDA	Perfluoroundecanoic acid
PHA	Provisional health advisories
PLE	Pressurized liquid extraction
POPs	Persistent organic pollutants
PTFE	Polytetrafluoroethylene
QQQ	Triple quadrupole
QTOF	Quadrupole-time of flight
QuEChERS	Quick easy cheap rugged safe
S-diPAPs	Disubstituted thioether phosphate surfactants
SPE	Solid-phase extraction
TDI	Tolerable daily intake
THPFOS	Tetrahydroperfluorooctane sulfonate
UHPLC	Ultra-high-performance liquid chromatography
US EPA	United States Environmental Protection Agency

REFERENCES

1. Haug, L. S., Thomsen, C., and Becher, G. A. 2009. A sensitive method for determination of a broad range of perfluorinated compounds in serum suitable for large-scale human biomonitoring. *J. Chromatogr. A* 1216:385–393.

2. Chang, Y-C., Chen, W-L., Bai, F-Y., Chen, P-C., Wang, G-S., and Chen, C-Y. 2012. Determination of perfluorinated chemicals in food and drinking water using high-flow solid-phase extraction and ultra-high performance liquid chromatography/tandem mass spectrometry. *Anal. Bioanal. Chem.* 402:1315–1325.
3. Yamashita, N., Taniyasu, S., Petrick, G., Wei, S., Gamo, T., Lam, P. K. S., and Kannan, K. 2008. Perfluorinated acids as novel chemical tracers of global circulation of ocean waters. *Chemosphere* 70:1247–1255.
4. Teng, J., Tang, S., and Ou, S. 2009. Determination of perfluorooctanesulfonate and perfluorooctanoate in water samples by SPE-HPLC/electrospray ion trap mass spectrometry. *Microchem. J.* 93:55–59.
5. Zhao, X., Li, J., Shi, Y., Cai, Y., Mou, S., and Jiang G. 2007. Determination of perfluorinated compounds in wastewater and river water samples by mixed hemimicelle-based solid-phase extraction before liquid chromatography-electrospray tandem mass spectrometry detection. *J. Chromatogr. A* 1154:52–59.
6. Loos, R., Locoro, G., Huber, T., Wollgast, J., Christoph, E. H, De Jager, A., Gawlik, B. M., Hanke, G., Umlauf, G., and Zaldivar, J. M. 2008. Analysis of perfluorooctanoate (PFOA) and other perfluorinated compounds (PFCs) in the River Po watershed in N-Italy. *Chemosphere* 71:306–313.
7. Yu, J., Hu, J., Tanaka, S., and Fujii, S. 2009. Perfluorooctane sulfonate (PFOS) and perfluorooctanoic acid (PFOA) in sewage treatment plants. *Water Res.* 43:2399–2408.
8. Furdui, V. I., Crozier, P. W., Reiner, E. J., and Mabury, S. A. 2008. Trace level determination of perfluorinated compounds in water by direct injection. *Chemosphere* 73:S24–S30.
9. Yoo, H., Washington, J. W., Jenkins, T. M., and Libelo, E. L. 2009. Analysis of perfluorinated chemicals in sludge: Method development and initial results. *J. Chromatogr. A* 1216:7831–7839.
10. Mc Murdo, C. J., Ellis, D. A., Webster, E., Butler, J., Christensen, R. D., and Reid, L. K. 2008. Aerosol enrichment of the surfactant PFO and mediation of the water—Air transport of gaseous PFOA. *Environ. Sci. Technol.* 42:3969–3974.
11. Gosetti, F., Chiuminatto, U., Zampieri, D., Mazzucco, E., Robotti, E., Calabrese, G., Gennaro, M. C., and Marengo, E. 2010. Determination of perfluorochemicals in biological, environmental and food samples by an automated on-line solid phase extraction ultra high performance liquid chromatography tandem mass spectrometry method. *J. Chromatogr. A* 1217:7864–7872.
12. Yang, L., Yu, W., Yan, X., and Deng, C. 2012. Decyl-perfluorinated magnetic mesoporous microspheres for extraction and analysis perfluorinated compounds in water using ultrahigh-performance liquid chromatography–mass spectrometry. *J. Sep. Sci.* 35:2629–2636.
13. Onghena, M., Moliner-Martinez, Y., Picó, Y., Campíns-Falcó, P., and Barceló, D. 2012. Analysis of 18 perfluorinated compounds in river waters: Comparison of high performance liquid chromatography-tandem mass spectrometry, ultra-high-performance liquid chromatography-tandem mass spectrometry and capillary liquid chromatography-mass spectrometry. *J. Chromatogr. A* 1244:88–97.
14. Gomez, C., Vicente, J., Echavarri-Erasun, B., Porte, C., and Lacorte, S. 2011. Occurrence of perfluorinated compounds in water, sediment and mussels from the Cantabrian Sea (North Spain). *Mar. Pollut. Bull.* 62:948–955.
15. Van Leeuwen, S. P. J., Swart, C. P., van der Veen, I., and de Boer, J. 2009. Significant improvements in the analysis of perfluorinated compounds in water and fish: Results from an interlaboratory method evaluation study. *J. Chromatogr. A* 1216:401–409.
16. Gebbink, W. A., Hebert, C. E., and Letcher, R. J. 2009. Perfluorinated carboxylates and sulfonates and precursor compounds in Herring Gull eggs from colonies spanning the Laurentian Great Lakes of North America. *Environ. Sci. Technol.* 43:7443–7449.

17. Domingo, J. L., Jogsten, I. E., Perello, G., Nadal, M., van Bavel, B., and Kärrman, A. 2012. Human exposure to perfluorinated compounds in Catalonia, Spain: Contribution of drinking water and fish and shellfish. *J. Agric. Food Chem.* 60:4408–4415.
18. Valsecchi, S., Rusconi, M., and Polesello, S. 2013. Determination of perfluorinated compounds in aquatic organisms: A review. *Anal. Bioanal. Chem.* 405:143–157.
19. Olsen, G. W., Mair, D. C., Reagen, W. K., Ellefson, M. E., Ehresman, D. J., Butenhoff, J. L., and Zobel, L. R. 2007. Preliminary evidence of a decline in perfluorooctanesulfonate (PFOS) and perfluorooctanoate (PFOA) concentrations in American Red Cross blood donors. *Chemosphere* 68:105–111.
20. Yeung, L. W. Y., Taniyasu, S., Kannan, K., Xu, D. Z. Y., Guruge, K. S., Lam, P. K. S., and Yamashita, N. 2009. An analytical method for the determination of perfluorinated compounds in whole blood using acetonitrile and solid phase extraction methods. *J. Chromatogr. A* 1216:4950–4956.
21. Lindström, G., Kärrman, A., and van Bavel, B. 2009. Accuracy and precision in the determination of perfluorinated chemicals in human blood verified by interlaboratory comparisons. *J. Chromatogr. A* 1216:394–400.
22. Toms, L. M. L., Calafat, A. M., Kato, K., Thompson, J., Harden, F., Hobson, P., Sjodin, A., and Mueller, J. F. 2009. Polyfluoroalkyl chemicals in pooled blood serum from infants, children, and adults in Australia. *Environ. Sci. Technol.* 43:4194–4199.
23. Henderson, W. M., Weber, E. J., Duirk, S. E., Washington, J. W., and Smith, M. A. 2007. Quantification of fluorotelomer-based chemicals in mammalian matrices by monitoring perfluoroalkyl chain fragments with GC/MS. *J. Chromatogr. B* 846:155–161.
24. Rylander, C., Duong, T. P., Odland, J. O., and Sandanger, T. M. 2009. Perfluorinated compounds in delivering women from south central Vietnam. *J. Environ. Monit.* 11:2002–2008.
25. Kärrman, A., Domingo, J. L., Llebaria, X., Nadal, M., Bigas, E., van Bavel, B., and Lindstrom, G. 2010. Biomonitoring perfluorinated compounds in Catalonia, Spain: Concentrations and trends in human liver and milk samples. *Environ. Sci. Pollut. Res.* 17:750–758.
26. Qiu, Y., He, J., and Shi, H. 2010. Perfluorocarboxylic acids (PFCAs) and perfluoroalkyl sulfonates (PFASs) in surface and tap water around Lake Taihu in China. *Front. Environ. Sci. Engin. China* 4:301–310.
27. Lin, A. Y-C., Panchangam, S. C., and Ciou, P-S. 2010. High levels of perfluorochemicals in Taiwan's wastewater treatment plants and downstream rivers pose great risk to local aquatic ecosystems. *Chemosphere* 80:1167–1174.
28. Takagi, S., Adachi, F., and Miyano, K. 2008. Perfluorooctanesulfonate and perfluorooctanoate in raw and treated tap water from Osaka, Japan. *Chemosphere* 72:1409–1412.
29. Picó, Y., Farré, M., Llorca, M., and Barceló, D. 2011. Perfluorinated compounds in food: A global perspective. *Crit. Rev. Food Sci. Nutr.* 51:605–625.
30. Cornelis, C., D'Hollander, W., Roosens, L., Covaci, A., Smolders, R., Van Den Heuvel, R., Govarts, E., Van Campenhout., K., Reynders, H., and Bervoets, L. 2012. First assessment of population exposure to perfluorinated compounds in Flanders, Belgium. *Chemosphere* 86:308–314.
31. Jogsten, I. E., Perello, G., Llebaria, X., Bigas, E., Marti-Cid, R. Karrman, A., and Domingo, J. L. 2009. Exposure to perfluorinated compounds in Catalonia, Spain, through consumption of various raw and cooked foodstuffs, including packaged food. *Food Cosmet. Toxicol.* 47:1577–1583.
32. Trier, X., Granby, K., and Christensen, J. H. 2011. Polyfluorinated surfactants (PFS) in paper and board coatings for food packaging. *Environ. Sci. Pollut. Res.* 18:1108–1120.
33. Andersen, M. E., Butenhoff, J. L., Chang, S-C., Farrar, D.G., Kennedy, G. L., Lau, C., Olsen, G. W., Seed, J., and Wallacekj, K. B. 2008. Perfluoroalkyl acids and related chemistries—Toxicokinetics and modes of action. *Toxicol. Sci.* 102:3–14.

34. Martínez-Moral, M. P. and Tena, M. T. 2012. Determination of perfluorocompounds in popcorn packaging by pressurized liquid extraction and ultra-performance liquid chromatography-tandem mass spectrometry. *Talanta* 101:104–109.
35. Domingo, J. L., Jogsten, I. E., Eriksson, U., Martorell, I., Perello. G., Nadal, M., and van Bavel, B. 2012. Human dietary exposure to perfluoroalkyl substances in Catalonia, Spain. Temporal trend. *Food Chem.* 135:1575–1582.
36. Ahrens, L., Plassmann, M., Xie, Z., and Ebinghaus, R. 2009. Determination of polyfluoroalkyl compounds in water and suspended particulate matter in the river Elbe and North Sea, Germany. *Front. Environ. Sci. Eng. China* 3:152–170.
37. Saito, K., Uemura, E., Ishizaki, A., and Kataoka, H. 2010. Determination of perfluorooctanoic acid and perfluorooctane sulfonate by automated in-tube solid-phase microextraction coupled with liquid chromatography-mass spectrometry. *Anal. Chim. Acta* 658:141–146.
38. Trier, X., Granby, K., and Christensen, J. H. 2011. Tools to discover anionic and nonionic polyfluorinated alkyl surfactants by liquid chromatography electrospray ionisation mass spectrometry. *J. Chromatogr. A* 1218:7094–7104.
39. US Environmental Protection Agency. 2002. Perfluoroalkyl sulfonates; significant new use rule, *Federal Register*, 67:72854–67.
40. http://www.epa.gov/oppt/pfoa/ (last accessed on April 2013).
41. European Directive 2006/122/EC relating to restrictions on the marketing and use of certain dangerous substances and preparations (perfluorooctane sulfonates), *Off. J. Eur. Union* L372, 2006:32–34.
42. Maximum acceptable concentrations of perfluorooctane (PFOS) and perfluorooctanoic acid (PFOA) in drinking water by Health Protection Agency (HPA), 2007, Available at http://www.hpa.org.uk/webc/HPAwebFile/HPAweb_C/1194947397222 (last accessed on April 2013).
43. Fact Sheet "Final Third Drinking Water Contaminant Candidate List (CCL 3)" presented by US EPA. Available at http://www.epa.gov/ogwdw/ccl/pdfs/ccl3_docs/fs_cc3_final.pdf (last accessed on April 2013).
44. Ericson, I., Domingo, J. L., and Nadal, M. 2009. Levels of perfluorinated chemicals in municipal drinking water from Catalonia, Spain: Public health implications. *Arch. Environ. Contam. Toxicol.* 57:631–638.
45. European Commission Recommendation 2010/161/UE on the monitoring of perfluoroalkylated substances in food, *Off. J. Eur. Union* L68, 2010:22–23.
46. European Food Safety Authority. 2012. Perfluoroalkylated substances in food: Occurrence and dietary exposure. *EFSA J.* 6:2743–2798.
47. Report from the Commission to the European Parliament and the council on the priority substances in the field of water policy, 2012.
48. Lacina, O., Hradkova, P., Pulkrabova, J., and Hajslova, J. 2011. Simple, high throughput ultra-high performance liquid chromatography/tandem mass spectrometry trace analysis of perfluorinated alkylated substances in food of animal origin: Milk and fish. *J. Chromatogr. A* 1218:4312–4321.
49. Vestergren, R., Ullah, S., Cousins, I. T., and Berger, U. 2012. A matrix effect-free method for reliable quantification of perfluoroalkyl carboxylic acids and perfluoroalkane sulfonic acids at low parts per trillion levels in dietary samples. *J. Chromatogr. A* 1237:64–71.
50. Fernández-Sanjuan, M., Meyer, J., Damásio, J., Faria, M., Barata, C., and Lacorte, S. 2010. Screening of perfluorinated chemicals (PFCs) in various aquatic organisms. *Anal. Bioanal. Chem.* 398:1447–1456.
51. Ding, H., Peng, H., Yang, M., and Hu, J. 2012. Simultaneous determination of mono- and disubstituted polyfluoroalkyl phosphates in drinking water by liquid chromatography-electrospray tandem mass spectrometry. *J. Chromatogr. A* 1227:245–252.

52. Barbarossa, A., Masetti, R., Gazzotti, T., Zama, D., Astolfi, A., Veyrand, B., Pession, A., and Pagliuca, G. 2013. Perfluoroalkyl substances in human milk: A first survey in Italy. *Environ. Int.* 51:27–30.
53. Tittlemier, S. A. and Braekevelt, E. 2011. Analysis of polyfluorinated compounds in foods. *Anal. Bioanal. Chem.* 399:221–227.
54. Wu, Y., Wang, Y., Li, J., Zhao, Y., Guo, F., Liu, J., and Cai, Z. 2012. Perfluorinated compounds in seafood from coastal areas in China. *Environ. Int.* 42:67–71.
55. Poothong, S., Boontanon, S. K., and Boontanon, N. 2012. Determination of perfluorooctane sulfonate and perfluorooctanoic acid in food packaging using liquid chromatography coupled with tandem mass spectrometry. *J. Hazard. Mater.* 205:139–143.
56. Labadie, P. and Chevreuil, M. 2011. Biogeochemical dynamics of perfluorinated alkyl acids and sulfonates in the River Seine (Paris, France) under contrasting hydrological conditions. *Environ. Pollut.* 159:391–397.
57. Pozo, O. J., Sancho, J. V., Ibanez, M., Hernandez, F., and Niessen, W. M. A. 2006. Confirmation of organic micropollutants detected in environmental samples by liquid chromatography tandem mass spectrometry: Achievements and pitfalls. *Trends Anal. Chem.* 25:1030–1042.
58. Kaufmann, A. and Butcher, P. 2006. Strategies to avoid false negative findings in residue analysis using liquid chromatography coupled to time-of-flight mass spectrometry. *Rapid Commun. Mass Spectrom.* 20:3566–72.
59. Gallart-Ayala, H., Nuñez, O., Moyano, E., Galceran, M. T., and Martins, C. P. B. 2011. Preventing false negatives with high-resolution mass spectrometry: The benzophenone case. *Rapid Commun. Mass Spectrom.* 25:3161–3166.
60. Standard XXXVI. Paper and board for food contact. Federal Institute for Risk Assessment, Germany. Available from: http://bfr.zadi.de/kse/faces/resources/pdf/360-english.pdf (last accessed on April 2013).
61. Forrest, M. J. 2005. Coatings and inks for food contact materials. In Rapra Review Report n. 186, Smithers Rapra Technology, Shrewsbury, United Kingdom, ISBN: 978-1-84735-079-4.
62. Begley, T. H., White, K., Honigfort, P., Twaroski, M. L., Neches R., and Walker, R. A. 2005. Perfluorochemicals: Potential sources of and migration from food packaging. *Food Addit. Contam.* 22:1023–1031.
63. Sinclair, E., Kim, S. K., Akinleye, H. B., and Kannan, K. 2007. Quantitation of gas-phase perfluoroalkyl surfactants and fluorotelomer alcohols released from nonstick cookware and microwave popcorn bags. *Environ. Sci. Technol.* 41:1180–1185.
64. Begley, T. H., Hsu, W., Noonan, G. O., and Diachenko G. 2008. Migration of fluorochemical paper additives from food-contact paper into foods and food simulants. *Food Addit. Contam.* 25: 384–390.
65. Wille, K., De Brabander, H. F., De Wulf, E., Van Caeter, P., Janssen, C. R., and Vanhaecke, L. 2012. Coupled chromatographic and mass-spectrometric techniques for the analysis of emerging pollutants in the aquatic environment. *Trends Anal. Chem.* 35:87–108.
66. Wille, K., Kiebooms, J. A. L., Claessens, M., Rappe, K., Vanden Bussche, J., Noppe, H., Van Praet, N., De Wulf, E., Van Caeter, P., Janssen, C. R., De Brabander, H. F., and Vanhaecke, L. 2011. Development of analytical strategies using U-HPLC-MS/MS and LC-ToF-MS for the quantification of micropollutants in marine organisms. *Anal. Bioanal. Chem.* 400:1459–1472.
67. Ericson, I., Gómez, M., Nadal, M., van Bavel, B., Lindström, G., and Domingo, J. L. 2007. Perfluorinated chemicals in blood of residents in Catalonia (Spain) in relation to age and gender: A pilot study. *Environ. Int.* 33:616–623.

13 Determination of Acrylamide in Foodstuffs Using UPLC–MS

Ayman Abdel Ghfar and Ibrahim Hotan Alsohaimi

CONTENTS

13.1 INTRODUCTION

Acrylamide is one of the latest discovered neurotoxic and carcinogenic substances in food. Upon single exposure, acrylamide is toxic or harmful by all routes of administration [1,2]. Acrylamide has been added to the list of food-borne toxicants since 2002, when the Swedish National Food Administration found relevant amounts of acrylamide in several heat-treated, carbohydrate-rich foods such as potato chips, coffee, and bread [3]. It has been widely used since the last century for various chemical and environmental applications [4]. Some of the common uses of acrylamide are in the paper, dyes, cosmetics, and toiletry industry. Acrylamides have also been used as flocculants for clarifying drinking water, and for waste water treatment. It is produced commercially as an intermediate in the production and synthesis of polyacrylamides [2]. They are also a component of tobacco smoke, which gave the earliest indication that it can be formed by heating of biological material [5]. Acrylamide can also be present in a variety of food cooked at high temperature. For example, the daily mean intake of acrylamide present in some foods and coffee in a Norwegian subpopulation have been estimated to be 0.49 and 0.46 g per kg body weight in males and females, respectively [6]. Acrylamide is formed during frying, roasting, and baking and is not typically found in boiled or microwaved foods. The highest acrylamide levels have been found in fried potato products, bread and bakery wares, and coffee [7]. All the same, a great variability in acrylamide level between different products of each food category as well as between different brands of the same product has been reported. The difference in the concentration of precursors (free asparagine and reducing sugars) in raw materials, difference in food composition, and in process conditions applied can easily explain the observed variability [8]. Moreover, the actual acrylamide content of a food as it is eaten can largely vary according to domestic cooking conditions. Estimates of dietary acrylamide intake have been made for populations in many countries. A great variability between populations has been found according to a population's eating habits and the way the foods are processed and prepared. Dybing et al. [9] reported an average daily intake for adults close to 0.5 mg/kg body wt, with 95th percentile values of about 1 mg/kg body wt. The World Health Organization (WHO) estimates a daily dietary intake of acrylamide in the range of 0.3–2.0 mg/kg body wt for the general population and up to 5.1 mg/kg body wt for the 99th percentile consumers [10].

13.1.1 Chemistry of Acrylamide

Acrylamide (CH_2=CH–CO–NH_2; 2-propenamide; CAS RN79-06-1) is a colorless and odorless white crystalline solid with a molecular weight of 71.08, a melting point of 84.5°C, low vapor pressure of 0.007 mmHg at 25°C, and a boiling point at 136°C of 3.3 kPa/25 mmHg. Acrylamide is soluble in water, acetone, and ethanol; however, it is not soluble in nonpolar solvents. These properties provide a high mobility in soil and groundwater to acrylamide [11–13]. Acrylamide has been used as an industrial chemical since the 1950s and is produced from the hydration of acrylonitrile. Acrylamide is also known as acrylic amide, ethylene carboxamide, vinyl amide, or 2-propenamide. The main use of acrylamide is as a chemical intermediate for the production of polyacrylamides. Monomeric acrylamide readily participates in radical-initiated polymerization reactions, whose products, polyacrylamides, form the basis of most of its industrial applications [14]. Acrylamide improves the aqueous solubility, adhesion, and cross-linking of polymers that are well known by population due to myriad types of uses in our society. The primary use of polyacrylamide is to strengthen paper, but polyacrylamides are also utilized in the synthesis of dyes; in copolymers for contact lenses; as well as in construction of dam foundations, tunnels, and sewers. Polymers are used as additives for water treatment, enhancers of oil recovery, flocculants, papermaking aids, thickeners, soil conditioning agents, sewage and waste treatment, ore processing, and permanent-press fabrics. Polyacrylamides are also applied in formulations of several types of personal care and grooming products, such as lotions, cosmetics, deodorants, soaps, and shampoos. Further uses are in oil well drilling fluids, for soil stabilization, as dye acceptors, as polymers for promoting adhesion, for increasing the softening point and solvent resistance of resins, as components of photopolymerizable systems, and as cross-linking agents in vinyl polymers. However, other applications of polyacrylamide are in the biomedical, genetic engineering, and research fields, like the separation of proteins by gel electrophoresis [2]. Besides its industrial applications, acrylamide is also present in tobacco smoke, in amounts of 1–2 mg per cigarette [15,16]. Polyacrylamide contains up to 0.1% free acrylamide monomer. Then, low amounts of acrylamide might also migrate from food packaging material into packaged foodstuff [17]. The specific migration limit for acrylamide from materials that come into contact with foodstuffs was defined to not be detectable, with a limit of detection (LOD) of 10 mg/kg [18,19].

13.2 CONTAMINATION OF FOODSTUFFS WITH ACRYLAMIDE

Acrylamide is a heat-induced contaminant naturally formed during home cooking and industrial processing of many foods consumed daily around the world. French fries, potato chips, bread, cookies, and coffee exert the highest contribution to dietary exposure of acrylamide to humans. Furthermore, food safety international bodies and industrial sectors are very active in implementing strategies to minimize its formation during roasting, baking, frying, toasting, and so on. Given the prevalence of acrylamide in the human diet and its toxicological effects, it is a general public health concern to determine the risk of dietary intake of acrylamide. However, associations between dietary acrylamide exposure and increased risk of different cancers are somewhat controversial and do not have a direct extrapolation to the global population. Accordingly, further long-term studies with a general view are ongoing to clarify the risk scenario and to improve the methodology to detect small increases in cancer incidence [20].

13.2.1 Maillard Reaction Acrylamide Production

Recently, after the discovery of the chemical mechanism governing this food-related contaminant, acrylamide production has been described through a series of reactions known as Maillard reaction, between an amino acid, primarily asparagines, and a reducing sugar such as fructose or glucose [21–25] (Figure 13.1). The amino acid

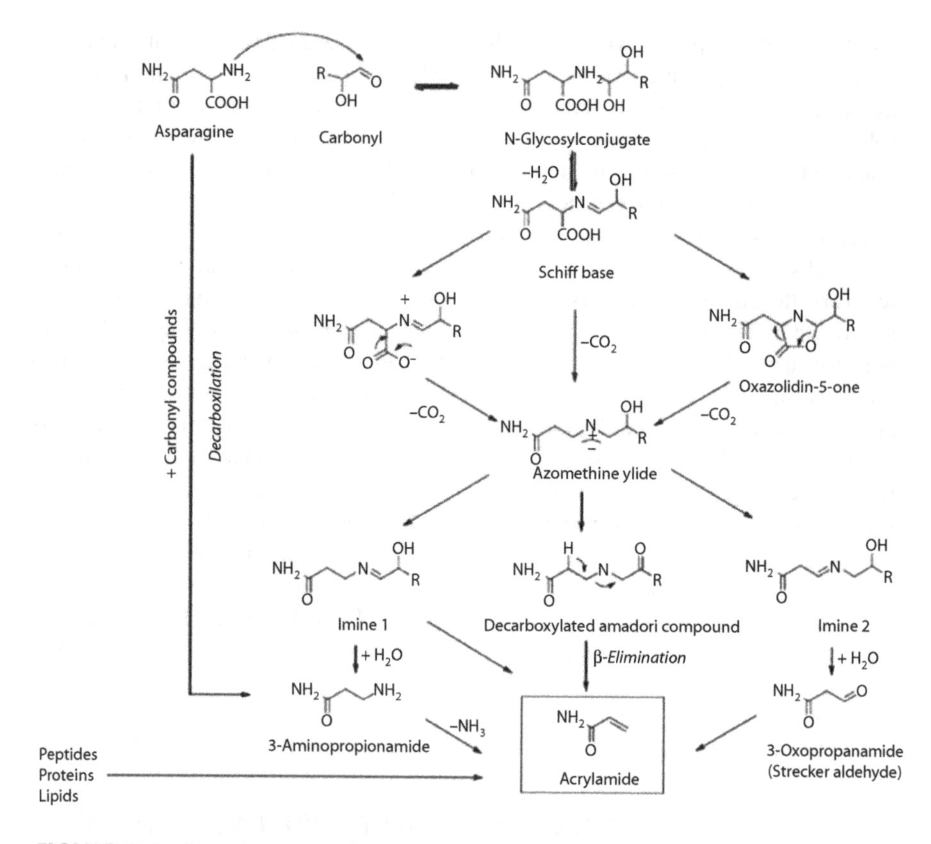

FIGURE 13.1 Reaction scheme for acrylamide formation in the Maillard reaction. (Adapted from Buhlert, J. et al. 2006. *Lett. Org. Chem.* 3:356–357. With permission.)

asparagines was first suspected and later confirmed to be necessary for the formation of acrylamide, as it furnishes the backbone of the acrylamide molecule. Later, additional formation mechanisms, for example, from peptides, proteins, lipids, and biogenic amines, were identified [26–29]. Briefly, acrylamide formation begins at temperatures around 120°C and peaks between 160°C and 180°C [30]. Thus, acrylamide is formed during frying, roasting, and baking and is not typically found in boiled or microwaved food. The highest levels appear in fried and roasted potato products and in cereal products such as breads, crackers, and breakfast cereals. Kinetic studies in model systems and foods clearly demonstrated the strong relationship between processing conditions (time and temperature), moisture and pH, and acrylamide. The formation of acrylamide becomes relevant at temperatures higher than 120°C, and at prolonged heating conditions above 170°C there is a balance between the rates of formation and of loss. Acrylamide losses are due to evaporation, polymerization, or reaction with other food components. Acrylamide is able to react via Michael addition with nucleophilic sites such as amino or thiol groups present in proteins. The almost exclusive formation of acrylamide from asparagines could explain the selective occurrence of acrylamide in certain food products that are rich in concentration of the dominant free amino acid sparagines such as potatoes and cereals. This is the case of potato processing, wherein asparagines comprise nearly 39% (59–143 mmol/kg) of the total free amino acids, and reducing sugar content can vary up to 5.0% of fresh weight. In the case of cereals, the content of free sparagines in wheat flour ranged from 0.15 to 0.4 g/kg, being up to 1.48 g/kg higher in wheat bran [27]. Shortly after discovering, it was established that the major pathway for acrylamide formation in foods is the Maillard reaction with free sparagines as the key amino precursor [18,19]. Other minor reaction routes for acrylamide formation in foods have been postulated, from acrolein and acrylic acid [28], from wheat gluten [26], and by deamination of 3-aminopropionamide [23]. This rapid identification of the main route of acrylamide formation during thermal processing of food was critical to launch the different mitigation strategies as offered today to industry. The Maillard reaction has also been known for producing other mutagenic compounds such as some dicarbonyl compounds (e.g., acrolein and glyoxal), furans (e.g., furfural and 5-sulfooxylmethylfurfural), heterocyclic amines, pyrroles (e.g., 1-nitro-2-acetylpyrrole), dithianes (e.g., 1,3-dithiane), thiazoles, and thiazolidines (e.g., 2-(1,2,3,4,5-pentahydroxy)-pentylthiazolidine). This background of knowledge from the scientific community was determinant in the rapid identification of the predominant pathway of formation in foodstuffs, and later for searching for ways of mitigation. Any information from the variables affecting the formation of acrylamide in foods could be of great interest, as it may open new opportunities for its mitigation. Because the precursors of acrylamide formation are common sugars and amino acids, it is impossible to eliminate them from our foods to avoid the formation of acrylamide, and alternatives should be studied. Accordingly, the structure-specific health effects of the harmful and beneficial compounds formed from the Maillard reaction can be determined, and the food processing technologies can be optimized toward a more selective formation of the health-beneficial ones. Upon lowering the formation of acrylamide in foodstuffs, the concomitant loss of valuable Maillard components, such as flavor or colored compounds, should also be taken into account [20].

13.3 ACRYLAMIDE TOXICITY

From a chemical point of view, acrylamide is a reactive electrophile, and due to its α-, β-unsaturated structure can react with nucleophiles such as amines, carboxylates, and thiols that are commonly found on biological molecules like DNA [29]. Acrylamide is biotransformed *in vivo* to its epoxide, glycidamide ($C_3H_5NO_2$; CAS RN 5694-00-8), which is genotoxic in a variety of *in vitro* and *in vivo* test systems [31]. Glycidamide is also known as glycidic acid amide, oxirane-2-carboxamide, or 2,3-epoxypropionamide. However, acrylamide is much less reactive than glycidamide toward DNA [32]. This fact, together with the genotoxicity of glycidamide, has led to the assumption that glycidamide is the genotoxic agent and probably also the cancer risk increasing factor in acrylamide exposure [33]. The biotransformation process by which acrylamide is converted into glycidamide is not only plausible in animals, but can be readily demonstrated to occur efficiently in both human and rodent tissues. It was noted in passing by the IARC working group that "acrylamide is not known to occur as a natural product" [34]. Long-term exposure to acrylamide may cause damage to the nervous system in both humans and animals to a certain extent. Meanwhile, acrylamide is also regarded as a potentially genetic and reproductive toxin with mutagenic and carcinogenic properties in both *in vitro* and *in vivo* studies [35,36]. However, acrylamide has not been known to be produced from the degradation of polyacrylamide gels in biomedical research applications [37]. The major exposure to the population from polyacrylamide comes from cosmetics, which might contain up to 2% of the gel. However, polyacrylamide is not considered to be harmful to humans. Acrylamide monomer, however, is described as having damaging effects in several aspects, which might become apparent after a delay of months or even years. Toxicological studies suggested that acrylamide vapors irritate the eyes and skin and cause paralysis of the cerebrospinal system, and its occupational exposure limit is set to 0.3 mg/m^3 [38,39]. For the general public, a potential source of exposure had only been seen by drinking water that had been treated with polyacrylamide in a refining process [40]. In order to minimize the risk for the general population, a maximum tolerable level of 0.1 mg acrylamide/L water has been established within the European Union [41]. Elevated levels of acrylamide bound to the hemoglobin were found in workers exposed to the chemical grout. Through measurement of reaction products with protein hemoglobin in blood, it was shown that several of the tunnel workers had developed peripheral nerve symptoms similar to those reported for acrylamide poisoning [13,42]. However, unexpected amounts of acrylamide–hemoglobin adducts could be found in the unexposed people, living outside the contaminated area and used as controls. The observation of a regularly occurring high background level (~0.03 nmol/g globin) of adducts from acrylamide to N-terminal valine in hemoglobin in nonsmoking and occupationally unexposed control persons outside the leakage water indicated the existence of another general exposure source [43]. In order to identify the origin of acrylamide adducts in these nonexposed persons, researchers investigated a number of suspected sources, like food. The importance of acrylamide in food was mentioned for the first time by Tareke [44] who showed that rats feeding fried feed led to a large increase in the level of the hemoglobin adduct, which was concluded to be *N*-(2-carbamoyl-methyl)-valine. This finding showed that almost the entire population is exposed to acrylamide on a daily basis and that the major cause for the observed background adducts was the ingestion of heated starchy food. Background hemoglobin adduct levels in adult humans range from 12 to 50 fmol/mg of globin [9]. After the confirmation of the formation of acrylamide in starchy foods, a variety of many other food products containing acrylamide have been identified [45,46]. For some foods, the measured amounts exceeded amply the maximum allowable concentration in drinking water of 0.1 mg acrylamide/L in EU countries [43] as well as the WHO guideline value for the maximum safe level concentration of acrylamide at 0.5 mg/L [47] and, of course, the existing EU regulations on chemical migration from plastic packaging of 10 mg acrylamide/kg [18]. Based on these observations, certain food products were suspected of being a potential source of exposure to acrylamide. These findings attracted worldwide interest, because of toxicological relevance of acrylamide. Thus, the discovery that the compound is found extensively throughout the food supply caused alarm that dietary acrylamide could be an important human cancer risk factor [46]. This finding showed that almost the entire population is exposed to acrylamide on a daily basis and that the major cause for the observed background adducts was the ingestion of heated starchy food. After rapid confirmation, numerous research activities concerning the extent of exposure, origin of acrylamide in food, health risk to humans, and mitigation of acrylamide in food were initiated. Acrylamide has been shown to be neurotoxic in humans [48] and has been shown to induce tumors in laboratory rats [34,41]. On the basis of tests in animals, a Joint FAO/WHO Expert Committee on Food Additives (JECFA) concluded that current acrylamide levels in foods may indicate a human health concern and that cancer could be the most important adverse effect of acrylamide [49,50]. As a consequence, a huge number of multidisciplinary studies with a great mobilization of human and economic resources were started around the world. There is some confusion with the denomination "contaminant" to acrylamide, as it is a compound naturally formed in food during cooking. Acrylamide is named as a processing contaminant or neoformed contaminant. Long-term exposure to acrylamide may cause damage to the nervous system in both humans and animals to a certain extent. Meanwhile, acrylamide is also regarded as a potentially genetic and reproductive toxin with mutagenic and carcinogenic properties in both *in vitro* and *in vivo* studies [37,38]. Acrylamide monomer is described as having damaging effects in several aspects, which might become apparent after a delay of months or even years.

Toxicological studies suggested that acrylamide vapors irritate the eyes and skin and cause paralysis of the cerebrospinal system, and its occupational exposure limit is set to 0.3 mg/m^3 [2,40,41]. For the general public, a potential source of exposure had only been seen by drinking water that had been treated with polyacrylamide in a refining process [42]. In order to minimize the risk for the general population, a maximum tolerable level of 0.1 mg acrylamide/L water has been established within the European Union [22].

13.3.1 Neurotoxicity

The neurotoxicity of acrylamide in humans is well known from occupational and accidental exposures [51]. For instance, Calleman et al. [52] reported peripheral neuropathy symptoms to highly exposed workers in China. It is characterized by skeletal muscle weakness, numbness of hands and feet, and ataxia. Acrylamide has been shown to be toxic to both the central and the peripheral nervous system [53], although the nerve terminal is now considered to be the primary site of acrylamide action [54,55]. Acrylamide induces nerve terminal degeneration and has effects on the cerebral cortex, thalamus, and hippocampus [56]. In double-blind studies of factory workers, no neurotoxicity was found in workers exposed to less than 3.0 mg/kg/day as determined by biomonitoring [55]. A very recent study demonstrates structural and ultra structural evidence of neurotoxic effects of fried potato chips on rat postnatal development.

13.3.2 Reproductive Toxicity

Acrylamide administered to drinking water of rodents at doses ≥5 mg/kg bw/day resulted in significant decreases in number of live fetuses per litter [57]. At doses of

155 mg/kg bw/day or greater, signs of neurotoxicity and copulatory behavior were noted, as well as effects on sperm motility and morphology. The toxicities in male animals include degeneration of the epithelial cells of the seminiferous tubules, decreased number of sperm, and abnormal sperm, and resulted in decreased fertility rates and retarded development of pups [56]. These toxic effects may be attributed to the interfering effect of acrylamide on the kinesin motor proteins, which also exist in the flagella of sperm, resulting in the reduction in sperm motility and fertilization events [58]. The exposure levels are, however, far above the dietary acrylamide intake in order to pose such a risk. Furthermore, there is no evidence for adverse reproductive or developmental effects from exposure to acrylamide in the general population [59].

13.4 CARCINOGENICITY

Acrylamide is found extensively throughout the food supply, causing alarm that dietary acrylamide could be an important human cancer risk factor [46]. Acrylamide was classified as probably carcinogenic to humans (Group 2A) by the International Agency for Research on Cancer (IARC) [34]. This conclusion was mainly based on positive bioassay results in rodents and supported by the evidence that acrylamide is transformed in mammalian tissues to its more reactive genotoxic metabolite, glycidamide. Several evidences were considered to reach this decision: (i) formation of covalent adducts with DNA in mice and rats, (ii) formation of covalent adducts with hemoglobin in humans and rats, (iii) induction of gene mutations and chromosomal aberrations in germ cells of mice and chromosomal aberrations in germ cells of rats, (iv) induction of chromosomal aberrations in somatic cells of rodents in vivo, (v) induction of gene mutations and chromosomal aberrations in cultured cells in vitro, and (vi) induction of cell transformation in mouse cell lines. Moreover, acrylamide is currently classified as "reasonably anticipated to be a human carcinogen" by the National Toxicology Program [60]. Clear evidence of carcinogenic activity in laboratory animals is demonstrated by studies that are interpreted as showing a dose-related (i) increase of malignant neoplasms, (ii) increase of a combination of malignant and benign neoplasms, or (iii) marked increase of benign neoplasms if there is an indication from this or other studies of the ability of such tumors to progress to malignancy. Acrylamide has not been found to induce mutations in bacteria, but it induced sex-linked recessive lethal and somatic mutations in Drosophila. Substantial laboratory evidence on experimental rodents shows that acrylamide is carcinogenic, causing tumors at multiple sites such as lungs, skin, brain, mammary gland, thyroid gland, and uterus [61,62]. Acrylamide is clastogenic and mutagenic in mammalian cells [42]. It has been stated that oxidation of acrylamide to glycidamide appeared to be a prerequisite for genotoxicity of acrylamide, due to the higher reactivity of glycidamide to form adducts with DNA [63,64]. Although acrylamide and glycidamide reacted directly with hemoglobin, only glycidamide reacted largely with DNA to produce the N7–guanine adduct and to a much lesser extent the N3–adenine adduct [65]. Hence, acrylamide is not mutagenic in *Salmonella typhimurium* assays; in contrast, glycidamide is mutagenic in this assay. The important role of CYP2E1 in epoxidation of acrylamide to glycidamide and formation of glycidamide–DNA adducts has been demonstrated by using CYP2E1-null mice, and when such mice were exposed to acrylamide, higher levels of acrylamide adducts were observed compared with wild-type mice [65]. Additionally, DNA adducts can be regarded as biomarkers of a biologically active internal dose of acrylamide. Before use of such DNA adducts, analytical methods with improved sensitivity would be required, but also complete information on the stability over longer periods of time. Acrylamide itself does not show direct reactivity toward DNA. Since 1994, acrylamide has been classified by the IARC as probably carcinogen to humans [34].

13.5 REGULATIONS

For some foods, the measured amounts exceeded amply the maximum allowable concentration in drinking water of 0.1 mg acrylamide/L in EU countries [43], as well as the WHO guideline value for the maximum safe level concentration of acrylamide at 0.5 mg/L [66], and, of course, the existing EU regulation on chemical migration from plastic packaging of 10 mg acrylamide/kg [18]. Based on these observations, certain food products were suspected of being a potential source of exposure to acrylamide. An acrylamide legislative framework is a critical determinant of whether reliable analytical methods can be developed. It stipulates (i) sampling and monitoring plans, (ii) definition of maximum residue limits (MRLs) for tolerated food contaminants and residues and minimum required performance limits (MRPLs) for some of the testing procedures to detect banned substances, and (iii) the performance characteristics of analytical methods [50,67,68]. The development, optimization, and validation of suitable analytical methods are important elements for the determination of acrylamide and to improve the reliability of analysis methods applied to food samples and residue testing. Because of this, a short description of the situation and aims of this legislative set-up is obligatory. An acrylamide legislation is not coordinated throughout the world. However, well-known international bodies, the most representative of which is the Codex Alimentarius Commission established by FAO and WHO, develops science and risk-based food safety standards that are a reference in international trade and a model for countries to use in their legislation. They considered the margin of exposure values to be low and concluded that it may indicate a human health concern [69].

13.6 MITIGATION STRATEGIES

Mitigation steps include replacement of reducing sugars with sucrose and of ammonium bicarbonate with sodium bicarbonate, or changing in process conditions and/or technologies (changing of time–temperature of frying or baking, changing in the type of oven, prolonged fermentation, etc.) [70]. One of the most promising tools to control acrylamide content in heat-treated foods is the addition of the enzyme asparaginase. The enzyme asparaginase (L-asparagine amidohydrolase) is an enzyme able to catalyze the hydrolysis of asparagine in aspartic acid and ammonia, thus lowering the content of precursor asparagine. Asparaginase has been successfully applied at lab scale both to potato [22] and cereal-based products [71] with a percentage of reduction up to 85–90%. It has no effect on product taste and appearance and is already being used for some products at industrial scale [72]. Some preliminary results achieved at lab scale highlight that asparaginase pretreatment of green beans may represent a viable way to reduce acrylamide concentration in roasted coffee, as well. Up to now, two commercial products, Acrylaway® (asparaginase from *Aspergillus oryzae*) from Novozyme and PreventASe® (asparaginase from *Aspergillus niger*) from Dutch-based multinational life sciences and materials sciences company (DSM), are on the market for food applications. Generally recognized as safe (GRAS) status has been obtained from the US FDA for both types of asparaginases available, and JECFA also endorsed the conclusion that asparaginase does not represent a hazard to human health [73]. Nevertheless, the high cost of the enzyme may represent a serious constraint on its application on a large scale. It should be noted that most of the mitigation measures proposed so far were only tested at laboratory or at pilot scale. Therefore, for those mitigation measures it is not clearly known whether the percentage of reduction in acrylamide claimed at laboratory scale could ever be achievable in food processed at an industrial scale. It has also been emphasized that some mitigation strategies are associated with an increase in other risks or a loss in benefits. For example, prolonging yeast fermentation can efficiently reduce acrylamide concentration in bread, but it is also associated with an increase in the levels of 3-monochloropropandiol (3-MCPD), another neo-formed contaminant [51]. Similarly, replacement of ammonium bicarbonate with sodium bicarbonate as a rising agent for fine bakery products results in an increase of sodium intake [74]. There is a wide consensus that the actions aiming at lowering acrylamide content of foods should be accompanied by a risk–risk or risk–benefit analysis to elucidate all the side effects and their impact on human health. Some options are hardly feasible because of the negative effect they have on HMF content. In addition, some of the mitigation strategies that have been proposed bring about changes in organoleptic properties of foods (excessive browning as a result of glycine addition, generation of off-flavors, insufficient browning as result of changing in time–temperature profile, etc.) that can dramatically affect the final quality and consumers' acceptance [75]. This is a fundamental point for the future, considering that mitigation strategies are not useful if for sensorial reasons consumers do not like the "mitigated" products, giving their preference to the "conventional" ones. In that respect, knowledge of the kinetics of acrylamide accumulation in foods is of utmost importance. Acrylamide starts to form at a temperature >100°C after an initial lag phase, during which no acrylamide forms. Later on, the acrylamide concentration increases exponentially with time to a maximum concentration, after which it can decrease again because of the exhaustion of one of the reactants and/or by the elimination of acrylamide. Acrylamide possesses two functional groups, an amide group and the electron-deficient vinylic double bond, that makes it available for a wide range of reactions, including nucleophilic and Diels–Alder additions and radical reactions. Acrylamide may undergo Michael addition-type reactions to the vinylic double bond with nucleophiles, including amino and thiol groups of amino acids and proteins. On the other hand, the amide group can undergo many reactions, including hydrolysis, dehydration, alcoholysis, and condensation with aldehydes [16]. Many kinetic models have been proposed to describe acrylamide formation and elimination and to predict its final concentration in model systems and foods.

Single-response models based on overall empirical reaction kinetics have been extensively used. Acrylamide formation has thus been modeled as a first-order [76] or second-order reaction [77] according to reaction conditions and reactant concentrations. Acrylamide elimination has usually been modeled as first-order kinetics [78]. Totally empirical models such as those proposed by Corradini and Peleg [79] and Kolek et al. [80] have also been proposed and proven to satisfactorily describe acrylamide concentration in model systems [81] and potato chips [82]. Such empirical models are not based on an underlying chemical mechanism and extrapolation is thus not possible outside the region of variables for which the function has been derived. On the other hand, multiresponse models using acrylamide data supplemented with data on reaction precursors, intermediates, and end products including mechanistic insights in the chemistry involved have also been proposed [83]. With such model systems, not only acrylamide formation but also that of other relevant Maillard-related compounds can be modeled and the estimation of kinetic parameters is much more precise.

13.7 ANALYTICAL METHODS FOR ACRYLAMIDE DETERMINATION IN FOOD

13.7.1 Sample Pretreatment Methods

13.7.1.1 Liquid–Liquid Extraction

A promising approach is to extract the analyte into a polar organic solvent, such as ethyl acetate. Sanders et al. (2002) [84] have employed ethyl acetate to extract acrylamide from the aqueous phase (removing interfering constituents such as salt, sugars, starches, amino acids, etc.). The ethyl acetate extract can then be concentrated and analyzed by either liquid chromatography mass spectrometry (LC–MS) or GC–MS. In most cases, the LOD is significantly lowered, even approaching 10 μg/kg [22]. Similarly, an ethyl acetate extraction step can also be included after the SPE cleanup, providing a significant improvement in sensitivity, especially for more difficult matrices, such as cocoa powder and coffee.

13.7.1.2 Solid-Phase Extraction

The whole extraction and clean-up procedures generally summarized from many peer-reviewed papers before sample injection of GC–MS or LC–MS/MS analysis are shown in Figure 13.2. Water at room temperature has been used to extract acrylamide from many kinds of sample matrices in most analytic methods published to date, because acrylamide is a good hydrophilic small molecule [46,47]. Besides water as an extractant, methanol can also be used to extract acrylamide for the convenience of rotatory evaporation and concentration [85]. Young et al. [86] suggested that acrylamide could be extracted from sample matrices by using NaCl aqueous solution with a relatively high level in order, so that the emulsification process during sample pretreatment was obviously inhibited and the high recovery of analytes was demonstrated. Moreover, one of the laboratories that took part in the proficiency test about acrylamide used a mixture of water and acetone as extractant [87]. In addition, a research group of the National Institute of Health Sciences of Japan chose

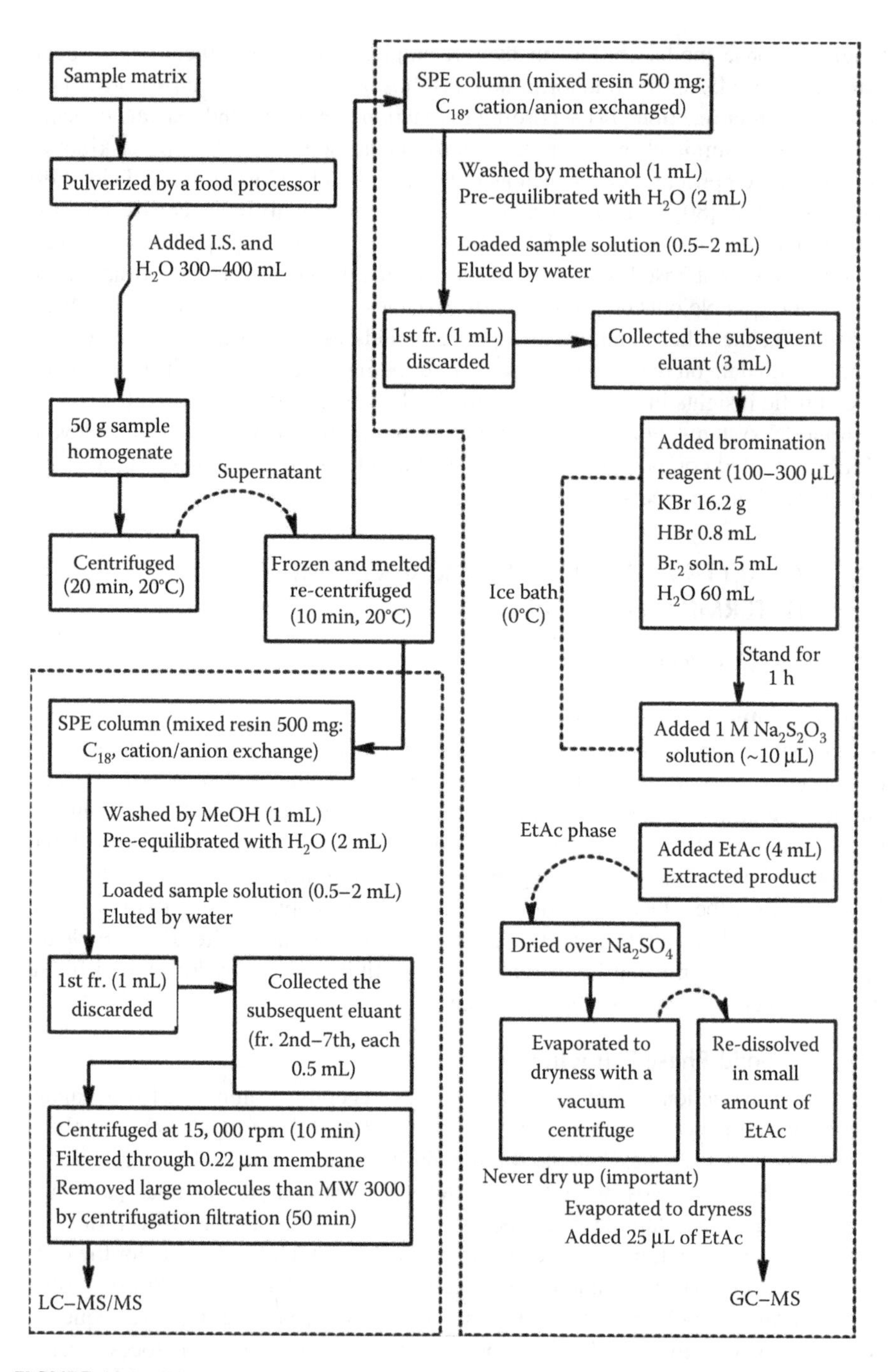

FIGURE 13.2 The whole extraction and clean-up pretreatment of acrylamide before GC–MS or LC–MS/MS. I.S., internal standard; fr., fraction; EtAc, ethyl acetate; MW, molecular weight. (Adapted from Zhang, Y., Zhang, G., and Zhang, Y. 2005. *J. Chromatogr. A* 1075:1–21. With permission.)

this solvent composition for acrylamide extraction [88]. Heating or ultrasonicating during the extraction step may as well be avoided because this may generate large amounts of slight particles that can saturate the SPE cartridges used in further clean-up steps and reduce the efficiency of clean-up and the operating life of SPE cartridges. However, water that had been previously heated to 80°C has been used with no extracting problems.

13.7.2 Chromatographic Techniques

The first method pertaining to the analysis of acrylamide in different cooked and processed foods was reported in May 2002 and is based on the use of isotope dilution LC–MS [89]. Since then, several analytical methods dealing with the analysis of acrylamide in cooked foods have been published in peer-reviewed journals or presented at international scientific meetings [90]. These methods are based mainly on MS as the determinative technique, coupled with a chromatographic step either by LC [64] or GC, the latter either after derivatization of the analyte [91], or, in a few cases, analysis of the compound directly [92]. The Working Group on Analytical Methods that convened during the recent meeting [93] of the European Workshop on Analytical Methods for the Determination of Acrylamide in Food [94], concluded that the majority of laboratories involved in acrylamide analysis use either GC–MS or LC–MS methods. The advantage of the LC–MS-based methods is that acrylamide can be analyzed without prior derivatization (e.g., bromination), which considerably simplifies and expedites the analysis.

13.7.2.1 Methods Based on Gas Chromatography–Mass Spectrometry Techniques

Assays employing GC–MS techniques are either based on bromination of the analyte or direct analysis without derivatization. The latter approach is less laborious and, in both reported cases, employs liquid–liquid extraction of the analyte. In the method reported by Biedermann et al. [94], the determinative step is either positive ion chemical ionization in the selected ion monitoring (SIM) mode or electron impact ionization, achieving an LOD of around 50 and <10 μg/kg, respectively, for potato products. Better sensitivity (level of quantitation, LOQ = 5 μg/kg) can be achieved in the tandem (MS/MS) mode using a high-resolution mass spectrometer. Although more tedious, the bromination of acrylamide to 2,3-dibromopropionamide has multiple advantages, which include (a) improved selectivity, (b) increased volatility, (c) removal of potentially interfering co-extractives, and (d) better sensitivity. Usually the ions *m/z* 150/152 $[CH_2\ CHBr\ CONH_2]+$ and *m/z* 106/108 $[CH_2\ CHBr]+$ are monitored in the SIM mode. Some analysts, however, choose to convert the rather labile di-bromo derivative into 2-bromopropenamide by treatment with triethylamine. This additional step avoids the risk of dehydrobromination in the injector or the ion source of the MS and has no impact on the selectivity or sensitivity of the method. In this case, the ions *m/z* 149/151 $[CH_2\ CBr\ CONH_2]+$ and *m/z* 106/108 are chosen in the SIM mode. Therefore, GC–MS after bromination is probably the best choice for the analysis of acrylamide in foods necessitating a detection level at or <10 μg/kg. A typical flow chart illustrating the individual steps in sample preparation and extraction for GC–MS analysis is shown in Pittet et al. [95]. A further advantage of this technique is that a relatively simple benchtop, GC–MS, can be employed for acrylamide analysis. Application of GC–MS/MS or coupling to a high-resolution MS would even further lower the detection limit for certain foods, approaching the range of 1–2 μg/kg.

13.7.2.2 Methods Based on Liquid Chromatography–Mass Spectrometry

The first LC–MS method for acrylamide in cooked foods was developed in early 2002 by Rosén and Hellenas [91] to verify the initial results procured in Sweden by GC–MS. The method essentially entailed extraction of the analyte with water, centrifugation, solid phase extraction over a Multimode (Isolute®) cartridge, filtering over a 0.22 μm syringe filter, and, subsequently, over a centrifuge spin filter (cut-off 3 kDa). Due to the low molecular weight of acrylamide (71 g/mol) and also its low mass fragment ions, confirmation of the analyte can be achieved with a three-stage mass spectrometer (monitoring of more than one characteristic mass transition). However, acrylamide is a very polar molecule with poor retention on conventional LC reversed-phase sorbents [64], and despite the use of a tandem mass spectrometer, more effort must, in most cases, be placed on the clean-up steps to avoid interference from co-extractives. Of the LC–MS methods communicated at different expert meetings, workshops, or published in the scientific literature, most are making use of SPE during the clean-up step. Acrylamide is difficult to bind actively to any of the conventional sorbents, but the major advantage of SPE is the retention of interfering matrix constituents. Therefore, enrichment or concentration of the analyte remains a challenge, and relatively low absolute recoveries have been recorded in the range of 62–74% in breakfast cereals and crackers, respectively [96]. Since the initial Swedish announcement, food industry laboratories have worked intensively on the development of LC–MS-based methods to determine acrylamide in processed and cooked foods [97]. Similar to the experiences of private and official food control laboratories, problems have been encountered in the analysis of difficult matrices due to interfering compounds in the characteristic acrylamide transitions (either for the internal standard or the analyte). A promising approach is to extract the analyte into a polar organic solvent, such as ethyl acetate. Sanders et al. [87] have employed ethyl acetate to extract acrylamide from the aqueous phase (removing interfering constituents such as salt, sugars, starches, amino acids, etc.). The ethyl acetate extract can then be concentrated and analyzed by either LC–MS or GC–MS. In most cases, the LOD is significantly lowered, even approaching 10 μg/kg [22]. Similarly, an ethyl acetate extraction step can also be included after the SPE clean-up, providing a significant improvement in sensitivity, especially for more difficult matrices, such as cocoa powder and coffee. Continuous progress is being made in optimizing LC–MS methods and reducing the quantification limits. Recently, Jezussek and Schieberle [98,99] have reported a promising method by derivatizing acrylamide with 2-mercaptobenzoic acid to the thioether and measuring the resulting adduct with a single-stage mass spectrometer. This method achieves an improvement in sensitivity of approximately 100-fold versus the nonderivatized analyte. Such approaches may potentially form the basis for the development of even more simple LC-UV methods for the determination of acrylamide in foods.

13.7.2.3 Online Methods-Proton Transfer Reaction MS

Proton transfer reaction mass spectrometry (PTR-MS) has been shown to be a suitable method for rapid and online measurements of volatile compounds of headspace samples [100] It combines a soft, sensitive, and efficient mode of chemical ionization, with a quadrupole mass filter. The headspace gas is continuously introduced into the drift tube, which contains a buffer gas and a controlled ion density of H_3O+. Volatile organic compounds (VOCs) that have proton affinities larger than water are ionized in the drift tube by proton transfer from H_3O+, that is, $VOC + H_3O + \rightarrow [VOC + H] + +H_2O$. The protonated VOCs are extracted from the drift tube by a small electric field and mass analyzed in the quadrupole mass spectrometer. The four key features of PTR-MS can be summarized as follows: (a) it is fast, and time-dependent variations of headspace profiles can be monitored with a time resolution of about 0.1 s; (b) the volatiles are not subjected to work-up or thermal stress and little fragmentation is induced by the ionization step, hence, mass spectral profiles closely reflect genuine headspace distributions; (c) mass spectral intensities can be transformed into absolute headspace concentrations; (d) it is not invasive. All these features make PTR-MS particularly suited to investigate fast dynamic processes, such as formation of aroma and volatile contaminants in Maillard reactions. The applicability of the PTR-MS approach for monitoring online the formation of acrylamide was evaluated in real food systems using thermally treated potatoes as an example [101]. The mass trace at *m/z* 72 indicated the presence of acrylamide in the headspace obtained by heating a potato at 170°C. The mass at *m/z* 72 was found to be homogeneous, without interference with other volatile compounds, using an offline coupling method. Retention index and EI mass spectrum were identical with those of the acrylamide reference compound, and only one peak with the mass at *m/z* 72 was detected by PTR-MS. The EI spectrum of the compound eluting 49.6 m was conclusively identified by the Wiley EI database as acrylamide. The formation of acrylamide on heating dried potato slices at 170°C for 70 m showed a rapid initial increase, followed by a broad maximum after 6–10 m of reaction time, and, subsequently, with a slow decline of the curve. However, the amounts of acrylamide in the headspace were very low compared to the Maillard model systems [101], which is most likely due to the lower concentration of the precursors (reducing sugars and asparagine), but also to the high polarity and low volatility of acrylamide. According to literature data [102], fresh potato contains about 1000 mg/kg of free asparagine. Taking into account the high water content of ca. 80% [103], an estimated 2.5 mg (19 μmol) of asparagine was available in 0.5 g dried potatoes for generating acrylamide. Despite the low precursor amounts in the experiment with potatoes, these data show that PTR-MS is sufficiently sensitive to monitor the formation of acrylamide under food processing conditions.

In summary, mainly two methods of analysis (LC–MS or GCMS) are used by laboratories worldwide, and based on the early indications of proficiency tests, it is difficult to say that one is more reliable than the other. Limits of quantification range from 30 to 50 μg/kg for LC–MS down to 10–30 μg/kg for GC–MS. However, it is quite clear that for the analysis of acrylamide at <30 μg/kg level, GC–MS after bromination is the best approach. This approach has the advantage of adequate sensitivity with multiple ion confirmation. A further advantage of this technique is that a relatively simple benchtop GC–MS can be employed for acrylamide analysis. Application of GC–MS/MS or coupling to a high-resolution MS would even further lower the detection limit of certain foods, approaching the range of 1–2 μg/kg [104].

13.7.2.4 LC–Time-of-Flight–MS/MS

Liquid chromatography–time-of-flight (TOF)–mass spectrometry has also been established as a valuable technique for the routine control of the wholesomeness of food. In this sense, TOF techniques can record an accurate full-scan spectrum throughout the acquisition range and have resulted as an excellent tool for the unequivocal target and nontarget identification and confirmation of food contaminants [105,106].

13.7.2.5 LC–Quadrupole Linear Ion Trap (QqLIT)–MS/MS

Recently introduced tandem mass spectrometers, having both features, such as quadrupole linear ion trap (QqLIT, LTQ or Q-trap), quadrupole time-of-flight (QqTOF), LTQ-Fourier transform ion cyclotron resonance mass spectrometry (FTICR-MS), and LTQ-Orbitrap, and so on, have allowed for the development of several new methods for acrylamide detection [107,108].

13.7.2.6 Capillary Zone Electrophoresis

A CZE method has been developed for the determination of acrylamide after derivatization with 2-mercaptobenzoic acid in foodstuffs products. The previously established derivatization procedure was improved by reducing several steps, allowing direct injection of AA derivative into the CZE system without the need to remove reagent excess. With this method, a LOD of 0.07 μg/mL was obtained, which involves a 10-fold enhancement when compared with that obtained by MEEKC. Good linearity ($r^2 > 0.999$) and run-to-run and day-today precisions (RSD lower than 5.8 and 11.2%, respectively) were achieved. The addition of a derivatization step did not negatively affect the precision of the established methodology. The results obtained show that the method can be used for quantitative purposes in foodstuffs products. The application of CZE for the determination of acrylamide in french fries, breakfast cereals, and biscuits, using external calibration and standard addition methods, has been demonstrated. As a consequence, external calibration can be successfully selected as a good strategy for the determination of AA in a large variety of samples. This less time-consuming method becomes especially appropriate for a routine screening of foodstuffs with a high risk of AA contamination due to their thermal processing [109].

13.8 UPLC–MS ANALYSIS OF ACRYLAMIDE IN FOOD SAMPLES

In today's global marketplace, the safety and quality of food products are of growing concern for consumers and governments, and analytical information, including surveillance data for both recognized and newly identified contaminants, is also essential. However, information about their occurrence in food is still (very) limited [110]. Against this background, LC–MS, traditionally an important part of the medical

TABLE 13.1
Common Acrylamide Contaminants Present in a Variety of Food Determined by LC–MS

Technique	Matrix	Extraction Method	LOQ (ng/g)	Comments	Ref.
HPLC–MS	Food stuffs (potato chips and biscuit)	Acetic acid after Carrez 1 and Carrez 2 solutions	5.0	Confirmatory and quantitative. Study the formation of acrylamide during cooking	[116]
LC–MS	Food stuffs (beef, chicken, biscuits, etc.)	Hexane and filtration through 0.45_m syringe filter	–	Confirmatory and quantitative. Study the formation of acrylamide during cooking	[116]
LC–QqQ–MS/MS	Roasted chestnuts and chestnut-based foods	Water and cleaned with multimode ENV + ® SPE and eluted with methanol	4–9	Confirmatory, quantitative, and study formation of acrylamide during roasting	[117]
LC–Qq-IT–MS/MS or LC–QqQ–MS/MS	Spanish products, potato chips, pastry products, sweet fritters ("churros"), Spanish omelet	Homogenized with water, clean up with Strata-XC SPE and a ENV+ SPE and elution with MeOH:H_2O (60:40)	2–6	Confirmatory and quantitative	[118]
LC–QqQ–MS/MS	Potato, coffee, cereals	Homogenized with water, clean up with ENV+ and elution with methanol	0.5	Confirmatory and quantitative	[119]
LC–QqQ–MS/MS	Chinese traditional carbohydrate-rich foods	Ethyl acetate clean up with SPE with Bond Elut Accutat mixed mode SPE column consisting of a strong cation and strong anion exchanges into one bed and analyte elution with methanol	4	Confirmatory, quantitative $^{13}C_3$-labeled acrylamide internal standard solution	[120]
LC–QqQ–MS/MS	Processes food (rice, bread, corn chips, potato chips)	C18 SPE and analyte elution with water	2	Confirmatory and quantitative	[121]
LC–APCI–MS	Potato and cereal-based foods	0.01 mM acetic acid in a vortex mixer and clean up with Oasis MCX SPE	–	Confirmatory and quantitative	[122]

laboratory, found a growing market from a new application—food safety testing [111]. LC–MS is particularly suited for the analysis of acrylamide in food samples, since it provides a large amount of information about a complex mixture, enabling the screening, confirmation, and quantitation of hundreds of components with one analysis [112,113]. These instruments are used to test other food safety issues, such as food authenticity and labeling accuracy [114,115]. Table 13.1 illustrates examples regarding major acrylamide contaminants in food determined by LC–MS. Triple quadrupole (QqQ) MS has been the cornerstone technique for screening and confirmation of food contaminants and residues [123]. The majority of current liquid chromatography–tandem mass spectrometry (LC–MS/MS)-based contaminants and residue analysis relies on the high sensitivity and selectivity of the selected reaction monitoring (SRM) mode of QqQ–MS/MS [124,125]. Modern instruments produce high signal-to-noise (S/N) ratios even when relying on short SRM dwell times and can be properly combined with ultra-performance liquid chromatography (UPLC). Conventional LC with C18 columns plays a dominant role for the determination of foodstuff, whereas UPLC along with sub-2 μm particle C18 columns reduces run time and improves sensitivity. Comparative analyses of acrylamide in potato chip extracts, which were analyzed in two alternative LC–MS/MS systems, documented the potential of UPLC to replace "classic" LC separation strategy [126]. The data generated in optimized systems employing either Acquity UPLC (Waters) or Alliance LC (Waters) hyphenated with Quattro Premier (Waters) MS detector

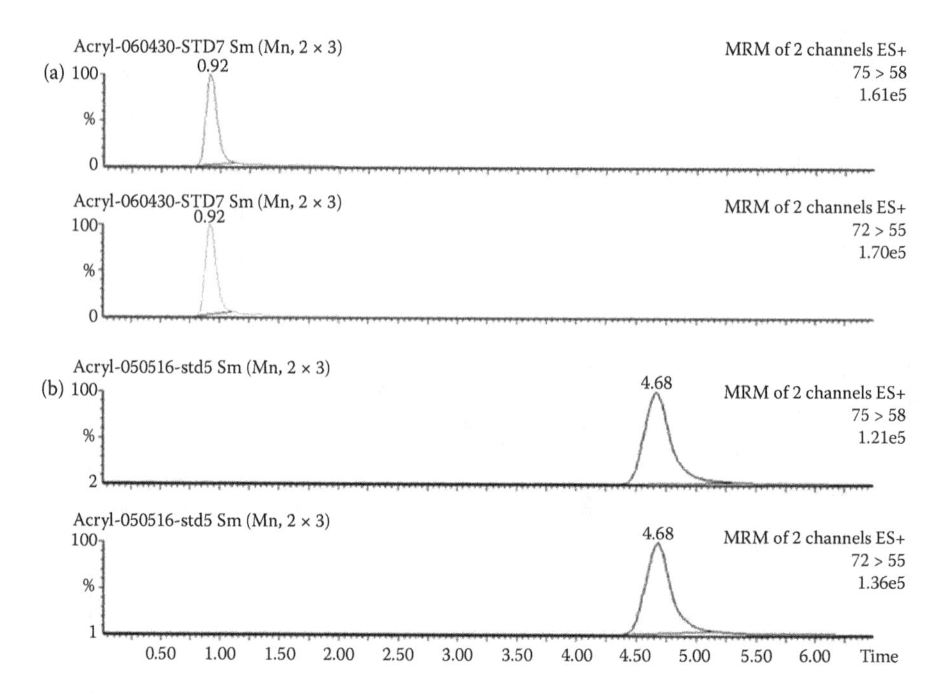

FIGURE 13.3 Comparison study on the determination of acrylamide by (a) UPLC–MS/MS and (b) HPLC–MS/MS. (Adapted from Zhang, Y. et al. 2007. *J. Chromatogr. A* 1142:194–198. With permission.)

(tandem quadrupole), showed that (i) the number of theoretical plates was for most analytes higher in a system employing LC, and with lower variability compared to UPLC, (ii) the values of height equivalent to the theoretical plate obtained in UPLC were mostly higher, however, their variability was also rather high, (iii) the analysis time in the system employing UPLC was reduced by more than 50% with similar analytical output, and (iv) UPLC provided significantly improved S/N followed by decreased LOQs for the majority of compounds [127]. Chromatogram illustrating LC–MS analysis of acrylamide in potato chips is shown in Figure 13.3. The reduced analysis time consequently resulted in significantly lower consumption of organic solvents [128].

13.9 CONCLUSION

This chapter emphasized number of recent techniques for the analysis of acrylamide in food. Despite presence of several analytical methods for acrylamide determination, these techniques still require further development. After review of this chapter, one can conclude that UPLC/MS/MS represents a powerful technique with high-resolution separation of acrylamide, short run time, and high specificity that excludes the matrix interferences of the extract.

ACKNOWLEDGMENTS

The authors would like to extend their sincere appreciation to King Saud University, Deanship of Scientific Research, College of Science Research Center for its supporting of this book chapter.

REFERENCES

1. European Commission. 2000. Risk assessment of acrylamide. (CAS No. 79–06-1, EINECS No. 201-173-7. 2003: 7–15.
2. Smith, E. A., and Oehme, F. W. 1991. Acrylamide and polyacrylamide: A review of production, use, environmental fate and neurotoxicity, *Rev. Environ. Health* 9: 215–228.
3. Swedish National Food Administration. 2002. Information about acrylamide in food. www.slv.seS.
4. Davis, L. N., Durkin, P. R., Howard, P. H., and Saxena, J. 1976. Investigation of selected potential environmental contaminants: Acrylamide. *EPA Technical Report, PB-257.* 704:1–147. Eugene, OR: EPA Press.
5. Bergmark, E. 1997. Hemoglobin adducts of acrylamide and acrylonitrile in laboratory workers smokers and non-smokers, *Chem. Res. Toxicol.* 10: 78–84.
6. Dybing, E., and Sanner, T. 2006. Risk assessment of acrylamide in foods, *Toxicol. Sci.* 75(1):7–15.
7. EFSA. 2009. Scientific report of EFSA prepared by data collection and exposure unit (DATEX) on Monitoring of acrylamide levels in food, *The EFSA Scientific Report*, 285:1–26.
8. Boon, P. E, de Mul, A., van der Voet, H., van Donkersgoed, G., Brette, M., and van Klaveren, J. D. 2005. Calculation of dietary exposure to acrylamide, *Mutat. Res.*, 580:143–155.
9. Dybing, E., Farmer, P. B., Andersen, M., Fennell, T. R., Lalljie, S. P., Muller, D. J., et al. 2005. Human exposure and internal dose assessments of acrylamide in food, *Food Chem. Toxicol.*, 43:365–410.

10. WHO. 2005. *Summary Report of the Sixty-Fourth Meeting of the Joint FAO/WHO Expert Committee on Food Additive (JECFA)*. Rome, Italy: The ILSI Press International Life Sciences Institute. Washington, DC, pp. 1–47.
11. Norris, N. V. 1967. Acrylamide, In: F. D. Snell, C. L. Hilton (Eds.), *Encyclopedia of Industrial Chemical Analysis*, Interscience, New York, pp. 160–168.
12. Ashoor, S. H., and Zent, J. B. 1984. Maillard browning of common amino acids and sugars, *J. Food Sci.* 49:1206–1207.
13. Eriksson, S. 2005. Acrylamide in food products: Identification, formation and analytical methodology, PhD thesis, Department of Environmental Chemistry, Stockholm University, Stockholm, Sweden.
14. Friedman, M. 2003. Chemistry, biochemistry, and safety of acrylamide. A review, *J. Agric. Food Chem.* 5:4504–4526.
15. Bergmark, E. Calleman, He, C. J. F., and Costa, L. G. 1993. Determination of hemoglobin adducts in humans occupationally exposed to acrylamide, *Toxicol. Appl. Pharmacol.* 120:45–54.
16. Urban, M., Kavvadias, D., Riedel, K., Scherer, G., and Tricker, A. R. 2006. Urinary mercapturic acids and a hemoglobin adduct for the dosimetry of acrylamide exposure in smokers and nonsmokers, *Inhal. Toxicol.*18:831–839.
17. Tritscher, A. M. 2004. Human health risk assessment of processing-related compounds in food, *Toxicol. Lett.* 149:177–186.
18. EEC, Commission Directive 92/39/EEC, amending Directive 90/128/EEC.1992. Relating to plastic materials and articles intended to come into contact with foodstuffs, *Off. J. Eur. Comm.* L330:21–29.
19. Commission Regulation, COMMISSION REGULATION (EU) No 10/2011 of 14 January 2011 on plastic materials and articles intended to come into contact with food, *Off. J. Eur. Union* L12:1–89.
20. Advances in Molecular Toxicology, Volume 6. 2012, http://dx.doi.org/10.1016/B978-0-444-59389-4.00005-7. Institute of Food Science, Technology and Nutrition (ICTAN-CSIC), Madrid, Spain. Elsevier B.V. ISSN 1872-0854.
21. Yaylayan, V. A., Wnorowski, A., and Locas Perez, C. 2003.Why asparagine needs carbohydrates to generate acrylamide, *J. Agric. Food Chem.* 51:1753–1757.
22. Zyzak, D. V., Sanders, R. A., Rtojanovic, M., Tallmadge, D. H., Eberhardt, B. L., Ewald, D. K., Gruber, D. C., Morsch, T. R., Strothers, M. A., Rizzi, G. P., and Villagran, A. D. 2003.Acrylamide formation mechanism in heated foods, *J. Agric. Food Chem.* 51:4782–4787.
23. Granvogl, M., Jezussek, M., Koehler, P., and Schieberle, P. 2004. Quantitation of 3-aminopropionamide in potatoes—A minor but potent precursor in acrylamide formation, *J. Agric. Food Chem.* 52:4751–4757.
24. Yaylayan, V. A. Perez Locas, C., Wnorowski, A., and O'Brien, J. 2004. The role of creatine in the generation of N-methylacrylamide: a new toxicant in cooked meat, *J. Agric. Food Chem.* 52:5559–5565.
25. Buhlert, J., Carle, R., Majer, Z., and Spitzner, D. 2006. Thermal degradation of peptides and formation of acrylamide, *Lett. Org. Chem.* 3:356–357.
26. Claus, A., Weisz, G. M., Schieber, A., and Carle, R. 2006. Pyrolytic acrylamide formation from purified wheat gluten and gluten-supplemented wheat bread rolls, *Mol. Nutr. Food Res.* 49: 87–93.
27. Noti, A., Biedermann-Brem, S., Biedermann, M., Grob, K., Albisser, P., and Realini, P. 2003. Storage of potatoes at low temperature should be avoided to prevent increased acrylamide formation during frying or roasting, *Mitt. Lebensm. Hyg.* 94:167–180.
28. Yasuhara, A., Tanaka, Y., Hengel, M., and Shibamoto, T. 2003. Gas chromatographic investigation of acrylamide formation in Browning model systems, *J. Agric. Food Chem.* 51:3999–4003.

29. Gold, B. G., and Schaumburg, H. H. 2000. Acrylamide, In: P. S. Spencer, H. H. Schaumburg (Eds.), *Experimental and Clinical Neurotoxicology*, second ed., Oxford University Press, New York, pp. 124–132.
30. Mottram, D. S., Wedzicha, B. L., and Dodson, A. T. 2012. Acrylamide is formed in the Maillard reaction, *Nature* 419:448–449.
31. Calleman, C. J. Bergmark, E., and Costa, L. G. 1990. Acrylamide is metabolized to glycidamide in the rat: Evidence from hemoglobin adduct formation, *Chem. Res. Toxicol.* 3:406–412.
32. Friedman, M. A., Dulak, L. H., and Stedham, M. A. 1995. A lifetime oncogenicity study in rats with acrylamide, *Fundam. Appl. Toxicol.* 27:95–105.
33. Marsh, G. M., Lucas, L. J., and Youk, A. O. 1999. Schall, Mortality patterns among workers exposed to acrylamide: 1994 follow up, *Occup. Environ. Med.* 56:181–190.
34. IARC (International Agency for Research on Cancer.1 1994. some industrial chemicals, IARC Monogr. Eval. *Carcinog. Risk Chem. Hum.* 60:389–433.
35. Gamboa da Costa, G., Churchwell, M. I., Hamilton, L. P., von Tungeln, L. S., Beland, F. A., Marques, M. M., and Doerge, D. R. 2003. DNA adduct formation from acrylamide via conversion to glycidamide in adult and neonatal mice, *Chem. Res. Toxicol.* 16:1328–1337.
36. Dearfield, K. L., Douglas, G. R., Ehling, U. H., Moore, M. M., Sega, G. A., and Brusick, D. J. 1995.Acrylamide: A review of its genotoxicity and an assessment of heritable genetic risk, *Mutat. Res.* 330:71–99.
37. Ahn, J. S., and Castle, L. 2003. Test for the depolymerization of polyacrylamides as a potential source of acrylamide in heated foods, *J. Agric. Food Chem.* 51:6715–6718.
38. Zhang, Y., Zhang, G., and Zhang, Y. 2005. Occurrence and analytical methods of acrylamide in heat-treated foods: Review and recent developments, *J. Chromatogr. A* 1075:1–21.
39. Johnson, K. A., Gorzinski, S. J., Bodner, K. M., Campbell, R. A., Wolf, C. H., Friedman, M. A., and Mast, R. W. 1986. Chronic toxicity and oncogenicity study on acrylamide incorporated in the drinking water of Fischer 344 rats, *Toxicol. Appl. Pharmacol.* 85:154–168.
40. Abramsson-Zetterberg, L. 2003. The dose–response relationship at very low doses of acrylamide is linear in the flow cytometer-based mouse micronucleus assay, *Mutat. Res.* 535:215–222.
41. EEC, Council Directive 98/83/EC on the quality of water intended for human consumption, *Off. J.* L330. 1998. 32–54.
42. Hagmar, L., Wirfa, E., Paulsoon, B., and Tornqvist, M. 2005. Differences in haemoglobin adduct levels of acrylamide in the general population with respect to dietary intake, smoking habits and gender, *Mutat. Res.* 580:157–165.
43. Albin, M., Tornqvist, M., Tinnerberg, H., Kautiainen, A., Eriksson, A., Magnusson, A. L., Gustavsson, C., Bjorkner, B., Isaksson, M., and Hagmar, L.1998. Results of halsounderso increase in housing Hallandsas—Possible exposure to emissions of Rhoca-Gil, besvaroch hemoglobin adducts of acrylamide. Rapport 1998-04-08 In Tunnel Commission, Kring Hallandsasen. Interim report of the Tunnel Commission, National Public Inquiries ((Governmental Official Report), 60, Stockholm: Government of Sweden, Office for Administrative Affairs.
44. Tareke, E., Rydberg, P., Karlsson, P., Eriksson, S., and Tornqvist, M. 2000. Acrylamide: A cooking carcinogen, *Chem. Res. Toxicol.* 13:517–522.
45. Becalski, A., Lau, B. P. Y., Lewis, D., and Seaman, S. W. 2003. Acrylamide in foods: Occurrence, sources, and modeling, *J. Agric. Food Chem.* 51:802–808.
46. Leung, K. S., Lin, A., Tsang, C. K., and Yeung, S. T. 2003. Acrylamide in Asian foods in Hong Kong, *Food Addit. Contam.* 20:1105–1113.
47. WHO, World Health Organization, Guidelines for Drinking-Water Quality, Fourth Edition. 2011. Available at http://whqlibdoc.who.int/publications/2011/9789241548151_eng.pdf.

48. He, F. S., Zhang, S. L., Wang, H. L., Li, G., Zhang, Z. M., Li, F. L., Dong, X. M., and Hu, F. R. 1989. Neurological and electroneuromyographic assessment of the adverse effects of acrylamide on occupationally exposed workers, *Scand. J. Work Environ. Health* 15:125–129.
49. WHO. 2002. FAO/WHO Consultation on The Health Implications of Acrylamide in Food. Summary Report of a Meeting Held in Geneva, 25–27 June 2002, Geneva, Switzerland: World Health Organization, http://www.who.int/foodsafety/publications/chem/acrylamide_june2002/en/.
50. Zeleny, R., Ulberth, F., Gowik, P., Polzer, J., van Ginkel, L. A., and Emons, H. 2006. Developing new reference materials for effective veterinary drug-residue testing in food-producing animals, *Trends Anal. Chem.* 25:927–936.
51. Hamlet, C. G., and Sadd, P. A. 2005. Effects of yeast stress and pH on 3-monochloropropanediol (3-MCPD)-producing reactions in model dough systems, *Food Additives Contam.*, 22:616–623.
52. Calleman, C. J., Wu, Y., He, F., Tian, G., Bergmark, E., Zhang, S., Deng, H., Wang, Y., Crofton, K. M., Fennell, T., and Costa, L. G. 1994. Relationships between biomarkers of exposure and neurological effects in a group of workers exposed to acrylamide, *Toxicol. Appl. Pharmacol.* 126:361–371.
53. Calleman, C. J. 1996. The metabolism and pharmacokinetics of acrylamide: Implications for mechanisms of toxicity and human risk estimation, *Drug Metab. Rev*. 28:527–590.
54. LoPachin, R. M., Schwarcz, A. I., Gaughan, C. L. Mansukhani, S., and Das, S. 2003. *In vivo* and *in vitro* effects of acrylamide on synaptosomal neurotransmitter uptake and release, *Neurotoxicology* 25:349–363.
55. LoPachin, R. M. 2004. The changing view of acrylamide neurotoxicity, *Neurotoxicology* 25:617–630.
56. Hashimoto, K., and Tanii, H. 1985. Mutagenicity of acrylamide and its analogues in *Salmonella typhimurium*, *Mutat. Res.* 158:129–133.
57. Shipp, A., Lawrence, G., Gentry, R., McDonald, T., Bartow, H., Bounds, J., Macdonald, N., Clewell, H., Allen, B., and van Landingham, C. 2006. Acrylamide: Review of toxicity data and dose–response analyses for cancer and noncancer effects, *Crit. Rev. Toxicol.* 36:481–608.
58. Tyl, R. W., and Friedman, M. A. 2003. Effects of acrylamide on rodent reproductive performance, *Reprod. Toxicol.* 17:1–13.
59. NTP-CERHR Monograph on the Potential Human Reproductive and Developmental. Effects of Acrylamide. 2005 Center for the Evaluation of Risks to Human Reproduction, National Toxicology Program, Research Triangle Park, NC, NIH Publication No. 05-4472, pp. i-III-76.
60. National Toxicology Program, Report on carcinogens, Twelfth Edition, 2011 pp. 25–28.
61. Rice, J. M. 1995. The carcinogenicity of acrylamide, *Mutat. Res.* 580:3–20.
62. Besaratinia, A., and Pfeifer, G. P. 2005. DNA adduction and mutagenic properties of acrylamide, *Mutat. Res.* 580:31–40.
63. Tareke, E., Twaddle, N. C., McDaniel, L. P., Churchwell, M. I., Young, J. F., and Doerge, D. R. 2006. Relationships between biomarkers of exposure and toxicokinetics in Fischer 344 rats and B6C3F1 mice administered single doses of acrylamide and glycidamide and multiple doses of acrylamide, *Toxicol. Appl. Pharmacol.* 217:63–75.
64. Ahn, J. S., Castle, L., Clarke, D. B., Lloyd, A. S., Philo, M. R., and Speck, D. R. 2002. Verification of the findings of acrylamide in heated foods, *Food Addit. Contam.*, 19:1116–1124.
65. Ghanayem, B. I., McDaniel, L. P., Churchwell, M. I., Twaddle, N. C., Snyder, R., Fennell, T. R., and Doerge, D. R. 2005. Role of CYP2E1 in the epoxidation of acrylamide to glycidamide and formation of DNA and hemoglobin adducts, *Toxicol. Sci.* 88:311–318.

66. WHO, World Health Organization, Guidelines for Drinking-Water Quality, Fourth Edition. 2011. Available at http://whqlibdoc.who.int/publications/2011/9789241548151_eng.pdf.
67. Marvin, H. J. P., Kleter, G. A., Frewer, L. J., Cope, S., Wentholt, M. T. A., and Rowe, G. 2009. A working procedure for identifying emerging food safety issues at an early stage: Implications for European and international risk management practice, *Food Control*. 20:345–356.
68. Schillhorn van Veen, T. W. 2005. International trade and food safety in developing countries. *Food Control* 16:491–496.
69. FAO/WHO Food Standards, FAO and WHO, Rome, Italy, 2009.
70. Anese, M., Suman, M., and Nicoli, M. C. 2010. Acrylamide removal from heated foods, *Food Chem.*, 119:791–794.
71. Capuano, E., Ferrigno, A., Acampa, I., Ait-Ameur, L., and Fogliano, V. 2008. Characterization of Maillard rection in bread crisps, *Eur. Food Res. Technol.*, 228:311–319.
72. Ciaa-Confederation of the Food and Drink Industries of the EU. 2009. CIAA acrylamide "toolbox"- REV 12-February 2009. www.ciaa.be/documents/acrylamide.
73. JECFA. 2007. Joint FAO/WHO expert committee on food additives (JECFA): Report on 68th meeting. Geneva.
74. Cook, N. R., Cutler, J. A., Obarzanek, E., Buring, J. E., Rexrode, K. M., Kumanyika, S. K., et al. 2007. Long term effects of dietary sodium reduction on cardiovascular disease outcomes: Observational follow-up of the trials of hypertension prevention (TOHP) 1, *Br. Med. J.* 3347599:885–888.
75. Capuano, E., Ferrigno, A., Acampa, I., Serpen, A., Açar, Ö. Ç., Gökmen, V., et al. 2009. Effect of flour type on Maillard reaction and acrylamide formation during toasting of bread crisp model systems and mitigation strategies, *Food Res. Int.* 42:1295–1302.
76. Gökmen, V., and Senyuva, H. Z. 2006a. A simplified approach for the kinetic characterization of acrylamide formation in fructose-asparagine model system, *Food Additives Contaminants*, 23(4):348–354.
77. De Vleeschouwer, K., Van der Plancken, I., Van Loey, A., and Hendrickx, M. 2006. Impact of pH on the kinetics of acrylamide formation/elimination reactions in model systems, *J. Agric. Food Chem.*, 54:7847–7855.
78. Biedermann, M., Biedermann-Brem, S., Noti, A., and Grob, K. 2002. Methods for determining the potential of acrylamide formation and its elimination in raw materials for food preparation, such as potatoes, *Mitteilungen aus Lebensmitteluntersuchung und Hygiene*, 93:653–667.
79. Corradini, M. G., and Peleg, M. 2006. Linear and non-linear kinetics in the synthesis and degradation of acrylamide in foods and model systems, *Critical Reviews in Food Science and Nutrition,* 46:489–517.
80. Kolek, E., Simon, P., and Simko, P. 2007. Nonisothermal kinetics of acrylamide elimination and its acceleration by table salt—A model study, *Journal of Food Science*, 72(6), E341–E344.
81. Zhang, Y., and Zhang, Y. 2008. Effect of natural antioxidants on kinetic behaviour of acrylamide formation and elimination in low-moisture asparagine-glucose model system, *J. Food Eng.*, 85:105–115.
82. Knol, J., Viklund, G. Å. I., Linssen, J. P. H., Sjöholm, I. M., Skog, K. I., et al. 2008. A study on the use of empirical models to predict the formation of acrylamide in potato crisps, *Mol. Nutr. Food Res.*, 52:313–321.
83. DeVleeschouwer, K., Van der Plancken, I., Van Loey, A., and Hendrickx, M. 2009a. Role of precursors on the kinetics of acrylamide formation and elimination under low-moisture conditions using a multiresponse approach—Part I: Effect of the type of sugar *Food Chem.*, 114:116–126.

84. Sanders, R. A., Zyzak, D. V., Stojanovic, M., Tallmadge, D. H., Eberhart, B. L., and Ewald, D. K. 2002. An LC/MS acrylamide method and it's use in investigating the role of asparagine. *Presentation at the Annual AOAC International Meeting*, Los Angeles, CA, September 22–26.
85. Zhong, C. Y. Chen, D. Z., and Xi, X. L. 2004. *Guangdong Chem. Eng*. 46:15.
86. Young, M. S., Jenkins, K. M., and Mallet, C. R. 2004. Solid-phase extraction and cleanup procedures for determination of acrylamide in fried potato products by liquid chromatography/mass spectrometry, *J. AOAC Int.* 87:961–964.
87. Fauhl, C., Klaffke, H., Mathar, W., Palvinskas, R., and Wit-tkowski, R. 2002. Acrylamide Interlaboratory Study 2002,, http://www. bfr.bund.de/cms/detail.php?template=internet de index js.
88. Takatsuki, S., Nemoto, S., Sasaki, K., and Maitani, T. J. 2003. *Food Hyg. Soc. Jpn*. 44:89.
89. Rosén, J., and Hellenäs, K.-E. 2002. Analysis of acrylamide in cooked foods by liquid chromatography tandem mass spectrometry, *The Analyst*, 127:880–882.
90. Wenzl, T., de la Calle, B., and Anklam, E. 2003. Analytical methods for the determination of acrylamide in food products: A review, *Food Addit. Contam.*, **20**:885–902.
91. Gertz, C., and Klostermann, S. 2002. Analysis of acrylamide and mechanisms of its formation in deep-fried products, *Eur. J. Lipid Sci. Technol.*, 104:762–771.
92. Biedermann, M., Biedermann-Brem, S., Noti, A., Grob, K., Egli, P., and M¨andli, H. 2002a. Two GC-MS methods for the analysis of acrylamide in foods, *Mitt. Lebensm. Hyg.* 93:638–652.
93. JIFSAN Infonet www.acrylamide-food.org/JIFSAN/NCFST Workshop "Acrylamide in Food, scientific issues, uncer-tainties, and research strategies," 28–30th October 2002. Rosemont, USA. www.jifsan.umd.edu/acrylamide/acrylamideworkshop.html
94. Clarke, D. B., Kelly, J., and Wilson, L. A. 2002. Assessment of performance of laboratories in determining acrylamide in crisp bread, *J. AOAC Int.*, 85:1370–1373.
95. Pittet, A., Périsset, A., and Oberson, J.-M. 2003. Trace level detection of acrylamide in cereal-based foods by gas chromatography mass spectrometry, *J. Chrom. A.*, 1035:123–130.
96. Riediker, S., and Stadler, R. H. 2003. Analysis of acrylamide in food using isotope-dilution liquid chromatography coupled with electrospray ionization tandem mass spectrometry, *J. Chrom. A.*, 1020:121–130.
97. Roach, J. A. G., Andrzejewski, D., Gay, M. L., Nortrup, D., and Musser, S. M. 2003. Rugged LC-MS/MS survey analysis for acrylamide in foods, *J. Agric. Food Chem.*, 51:7547–7554.
98. Jezussek, M., and Schieberle, P. 2003a. Entwicklung neuer Methoden zur Bestimmung von Acrylamid. *Lebensmittelchemie*, 57(73–104):85–86.
99. Jezussek, M., and Schieberle, P. 2003b. A new LC/MS-method for the quantitation of acrylamide based on a stable isotope dilution assay and derivatization with 2-mercaptobenzoic acid. Comparison with two GC/MS methods, *J. Agric. Food Chem.*, 51:7866–7871.
100. Yeretzian, C., Jordan, A., Brevard, H., and Lindinger, W. 2002. Time-resolved headspace analysis by proton-transfer-reaction mass-spectrometry. In *Flavor Release.* Eds. Roberts, D. D., Taylor, A. J. ACS Symposium Series 763. American Chemical Society: Washington, DC, pp. 58–72.
101. Pollien, P., Lindinger, C., Yeretzian C., and Blank, I. 2003. Proton transfer reaction mass spectrometry is a suitable tool for on-line monitoring of acrylamide in food and Maillard systems, *Anal. Chem.*, 75:5488–5494.
102. Martin, F. L., and Ames, J. M. 2001. Formation of Strecker aldehydes and pyrazines in a fried potato model system, *J. Agric. Food Chem.*, 49:3885–3892.
103. Souci, S. W., Fachmann, W., and Kraut, H. 2000. In: *Food Composition and Nutrition Tables*, pp. 639–641, 6th edn., CRC Press: Boca Raton, FL.

104. Taeymans, D. W. John. 2004., A review of acrylamide: An industry perspective on research, analysis, formation, and control, *Critical Reviews in Food Science and Nutrition*, 44:323–347.
105. García-Reyes, J. F., Ferrer, C., Gomez-Ramos, M. J., Molina-Diaz A. and Fernandez-Alba A.R. 2007. Determination of pesticide residues in olive oil and olives, *TrAC—Trends Anal. Chem.* 26:239–251.
106. Garcia-Reyes, J. F. M. D., Hernando, A., Molina-Diaz, A. R., Fernandez-Alba. 2007. Comprehensive screening of target, non target and unknown pesticides in food by LC-TOF-MS, *Trends Anal. Chem.* 26:828–841.
107. Barcel, D., and Petrovic, M. 2007. Challenges and achievements of LC-MS in environmental analysis: 25 years on, *Trends Anal. Chem.* 26:2–11.
108. Fernandez-Alba, A. R., and Garcia-Reyes, J. F. 2008. Large-scale multi-residue methods for pesticides and their degradation products in food by advanced LC-MS, *Trends Anal. Chem.* 27:973–990.
109. Bermudo, E., Nunez, O., and Puignou, L. 2006. Analysis of acrylamide in food samples by capillary zone electrophoresis, *J. Chromatogr. A.*1120:199–204.
110. International Commission on Microbiological Specifications for Foods (ICMSF). 2006. *Food Control* 17,825.
111. Hobbs, J. E., Fearne, A., and Spriggs, J. 2002. Incentive structures for food safety and quality assurance: An international comparison, *Food Control* 13:77–81.
112. Nunez, O., Moyano, E., and Galceran, M. T. 2005. LC-MS/MS analysis of organic toxics in food, *Trends Anal. Chem.* 24:683–703.
113. Pico, Y., Font, G., Ruiz, M. J., and Fernandez, M. 2006. Control of Pesticide residues by liquid chromatography-mass spectrometry to ensure food safety, *Mass Spectrom. Rev.* 25:917–960.
114. Monaci, L., and Visconti, A. 2009. Mass spectrometry-based proteomics methods for analysis of food allergens, *Trends Anal. Chem.* 28:581–591.
115. Westenbrink, S., Oseredczuk, M., Castanheira, I., and Roe, M. Food composition databases: The EuroFIR approach to develop tools to assure the quality of the data compilation process. *Food Chem.* 113: 759–67.
116. Kaplan, O., Kaya, G., Ozcan, C., Ince, M., and Yaman, M. 2009. Acrylamide concentrations in grilled foodstuffs of Turkish kitchen by high performance liquid chromatography-mass spectrometry. *Microchem. J.* 932009:173–179.
117. Karasek, L., Wenzl, T., and Anklam, E. 2009 Determination of acrylamide in roasted chestnuts and chestnut-based foods by isotope dilution HPLC-MS/MS. *Food Chem.* 114:1555–1558.
118. Bermudo, E., Moyano, E., Puignou, L., and Galceran, M. T. 2008. Liquid chromatography coupled to tandem mass spectrometry for the analysis of acrylamide in typical Spanish products. *Talanta* 76:389–394.
119. Rosqn, J., Nyman, A., and Hellen, K. E. 2007. Retention studies of acrylamide for the design of a robust liquid chromatography-tandem mass spectrometry method for food analysis, *J Chromatogr. A.* 1172:19–24.
120. Zhang, Y., Ren, Y., Zhao, H., and Zhang, Y. 2007. Determination of acrylamide in Chinese traditional carbohydrate-rich foods using gas chromatography with micro-electron capture detector and isotope dilution liquid chromatography combined with electrospray ionization tandem mass spectrometry, *Anal. Chim. Acta.* 584:322–332.
121. Kim, C. T., Hwang, E. S., and Lee, H. J. 2007. An improved LC-MS/MS method for the quantitation of acrylamide in processed foods, *Food Chem.* 101:2007 401.
122. Wenzl, T., Karasek, L., Rosen, J., Hellenaes, K. E., Crews, C., Castle, L., and Anklam, E. 2006. Collaborative study of two methods, one based on high performance liquid chromatography and on gas chromatography–tandem mass spectrometry for the determination of acrylamide in bakery and potato products, *J. Chromatogr.* A.1132:211.

123. Debrauwer, L., Chevolleau, S., Zalko, D., Paris, A., and Tulliez, J. 2005. *Sci. Alim.* 25:273.
124. Songsermsakul, P., and Razzazi-Fazeli, E. 2008. A Review of recent trends in applications of liquid chromatography-mass spectrometry for determination of mycotoxins. *J. Liquid Chromatogr. Rel. Technol.* 31:1641–1686.
125. G. S. Shephard, 2008. Determination of mycotoxins in human foods, *Chem. Soc. Rev.* 37:2468.
126. Kovalczuk, T., Jech, M., Poustka, J., and Hajslova, J. 2006. Ultra-performance liquid chromatography-tandem mass spectrometry: A novel challenge in multiresidue pesticide analysis in food, *Anal. Chim. Acta* 577(1):8–17.
127. Ashok Kumar Malik, Cristina Blasco, and Yolanda Picó, 2010. Liquid chromatography–mass spectrometry in food safety. *J. Chromatogr. A*, 1217:4018–4040.
128. Zhang, Y., Jiao, J., Cai, Z., Zhang, Y., and Ren, Y. 2007. An improved method validation for rapid determination of acrylamide in foods by ultra-performance liquid chromatography combined with tandem mass spectrometry, *J. Chromatogr. A* 1142:194–198.

14 Determination of Procyanidins and Alkaloids in Cocoa and Biological Samples by Ultra-High-Performance Liquid Chromatography Coupled to Tandem Mass Spectrometry

Alba Macià, Maria José Motilva, and Nàdia Ortega

CONTENTS

14.1 INTRODUCTION

In recent years, the studies of cocoa and cocoa products (cocoa powder, dark chocolate, and cocoa liquor) and their related products have become an area of interest given their health-promoting properties due to the presence of bioactive compounds such as the flavanols (catechins and proanthocyanins), the flavonoids subgroup, and alkaloids (methylxanthines). Several *in vitro* and *in vivo* studies have suggested that polyphenols may protect against many degenerative diseases. A relation has been found in some of these between the consumption of cocoa derivatives and their cardiovascular effect [1,2].

These minor components of cocoa have become an intense focus of research interest because of their demonstrated beneficial effect on health. The interest has been recently reinforced by the scientific opinion on the substantiation of a health claim related to cocoa flavanols and the maintenance of normal endothelium-dependent vasodilation issued by the European Food Safety Authority (EFSA) Panel on Dietetic Products, Nutrition and Allergies (NDA) [3]. In order to obtain the claimed effect, 200 mg of cocoa flavanols should be consumed daily. This amount could be provided by 2.5 g of high-flavanol cocoa powder or 10 g of high-flavanol dark chocolate. This reinforces the scientific interest in the development of sensitive and rapid analytical methods for determining these minor components in cocoa, cocoa products, and chocolate. Another important aspect is related to the development of analytical methodologies for sensitive and selective identification and quantification of the main metabolites of cocoa polyphenols in biological samples in order to understand their bioavailability and their contribution to health.

However, the extraction, separation, identification, and analysis of these minor compounds of cocoa is still challenging due to their chemical complexity and the complexity of the cocoa process. These processes include fermenting and drying cocoa beans to develop the characteristic flavors, and the complex cocoa process in chocolate production, including roasting, grinding, alkalizing, and conching. These steps lead to a change in the phenolic composition of the cocoa bean, which increases the complexity of these minor fractions and requires sensitive and selective methods to analyze them.

14.1.1 Cocoa Flavanols: Catechins and Proanthocyanidins

Phenolic compounds or polyphenols constitute one of the most numerous and ubiquitous groups of plant phytochemicals in our diet. Phenolic compounds are organic chemicals characterized by the presence of at least one aromatic ring with one or more hydroxyl groups attached. The term "polyphenols" should be used to define phenolic compounds with more than one phenolic ring and devoid of any nitrogen-based functional group in their most basic structural expression [4]. Polyphenols are classified into families according to the differences on their carbon skeleton number and arrangement of their carbon atoms, ranging from simple small, single aromatic ring structures to the complex and heavy condensed tannins [5–7].

Dietary flavonoids, commonly present in the epidermis of leaves and the skin of fruits, are the most numerous and widespread phenolic compounds. Flavonoids are

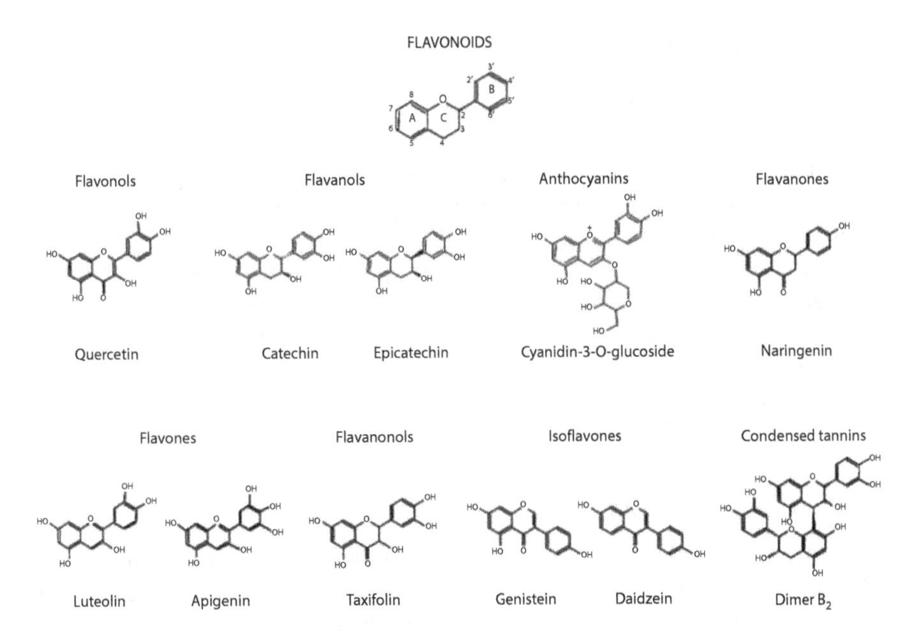

FIGURE 14.1 Classification, chemical structures, and some examples of flavonoids.

characterized by a C_{15} phenylchromane core, composed of two aromatic rings linked by a three-carbon bridge (C_6–C_3–C_6) (Figure 14.1) [8,9] and are subclassified into flavonols (e.g., quercetin), flavanols (e.g., catechin and epicatechin), anthocyanins (e.g., cyanidin-3-*O*-glucoside), flavanones (e.g., naringenin), flavones (e.g., luteolin and apigenin), flavanonols (e.g., taxifolin), and isoflavones (e.g., genistein or daidzein) [10–13], which are sometimes classified into an independent subcategory apart from flavonoids [7]. The majority of flavonoids present a hydroxylation pattern, usually in the 4′, 5-, and 7-position (Figure 14.1), or a glycosilation pattern that reflects a biological strategy in plant cells to increase their water solubility. The presence of methyl groups or isopentyl units may give a lipophilic character to flavonoid molecules [8].

Condensed tannins or proanthocyanidins are high-molecular-weight polymers. The monomeric unit is a flavan-3-ol (e.g., catechin and epicatechin), with a flavan-3,4-diol as its precursor (Figure 14.1). Oxidative condensation occurs between carbon C-4 in the heterocycle and carbons C-6 or C-8 of adjacent units [14]. However, most of the literature on the condensed tannin contents refers only to oligomeric proanthocyanidins (dimers, trimers, and tetramers) because of the difficulty of analyzing highly polymerized molecules. Proanthocyanidins, however, can occur as polymers with a degree of polymerization of 50 and more.

The polyphenols in cocoa (*Theobroma cacao* L.) and cocoa products can be attributed mainly to procyanidins (flavan-3-ols) (37%) [15]. The procyanidins identified range in size from monomers (catechin and epicatechin) to long polymers (with a degree of polymerization higher than 10 or decamers). The concentrations of polyphenols in cocoa products depend on their origin [16] and the cocoa processing conditions [14,17]. Epicatechin has been identified as the main monomer procyanidin in cocoa products. However, the epicatechin content varies according to the manufacturing process due to epimerization of this compound to catechin [18]. Alkalization (or dutching) of cocoa powder also influences the procyanidin contents [19,20]. Given the above-mentioned factors, the nature and especially the concentration of procyanidins in cocoa products may vary. For instance, Adamson et al. [19] quantified different amounts depending on the cocoa product. They found 19.4 mg of procyanidins per gram of cocoa liquor, 1.7 mg of procyanidins per gram of chocolate with high cocoa content, and 0.7 mg of procyanidins per gram of milk chocolate. A recent study also identified and quantified small quantities of flavonoids (quercetin, kaempferol, naringenin, luteolin, apigenin, and their glycoside forms) and phenolic acids (caffeic, coumaric, protocatechuic, and ferulic acids) in cocoa beans, cocoa nibs, cocoa liquor, and cocoa powder [20].

14.1.2 Cocoa Alkaloids

Similar to polyphenols, alkaloids are products of the secondary metabolism of plants, which have been identified in hundreds of plant species with great structural diversity [21]. Methylxanthines derived from purine nucleotides are known collectively as purine alkaloids. Caffeine, theophylline, and theobromine alkaloids are methylated xanthine derivatives. Figure 14.2 shows their chemical structures and, as can be observed, these structures differ only in the number and the position of one methyl substituent and/or hydrogen atom around the xanthine ring system.

These three alkaloids are well-known compounds, as they are present in everyday foods and beverages, such as cocoa, drinking chocolate, tea, and cola, as well as in pharmaceutical products. In cocoa products, theobromine is the major alkaloid, followed by caffeine, which is found in small quantities. Slight traces of theophylline have been identified; therefore, it is not considered relevant and is not reviewed in the cocoa products [22,23]. The range of methylxanthines also depends on variables such as the origin of the cocoa beans, the fermentation process, and the cocoa production process. Thus, for instance, defatted cocoa beans may contain about 4% and 0.2% of theobromine and caffeine, respectively [24]. Regarding their physiological effects in humans, methylxanthines have been related to various body systems, mainly the central nervous system, but also the cardiovascular, renal, and respiratory systems [25,26].

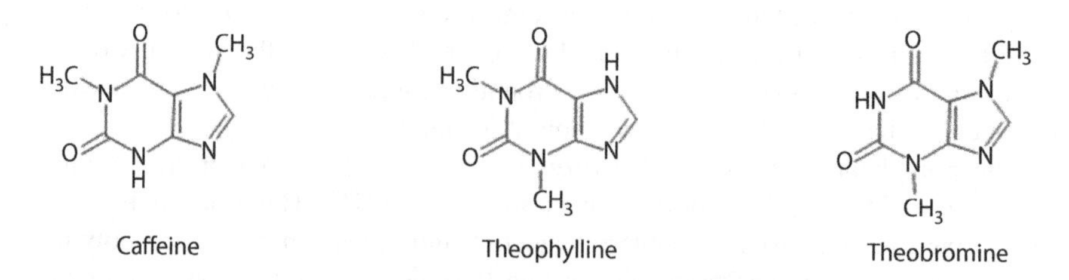

FIGURE 14.2 Chemical structures of alkaloids (methylated xanthine derivatives).

In the following sections, the sample pretreatment techniques and the UHPLC–MS/MS conditions for determining procyanidins and alkaloids in cocoa samples and their metabolites in biological samples are reported.

14.2 SAMPLE PRETREATMENT

In order to determine the procyanidins and alkaloids in cocoa and their metabolites in biological samples, a sample has to be pretreated prior to chromatographic analysis to extract the target compounds and clean up the sample matrix. For the analysis of biological samples, it is very important to concentrate the generated procyanidin and alkaloid metabolites, as these are present at low concentrations in these complex matrices. In addition, cleaning up the cocoa and biological samples is essential in order to remove the interference compounds, as these could suppress the ionization.

Different sample pretreatment strategies have been reported depending on whether the sample matrix is a cocoa or biological sample. These are examined next.

14.2.1 Cocoa Samples

For the analysis of cocoa samples, the beans were ground using a mortar and pestle [27] or a laboratory mill [28,29] to obtain a homogeneous material. Then, to analyze procyanidins and alkaloids, these were first defatted between three and four times with *n*-hexane to remove the fat from the cocoa matrix. Second, an off-line liquid–liquid extraction (LLE) was performed to extract the procyanidins and alkaloids from this matrix. Various solutions formed by acetone were used as the extraction solutions. These were acetone/water (70/30, v/v) [27], acetone/water/acetic acid (70/28/2, v/v/v) [18], and acetone/water/acetic acid (70/29.5/0.5, v/v/v) [28,29] (see Table 14.1). Afterwards, the extracts were sonicated [18] or vortexed [28,29] and centrifuged. The supernatants were combined from the centrifuged tubes and filtered with glass wool [28,27]. Finally, the organic solvent of these extracts was removed by rotary evaporation at a temperature between 30°C and 40°C.

This sample pretreated is the only one reported in the literature for determining procyanidins in cocoa samples. This is tedious and uses a large volume of solvent, both *n*-hexane and acetone, and takes a long time to extract these target compounds. This time required for defatting and extracting is much higher than in UHPLC separations. Normally, the use of this cocoa pretreatment requires 1 or 2 h or more to extract the compounds, while the UHPLC separations only require a few minutes. Further studies should be carried out in order to improve the cocoa sample pretreatment for the analysis of procyanidins.

14.2.2 Biological Samples

To determine procyanidins, alkaloids, and their generated metabolites in biological samples (fluids and tissues), it is necessary to preconcentrate them given that they are found at low concentrations. Additionally, only a small volume of the biological sample is usually available.

TABLE 14.1
UHPLC–MS/MS Methodologies for Determining Procyanidins and Alkaloids in Cocoa Samples

Cocoa Samples	Extraction Solution	Analytical Column	Mobile Phase	MS Analyzer	Analysis Time (min)	Validation	Identification/ Quantification	Ref.
Chocolates	Acetone/water/ acetic acid (70/28/2, v/v/v)	BEH C_{18} (50 × 2.1 mm, 1.7 μm)	(A) water/THF/TFA (98/2/0.1, v/v/v) (B) 0.1% Formic acid in acetonitrile	QTOF	8	NIST Certified Reference Material (CRM)	Quantification	[18]
Cocoa beans Cocoa nibs Cocoa liquor Cocoa powder	Acetone/water/ acetic acid (70/29.5/0.5, v/v/v)	HSS T3 (100 × 2.1 mm, 1.7 μm)	(A) 0.2% Acetic acid (B) Acetonitrile	QqQ	12.5	—	Quantification	[28]
Cocoa beans	Acetone/water (70/30, v/v)	Comprehensive 2D Zorbax SB C_{18} (50 × 4.6 mm, 1.8 μm)	(A) 0.1% Formic acid (B) Acetonitrile	QTOF	20	—	Identification	[27]
Cocoa nibs	Acetone/water/ acetic acid (70/29.5/0.5, v/v/v)	HSS T3 (100 × 2.1 mm, 1.7 μm)	(A) 0.2% Acetic acid (B) Acetonitrile	QqQ	12.5	Spiking the extraction solution with different procyanidin and alkaloid concentrations	Quantification	[29]

Note: THF, tetrahydrofuran.

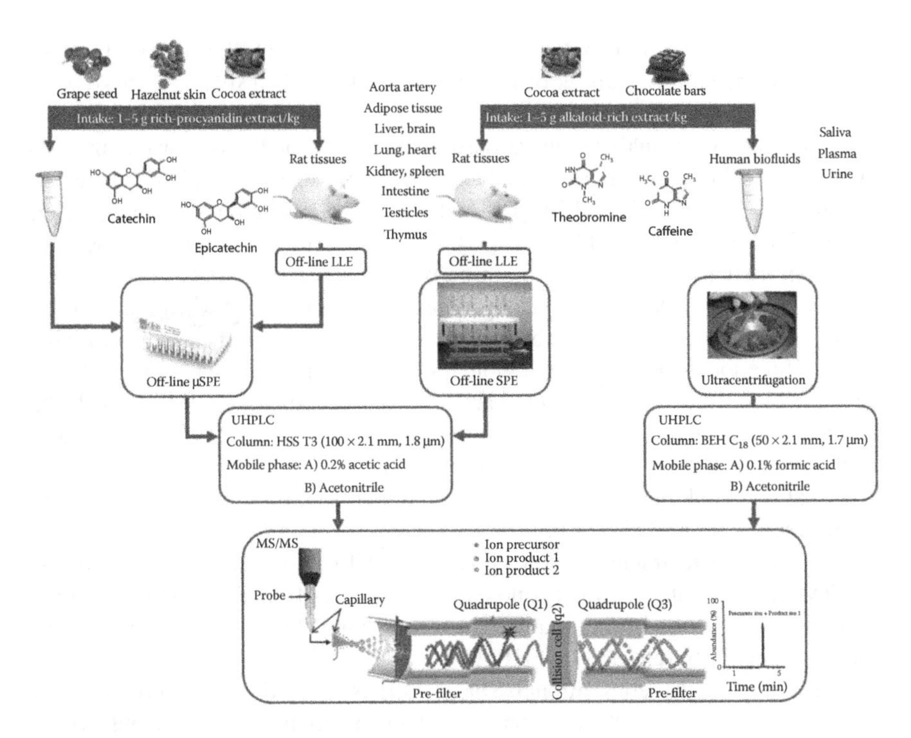

FIGURE 14.3 Schematic representation of the sample pretreatment strategies for determining procyanidins and alkaloids and their metabolites in biological samples by UHPLC–MS/MS.

Figure 14.3 shows a schematic representation of the different sample pretreatment strategies reported in the literature for determining procyanidins and alkaloids and their metabolites in biological samples when UHPLC–MS/MS is the analytical separation technique and the detection system.

A range of sample pretreatment strategies have been reported in the literature depending on whether the compounds analyzed are procyanidins or alkaloids, and also the complexity of the biological matrix, either plasma or tissue. Prior to the sample pretreatment, the biological samples (plasma and tissues) are mixed with phosphoric acid in order to break possible polyphenols–plasmatic or tissue protein linkages [30–39].

The sample pretreatment used to determine procyanidins and their metabolites in plasma samples (from rats) is off-line solid-phase extraction (SPE) with hydrophilic–lipophilic balanced (HLB) copolymer. Both cartridges (60 mg) [31,32] and microelution SPE plates (µSPEs) (2 mg) [33,36,37] have been used as the device format. For the analysis of tissues (also from rats), such as the liver, brain, aorta artery, and adipose tissue, these were first freeze-dried and then pretreated by off-line LLE (with the solution made up of methanol, water, and phosphoric acid) followed by µSPE [35–37]. The use of two sample pretreatments (LLE and µSPE) is due to the complexity of the tissue sample compared with the biofluid plasma sample. The extraction recoveries (%Rs) for determining the procyanidins catechin, epicatechin, dimer B_2, and trimer in the plasma samples by µSPE [33] and SPE [31] were higher than 83%, with the exception of the trimer, which was 65% in SPE [31]. Similar %Rs were obtained from the analysis of tissue samples by LLE-μSPE, and these were higher than 81%.

The alkaloids were analyzed in rat tissues and human biofluids. For the analysis of tissues (liver, brain, aorta artery, and adipose tissue), these were also first freeze-dried and then pretreated by off-line LLE (also with the solution of methanol, water, and phosphoric acid) followed by SPE [35]. With this sample pretreatment, the %Rs of caffeine, theophylline, and theobromine were between 62% and 100%, according to the tissue analyzed [35]. To analyze human biofluids, such as saliva, plasma, and urine, these were pretreated by sample dilution with commercially available ultracentrifugation devices. In this sample pretreatment, the biofluids were first centrifuged at high speed to remove any particulate matter from the fluid. Second, the supernatant was filtered in a 10-kDa molecular weight cutoff (MWCO) filter and centrifuged at high speed. Finally, the low-MW filtrate was diluted 20-fold [40]. The %Rs of caffeine and theobromine in these biofluids from this sample pretreatment were between 73% and 118% [40].

This sample pretreatment dilution with ultracentrifugation devices [40] and μSPE [33] allows the procyanidins and alkaloids to be extracted from the biological matrix in a short analysis time of less than 15 min. Until now, these two sample pretreatments have only been reported in UHPLC.

Compared with the conventional SPE, μSPE is a rapid sample pretreatment technique because the postextraction solvent evaporation and the reconstitution steps, necessary to concentrate the analytes in SPE with cartridges, are eliminated. The device format of μSPE is a 96-well plate and this technique allows the sample throughput to be increased. Therefore, the use of μSPE plates instead of SPE cartridges as the device format has great potential in terms of speed and sample throughput. Another additional advantage of this sample pretreatment technique is that a small amount of biological sample can be analyzed, up to 350 μL of sample in comparison with the 1000 μL usually used in SPE.

14.3 UHPLC–MS/MS METHODS

Over the last few years, due to improvements in the packing material of the chromatographic columns, the use of UHPLC technique has improved the reported HPLC methodologies in terms of peak efficiency, peak resolution, speed, sensitivity, and solvent consumption. However, this technique must operate at high backpressures [41–43]. Another advantage of the UHPLC technique is that, given this high peak resolution, when it is combined with MS as the detector system, it is less susceptible to matrix effects (MEs), one of the main problems when the MS is used to quantify minor components in food samples. This is because an efficient UHPLC separation may contribute to a reduction in ion suppression, when this is only produced by the coelution of two different compounds [44].

The following sections examine the chromatographic UHPLC separations and MS conditions used to determine procyanidins and alkaloids in cocoa samples and also procyanidins, alkaloids, and their metabolites in biological samples obtained after procyanidin-rich or alkaloid-rich extract consumptions. Tables 14.1 and 14.2 show a

TABLE 14.2
Quality Parameters for Determining Procyanidins and Alkaloids in Biological Samples by UHPLC–MS/MS

Compounds	Mobile Phase	Stationary Phase	Biological Sample	Sample Pretreatment	%R	LOD (nM)	LOQ (nM)	Ref.
Procyanidins								
Catechin, epicatechin, dimer B_2, trimer	0.2% Acetic acid Acetonitrile	HSS T3 (100 × 2.1 mm, 1.8 µm)	Rat plasma	Off-line SPE	65–102	3–800	10–980	[31]
Catechin, epicatechin, dimer B_2, trimer	0.2% Acetic acid Acetonitrile	HSS T3 (100 × 2.1 mm, 1.8 µm)	Rat plasma	Off-line µSPE	83–100	7–700	10–1380	[33]
Catechin, epicatechin, dimer B_2	0.2% Acetic acid Acetonitrile	HSS T3 (100 × 2.1 mm, 1.8 µm)	Rat liver	Off-line LLE-µSPE	85–100	0.4–0.08[a]	0.2–1.0[a]	[35]
Catechin, epicatechin, dimer B_2	0.2% Acetic acid Acetonitrile	HSS T3 (100 × 2.1 mm, 1.8 µm)	Rat brain	Off-line LLE-µSPE	84–100	0.3–0.06[a]	0.2–0.9[a]	[35]
Catechin, epicatechin, dimer B_2	0.2% Acetic acid Acetonitrile	HSS T3 (100 × 2.1 mm, 1.8 µm)	Rat aorta artery	Off-line LLE-µSPE	100	3.0–0.07[a]	0.4–11.4[a]	[35]
Catechin, epicatechin, dimer B_2	0.2% Acetic acid Acetonitrile	HSS T3 (100 × 2.1 mm, 1.8 µm)	Rat adipose tissue	Off-line LLE-µSPE	100	1.6–0.03[a]	0.1–6.1[a]	[35]
Alkaloids								
Theophylline, theobromine, caffeine	0.2% Acetic acid Acetonitrile	HSS T3 (100 × 2.1 mm, 1.8 µm)	Liver	Off-line LLE-SPE	75–82	0.4–0.9[a]	1.0–3.3[a]	[35]
Theophylline, theobromine, caffeine	0.2% Acetic acid Acetonitrile	HSS T3 (100 × 2.1 mm, 1.8 µm)	Rat brain	Off-line LLE-SPE	57–92	0.3–1.1[a]	0.9–4.6[a]	[35]
Theophylline, theobromine, caffeine	0.2% Acetic acid Acetonitrile	HSS T3 (100 × 2.1 mm, 1.8 µm)	Rat aorta artery	Off-line LLE-SPE	62–100	1.1–13.3[a]	3.0–35.8[a]	[35]

continued

TABLE 14.2 (continued)
Quality Parameters for Determining Procyanidins and Alkaloids in Biological Samples by UHPLC–MS/MS

Compounds	Mobile Phase	Stationary Phase	Biological Sample	Sample Pretreatment	%R	LOD (nM)	LOQ (nM)	Ref.
Theophylline, theobromine, caffeine	0.2% Acetic acid Acetonitrile	HSS T3 (100 × 2.1 mm, 1.8 μm)	Rat adipose tissue	Off-line LLE-SPE	63–79	1.1–5.7[a]	3.4–19.0[a]	[35]
Theobromine, caffeine	0.1% Formic acid Acetonitrile	BEH C_{18} (50 × 2.1 mm, 1.7 μm)	Human saliva	Dilution and ultracentrifugation	95–100	1250	2500	[40]
Theobromine, caffeine	0.1% Formic acid Acetonitrile	BEH C_{18} (50 × 2.1 mm, 1.7 μm)	Human plasma	Dilution and ultracentrifugation	94–95	1250	2500	[40]
Theobromine, caffeine	0.1% Formic acid Acetonitrile	BEH C_{18} (50 × 2.1 mm, 1.7 μm)	Human urine	Dilution and ultracentrifugation	111–112	1250	2500	[40]

[a] nmol/g tissue.

summary of the UHPLC–MS/MS methods reported in the literature for determining, respectively, procyanidins and alkaloids in cocoa and in biological samples.

14.3.1 Chromatographic Separations

The reversed-phase mode was used to determine procyanidins and alkaloids in cocoa and in biological samples by the UHPLC technique. The reported UHPLC columns were the ethylene bridged hybrid (BEH) C_{18} [18,40] and the high-strength silica (HSS) T3 [28,29–39] from Waters, and the Zorbax SB C_{18} [27] from Agilent. These UHPLC columns have an internal diameter of 2.1 mm, a length between 50 and 100 mm, and a particle size between 1.7 and 1.8 μm. The Zorbax SB C_{18} column, with an inner diameter of 4.6 mm, was an exception [27]. The most widely used is the HSS T3, and its particles are characterized by being 100% silica. It was reported that most compounds are more strongly retained in HSS T3 columns than in BEH C_{18} columns [45].

Unlike a UHPLC system, in the HPLC system, both the reversed-phase [46,47] and the normal-phase [29,48–50] mode have reportedly been used to determine procyanidins and alkaloids in cocoa samples. This is explained because, until now, no UHPLC columns with polar stationary phase (normal-phase mode) with sub-2 μm have been developed. Initial reports for the determination of procyanidins and alkaloids in cocoa samples by HPLC technique were based on the normal-phase mode, and monomer procyanidins through decamer procyanidins were resolved. In this normal phase, two different stationary phases were reported, an unmodified silica column [29,50] and a dihydroxypropyl-bonded silica [48,49].

The particles in the UHPLC columns are reported to be totally porous silica. Recently, a new generation of analytical columns, known as core–shell columns, has been developed. These columns have a solid core and a porous shell, and their particles have an inner diameter of 2.6 μm. The reduced depth of the outer porous layer limits the diffusion path of the analytes and the mass transfer resistance and the peak broadening are minimized [51]. The great advantage of these columns is that the high peak efficiency and high separations in a reduced analysis time are also achieved, but using the conventional HPLC system because no excessive backpressure is generated. Magi et al. [52] used a Kinetex C_{18} (from Phenomenex) column to determine procyanidins in cocoa liquors. The internal diameter and length of this analytical column are 2.1 and 100 mm, respectively. These columns seem to be very promising and could be a good competitor for UHPLC columns.

The mobile phase used in the UHPLC separations for the analysis of procyanidins and alkaloids is acidified water and an organic modifier solvent as phases A and B, respectively. The pH of phase A of the mobile phase is normally kept below three by the addition of a small amount of formic acid [27,40], acetic acid [28,29–39], and trifluoroacetic acid (TFA) [18]. Its concentration is kept as low as possible in order to ensure satisfactory ionization, and it is between 0.1% and 0.2%. Marti et al. [33] demonstrated that as the percentage of acetic acid was increased from 0.2% to 10% in the mobile phase, the sensitivity of the procyanidins decreased significantly. This fact was explained because the high percentage of acetic acid suppressed the ionization of these compounds and no good ionization was obtained.

Acetonitrile was chosen as phase B in the mobile phase, and the separation of the procyanidin and alkaloids compounds was accomplished by gradient elution [18,28,29–40].

By using these chromatographic conditions, the procyanidins are eluted according to their degree of polymerization, from monomers to decamers, in the analysis of cocoa samples. The procyanidins with a degree of polymerization higher than 10, also known as procyanidin polymers, elute after decamers as a single peak. At present, any LC methods for analyzing or separating this chromatographic peak have been reported. With regard to alkaloids, caffeine and theobromine are eluted according to their polarity, first theobromine and then caffeine, and, additionally, are always eluted before the procyanidins. For the analysis of biological samples, the procyanidin metabolites are also eluted according to their polarity. First, catechin and epicatechin in its glucuronide forms, then methyl catechin and methyl epicatechin in its glucuronide forms, then catechin and epicatechin in its sulfate forms, and, finally, methyl catechin and methyl epicatechin in its sulfate forms.

The total analysis time for the determination of the studied compounds under these UHPLC conditions was between 12.5 min [18,28,29] and 20 min [27] for the analysis of cocoa samples, and between 3.1 min [40] and 12.5 min [30–39] for the analysis of biological samples. Although the procyanidins and alkaloids and their metabolites were eluted in a shorter analysis time, some time was necessary to clean up the column with phase B and reequilibrate the column to return to the initial conditions.

In contrast, when these compounds are analyzed by HPLC–MS/MS, high resolution is required and longer analysis time is reported. The high resolution is required to resolve the chromatographic peaks of different procyanidins related to this degree of polymerization for the analysis of cocoa samples, and the different procyanidins metabolites generated for the analysis of biological samples. In HPLC, the resolution can be achieved by increasing the column length up to 250 mm. However, as a consequence, the retention time and band broadening are also increased. The long analysis time, which is higher than an hour per sample, is an important limitation in research or quality control, when a high number of samples have to be analyzed.

To analyze cocoa samples by HPLC, in addition to the long analysis time reported for the analysis of procyanidins with a higher degree of polymerization, when the normal-phase mode was used, the monomers catechin and epicatechin were not resolved. In addition, the use of the unmodified silica column in this normal-phase mode requires the use of toxic chlorinated solvents that are considered an ecological hazard [53].

In addition to the advantages of the UHPLC technique over HPLC, including its higher peak efficiency, higher peak resolution, and speed, the mobile phase consumption is also reduced. As a consequence of the lower inner diameter and length of UHPLC columns, the flow rate in UHPLC is 0.4 mL/min, while in HPLC, it is 1 mL/min. This lower flow rate in UHPLC results in a lower organic solvent (mobile phase) consumption, around 80% less in HPLC, and, therefore, UHPLC could also be considered cost-effective. Additionally, the sample volume injection in UHPLC is also reduced in comparison with the conventional HPLC, it being around 3 and 20 μL, respectively.

These different advantages of the UHPLC technique compared with HPLC were reported by Ortega et al. [29] for determining procyanidins and alkaloids in cocoa nibs. These authors used the HSS T3 in the reversed-phase mode (100 × 2.1 mm, 1.7 μm) as the UHPLC column and the Luna Silica in the normal-phase mode (2504.6 mm, 5 μm) as the HPLC column. By means of the UHPLC system, the analysis time was reduced sevenfold, from 80 min in the conventional HPLC to 12.5 min in UHPLC. The efficiency of the peaks in UHPLC–MS/MS allowed the nonamer procyanidins to be identified while HPLC–MS/MS allowed identification up to the heptamers. The quantification limits (LOQs) of catechin, epicatechin, and dimer B_2 in the UHPLC system were between 10 and 90 μg/L, lower than in HPLC, where they were between 20 and 200 μg/L.

Additionally, in this study, several differences were observed between these two chromatographic modes, the normal phase (HPLC) and reversed-phase (UHPLC) for the analysis of procyanidins and alkaloids. First, the monomers catechin and epicatechin were not resolved very well in the normal phase, and these compounds coeluted in a single peak. These two monomers were completely separated by using the reversed phase. Second, the elution order of the procyanidins and alkaloids was changed, as was expected. In the reserved-phase mode, the procyanidins and alkaloids were eluted according to their polarity and molecular size. Therefore, caffeine eluted after theobromine, and epicatechin eluted after catechin. In contrast, in the normal-phase mode, these compounds were eluted according to their apolarity and increasing size. As a consequence, caffeine eluted before theobromine, and epicatechin eluted before catechin. In all the analyses, the alkaloids eluted before the monomer procyanidins.

Apart from these UHPLC–MS/MS methodologies, which were based on a single-dimension (1D-LC) [18,28,29] system, UHPLC–MS/MS technique has also been applied by using a comprehensive two-dimensional LC (2D-LC) system to determine procyanidins in cocoa samples [27]. Typically, this system combines two analytical columns with different separation mechanisms in order to achieve better separation. The overall analytical time is determined by the speed of the second column, and this speed should be as fast as possible and with good resolution. In order to achieve this fast speed, short columns packed with small particles, such as UHPLC columns, are used. Therefore, a combination of HPLC (first column) with UHPLC (second column) offers a good configuration for a comprehensive 2D system. Kalili et al. [27] used the Zorbax SB-C_{18} (50 × 4.6 mm, 1.8 μm) as the second column (UHPLC column). With this methodology, higher resolution was obtained by a comprehensive 2D-LC in comparison to a single 1D-LC. Additionally, higher screening and selectivity was obtained by the use of the quadrupole-time-of-flight (Q-TOF) detection system. This 2D-LC system has only been applied to characterize and identify procyanidins in cocoa samples (see Table 14.1).

14.3.2 Mass Spectrometry

With the aim of determining procyanidins with their different degree of polymerization (from monomers to decamers) in cocoa samples, and determining procyanidins and their metabolites in biological samples, after an intake of a procyanidin or alkaloid-rich extract, the analytical method must be sensitive, selective, and reliable. This is due to the low concentrations in which compounds are found in complex biological matrices (biofluids and tissues), and the different chromatographic peaks of the procyanidins in their different degrees of polymerization in complex food samples (cocoa samples).

UHPLC combined with tandem MS has a great potential in terms of sensitivity and selectivity, and this UHPLC–MS/MS technique has been successfully applied to determine procyanidins and alkaloids in cocoa samples [18,28,27,29] and their metabolites in biological samples [30–40]. Table 14.1 shows a summary of the different applications of the UHPLC–MS/MS technique for the analysis of cocoa samples, and Tables 14.3 and 14.4 for the determination of procyanidins and alkaloids, respectively, in biological samples.

These applications have been reported by using the Acquity triple quadrupole detector (TQD) [28–39], the Micromass® Quattro Premier triple quadrupole (QqQ) system [40], the Micromass Q-TOF II [18], and the Q-TOF Ultima [27] from Waters. As can be seen in these tandem MS equipments, two different analyzers have been reported, QqQ [28–40], the same as the two analyzers, and Q-TOF [18,27] as the hybrid system. The QqQ analyzer was the only one reported for the analysis of biological samples [30–40]. Generally, when the QqQ analyzer was used, quantification studies were performed due to their higher sensitivity in the selected ion monitoring (SRM) mode, and when the Q-TOF was used, identification or characterization was done due to its higher power for confirmation purposes.

Electrospray ionization (ESI) was used as the ionization technique. The procyanidins in the negative mode and the alkaloids in the positive mode were analyzed in a single analysis.

The determination of procyanidins in cocoa samples using tandem MS requires the use of single, double, and triple ions to ionize according to the degree of polymerization (molecular weight). Singly charged molecular ions, $[M–H]^-$, were the most abundant for the analysis of monomers until tetramers at *m/z* 289.1, 577.3, 865.5, and 115.7, respectively. For the analysis of pentamers and hexamers, the doubly charged ions at *m/z* 720.4 and 864.5 were the predominant ions. In addition, doubly charged ions were the most abundant for the analysis of heptamers through nonamers as well as for dodecamers, while triply charged species were reported to be the most intense ions for decamers and undecamers [28,29,46,48,50].

On the other hand, different strategies were reported for the identification of the different procyanidin metabolites generated in biological samples using this detector system. First, the full-scan mode by MS mode was applied in order to know the pseudomolecular ion (precursor ion), and different cone voltages were applied. Second, the daughter scan and neutral loss scan modes were applied in tandem MS. With these two modes (daughter scan mode and neutral loss mode), selective fragments of each precursor ion were obtained by applying different collision energies, which generated fragments (product ions). Additionally, these two modes have been reported to be excellent tools for verifying structural information about the compounds when standards are not available. In daughter scan experiments, the daughter or product ions are produced by collision-activated dissociation of the selected precursor ion in the collision cell, usually with argon. Neutral loss scans at *m/z* 176, 80, and 14 were

TABLE 14.3
Application of UHPLC–MS/MS for Determining Procyanidins and Their Metabolites in Biological Samples Obtained at Different Times after the Consumption of Different Procyanidin

Procyanidin Extract	Dose (g/kg)	Biological Sample	Time after Consumption	Sample Pretreatment	Main Generated Metabolites	Concentration (μM)	Ref
Grape seed	1	Rat plasma	2 h	Off-line SPE	Trimer	8.55	
					Catechin glucuronide	23.90	
					Epicatechin glucuronide	20.57	
					Methyl catechin glucuronide	13.75	
					Methyl epicatechin glucuronide	9.06	[31]
Grape pomace	5	Rat plasma	4 h	Off-line μSPE	Catechin glucuronide	33.78	
					Epicatechin glucuronide	85.67	
					Methyl catechin glucuronide	7.26	
					Methyl epicatechin glucuronide	16.27	[33]
Hazelnut skin	0.5	Rat plasma	2 h	Off-line μSPE	Catechin glucuronide	0.30	
					Methyl catechin glucuronide	0.30	[40]
Grape pomace	1	Rat liver	4 h	Off-line LLE-μSPE	Epicatechin	13.6[a]	
					Methyl epicatechin glucuronide	13.5[a]	
					Catechin sulfate	16.1[a]	
					Epicatechin sulfate	14.0[a]	
					Methyl catechin sulfate	32.8[a]	
					Methyl epicatechin sulfate	30.3[a]	[35]
Grape pomace	1	Rat brain	4 h	Off-line LLE-μSPE	Catechin glucuronide	2.12[a]	
					Epicatechin glucuronide	5.48[a]	
					Methyl catechin glucuronide	1.87[a]	
					Methyl epicatechin glucuronide	1.60[a]	[35]
Grape pomace	1	Rat aorta artery	4 h	Off-line LLE-μSPE	Dimer	1.05[a]	[35]

continued

TABLE 14.3 (continued)
Application of UHPLC–MS/MS for Determining Procyanidins and Their Metabolites in Biological Samples Obtained at Different Times after the Consumption of Different Procyanidin

Procyanidin Extract	Dose (g/kg)	Biological Sample	Time after Consumption	Sample Pretreatment	Main Generated Metabolites	Concentration (μM)	Ref
Grape pomace	1	Rat adipose tissue	4 h	Off-line LLE-μSPE	Dimer	0.17[a]	[35]
Hazelnut skin	0.5	Rat liver	2 h	Off-line LLE-μSPE	Methyl catechin glucuronide	8.00[a]	[40]
Hazelnut skin	0.5	Rat brain	2 h	Off-line LLE-μSPE	Methyl catechin sulfate	6.40[a]	[40]
Hazelnut skin	5	Rat plasma	2 h	Off-line μSPE	Catechin glucuronide	1763	[36]
					Epicatechin glucuronide	154	
					Methyl catechin glucuronide	1103	
					Trimer	1748	
Hazelnut skin	5	Rat thymus	2 h	Off-line LLE-μSPE	Methyl catechin glucuronide	2.7[a]	[36]
Hazelnut skin	5	Rat intestine	2 h	Off-line LLE-μSPE	Catechin glucuronide	42[a]	
					Methyl catechin glucuronide	218[a]	
					Dimer	27[a]	
					Trimer	6.8[a]	[36]
Hazelnut skin	5	Rat lung	2 h	Off-line LLE-μSPE	Epicatechin	59[a]	
					Catechin glucuronide	19[a]	
					Methyl catechin glucuronide	23[a]	[36]
Hazelnut skin	5	Rat kidney	2 h	Off-line LLE-μSPE	Catechin glucuronide	5.1[a]	
					Methyl catechin glucuronide	13[a]	
					Methyl catechin sulfate	1.8[a]	[36]
Hazelnut skin	5	Rat spleen	2 h	Off-line LLE-μSPE	Methyl catechin glucuronide	1.5[a]	[36]
Hazelnut skin	5	Rat testicle	2 h	Off-line LLE-μSPE	Catechin glucuronide	2.2[a]	
					Methyl catechin glucuronide	2.3[a]	[36]

Grape seed	5–50	Rat plasma	21 days	Off-line µSPE	Catechin glucuronide	1.5–21	
					Epicatechin glucuronide	0.40–6.7	
					Methyl catechin glucuronide	1.7–11	
					Methyl epicatechin glucuronide	0.28–1.5	[37]
Grape seed	5–50	Rat muscle	21 days	Off-line LLE-µSPE	Catechin glucuronide	15–28[a]	
					Epicatechin glucuronide	2.0–26[a]	[37]
Grape seed	5–50	Rat liver	21 days	Off-line LLE-µSPE	Methyl catechin glucuronide	n.d.–78[a]	
					Methyl epicatechin glucuronide	n.d.–51[a]	[37]
Grape seed	5–50	Rat brown adipose tissue	21 days	Off-line LLE-µSPE	Catechin glucuronide	n.q.–44[a]	
					Epicatechin glucuronide	n.d.–57[a]	
					Methyl catechin glucuronide	n.d.–29[a]	[37]
Grape seed	5–50	Rat mesenteric adipose tissue	21 days	Off-line LLE-µSPE	Catechin glucuronide	n.q.–31[a]	
					Epicatechin glucuronide	8.9.–56[a]	
					Methyl catechin glucuronide	n.d.–16[a]	
					Methyl epicatechin glucuronide	n.d.–46[a]	[37]
Grape seed	5–50	Rat brown adipose tissue	21 days	Off-line LLE-µSPE	Catechin glucuronide	n.d.–41[a]	
					Epicatechin glucuronide	n.d.–37[a]	
					Methyl catechin glucuronide	n.d.–34[a]	
					Methyl epicatechin glucuronide	n.d.–35[a]	[37]

[a] nmol analyte/g tissue.
n.d., not detected; n.q., not quantified.

TABLE 14.4
Application of UHPLC–MS/MS for Determining Alkaloids in Biological Samples Obtained at Different Times after the Consumption of Cocoa and Chocolate

Alkaloid Extract	Dose	Biological Sample	Time after Consumption (h)	Sample Pretreatment	Main Generated Metabolites	Concentration (μM)	Ref.
Nuts-cocoa cream	1.5 g/kg	Rat large intestine	18	Off-line LLE	Caffeine	95[a]	[34]
Cocoa	1 g/kg	Rat liver	4	Off-line LLE-SPE	Theobromine	3.82[a]	[35]
					Caffeine	5.24[a]	
Cocoa	1 g/kg	Rat brain	4	Off-line LLE-SPE	Theobromine	25.6[a]	[35]
					Caffeine	2.36[a]	
Cocoa	1 g/kg	Rat aorta artery	4	Off-line LLE-SPE	Theobromine	289[a]	[35]
					Caffeine	27.2[a]	
Cocoa	1 g/kg	Rat adipose tissue	4	Off-line LLE-SPE	Theobromine	n.q.	[35]
					Caffeine	n.q.	
Chocolate bars	41 g	Human saliva	1.5	Dilution and ultracentrifugation	Theobromine	27.2–39.3	[40]
					Caffeine	3.4–12.5	
Chocolate bars	41 g	Human plasma	1.5	Dilution and ultracentrifugation	Theobromine	43.2–67.5	[40]
					Caffeine	4.6–25.3	
Chocolate bars	41 g	Human urine	1.5	Dilution and ultracentrifugation	Theobromine	131.9–449.4	[40]
					Caffeine	4.2–24.7	

[a] nmol analyte/g tissue.
n.d., not detected; n.q., not quantified.

used to characterize the glucuronide, sulfate, and methyl conjugate forms, respectively. Then, once these generated metabolites were identified, they were quantified and the mode SRM was used as the most sensitive. In all the reported UHPLC–MS/MS methodologies for determining procyanidins and alkaloids and their metabolites, the analyzer used was QqQ.

Another strategy reported for confirming the presence of the conjugated forms of procyanidins and their metabolites is to perform an enzymatic treatment of β-glucuronidase and sulfatase before the sample analysis, thus generating the respective free forms. However, this strategy has only been applied in HPLC–MS/MS [54–57].

As it is very well known, the ionization process is susceptible to matrix signal suppression or enhancements when the ESI as the ionization technique is used, and it can severely compromise the quantitative results. The UHPLC–MS/MS instrumental response from a standard can differ significantly from matrix samples. This occurrence is named ME. In the reported UHPLC–MS/MS, the ME was also evaluated for determining procyanidins and alkaloids in cocoa samples [29] and biological samples [31,33,35]. Either a positive or negative effect was observed, lower than 17%, which means an increase or decrease in the detector response, respectively. Nevertheless, this effect was considered small. Owing to the complexity of the biological sample matrix, the sample preparation and the use of an exhaustive sample extraction step is very important to maintain high sensitivity and signal reproducibility to qualitatively and quantitatively determine, for instance, procyanidins and alkaloids and their metabolites, at very low concentration levels.

14.3.3 High-Resolution Mass Spectrometry

Owing to its sensitivity, mass accuracy, and rapid full scanning mode, high-resolution MS (HR-MS), such as TOF, Q-TOF, and Orbitrap, has emerged as an ideal tool for profiling complex samples. Only two studies for the analysis of procyanidins and alkaloids by UHPLC and using Q-TOF as the HR-MS have been reported in the literature [18,27].

Cooper et al. [18] and Kalili et al. [27] used a UHPLC with a Q-TOF system as the HR-MS and a hybrid system to analyze procyanidins in cocoa samples. By using these systems, higher screening and selectivity was obtained. As is very well known, this hybrid system (Q-TOF) combines the advantages of both different analyzers, and offers a great potential for the screening, selectivity, confirmation, and, above all, the structural elucidation of unknown compounds in complex matrices, such as cocoa samples, due to the mass accuracy of a characteristic fragment ion in studied compounds.

14.4 ANALYSIS OF PROCYANIDINS AND ALKALOIDS IN COCOA AND BIOLOGICAL SAMPLES

14.4.1 Cocoa Samples

The quantification of procyanidins is complex due to the lack of commercial procyanidin standards. At present, only monomers (catechin and epicatechin), dimers

(B_1 and B_2) and, recently, trimers (C_1) are commercially available, but dimer B_1 and trimer C_1 are very expensive. Therefore, the procyanidins, which are not commercially available, were tentatively quantified compared with catechin or epicatechin. As examples, dimer procyanidins were tentatively quantified by using the calibration curve of the dimer B_2 [29] or epicatechin [18], and the procyanidins with a degree of polymerization higher than 2 were tentatively quantified by using the calibration curve of catechin [29] or epicatechin [18]. Catechin, epicatechin, and the caffeine and teobromine alkaloids were quantified by using their own calibration curves.

In order to obtain reliable results for quantification purposes, the quality parameters of the methods developed need to be studied. Only two developed UHPLC–MS/MS methods have been validated in the literature. One study was validated by using the National Institute of Standards and Technology (NIST) Certified Reference Material (CRM) [18], and the second by spiking the extraction solution with different concentrations of procyanidin and alkaloid [29]. The LOQs obtained for catechin and epicatechin were 11.2 and 65.5 μg/g of chocolate when the NIST CRM was used [18]. On the other hand, when the extraction solution was spiked with the target analytes standards, the detection limits (LODs), and LOQs of the procyanidins (catechin, epicatechin, and the dimer B_2) were between 7 and 30 μg/L and 10 and 90 μg/L, respectively, and for the caffeine and teobromine alkaloids, these values were below 30 and 100 μg/L, respectively [26].

Different cocoa procyanidin extracts obtained from cocoa beans, cocoa nibs, cocoa liquor, and cocoa powder were analyzed, and their concentrations of procyanidins ranged from 6 to 30 mg/g of lyophilized extract [28,29]. These cocoa sources corresponded to the Forastero variety from Ghana (West Africa). The cocoa extract from the nibs was the most abundant in procyanidins, followed by the extracts from bean and cocoa liquor, and the lower procyanidin concentration was observed in the extract from cocoa power [28].

The quantities of the monomers, catechin and epicatechin, were between 3 and 8 mg/g in the cocoa extracts, beans, nibs, liquor, and powder, respectively [28,29]. Epicatechin was detected as the main monomer for the bean, nib, and liquor cocoa extract, and this result was in line with another study reported in the literature [18]. In contrast, catechin was the most abundant monomer in cocoa powder [28]. As expected, the concentrations of these monomer compounds (catechin and epicatechin) were lower for the analysis of chocolates at a value of around 1 mg/g [18].

The quantification of the procyanidins with a degree of polymerization higher than 2 also varied according to the cocoa source analyzed. For instance, cocoa powder showed the lowest content compared with the other cocoa sources studied, which could be related to the manufacturing process comprising oxidation steps and condensation reactions. Lacueva et al. [58] suggested a 60% loss of the flavonoid content during the process of alkalinization to obtain the cocoa powder.

With regard to alkaloids analysis, theobromine was the main compound found in cocoa extracts. This is in agreement with this compound being the most important methylxanthine found in products from the cocoa tree, *Theobroma cocao* [22]. Theobromine was also detected at higher concentrations than caffeine. These were around 42 and 3.5 mg/g lyophilized cocoa extracts (beans, nibs, liquor, and powder) [28].

14.4.2 Biological Samples

Different strategies have been proposed to quantify the procyanidin metabolites, generated during digestive absorption and hepatic metabolism, because standards for these are not available commercially. These strategies are based on a tentative quantification [30–33,35,36,38,39] or an isolation of procyanidin-conjugated metabolites by semipreparative LC [31].

The strategy most widely used is the tentative quantification of the procyanidin metabolites with respect to others with similar procyanidin structures. For example, the catechin and epicatechin metabolites were tentatively quantified with its aglycone, catechin and epicatechin, respectively. In UHPLC, the elution order of the catechin and epicatechin metabolites was considered to be the same as the elution order of catechin (first) and epicatechin (second) in their free forms. The different dimer isomer procyanidins were tentatively quantified with regard to the monomer catechin [30] or the commercial dimer B_2 [31–33,35,36,38].

Although the tentative quantification was the strategy most widely used to quantify procyanidins, when the MS detector system is used, it is very important to quantify each analyte with its own calibration curve, because the ionization can vary depending on the molecular structure. Serra et al. [31] compared the results obtained from quantifying the trimer procyanidin in the plasma sample, according to whether this compound was quantified by using the calibration curve of catechin, epicatechin, dimer B_2, or trimer. In this report, the authors isolated the trimer procyanidin from the cocoa extract by semipreparative HPLC. A great difference in concentration was shown depending on the calibration curve used to quantify the trimer procyanidin. The concentration of the oligomer trimer in the plasma was 8.55 μM, when this was quantified by its own calibration curve, and between 0.05 and 0.06 μM, when this was quantified with the calibration curve of catechin, epicatechin, or dimer B_2.

Another strategy reported for the quantification of procyanidin metabolites in biological samples by HPLC–MS/MS by different authors is based on quantifying the generated aglycones after an enzymatic treatment of β-glucuronidase and sulfatase [54–57]. As mentioned above, the quantification of these procyanidin metabolites in these samples is complex. Therefore, the application of an enzymatic treatment before sample pretreatment and its chromatographic analysis to tentatively quantify in reference to the respective free forms could be justified. However, the enzymatic treatment could be more an additional confirmation in the identification of the conjugate forms of procyanidins than an accurate quantification. The time and temperature conditions during the enzymatic treatment could result in a loss of the free procyanidins, leading to underquantification. Nowadays, this strategy is not applied in UHPLC–MS/MS.

With the aim of obtaining accurate, reliable, and robust results for determining procyanidins, alkaloids, and their metabolites in biological samples by UHPLC–MS/MS, the different quality parameters of the developed methods were studied. These parameters were linearity, repeatability, accuracy, LOD, and LOQ. These studies were carried out by spiking organic solvent [30,32,34,36] or a blank biological matrix [31–33,35,37–40] with different procyanidin and alkaloid concentrations. The blank biological matrix was obtained under fasting conditions. Table 14.2 shows the LODs and LOQs reported for the different UHPLC–MS/MS methodologies for determining these compounds in biological samples.

The reported LODs and LOQs for procyanidins (catechin, epicatechin, dimer, and trimer) in plasma samples that were lower than 4 and 10 nM, except for the trimers, which were 800 and 980 nM, respectively, using the off-line SPE-UHPLC–MS/MS [31]. Similar results were obtained when the off-line μSPE-UHPLC–MS/MS was used to analyze these procyanidin compounds in the biological sample [33]. This demonstrated the high capacity of the μSPE sample pretreatment technique, where the LODs and LOQs were similar in comparison with SPE. In contrast, the preconcentration factor obtained in SPE was 10, in comparison with obtained in μSPE, which was 3.5.

By comparing the LODs obtained by UHPLC–MS/MS [31] and HPLC–MS/MS [59,60] for determining these compounds in plasma samples using the sample pretreatment off-line SPE, the LODs in UHPLC were between three- and fourfold lower than in HPLC. For example, the LODs of epicatechin and the dimer B_2 in the UHPLC system were 4 nM [31] and 3 nM [31], respectively. In contrast, the LODs of these compounds with HPLC were 14 nM [60] and 10 nM [59]. Thus, these results demonstrate the higher peak efficiency and sensitivity of UHPLC compared with the HPLC system.

To determine procyanidins by UHPLC–MS/MS in rat tissue samples, such as liver, brain, aorta artery, and adipose tissue, the reported LODs and LOQs were at low nmol/g tissue levels. For example, the LODs for catechin, epicatechin, and dimer B_2 in rat liver were 1.0, 1.0, and 0.2 nmols/g, respectively [35].

With regard to the analysis of alkaloids, theobromine and caffeine, their LODs and LOQs were at low μM in the analyses of saliva, plasma, and urine [40], and at low nmol/g for the analysis of tissues samples [35]. For example, the LODs for theobromine and caffeine in rat liver were 3.3 and 1.5 nmols/g, respectively [35], and 1.25 μM for urine, plasma, and saliva [40]. When these results (LODs and LOQs) were compared with the ones reported for HPLC–MS/MS for the analysis of urine, the LODs and LOQs were similar or slightly higher in UHPLC [61]. In contrast, the LODs and LOQs from UHPLC–MS/MS could not be compared for the analysis of tissue samples with HPLC–MS/MS because there are no values reported in the literature.

These UHPLC–MS/MS methodologies were applied to determining procyanidins, alkaloids, and their generated metabolites in order to study their bioavailability and metabolism (Tables 14.3 and 14.4). In these reports, the biological samples (saliva [40], plasma [31–33,36,37,39,40], tissues [34–37,39], and urine [40]) were analyzed after an acute or long-term intake of procyanidin- or alkaloid-rich extracts. These extracts were grape seed [31,32,37], grape pomace [33], nut-cocoa cream [34], and hazelnut skin [36,37] as procyanidin-rich extracts, and cocoa extract [35] and chocolate bars [40] as alkaloid-rich extracts, wherein the doses ranged from 1 to 5 g/kg. After the intake of these extracts, the biological samples (saliva, plasma, urine, and tissues) were analyzed at a single time, such as 1.5 h [40], 2 h [31,36], or 4 h [33,35], or at different times, from 1 to 18 h [32,39]. In the latter case, a pharmacokinetic curve of each procyanidin metabolite generated was obtained. Longer times, such as 24 h, were reported for the analysis of the large intestine, including the intestinal contents after a single dose of a nut-cocoa cream. In this study, the metabolic pathways of the colonic metabolism of procyanidins and alkaloids were studied [34].

Then, different procyanidin metabolites were generated, these being catechin and epicatechin in its glucuronide forms, catechin and epicatechin in its sulfate forms, methyl-catechin and methyl-epicatechin in its glucuronide forms, and methyl-catechin and methyl-epicatechin in its sulfate forms [31–33,35–39]. In some studies, the aglycones, catechin, epicatechin, dimer, and trimer were also identified [31,33,35,36]. The main metabolites generated from the procyanidin-rich extract intake are shown in Table 14.3.

These metabolites were generated during digestion and intestinal absorption as a result of phase II enzymes and were converted into glucuronidated and sulfated metabolites as well as methylated metabolites. The procyanidins, which were not absorbed in the small intestine, passed into the large intestine where they were degraded by the colonic microflora into phenolic acids, which could be absorbed into the circulatory system and subjected to phase II metabolism prior to excretion. Serra et al. [34] used a colonic fermentation model with rat colonic microflora and reported that the phenylacetic acid and its hydroxylated forms were the main fermentation products, and hydroxy-γ-valerolactone was specifically from procyanidin compounds.

Apart from these *in vivo* applications, a coculture system on hepatic cells was also reported for assaying the bioactivity of a procyanidin extract. In this report, different procyanidin metabolites were identified, these being catechin and epicatechin in its sulfate forms; methyl-catechin and methyl-epicatechin in its sulfate forms; epicatechin glucuronide, methyl-catechin, and methyl-epicatechin in its glucuronide forms; and the aglycones catechin and epicatechin, epicatechin in its gallate forms, dimer and trimer [38].

It has been reported that the absorption of dietary procyanidins may be influenced by the matrix within which these compounds are consumed [62]. Ortega et al. [30] and Serra et al. [32] studied the *in vitro* bioavailability of procyanidins in two cocoa matrices with different fat content and in a sample matrix that contained a carbohydrate-rich food, respectively, using a UHPLC–MS/MS method.

Regarding alkaloid metabolism, only caffeine and theobromine were found to be the main metabolites in the biological samples in all the UHPLC–MS/MS methods (see Table 14.4) [31,34,35,40].

The concentrations of the procyanidin metabolites generated in the plasma samples were in hundreds of nM after an acute intake of 0.5 mg/g hazelnut skin extract [40], ranged from 85 μM to low μM after an acute intake of 1 and 5 g/kg [31,33], or a long-term intake of 50 mg/kg [37] of a grape seed phenolic extract. Higher procyanidin concentrations were found when 5 g/kg of hazelnut skin extract was ingested, and these were from hundreds to thousands of μM [36]. These results are shown in Table 14.3.

Generally, catechin and epicatechin in its glucuronide forms were reported as the most abundant procyanidin metabolites after the intake of grape seed [31,37], grape pomace, [33] and hazelnut skin [36,40] extract. Apart from these two metabolites, trimer [36] and methyl catechin glucuronide [36,37,39] were also the most important when the intake was hazelnut skin extract, and methyl catechin glucuronide in the case of the long-term intake of grape seed extract [37].

The same procyanidin metabolites identified in the plasma samples were identified in rat tissues such as the liver, brain, lung, kidney, testicles, spleen, heart, thymus, intestine, aorta artery, and adipose tissue between 2 and 4 h after an acute intake of grape pomace extract [35] and hazelnut extract [36,39]. The concentrations of these procyanidin metabolites were at low nmol/g tissue, with the exception of the metabolite methyl catechin glucuronide, which was higher at 218 nmol/g in the intestine tissue [36].

Among the alkaloid metabolites, only caffeine and theobromine were analyzed, and as their standard compounds are commercially available, these were quantified with regard to their own calibration curves. Caffeine and theobromine were detected in rat tissues (liver, brain, aorta artery, and adipose tissue) after an acute intake of 1 g/kg of cocoa extract [35], and in saliva, plasma, and urine after an acute intake of 41 g of a chocolate bar [40]. Their concentrations were from low to hundreds of nmols/g for the analysis of tissues, and in the range of μM for the analysis of biofluids (saliva, plasma, and urine). The aorta artery was the tissue in which theobromine was the most abundant, at a concentration of 289 nmols/g aorta artery [35], and this alkaloid, which was also more abundant than caffeine, was detected at 32, 60, and 260 μM in saliva, plasma, and urine, respectively [40]. Caffeine and theobromine were also identified in the large intestine 18 h after an acute intake of 1.5 g/kg of nuts cocoa-cream. Their respective concentrations were 95 and 0.24 nmols/g large intestine [34]. These results are shown in Table 14.4.

14.5 SUMMARY AND CONCLUSIONS

The continuous improvements in the chromatographic packing material in the LC columns, such as the totally porous silica sub-2 μm UHPLC columns and, recently, the core–shell columns with 2.6 μm as the particle size, are opening new possibilities for determining procyanidins and alkaloids in food samples, such as cocoa samples and cocoa products, and for determining procyanidins, alkaloids, and their metabolites in biological samples. By means of these columns, higher peak efficiency and higher peak resolution are obtained in shorter analysis times in comparison with the HPLC columns. The core–shell columns seem to be very promising because these can be used in conventional HPLC equipment, so no excessive backpressure is generated. Nevertheless, nowadays, sub-2 μm UHPLC columns are chosen instead of the core–shell columns because better results are reported.

Despite the important advances in fast LC (UHPLC), food and biological matrices are very complex, and although multiresidue methods with minimal sample manipulation are demanded, sample extraction and clean-up pretreatments must be carefully developed to reduce the total analysis time and consumption of organic solvents.

The use of the μSPE as the sample pretreatment technique using plates as the device format instead of cartridges, reported for determining procyanidin metabolites in biological samples, is a fast technique that requires loading a lower sample volume. This speed in the extraction of the target procyanidin compounds and their metabolites and the clean-up of the biological matrix is achieved by avoiding the postextraction solvent evaporation and reconstitution steps.

In contrast, the only reported sample pretreatment technique for determining procyanidin and alkaloids in cocoa samples is off-line LLE, first using *n*-hexane to defat the cocoa matrix, and, second, an extraction solution of acetone/water or acetone/water/acetic acid to isolate the procyanidins and alkaloids. This extraction technique is tedious and needs a large volume of solvent and a long time to extract these compounds. The time required for LLE is extremely long compared to the UHPLC separations. Normally, the use of LLE requires 1 or 2 h or more to extract the compounds, while the UHPLC separations only require a few minutes. This aspect should be improved and further studies should be carried out for the sample pretreatment of procyanidins in cocoa samples.

ABBREVIATIONS

1D-LC	single-dimension liquid chromatography
2D-LC	two-dimension liquid chromatography
BEH	ethylene bridged hybrid
CRM	Certified Reference Material
ESI	electrospray ionization
HPLC	high-performance liquid chromatography
HR-MS	high-resolution mass spectrometry
HSS	high-strength silica
LLE	liquid–liquid extraction
LOD	detection limit
LOQ	quantification limit
ME	matrix effect
MS	mass spectrometry
MS/MS	tandem mass spectrometry
MWCO	molecular weight cutoff
QqQ	triple quadrupole
Q-TOF	quadrupole-time-of-flight
R (%)	extraction recovery
SPE	solid-phase extraction
SRM	selected reaction monitoring
TFA	trifluoroacetic acid
TQD	triple quadrupole detector
UHPLC	ultra-high-performance liquid chromatography
µSPE	microelution solid-phase extraction plates

ACKNOWLEDGMENTS

This work was supported by the Spanish Ministry of Education and Science financing project AGL2009-13517-C13-02. The present study was supported by the CENIT program from the Spanish Minister of Industry and by a consortium of companies led by La Morella Nuts S.A. (Reus, Catalonia, Spain).

REFERENCES

1. Keen, C. L., Holt, R. R., Oteiza, P. I., Fraga, C. G., and Schmitz, H. H. 2005. Cocoa antioxidants and cardiovascular health. *Am. J. Clin. Nutr.* 81:298S–303S.
2. Cooper, K. A., Donovan, J. L., Waterhouse, A. I., and Williamson, G. 2008. Cocoa and health: A decade of research. *Br. J. Nutr.* 99:1–11.
3. EFSA Panel on Dietetic Products, Nutrition and Allergies (NDA). 2012. Scientific opinion on the substantiation of a health claim related to cocoa flavanols and maintenance of normal endothelium-dependent vasodilation pursuant to Article 13(5) of Regulation (EC) No 1924/2006. 2012. *EFSA J.* 10:2809–2830.
4. Quideau, S., Deffieux, D., Douat-Casassus, C., and Pouységu, L. 2011. Plant polyphenols: Chemical properties, biological activities, and synthesis. *Angew. Chem. Int. Ed.* 50:586–621.
5. Harbone, J. B. 1989. *Methods in Plant Biochemistry I: Plant Phenolics*. Chapman and Hall, Academic Press, London, UK.
6. Seabra, R. M., Andrade, P. B., Valentao, P., Fernandes, E., Carvalho, F., and Bastos, M. L. 2006. Anti-oxidant compounds extracted from several plant materials. In: *Biomaterials from Aquatic and Terrestrial Organisms*. Finngerman, M., and Nagabhushana, R. (Eds.) Science Publishers, Enfield, NH, USA. pp. 115–174.
7. González-Castejón, M. and Rodriguez-Casado, A. 2011. Dietary phytochemicals and their potential effects on obesity: A review. *Pharmacol. Res.* 64:438–455.
8. Crozier, A., Jaganath, I. B., and Cliffort, M. N. 2009. Dietary phenolics: Chemistry, bioavailability and effects on health. *Nat. Prod. Rep.* 26:1001–1043.
9. Passamonti, S., Terdoslavich, M., Francia, R., Vanzo, A., Tramer, F., Braidot, E., Petrussa, E., and Vianello, A. 2009. Bioavailability of flavonoids: A review of their membrane transport and the function of biotranslocase in animal and plant organisms. *Curr. Drug Metab.* 10:369–394.
10. Hallman, P. C. H. and Katan, M. B. 1997. Absorption, metabolism and health effects of dietary flavonoids in man. *Biomed. Pharmother.* 51:305–316.
11. Bravo, L. 1998. Polyphenols: Chemistry, dietary sources, metabolism and nutritional significance. *Nutr. Rev.* 56:317–333.
12. Harborne, J. B. and Baxter, H. 1999. *The Dadbook of Natural flavonoids, Vol. 2. US Department of Agriculture. USDA Database for Flavonoid Content of Selected Foods-2003.* Wiley, West Sussex, UK.
13. Williams, R. J., Spencer, J. P. E., and Rice-Evans, C. 2004. Flavonoids: Antioxidants or signaling molecules? *Free Rad. Biol. Med.* 36:838–849.
14. Wollgast, J. and Anklam, E. 2000. Polyphenols in chocolate: Is there a contribution to human health? *Food Res. Int.* 33:449–459.
15. Wollgast, J. and Anklam, E. 2000. Review on polyphenols in *Theobroma cacao*: Changes in composition during the manufacture of chocolate and methodology for identification and quantification. *Food Res. Int.* 33:423–447.
16. Counet, C., Ouwerx, C., Rosoux, D., and Collin, S. 2004. Relationship between procyanidin and flavor contents of cocoa liquors from different origins. *J. Agric. Food Chem.* 52:6243–6249.
17. Summa, C., Raposo, F. C., McCourt, J., Scalzo, R. L., Wagner, K.-H., El-Madfa, I., and Anklam, E. 2006. Effect of roasting on the radicals scavenging activity of cocoa beans. *Eur. Food Res. Technol.* 222:368–375.
18. Cooper, K. A., Campos-Giménez, E., Alvarez, D. J., Nagy, K., Donovan, J. L., and Williamson, G. 2007. Rapid reversed phase ultra-performance liquid chromatography analysis of the major cocoa polyphenols and inter-relation ships of their concentrations in chocolate. *J. Agric. Food Chem.* 55:2841–2847.

19. Adamson, G. E., Lazarus, S. A., Mitchell, A. E., Prior, R. L., Cao, G., Jacobs, P. H., Kremers, B. G. et al. 1999. HPLC method for the quantification of procyanidins in cocoa and chocolate samples and correlation to total antioxidant capacity. *J. Agric. Food Chem.* 47:4184–4188.
20. Natsume, M., Osakabe, N., Yamagishi, M., Takizawa, T., Nakamura, T., Miyatake, H., Hatano, T., and Yoshida, T. 2000. Analyses of polyphenols in cacao liquor, cocoa, and chocolate by normal-phase and reversed-phase HPLC. *Biosci. Biotechnol. Biochem.* 64:2581–2587.
21. Zulak, K. G., Liscome, D. K., Ashihara, H., and Facchini, P. J. 2006. Alkaloids. In: Crozier, A., Clifford, M. N., and Ashihara, H. (Eds). *Plant Secondary Metabolites: Occurrence Structure, and Role in the Human Diet.* Blackwell, Oxford. pp. 102–136.
22. Matissek, R. 1997. Evaluation of xanthine derivatives in chocolate—Nutritional and chemical aspects. *Eur. Food Res. Techn.* 205:175–184.
23. Caudle, A. G. and Bell, L. N. 2000. Caffeine and theobromine contents of ready-to-eat chocolate cereals. *J. Am. Diet. Assoc.* 100:690–692.
24. Timbie, D. J., Sechrist, L., and Keeney, P. G. 1978. Application of high pressure liquid chromatographic to the study of variables affecting theobromine and caffeine concentrations in cocoa beans. *J. Food Sci.* 43:560–565.
25. Nehlig, A., Daval, J.-L., and Debry, G. 1992. Caffeine and the central nervous system: Mechanisms of action, biochemical, metabolic and psychostimulant effects. *Brain Res. Rev.* 17:139–169.
26. Spiller, G. A. 1998. Basic metabolism and physiological effects of the methylxanthines. In: G. Spiller (Ed.) *Caffeine*. CRC Press, Washington DC. pp. 225–232.
27. Kalili, K. M. and Villiers, A. 2009. Off-line comprehensive two-dimensional hydrophilic interaction x reversed phase liquid chromatography analysis of procyanidins. *J. Chromatogr. A* 1216:6274–6284.
28. Ortega, N., Romero, M. P., Macià, A., Reguant, J., Anglès, N., Morelló, J. R., and Motilva, M. J. 2008. Obtention and characterization of phenolic extracts from different cocoa sources. *J. Agric. Food Chem.* 56:9621–9627.
29. Ortega, N., Romero, M. P., Macià, A., Reguant, J., Anglès, N., Morelló, J. R., and Motilva, M. J. 2010. Comparative study of UPLC-MS/MS and HPLC-MS/MS to determine procyanidins and alkaloids in cocoa samples. *J. Food Comp. Anal.* 23:298–305.
30. Ortega, N., Reguant, J., Romero, M. P., Macià, A., and Motilva, M. J. 2009. Effect of fat content on the digestibility and bioaccessibility of cocoa polyphenol by an *in vitro* digestion model. *J. Agric. Food Chem.* 57:5743–5749.
31. Serra, A., Macià, A., Romero, M. P., Salvadó, M. J., Bustos, M., Fernández-Larrea, J., and Motilva, M. J. 2009. Determination of procyanidins and their metabolites in plasma samples by improved liquid-chromatography-tandem mass spectrometry. *J. Chromatogr. B* 877:1169–1176.
32. Serra, A., Macià, A., Romero, M. P., Valls, J., Bladé, C., Arola, Ll., and Motilva, M. J. 2010. Bioavailability of procyanidin dimers and trimmers and matrix food effects in *in vitro* and *in vivo* models. *Br. J. Nutr.* 103:944–952.
33. Martí, M. P., Pantaleón, A., Rozek, A., Soler, A., Valls, J., Macià, A., Romero, M. P., and Motilva, M. J. 2010. Rapid analysis of procyanidins and anthocyanidins in plasma by microelution SPE and ultra-HPLC. *J. Sep. Sci.* 33:2841–2853.
34. Serra, A., Macià, A., Romero, M. P., Anglés, N., Morelló, J. R., and Motilva, M. J. 2011. Metabolic pathways of the colonic metabolism of procyanidins (monomers and dimers) and alkaloids. *Food Chem.* 126:1127–1137.
35. Serra, A., Macià, A., Romero, M. P., Piñol, C., and Motilva, M. P. 2011. Rapid methods to determine procyanidins, anthocyanidins, theobromine and caffeine in rat tissues by liquid chromatography-tandem mass spectrometry. *J. Chromatogr. B* 879:1519–1528.

36. Serra, A., Macià, A., Romero, M. P., Anglés, N., Morelló, J. R., and Motilva, M. J. 2011. Distribution of procyanidins and their metabolites in rat plasma and tissues after an acute intake of hazelnut extract. *Food Funct.* 2:562–568.
37. Serra, A., Bladé, C., Arola, L., Macià, A., and Motilva, M. J. 2013. Flavanol metabolites distribute in visceral adipose depots after long-term intake of grape seed proanthocyanidin extract in rats. *J. Br. Nutr.* 110:1411–1420.
38. Castell-Auví, A., Motilva, M. J., Macià, A., Torrell, H., Bladé, C., Pinent, M., Arola, Ll., and Ardévol, A. 2010. Organotypic co-culture system to study plant extract bioactivity on hepatocytes. *Food Chem.* 122:775–781.
39. Serra, A., Macià, A., Rubió, L., Anglès, N., Ortega, N., Morelló, J. R., Romero, M. P., and Motilva, M. J. 2013. Distribution of procyanidins and their metabolites in rat plasma and tissues in relation to ingestion of procyanidin-enriched or procyanidin-rich cocoa creams. *Eur. J. Nutr.* 52:1029–1038.
40. Ptolemy, A. S., Tzioumis, E., Thomke, A., Rifai, S., and Kellogg, M. 2010. Quantification of theobromine and caffeine in saliva, plasma and urine via liquid chromatography–tandem mass spectrometry: A single analytical protocol applicable to cocoa intervention studies. *J. Chromatogr. B* 878:409–416.
41. Swartz, M. E. 2005. Ultra-performance liquid chromatography (UPLC): An introduction. *LC-GC North America* 23:8–14.
42. Swartz, M. E. 2005. UPLC™: An introduction and review. *J. Liq. Chromatogr. Rel. Technol.* 28:1253–1263.
43. Mazzeo, J. R., Neue, U. D., Lele, M., and Plumb, R. S. 2005. Advancing LC performance with smaller particles and higher pressure. *Anal. Chem.* 77:460A–467A.
44. Guillarme, D., Schappler, J., Rudaz, S., and Veuthey, J. L. 2010. Coupling ultra-high-performance liquid chromatography with mass spectrometry. *Trends Anal. Chem.* 29:15–27.
45. http://www.waters.com/waters/home.htm. February 2013.
46. Wollgast, J., Pallaroni, L., Agazzi, M. E., and Anklam, E. 2001. Analysis of procyanidins in chocolate by reversed-phase high-performance liquid chromatography with electrospray ionization mass spectrometric and tandem mass spectrometric detection. *J. Chromatogr. A* 926:211–220.
47. Sánchez-Rabaneda, F., Jáuregui, O., Casals, I., Andrés-Lacueva, C., Izquierdo-Pulido, M., and Lamuela-Raventós, R. M. 2003. Liquid-chromatography/electrospray ionization tandem mass spectrometry study of the phenolic composition of cocoa (*Theobroma cacao*). *J. Mass Spectrom.* 38:35–42.
48. Pereira-Caro, G., Borges, G., Nagai, C., Jackson, M. C., Yokota, T., Crozier, A., and Ashihara, H. 2013. Profiles of phenolic compounds and purine alkaloids during the development of seeds of *Theobroma cacao* cv. Trinitario. *J. Agric. Food Chem.* 61:427–434.
49. Robbins, R. J., Leonczak, J., Li, J., Christopher, J., Collins, T., Uribe, C. K., and Schmitz, H. H. 2012. Determination of flavanol and procyanidin (by degree of polymerization 1–10) content of chocolate, cocoa liquors, powder(S), and cocoa flavanol extracts by normal phase high-performance liquid chromatography: collaborative study. *J. AOAC Int.* 95:1153–1160.
50. Hammerstone, J. F., Lazarus, S. A., Mitchell, A. E., Rucker, R., and Schmitz, H. H. 1999. Identification of procyanidins in cocoa (*Theobroma cacao*) and chocolate using high-performance liquid chromatography-mass spectrometry. *J. Agric. Food Chem.* 47:490–496.
51. Fekete, S., Erzsébet, E., and Felete, J. 2010. Fast liquid chromatography of core-shell and very fine particles. *J. Chromatogr. A* 1228:57–71.
52. Magí, M., Bono, L., and Di Carro, M. 2012. Characterization of cocoa liquors by GC-MS and LC-MS/MS: Focus on alkylpyrazines and flavanols. *J. Mass Spectrom.* 47:1191–97.

53. Gu, L., Kelm, M., Hammerstone, J. F., Beecher, G., Cunningham, D., Vannozi, S., and Prior, R. L. 2002. Fractionation of polymeric procyanidins from lowbush blueberry and quantification of procyanidins in selected foods with an optimized normal-phase HPLC-MS fluorescent detection method. *J. Agric. Food Chem.* 50:4852–4860.
54. Shoji, T., Masumoto, S., Moriichi, N., Akiyama, H., Kanda, T., Ohtake, Y., and Goda, Y. 2006. Apple procyanidin oligomers absorption in rats after oral administration: Analysis of procyanidins in plasma using the Porter method and HPLC-MS/MS. *J. Agric. Food Chem.* 54:884–892.
55. Urpi-Sardà, M., Monagas, M., Khan, N., Lamuela-Raventós, R. M., Santos-Buelga, C., Sacanella, E., Castell, M., Permanyer, J., and Andrés-Lacueva, C. 2009. Epicatechin procyanidins and phenolic microbial metabolites after cocoa intake in humans and rats. *Anal. Bioanal. Chem.* 394:1545–1556.
56. Ottaviani, J., Momma, T., Kuhnle, G., Keen, C., and Schroeter, H. 2012. Structurally related (-)-epicatechin metabolites in humans: Assessment using de novo chemically synthesized authentic standards. *Free Rad. Biol. Med.* 52:1403–1412.
57. Li, S., Sui, Y., Xiao, J., Wu, Q., Hu, B., Xie, B., and Sun, Z. 2013. Absorption and urinary excretion of A-type procyanidin oligomers from *Litchi chinensis* pericarp in rats by selected ion monitoring liquid chromatography-mass spectrometry. *Food Chem.* 138:1536–1542.
58. Andrés-Lacueva, C., Monagas, M., Khan, N., Izquierdo-Pulido, M., Urpi-Sardà, M., Permanyer, J., and Lamuela-Raventós, R. M. 2008. Flavanol and flavonol contents of cocoa powder products: Influence of the manufacturing process. *J. Agric. Food Chem.* 56:3111–3117.
59. Holt, R., Lazarus, S., Sullards, M., Zhu, Q., Schramm, D., Hammerstone, J., Fraga, C., Schmitz, H., and Keen, C. 2002. Procyanidin dimer B2 [epicatecin-(4(-8)-epicatechin] in human plasma after the consumption of a flavanol-rich cocoa. *Am. J. Clin. Nutr.* 76:798–804.
60. Roura, E., Andrés-Lacueva, C., Jáuregui, O., Badia, E., Estruch, R., Izquierdo-Pulido, M., and Lamuela-Raventós, R. 2005. Rapid liquid-chromatography tandem mass spectrometry assay to quantify plasma epicatecin metabolites after ingestion of a standard portion of cocoa beverage in humans. *J. Agric. Food Chem.* 53:6190–6194.
61. Schneider, H., Ma, L., and Glatt, H. 2003. Extractionless method for the determination of urinary caffeine metabolites using high-performance liquid chromatography coupled with tandem mass spectrometry. *J. Chromatogr. B* 789:227–237.
62. Clifford, M. and Brown, J. E. 2006. Dietary flavonoids and health-broadening the perspective. In: Andersen, R. M. and Markham, K. R. (Eds.) *Flavonoids: Chemistry, Biochemistry and Applications*. Taylor & Francis Group, CRC Press, New York. pp. 319–370.

15 Ultra-Performance Liquid Chromatography–Mass Spectrometry and Its Application toward the Determination of Lactose Content in Milk

Mahamudur Islam

CONTENTS

15.1 INTRODUCTION

The physical characteristics of milk are highly complex, as milk is composed of an intricate mixture of fat globules and protein (casein, whey) in an aqueous solution of lactose, minerals, and other minor constituents. Milk's physical characteristics are affected by several factors, including the composition and processing of milk. Measurement of milk's physical properties is used in processing, to determine the concentration of milk's components, and to evaluate the quality of milk products. Lactose is the major carbohydrate in milk. Measurement of lactose in milk is important because it contributes to the sensory and functional properties of milk. Lactose (β-D-galactopyranosyl-(1 → 4)-D-glucose) is a disaccharide sugar that is found most notably in cow milk (but is in higher concentration in human milk, and is also found in those of other mammals) and is formed from galactose and glucose. Lactose (Figure 15.1) was discovered in milk in 1619 by Fabriccio Bartoletti, and identified as a sugar in 1780 by Carl Wilhelm Scheele. Lactose makes up around 2~8% of milk (by weight), although the amount varies among species and individuals. It is extracted from sweet or sour whey. The name comes from *lac*, or *lactis*, the Latin word for milk, plus the "ose" ending used to name sugars. Lactose is the main carbohydrate in dairy products. This disaccharide is composed of glucose and galactose and is the only saccharide synthesized by mammals. Lactose plays an important role in the formation of the neural system and the growth of skin (texture), bone skeleton, and cartilage in infants. It also prevents rickets and saprodontia [1]. The need to quantify lactose came up with the knowledge of its importance in the human diet. It has been reported that cow mastitis can cause a reduction of milk yield and also of its nutritive value. Determination of the lactose content is one of the methods used to evaluate whether milk is acceptable for human consumption [2]. Moreover, the precise control of the amount of lactose in dairy food products is vital, as many people are intolerant to this carbohydrate. Therefore, the precise determination of lactose during the production process, as well as in the final product, is fundamental for the food industry. It also has economic value since the price of milk is based on

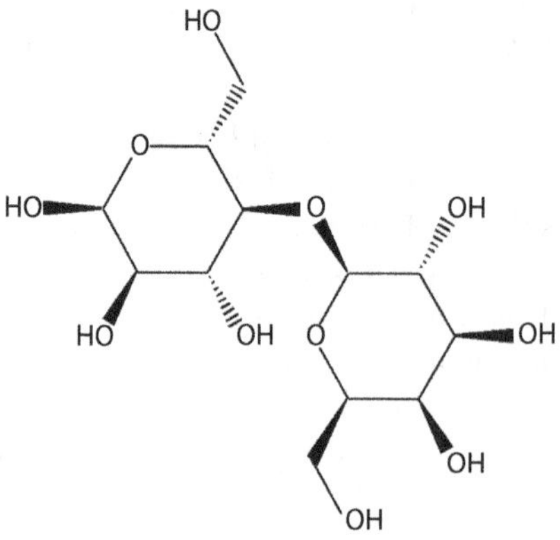

FIGURE 15.1 Structure of lactose.

milk solids content. The lactose content of cow milk can vary from 3.8% to 5.3% (38,000–53,000 μg/mL). Modern 1% and 2% milk have higher levels of lactose. The Association of Official Agricultural Chemists (AOAC) official method (984.15) for lactose in milk is both complex and time consuming. It involves enzymatic hydrolysis of lactose to glucose and galactose at pH 6.6 by β-galactosidase [3]. Subsequent oxidation of the β-galactose released to galactonic acid at pH 8.6, as catalyzed by β-galactose dehydrogenase, then occurs with concomitant reduction of nicotinamide adenine dinucleotide (NAD^+). The amount of reduced NAD formed is measured at 340 nm and is proportional to the lactose content. The method requires seven different reagents, two of which must be prepared weekly.

Various methods have been used for lactose determination, such as polarimetry, gravimetry, spectrophotometry, high-performance liquid chromatography (HPLC), and infrared spectroscopy (IR) [4–7]. The classic method for the determination of saccharides is chemical analysis as well as the enzyme catalyzed method and spectrophotometry [8–10]. The major disadvantage of these three methods consists of the difficulty in evaluating different saccharides simultaneously. One of the methods commonly used in saccharide analysis is gas chromatography (GC), which has been used in the study of milk [11,12], dried skim milk [11], and pasteurized milk [13]. Although GC is a sensitive method for saccharide analysis, sample preparation is laborious because saccharides must be derivatized before GC analysis. The procedure is too tedious to be used routinely. In the last decade, several types of enzymatic biosensors for lactose determination have been developed. The first generation of lactose biosensors was based on the detection of the formed hydrogen peroxide by using galactosidase in combination with glucose oxidase [14–15]. In the second generation, redox mediators are bound to the electrode in order to enhance the electron transfer between the active site of the enzyme and the electrode [16]; these include galactosidase, glucose oxidase [17–22], and horseradish peroxidase [23]. The third-generation biosensors are based on a direct electron transfer (DET) between the redox enzyme and the electrode, thus allowing a more simple design of the biosensor and reducing the number of components [24]. Examples of enzymes employed in third-generation biosensors are horseradish peroxidase [25], cellobiose dehydrogenase [26], superoxide dismutase [27], and alcohol PQQ-dehydrogenase [28]. They are to a lesser or higher degree time consuming, expensive, and dependent on sample pretreatment, and require considerable technical skills. Therefore, it is of great importance to find a rapid, simple, and robust method to quantify lactose in milk.

For many years, researchers have looked at HPLC as a way to speed up analyses. Smaller columns and faster flow rates (among other parameters) have been used. Elevated temperature, having the dual advantages of lowering viscosity and increasing mass transfer by increasing the diffusivity of the analytes, has also been investigated. HPLC with different types of detectors has been used as a reference lactose determination method because it is direct, automated, and able to differentiate between carbohydrates [29–32]. The most common HPLC method obtains measurements using a refractive index detector with differentiation among various monosaccharides, disaccharides, and oligosaccharides of milk [33–36]. Furthermore, both mono- and disaccharides were detected in both positive and negative modes using tandem mass spectrometry (MS) [37]. Mulroney et al. demonstrated that the measurement of β-linked glucose disaccharides in the negative ion mode provided more structural information than in the positive ion mode because deprotonation of the anomeric hydroxyl causes fragmentation [31]. Moreover, HPLC combined with a corona-charged aerosol detector (CAD) was presented at the PITTCON conference as a new lactose determination method. The benefits of CAD are that it provides a consistent response independent from the chemical structure of a compound, and that it is simply easy to utilize. However, CAD is not a mass specific unlike MS, so it may not be sensitive to detect coeluting compounds. Thus, the need for an accurate and reliable method of lactose analysis has become essential for quantity and safety assurance. However, using conventional particle sizes and pressures, limitations are soon reached and compromises must be made, sacrificing resolution. HPLC is an approved technique and has been used in laboratories worldwide over the past 30 years. Ultra-performance liquid chromatography (UPLC) can be regarded as a new invention for liquid chromatography (LC). UPLC brings dramatic improvements in sensitivity, resolution, and speed of analysis. It has instrumentation that operates at high pressure than that used in HPLC, and this system uses fine particles (less than 2.5 μm) and mobile phases at high linear velocity, which decreases the length of the column, reduces solvent consumption, and saves time. According to the van Deemter equation, as the particle size decreases to less than 2.5 μm, there is a significant gain in efficiency, while the efficiency does not diminish at increased flow rates or linear velocities [38]. Therefore, by using smaller particles, speed and peak capacity (number of peaks resolved per unit time in gradient separations) can be extended to new limits [38]. The technology takes full advantage of chromatographic principles to run separations using columns packed with smaller particles (less than 2.5 μm) and/or higher flow rates for increased speed. This gives superior resolution and sensitivity [38]. Nowadays, in industrial areas, UPLC is used for some of the most recent work in the field.

The determination of lactose by UPLC–MS/MS in negative ion mode has been found in this regard, since it is convenient and reliable. Recently, lactose, maltose, and sucrose in vegetable matrices were detected by U-HPLC–MS/MS in the negative selective reaction monitoring (SRM) mode using only mother ions for quantification [39]. Ultra-performance liquid chromatography mass spectrometry (UPLC–MS/MS) takes advantage of technological strides made in particle chemistry performance, system optimization, detector design, and data processing and control. Using sub-2 mm particles and mobile phases at high linear velocities, and instrumentation that operates at higher pressures than those used in HPLC, dramatic increases in resolution, sensitivity, and speed of analysis can be obtained. This new category of analytical separation science retains the practicality and principles of HPLC while creating a step function improvement in chromatographic performance. This chapter describes a simple, rapid, and accurate method for the determination of lactose in milk using standard UPLC–MS/MS in negative ion mode. The method was validated with respect to specificity, linearity, accuracy, precision, and sensitivity.

15.2 SAMPLE TREATMENT

The sampling procedures differ according to the nature of the material and the purpose for which it is needed. The sample should be such that it should be truly representative of the bulk. Different procedures are followed in collecting a sample of milk. The operations involved are mixing and then immediate taking of the sample. Mixing can be done by using a plunger, which should be moved up and down the milk vigorously about 10 times. If it is in a small container, it should be poured from one vessel to the other and shaking it. Care should be taken to avoid fat separation. If samples of milk of individual cows are needed, it may be done in the weighing room.

The determination of lactose content by UPLC–MS/MS is a simple, rapid, accurate, and validated method. The method uses UPLC separation and direct measurement of lactose by tandem quadrupole detector. The determination involves a simple 150,000-fold dilution of milk and analysis is completed in 5 min. The range of the method is from 0.00001% to 10% lactose, allowing the accurate determination of this carbohydrate in lactose-reduced, fat-free milk. The detection limit of the method is very low. The method demonstrates good accuracy and precision and is suitable for the routine determination of lactose in milk.

One-lower and one-higher concentrations of lactose containing cow milk (milk from the certified laboratories with known lactose) are chosen as quality control samples. Cow milk samples are diluted to 150,000-fold to meet optimal concentrations that UPLC–MS/MS can detect with high sensitivity and specificity. Cow milk is first diluted to 200,000-fold and is used as the baseline to conserve the matrix effect for the development of calibration curves. Five different lactose concentrations are produced by adding different lactose concentrations in cow milk. The aqueous lactose solution is prepared by adding 1.1 mg lactose monohydrate (MW 360.32 g/mol) in 100 mL of water, and 2, 5, 10, 15, and 20 μL of the aqueous lactose solution is added to 198, 195, 190, 185, and 180 μL of diluted cow milk, respectively. As the internal standard, a 100 μL labeled lactose (26.86 μmol/L) is subsequently added, resulting in a total dilution of milk 300,000-fold. One milk sample is prepared without adding any external lactose so that the lactose concentration itself within the cow milk can be determined.

15.3 UPLC–MS/MS METHODS

The UPLC–MS/MS is based on the principle of use of stationary phase, consisting of particles less than 2.5 μm (while HPLC columns are typically filled with particles of 3–5 μm). The underlying principles of this evolution are governed by the van Deemter equation, which is an empirical formula that describes the relationship between linear velocity (flow rate) and plate height (HETP or column efficiency) [38,40].

$$H = A + \frac{B}{v} + C \cdot v \tag{15.1}$$

where A, B, and C are constants and v is the linear velocity of the mobile phase. The A term is independent of velocity and represents "eddy" mixing. It is smallest when the packed column particles are small and uniform. The B term represents axial diffusion or the natural diffusion tendency of molecules. This effect is diminished at high flow rates, and so this term is divided by v. The C term is due to the kinetic resistance to equilibrium in the separation process. The kinetic resistance is the time lag involved in moving from the mobile phase to the packing stationary phase and back again. The greater the flow of the mobile phase, the more a molecule on the packing tends to lag behind molecules in the mobile phase. Thus the term is proportional to v.

Therefore, it is possible to increase throughput, and thus the speed of analysis without affecting the chromatographic performance. The advent of UPLC has demanded the development of a new instrumental system for LC, which can take advantage of the separation performance (by reducing dead volumes) and consistent with the pressures (8000–15,000 psi, compared with 2500 to 5000 psi in HPLC). Efficiency is proportional to column length and inversely proportional to the particle size [41]. Smaller particles provide increased efficiency as well as the ability to work at increased linear velocity without a loss of efficiency, providing both resolution and speed. Efficiency is the primary separation parameter behind UPLC since it relies on the same selectivity and retentivity as HPLC. In the fundamental resolution (R_S) equation [38], resolution is proportional to the square root of N.

$$R_S = \frac{\sqrt{N}}{4}\left(\frac{\alpha - 1}{\alpha}\right)\left(\frac{k}{k+1}\right) \tag{15.2}$$

$$N \propto \frac{1}{dp} \tag{15.3}$$

Since N is inversely proportional to particle size (dp), as the particle size is lowered by a factor of three, for example, from 5 μm (HPLC scale) to 1.7 μm (UPLC scale), N is increased by three and resolution by the square root of three, or 1.7. N is also inversely proportional to the square of the peak width. So, as the particle size decreases to increase N and subsequently R_S, an increase in sensitivity is obtained. Also, peak height is inversely proportional to the peak width.

$$H \propto \frac{1}{w} \tag{15.4}$$

An increase in sensitivity is obtained, since narrower peaks are taller. Narrower peaks also mean more peak capacity per unit time in gradient separations, desirable for many applications. As stated earlier, efficiency is proportional to the column length and inversely proportional to the particle size [38].

$$N \propto \frac{L}{dp} \tag{15.5}$$

Therefore, the column can be shortened by the same factor as the particle size without loss of resolution. Using a flow rate 3 times higher due to the smaller particles, and shortening the column by one-third (again due to the smaller particles), the separation is completed in 1/9 the time while maintaining resolution. The application of UPLC resulted in the detection of additional drug metabolites, superior separation, and improved spectral quality [42–43].

A commercially available nonporous, high iffiest small particle has poor retention and loading capacity due to low surface area. To maintain retention and loading capacity, novel porous particles that can withstand high pressures must be used. The promises of the van Deemter equation cannot be fulfilled without smaller particles than those traditionally used in HPLC. The design and development of sub-2 μm particles is a significant challenge, and researchers have been active in this area for some time to capitalize on their advantages [43,44]. Although high-efficiency, nonporous 1.5 μm particles are commercially available, they suffer from poor loading capacity and retention due to low surface area. To maintain retention and capacity similar to HPLC, UPLC must use novel porous particles that can withstand high pressures. Silica-based particles have good mechanical strength, but can suffer from a number of disadvantages, which include a limited pH range and tailing of basic analytes. Polymeric columns can overcome pH limitations, but they have their own issues, including low efficiencies and limited capacities.

In 2000, a first-generation hybrid chemistry that took advantage of the best of both the silica and polymeric column worlds was introduced. These columns were produced using a classical sol–gel synthesis that incorporates carbon in the form of methyl groups, these columns are mechanically strong. They are highly efficient and can operate at a wide range of pH. However, in order to provide the kind of enhanced mechanical stability required for UPLC, a second-generation bridged ethane hybrid (BEH) technology was developed [45]. These 1.7 μm particles derive their enhanced mechanical stability by bridging the methyl groups in the silica matrix. Packing 1.7 μm particles into reproducible and rugged columns was also a challenge that needed to be overcome. Requirements include a smoother interior surface of the column hardware, and redesigning the end frits to retain the small particles and resist clogging. Packed bed uniformity is also critical, especially if shorter columns are to maintain resolution while accomplishing the goal of faster separations. In addition, at high pressures, frictional heating of the mobile phase can be quite significant and must be considered [46]. With column diameters typically used in HPLC (3.0–4.6 mm), a consequence of frictional heating is the loss of performance due to temperature-induced nonuniform flow. To minimize the effects of frictional heating, smaller-diameter columns (1–2.1 mm) are typically used for UPLC [47,48].

Only small particles are not responsible for fast resolution and sensitivity. Some special instrumentation system should be designed. The special kind of system capable of delivering the pressures required to utilize the potential of UPLC has been reported in the literature [38,49]. Small particles alone do not make it possible to fulfill the promises of the van Deemter equation. Instrument technology also had to keep pace to truly take advantage of the increased speed, superior resolution, and sensitivity afforded by smaller particles. Standard HPLC technology (pumps, injectors, and detectors) simply does not have the horsepower to take full advantage of sub-2 μm particles.

Wu et al. described the design of injection valves and separation reproducibility, and the use of a carbon dioxide-enhanced slurry packing method on the capillary scale for the separation of some benzodiazepines, herbicides, and various pharmaceutical compounds [43,50]. Tolley et al. modified a commercially available HPLC system to operate at 17,500 psi and used 22 cm long capillaries packed with 1.5 μm

C18 modified particles for the analysis of proteins [49]. These reports illustrated that, to take full advantage of low dispersion and small-particle technology to achieve high peak capacity UPLC separations, a greater pressure range than that achievable by today's HPLC instrumentation was required. The calculated pressure drop at the optimum flow rate for maximum efficiency across a 15 cm long column packed with 1.7 μm particles is about 15,000 psi. Therefore, a pump capable of delivering solvent smoothly and reproducibly at these pressures, that can compensate for solvent compressibility, and can operate in both the gradient and isocratic separation modes, was required. Conventional injection valves, either automated or manual, are not designed and hardened to work at extreme pressure. To protect the column from experiencing extreme pressure fluctuations, the injection process must be relatively pulse free. The swept volume of the device also needs to be minimal to reduce potential band spreading. A fast injection cycle time is needed to fully capitalize on the speed afforded by UPLC, which, in turn, requires a high sample capacity. Low-volume injections with minimal carryover are also required to realize the increased sensitivity benefits.

With 1.7 μm particles, half-height peak widths of less than 1 s are obtained, posing significant challenges for the detector. In order to accurately and reproducibly integrate an analyte peak, the detector sampling rate must be high enough to capture enough data points across the peak. In addition, the detector cell must have minimal dispersion (volume) to preserve separation efficiency. Conceptually, the sensitivity increase for UPLC detection should be 2–3 times higher than with HPLC separations, depending on the detection technique that is used. Conventional absorbance-based optical detectors are concentration-sensitive detectors, and, for UPLC use, the flow cell volume would have to be reduced in standard UV/visible detectors to maintain concentration and signal, while avoiding Beer's law limitations.

In early 2004, the first commercially available UPLC system that embodied these requirements was described for the separation of various pharmaceutical-related small organic molecules, proteins, and peptides; it is called the Acquity UPLC–MS/MS system. The Acquity UPLC system consists of a binary solvent manager, sample manager (including the column heater), detector, and optional sample organizer. The binary solvent manager uses two individual serial flow pumps to deliver a parallel binary gradient mixed under high pressure [38]. Using UPLC, it is now possible to take full advantage of chromatographic principles to run separations using shorter columns, and/or higher flow rates for increased speed, with superior resolution and sensitivity. For some analyses, however, speed is of secondary importance; peak capacity and resolution take center stage. In these applications, the increased peak capacity (number of peaks resolved per unit time) of UPLC dramatically improves the quality of the data, resulting in a more definitive map.

15.3.1 Chromatographic Separations

Low-pressure LC (LPLC) is a standard method for purifying substrates from complicated matrices, providing a crude but generally effective method for isolating relatively large amounts of material. In addition, the method is relatively simple and robust, allowing for a wide range of applications. LPLC is often distinguished from medium-pressure LC (MPLC) by the pressure typically used. LPLC is traditionally between 0 and 5 bar, whereas MPLC is between 5 and 20 bar. In contrast, HPLC is operated between 20 and 400 bar, and ultra-performance liquid chromatography runs between 400 and 1200 bar. Each of these techniques offers unique strengths in terms of cost, compatible materials, speed, resolution, and loading capacity.

A nonexhaustive overview of methods used for lactose analysis is chromatographic separations. Traditionally, anion exchange chromatography, particularly high-pH anion exchange chromatography with pulsed amperometric detection, has been used for the analysis of lactose [51,52]. Using anion exchange columns, lactose can be separated, resulting in the separation of several isomers. However, previous separation of the neutral and acidic oligosaccharides may be required, resulting in doubled analysis times. A second mode of separation often used for lactose analysis is reverse-phase LC. Native lactose is not retained on reverse-phase material because of their hydrophilic properties, and therefore derivatization is required. Retention and separation of the lactose on reverse-phase LC thus depends mainly on the method of derivatization; some isomer separation was obtained so far, but no method has emerged that provides comprehensive isomer separation. Labeling with chromophoric active tags such as 1-phenyl-3-methyl-5-pyrazolone, 2-aminopyridine, and 2-aminobenzoic acid as well as perbenzoylation has been applied for the analysis of milk oligosaccharides. These labels served two major purposes. They provided a chromophore for detection with LC and a hydrophobic label to allow chromatographic separation in stationary phases such as C18, which do not normally retain or separate native lactose. Additional labels have been used in the analysis of other oligosaccharides [53] and may also be applicable for the analysis of milk oligosaccharides. More recently, milk oligosaccharides have been separated using hydrophilic interaction chromatography, a method that has already been applied extensively for the analysis of *N*- and *O*-glycans [54]. The oligosaccharides are labeled with 2-aminobenzamide using reductive amination to allow fluorescence detection, but retention is mostly based on the oliogosaccharide portion, and the elution order is mainly influenced by the number of monosaccharide residues. Electromigration-based separation techniques have also been applied in the analysis of lactose [55]. Using micellar electrokinetic chromatography, native sialylated milk oligosaccharides were separated.

15.3.2 Mass Spectrometry

Detection with MS is particularly important for the isolation of compounds that lack useful chromophores or are found only in complicated mixtures such as lactose [56]. The lack of a UV-absorbing chromophore in the saccharide structure limits the mode of detection. Refractive index detection requires precise control of the mobile phase and often does not meet the demands of trace-level analysis needed concerning sensitivity and selectivity. Even chemical derivatization, for example, postcolumn derivatization and enzymatic derivatization, which greatly improves the selectivity and sensitivity of a chromatographic detection system, does not meet the needed trace levels for the detection of the fine particle dose of lactose. The reason for this phenomenon is that the sample contains interfering factors, which make it impossible to obtain reproducible results. From 3 to 15 monosaccharides are linked through various glycosidic bonds to form a plethora of linkage-specific isomeric forms. Detection of human milk oligosaccharides (HMO) and other nonchromophores can be enhanced by derivatization with a chromophore or hydrophobic substituent [57,58]; however, biological assays such as cell binding and *in vitro* bacterial growth studies require compounds to be in their native (underivatized) states. The work by Thanawiroon and coworkers gives a good comparison of a chromatogram of oligosaccharides obtained by HPLC/MS versus HPLC/UV absorption [59]. The MS-based chromatogram showed more peaks and greater sensitivity because underivatized oligosaccharides do not efficiently absorb UV light [60].

15.3.3 High-Resolution Mass Spectrometry

All the analytical techniques described so far are based on lactose separation alone; however, structural confirmation can in such cases only be obtained based on standards. These standards are expensive and not readily available for lactose. Moreover, coelution/migration cannot be excluded. Because elution or migration is not perfectly identical in all runs, identification of the signals in each of the samples may be ambiguous. For better identification, coupling of the separation with MS has proven to be effective. Nano-LC porous graphitic carbon (PGC) chip time-of-flight (TOF) MS in the positive mode for the analysis of lactose has been introduced recently [61]. In this method, good separation is combined with unambiguous identification using MS. Lactose may be separated in one run, and, using a library containing retention time, mass, and fragmentation information, immediate identification is possible. Using this method, >200 milk oligosaccharide structures can be separated. Reduction of the reducing end of the oligosaccharides is necessary because the α- and β-anomers are separated on the PGC stationary phase. More recently, a method consisting of capillary electrophoresis (CE) with laser-induced fluorescence (LIF) coupled to MS was developed [62]. Although good separation can be achieved using offline CE with LIF using very fast runs, both resolution and separation times must be compromised when coupling CE to MS.

Coupling of MS to chromatographic techniques has always been desirable due to the sensitive and highly specific nature of MS compared to other chromatographic detectors. The coupling of GC to MS (GC–MS) was achieved in the 1950s with commercial instruments available from the 1970s. Relatively cheap and reliable GC–MS systems are now a feature of many clinical biochemistry laboratories and are indispensable in several areas where the analysis of complex mixtures and unambiguous identification is required, for example, screening urine samples for inborn errors of metabolism or drugs. The coupling of MS with LC (LC–MS) was an obvious extension, but progress in this area was limited for many years due to the relative incompatibility of existing MS ion sources with a continuous liquid stream. Several interfaces were developed but they were cumbersome to use and unreliable, so uptake by clinical laboratories was very limited. This situation changed with the development of the electrospray (ES) ion source by Fenn in the 1980s [63]. Manufacturers rapidly developed instruments equipped with ES sources, which had a great impact on protein and peptide biochemistry. Fenn was awarded the Nobel Prize in 2002 with Koichi Tanaka, who developed matrix-assisted laser desorption ionization, another extremely useful MS ionization technique for the analysis of biological molecules. By the mid-1990s, the price and performance of LC–MS instruments had improved to the extent that clinical biochemistry laboratories were able to take advantage of the new technology. Biochemical genetics was one of the first areas to do so, and the analysis of neonatal dried blood spot samples for a range of inborn errors of metabolism was a major early application [64]. There are a number of other clinical applications of LC–MS, and the technique is more generally applicable than GC–MS, owing to the broader range of biological molecules that can be analyzed and the greater use of LC separations in clinical laboratories. The reasons for choosing LC–MS over LC with conventional detectors are essentially the same as with GC–MS, namely, high specificity and the ability to handle complex mixtures.

15.3.4 Alternative Acquisition and Confirmatory Systems

Milk contains a highly complex mixture of neutral and acidic oligosaccharides; up to now, more than 100 different structures have been detected. MS has shown great utility for the detection of milk saccharides like lactose. Many different analytical methods using LC with MS detection have also been reported in the literature for the detection of lactose in milk with the application of mass analyzers. These methods use ES as an ion source, and in some cases tandem MS is also applied. Limits of detection below 10 ppb are achieved for compounds in milk using multiple analyzer configuration, that is, triple quadrupole instruments. The same sensitivity is obtained by utilizing an ion trap mass analyzer. This latter analyzer provides full scan MS/MS performance at full instrument sensitivity, in contrast to triple quadrupole MS/MS, which requires a selection of only a few ions to maximize sensitivity. Electrospray ionization mass spectrometry (ESI–MS) and MS/MS are valuable detection methods for underivatized lactose; however, the absence of distinctive ions and collision-induced dissociation (CID) fragmentation patterns does not allow discrimination of stereo isomers without good chromatographic resolution. Ultra-high-performance liquid chromatography–ESI (U-HPLC–ESI) approach, based on PGC columns, working at 5°C has also been developed to separate and detect the disaccharides in their anomeric forms as formate adducts obtained directly in-column by eluting with formate buffer/acetonitrile gradient mixtures. Paul trap has been utilized to monitor the adducts $[M + HCOO]^-$ at *m/z* 387 in ESI negative mode (MS^1) as well as the CID fragment ion $[M - H]^-$ at *m/z* 341 (MS^2) and uses MS^3 fragment ions at *m/z* 178 and 161 to confirm disaccharides in milk.

15.4 LEVELS OF LACTOSE IN MILK

Lactose, a disaccharide found in milk at concentrations from 4.5 to 5.0 g/100 mL, is absorbed after conversion to its monosaccharide units, glucose and galactose by lactase. The majority of milk consumed in India comes from cows; other animals, including goat, sheep, and buffalo, also produce milk. Liquid cow milk is available as nonfat, 1%, 2%, or whole milk, and contains 11 g lactose per 1 cup of milk,

TABLE 15.1
Amount of Lactose in Different Milk Samples

In 100 g of Different Milk	Amount of Lactose
Condensed milk	10% 12.5 g
Full-fat milk	3.5% 4.8 g
Full-fat milk powder	38.0 g
Goat milk	4.1 g
Low-fat milk	5.0 g
Low-fat milk powder	52.0 g
Sheep milk	4.8 g
Sweetened condensed milk	10.2 g
Sweetened condensed milk (light)	12.8 g
Yogurt milk based	3.5% 4.0 g

according to the University of Virginia Health System. There is a fall in lactose level in autumn. This is due to changes in the physiology and metabolism of the mammary gland. In late lactation, the lactose content in milk declines, coinciding with the decline in milk production. In an autumn-calving herd, where cows are calving over a 6-month period, there are no major declines in milk lactose, as it is balanced by cows at both ends of lactation. However, the problem is much more pronounced in a spring-calving herd, where all the cows have calved in a 12-week period and they enter into late lactation at the same time.

The lactose content of different types of milks is given in Table 15.1. Canned milks were originally produced to extend the shelf life of milk. To produce canned milks, water is removed, which concentrates the milk, therefore concentrating the lactose. Evaporated milk, also known as "dehydrated milk," removes about 60% of the water. It, therefore, contains 24 g of lactose per 1 cup serving. Evaporated milk can replace fresh milk when mixed in equal parts with water. Sweetened condensed milk, also available in cans, removes water and adds sugar to inhibit bacterial growth. Sweetened condensed milk contains 40 g of lactose per cup. Flavored milks, such as chocolate milk, also contain 11 g of lactose per cup. Removing the cream and water from milk produces nonfat dry milk, which still contains all the vitamins and minerals of fluid milk. By removing all of the water, dry milk contains more lactose than any other milk product. Often used in commercial products because it contributes less weight, dry milk contains 62 g lactose per 1 cup of un-reconstituted powder.

15.5 A VALIDATED METHOD FOR THE DETERMINATION OF LACTOSE IN MILK BY UPLC–MS/MS

A Waters Acquity UPLC coupled to an Acquity tandem quadrupole detector (Waters Corporation, Manchester, UK) was used for all analyses. Chromatographic separations were performed on Acquity UPLC BEH C18 1.7 μm column (2.1 × 100 mm column) at 45°C at a flow rate of 400 μL/min running a gradient 98% of "A" as initial setting for

0.8 min, 90% of "A" for 0.5 min, 60% of "A" for 1.8 min, and back to an initial setting of 97% of "A" for 1.9 min. Total run time was 5 min. The mobile phase "A" consisted of water with 0.1% formic acid and "B" of acetonitrile with 0.1% formic acid. Argon was used as collision gas and its gas flow was set to be 0.1 mL/min. A highly purified nitrogen was used as desolvation gas and its temperature and flow rate were 400°C and 650 L/h, respectively. The Waters Tandem Quadruple detector was operated in negative ion mode using the following settings: capillary voltage, source temperature, and cone voltage were set to be 2.00 kV, 150°C, and 22 V, respectively. The quantification was performed by monitoring the transitions of *m/z* 353 → *m/z* 167 for labeled lactose and *m/z* 341 → *m/z* 161 for the lactose.

Cow milk with known lactose concentrations was collected from the certified laboratories in India. Twenty-five different cow milk samples were collected from local dairy farms. [UL-$^{13}C_{12}$]-lactose monohydrate was purchased from Omicron Biochemicals Inc. Lactose monohydrate from Sigma Aldrich, water, acetonitrile, and formic acid were obtained as Optima LC/MS grade from Fisher Scientific.

The linearity of the calibration curve was ascertained by adding known concentrations of manufactured lactose to cow milk over the concentration range of interest. Intraday precision was determined by analyzing 15 replicates of low- and high-quality control samples. Interday precision was determined by analyzing quality control samples over 15 consecutive days, and the coefficient of variation for lactose over two levels was calculated. The low limit of quantification was conducted by additional dilution until indication of peak-to-peak signal-to-noise ratio as 5:1 was achieved. To determine the recovery rate, lactose was added to cow milk samples. Absolute recovery was indicated by a ratio of the observed value to the corresponding expected value.

15.5.1 Calibration Curve for Cow Milk

The linear regression of concentrations versus the signal intensity over the concentration range of 0.6–4.6 mg/L for lactose in diluted cow milk is shown in Figure 15.2. The linear regression equation of cow milk was $y = 142.5x–31.4$, where x and y indicated concentration and signal intensity, respectively. The calibration curve presented in Figure 15.2 showed an R^2 value of 0.988, when plotted with the concentration and response axes.

15.5.2 UPLC-MS/MS Spectra of Native and Labeled Lactose in Negative Ion Mode

The mother ions (*m/z* 353) of labeled lactose and (*m/z* 341) of native lactose readily fragments into daughter ions with the negative ion mode. The different daughter ions for labeled lactose were *m/z* 105, 149, and 166, and for native lactose were *m/z* 101, 143, and 160. Under the given experimental condition, *m/z* 160 and 166 were the most abundant fragments for labeled lactose and native lactose, respectively, in the full MS spectra. Therefore, the fragments with *m/z* of 160 and 166 were selected for the quantification of lactose in milk (Figure 15.3).

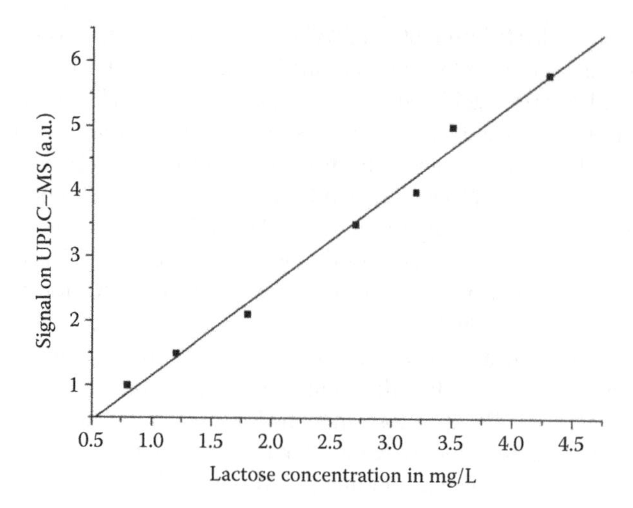

FIGURE 15.2 Calibration curves for lactose in milk via UPLC–MS/MS.

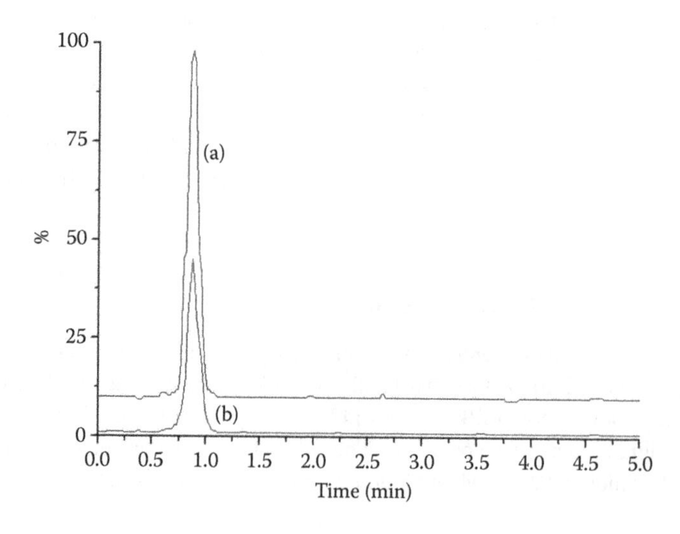

FIGURE 15.3 UPLC–MS/MS chromatogram of lactose standard (a) and of a real milk sample (b).

15.5.3 Validation of Determination of Lactose by UPLC–MS/MS

Intraday and interday precision were evaluated using quality control samples with low and high lactose content. An intraday precision study with low-lactose quality control samples was determined by analyzing 15 replicates per day, and the mean lactose content and coefficient of variation was found to be 54.6 g/L and 2.11%, respectively. Mean lactose content and coefficient of variation with high-lactose quality control samples was found to be 70.1 g/L and 1.88%, respectively. Interday precision was determined by analyzing quality control samples over 15 consecutive days, and the mean lactose content and coefficient of variation was found to be 53.2 g/L and 2.62%, respectively. Mean lactose content and coefficient of variation with high-lactose quality control samples was found to be 71.2 g/L and 2.12%, respectively. Coefficient of variation for intraday and interday was achieved, ranging from 1.88% to 2.11% and 2.12% to 2.62%, respectively, which fell into the expected range. The recovery rate was 96–102% by adding 26, 56, and 110 g/L of pure lactose to low-quality control samples. The limit of quantification was <5 ng/L.

15.5.4 Comparison of the Results Obtained by UPLC–MS/MS and Lactose Assay Kit

Twenty standard milk samples were collected from different certified laboratories. The accuracy of the lactose concentration was determined by choosing commercially available milk standards for cow milk, as well as an enzymatic lactose assay, which is regarded as an official method for lactose determination in the dairy industry. The lactose concentrations were also determined by the newly developed UPLC–MS/MS application. Lactose concentration obtained by UPLC–MS/MS and lactose assay kit was plotted against its standard lactose concentration and is presented in Figure 15.4. The mean regression equation indicated $y = 1.0276x - 0.7422$ with $r^2 = 0.982$, where x was the known lactose concentration of the milk from certified laboratories and y was the UPLC–MS/MS values. These results were confirmed with a lactose assay kit, which showed the mean regression equation as $y = 1.1156x - 5.4074$, with $r^2 = 0.894$. The mean difference of the measured values between UPLC–MS/MS and enzymatic

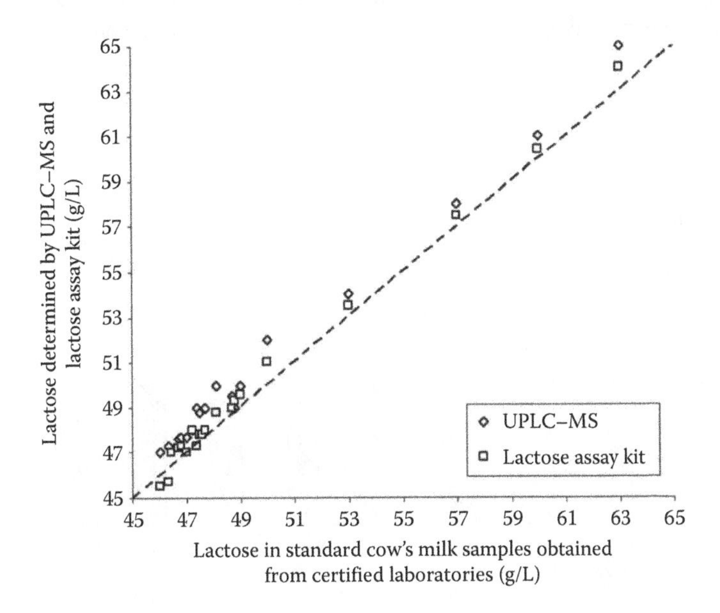

FIGURE 15.4 Comparison of concentrations of 20 known standard cow milk samples versus values obtained by UPLC–MS/MS and lactose assay kit.

lactose assay was 1.6 g/L indicating that the two methods in cow milk are in good agreement. The deviation of concentrations of the milk from certified laboratories and UPLC–MS/MS was 4.8%. It might be explained by cow milk samples from certified laboratories that were calculated by simply subtracting measured protein and fat contents from total solids corrected by a factor of 0.95. The lactose concentrations detected by enzymatic assay kit were 1.8% lower than the UPLC–MS/MS method [65]. Casadio et al. demonstrated in a validation study of a milk analyzer that calibration using an HPLC method showed lower lactose values than using the enzymatic lactose assay [66]. It is due to the fact that UPLC–MS/MS measures lactose directly but the enzymatic lactose assay measures galactose based on the hydrolysis of lactose into galactose and glucose by β-galactosidase [67,68]. The enzyme β-galactosidase is additionally able to convert oligosaccharides to galactose and glucose because of galactose–glucose-based oligosaccharides backbone structure [33,69]. We observed insignificant differences in values between UPLC–MS/MS and enzymatic lactose assay in cow milk since only small amounts of oligosaccharides are detectable in cow milk [68].

15.5.5 Determination of Lactose Content in Different Milk Samples Collected from Local Farms

Milk samples (five cow milk samples) were collected from different local dairy farms, and the lactose content was determined by the UPLC–MS/MS method. Milk samples were diluted to 150,000-fold to meet optimal concentration that UPLC–MS/MS can detect with high sensitivity and specificity. Packaged double-toned milk and flavored milk samples (like Amul Taaza, Amul Kool, and Amul Kool Café) were also collected and the lactose content was determined. Percentage lactose content of the milk

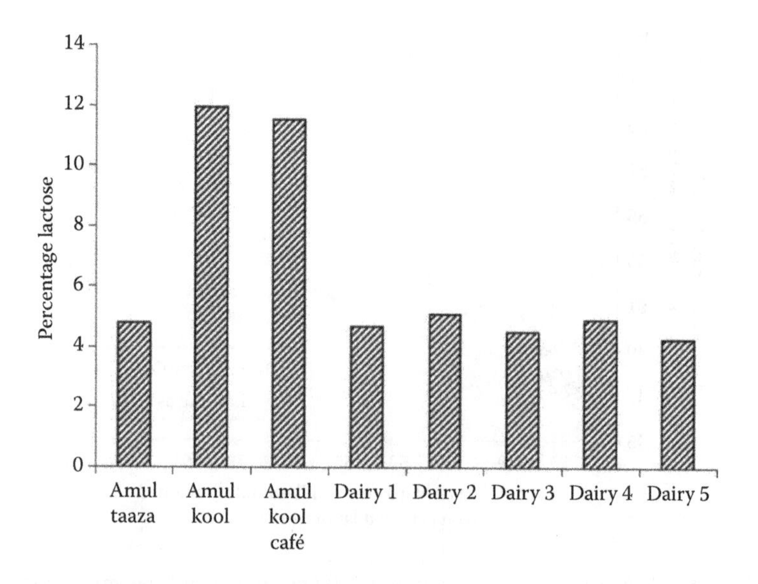

FIGURE 15.5 Lactose present in different milk samples.

samples are presented in Figure 15.5. It is evident from the figure that Amul Kool and Amul Kool Café (flavored milk) had high lactose content compared to that of Amul Taaza (double-toned milk) and milk samples collected from local dairies. The percentage of lactose present in Amul Taaza was found to be 4.8%. Amul Kool and Amul Kool Café (flavored milk) were found to contain 11.9% and 11.5%, respectively. A high result obtained for Amul Kool and Amul Kool Café (flavored milk) in the determination process by UPLC–MS/MS is possibly due to the presence of added sugar in the flavored milk samples. The UPLC–MS/MS method we used cannot separate among sucrose, maltose, and lactose since these disaccharides have the same molecular weight. Percentage of lactose in the milk samples collected from local farms varied in the range of 4.3–5.1%. Lactose present in local cow milk samples were within the normal range. Hence, the determination of lactose content in cow milk by UPLC–MS/MS is fast, reliable, and efficient. However, the method fails when applied for the determination of lactose in the milk samples containing other disaccharides.

Despite high sensitivity and specificity of lactose measurement via UPLC–MS/MS, the column we used cannot separate among sucrose, maltose, and lactose since these disaccharides have the same molecular weight. However, this fact can be disregarded for cow milk because 90% of their carbohydrates are composed of lactose, and they do not contain a significant amount of disaccharides other than lactose.

15.6 SUMMARY AND CONCLUSIONS

Lactose is the major carbohydrate in milk. Its measurement is important because it contributes to both the sensory and functional properties of milk. It also has economic value since the price of milk is based on milk solids content. The lactose content of cow milk can vary from 3.8% to 5.3% (g/100 mL) (38,000–53,000 μg/mL). Modern 1% and 2% milks have higher levels of lactose. Scientists are used to making compromises, and one of the most common scenarios involves sacrificing resolution for speed. With UPLC–MS/MS, increased resolution in shorter run times can generate more information faster without sacrifices. Acquity UPLC using 1.7 μm particles and a properly holistically designed system provide significantly more resolution (information) while reducing run times, and improve the sensitivity for the analyses of lactose in milk and many other compound types. At a time when many scientists have reached separation barriers with conventional HPLC, UPLC–MS/MS presents the possibility to extend and expand the utility of chromatography. The new technology in both chemistry and instrumentation boosts productivity by providing more information per unit of work, as UPLC–MS/MS fulfills the promise of increased resolution, speed, and sensitivity predicted for liquid chromatography.

ACKNOWLEDGMENTS

The author is thankful to Professor K. M. Purohit (director), Purushottam Institute of Engineering & Technology, Rourkela, India, for his necessary help and cooperation. The author is also thankful to Professor S. Nayak and Dr. P. C. Mishra, Department of Chemistry, Purushottam Institute of Engineering & Technology, Rourkela, India for providing necessary help in completing the chapter.

REFERENCES

1. Emmett, P. M. and Rogers, I. S. 1997. Properties of human milk and their relationship with maternal nutrition. *Early Human Development* 49:7–28.
2. Sharif, A., Ahmad, T., and Bilal, M. Q. 2007. Estimation of milk lactose and somatic cells for the diagnosis of sub-clinical mastitis in dairy buffaloes. *Int. J. Agric. Biol.* 9:267–270.
3. Essig, A. M. and Kleyn, D. H. 1983. Determination of lactose in milk: Comparison of methods. *J. Assoc. Off. Anal. Chem.* 66:1514–1516.
4. Smith, J. S., Villalobos, M.C., and Kotteman, C.M. 1986. Quantitative determination of sugars in various food products. *J. Food Sci.* 51:1373–1375.
5. Betschart, H. F. and Prenosil, J. E. 1984. High performance liquid chromatography analysis of the products of enzymatic lactose hydrolysis. *J. Chromatogr.* 299:498–502.
6. Yang, M. T., Milligan, L. P., and Mathison, G. W. 1981. Improved sugar separation by high-performance liquid chromatography using porous microparticles carbohydrate columns. *J. Chromatogr.* 209:316–322.
7. Tkac, J., Sturdik, E., and Gemeiner, P. 2000. Novel glucose non-interference biosensor for lactose detection based on galactose oxidase-peroxidase with and without co-immobilised galactosidase. *Analyst* 125:1285–1289.
8. Brereton, P., Hasnip, S., Bertrand, A., Wittkowski, R., and Guillou, C. 2003. Analytical methods for the determination of spirit drinks. *Trends Analyt. Chem.* 2:19–25.
9. Gonzáles, A. S. P., Naranjo, G. B., Malec, L. S., and Vigo, M. S. 2003. Available lysine, protein digestibility and lactulose in commercial infant formulas. *Int. Dairy J.* 13:95–99.
10. Salvador, L. D., Suganuma, T., Kitahara, K., Fukushige, Y., and Tanoue, H. 2002. Degradation of cell wall materials from sweet potato, cassava and potato by a bacterial protopectinase and terminal sugar analysis of the resulting solubilized products. *J. Biosci. Bioeng.* 93:64–72.
11. Troyano, E., Olano, A., Fernández-Díaz, M., and Sanz, J. 1994. Gas chromatographic analysis of free monosaccharides in milk. *Chromatographia* 32:379–382.
12. Olano, A., Calvo, M. M., and Corzo, N. 1989. Changes in the carbohydrate fraction of milk during heating processes. *Food Chem.* 31:259–265.
13. Ruas-Madiedo, P., delosReyes-Gavilan, C. G., Olano, A., and Villamiel, M. 2000. Influence of refrigeration and carbon dioxide addition to raw milk on microbial levels free monosaccharides and myoinositol content of raw and pasteurized milk. *Eur. Food Res. Technol.* 212:44–47.
14. Watanabe, E., Takagi, M., and Takei, S. 1991. Development of biosensors for the simultaneous determination of sucrose and glucose lactose and glucose, and starch and glucose. *Biotechnol. Bioeng.* 38:99–103.
15. Garcia, J. L., Lopez-Mungia, A., and Galindo, E. 1991. Modeling the non-steady-state response of an enzyme electrode for lactose. *Enzyme Microb. Technol.* 13:672–675.
16. Cass, A. E. G., Davis, G., and Francis, G. D. 1984. Ferrocene-mediated enzyme electrode for ampero-metric determination of glucose. *Anal. Chem.* 56:667–671.
17. Albery, J. W., Kalia, Y. N., and Magner, E. 1992. Amperometric enzyme electrodes. Part VI. Enzyme electrodes for sucrose and lactose. *J. Electroanal. Chem.* 325:83–93.
18. Gulce, H., Gulce, A., and Yildiz, A. 2002. A novel two-enzyme amperometric electrode for lactose determination. *Anal. Sci.* 18:147150.
19. Xu, Y., Guilbault, G. G., and Kuan, S. S. 1990. Fast responding lactose enzyme electrode. *Enzyme Microb. Technol.* 12:104–108.
20. Adanyi, N., Szabo, E. E., and Varadi, M. 1999. Multi-enzyme biosensor with amperometric detection for determination of lactose in milk and dairy products. *Eur. Food Res. Technol.* 209:206–226.

21. Rajendran, V. and Irudayaraj, J. 2002. Detection of glucose galactose and lactose in milk with a microdialysis-coupled flow injection amperometric sensor. *J. Dairy Sci.* 85:1357–1361.
22. Sharma, S. K., Singhal, R., Malhotra, B. D., Sehgal, N., and Kumar, A. 2004. Lactose biosensor based on Langmuir–Blodgett films of poly(3-hexyl thiophene). *Biosens. Bioelectron.* 20:651–657.
23. Eshkenazi, I., Maltz, E., Zion, B., and Rishpon, J. 2000. A three-cascaded-enzymes biosensor to determine lactose concentration in raw milk. *J. Dairy Sci.* 83:1939–1945.
24. Leger, C. and Bertrand, P. 2008. Direct electrochemistry of redox enzymes as a tool for mechanistic studies. *Chem. Rev.* 108:2379–2438.
25. Lindgren, A., Ruzgas, T., and Gorton, L. 2000. Biosensors based on novel peroxidases with improved properties indirect and mediated electron transfer. *Biosens. Bioelectron.* 15:491–497.
26. Lindgren, A., Gorton, L., and Ruzgas, T. 2001. Direct electron transfer of cellobiose dehydrogenase from various biological origins at gold and graphite electrodes. *Electroanal. Chem.* 496:76–81.
27. Ferapontova, E. E., Ruzgas, T., and Gorton, L. 2003. Direct electron transfer of heme- and molybdopterin cofactor-containing chicken liver sulfite oxidase on alkanethiol-modified gold electrodes. *Anal. Chem.* 75:4841–4850.
28. Razumiene, J., Niculescu, M., Ramanavieius, A., Laurinavieius, V., and Csoregi, E. 2002. Direct bioelectrocatalysis at carbon electrodes modified with quinohemoprotein alcohol dehydrogenase from *Gluconobacter*. *Electroanalysis* 14:43–49.
29. Weiskopf, A. S., Vouros, P., and Harvey, D. J. 1997. Characterization of oligosaccharide composition and structure by quadrupole ion trap mass spectrometry. *Rapid Commun. Mass Spectrom.* 11:1493–1504.
30. Nikolov, Z. L., Jakovljevic, J. B., and Boskov, Z. M. 1984. High performance liquid chromatographic separation of oligosaccharides using amine modified silica columns. *Starke* 36:97–100.
31. Mulroney, B., Traeger, J. C., and Stone, B. A. 1995. Determination of both linkage position and anomeric configuration in underivatized glucopyranosyl disaccharides by electrospray mass spectrometry. *J. Mass Spectrom.* 30:1277–1283.
32. Liu, Y., Urgaonkar, S. Verkade, J. G., and Armstrong, D. W. 2005. Separation and characterization of underivatized oligosaccharides using liquid chromatography and liquid chromatography–electrospray ionization mass spectrometry. *J. Chromatogr. A* 1079:146–152.
33. Coppa, G. V., Gabrielli, O., Pierani, P., Catassi, C., Carlucci, A., and Giorgi, P. L. 1993. Changes in carbohydrate composition in human milk over 4 months of lactation. *Pediatrics* 91:637–641.
34. Ball, G. F. M. 1990. The application of HPLC to the determination of low molecular weight sugars and polyhydric alcohols in foods: A review. *Food Chem.* 35:117–152.
35. Dunmire, D. L. and Otto, S. E. 1979. High pressure liquid chromatographic determination of sugars in various food products. *J. Assoc. Off. Anal. Chem.* 62:176–185.
36. Brons, C. and Olieman, C. 1983. Study of the high-performance liquid chromatographic separation of reducing sugars, applied to the determination of lactose in milk. *J. Chromatogr. A.* 259:79–86.
37. March, R. E. and Stadey, C. J. 2005. A tandem mass spectrometric study of saccharides at high mass resolution. *Rapid Commun. Mass Spectrom.* 19:805–812.
38. Swartz, M. E. 2005. UPLC: An introduction and review. *J. Liq. Chromatogr. Related Technol.* 28:1253–1263.
39. Gabbanini, S., Lucchi, E., Guidugli, F., Matera, R., and Valgimigli, L. 2010. Anomeric discrimination and rapid analysis of underivatized lactose, maltose, and sucrose in

vegetable matrices by U-HPLC–ESI-MS/MS using porous graphitic carbon. *J. Mass Spectrom.* 45:1012–1018.

40. van Deemter, J. J., Zuiderweg, F. J., and Klinkenberg, A. 1956. Longitudinal diffusion and resistance to mass transfer as causes of nonideality in chromatography. *Chem. Eng. Sci.* 5:271–289.
41. Lars, Y. and Honore, H. S. 2003. On-line turbulent-flow chromatography–high-performance liquid chromatography–mass spectrometry for fast sample preparation and quantitation. *J. Chromatogr. A.* 1020:59–67.
42. MacNair, J. E., Patel, K. D., and Jorgenson, J. W. 1999. Ultrahigh-pressure reversed-phase capillary liquid chromatography: Isocratic and gradient elution using columns packed with 1.0-µm particles. *Anal. Chem.* 71:700–708.
43. Wu, N., Lippert, J. A., and Lee, M. L. 2001. Practical aspects of ultrahigh pressure capillary liquid chromatography. *J. Chromotogr. A.* 911:1–12.
44. Unger, K. K., Kumar, D., and Grun, M. 2000. Synthesis of spherical porous silicas in the micron and submicron size range: Challenges and opportunities for miniaturized high-resolution chromatographic and electrokinetic separations. *J. Chromatogr. A.* 892:47–55.
45. Mazzeo, J. R., U. D. Neue, U. D., Kele, M., and Plumb, R. S. 2005. Advancing LC performance with smaller particles and higher pressure. *Anal. Chem.* 77:460–467.
46. Halasz, I., Endele, R., and Asshauer, J. 1975. Ultimate limits in high-pressure liquid chromatography. *J. Chromatogr.* 112:112, 37–60.
47. MacNair, J. E., Lewis, K. C., and Jorgenson, J. W. 1997. Ultrahigh-pressure reversed-phase liquid chromatography in packed capillary columns. *Anal. Chem.* 67:983–989.
48. Colon, L. A., Citron, J. M., Anspach, J. A., Fermier, A. M., and Swinney, K. A. 2004. Very high pressure HPLC with 1 mm id columns. *Analyst* 129:503–504.
49. Tolley, L., Jorgenson, J. W., and Mosely, M. A. 2001. Very high pressure gradient LC/MS/MS. *Anal. Chem.* 73:2985–2991.
50. Lippert, J. A., Xin, B., Wu, N., and Lee, M. L. 1999. Fast ultrahigh-pressure liquid chromatography: On-column UV and time-of-flight mass spectrometric detection. *J. Microcol. Sep.* 11:631–643.
51. Thurl, S., Henker, J., Siegel, M., Tovar, K., and Sawatzki, G. 1997. Detection of four human milk groups with respect to Lewis blood group dependent oligosaccharides. *Glycoconj J.* 14:795–799.
52. Thurl, S., Munzert, M., Henker, J., Boehm, G., Muller-Werner, B., Jelinek, J., and Stahl, B. 2010. Variation of human milk oligosaccharides in relation to milk groups and lactational periods. *Br. J. Nutr.* 104:1261–71.
53. Anumula, K. R. 2006. Advances in fluorescence derivatization methods for high-performance liquid chromatographic analysis of glycoprotein carbohydrates. *Anal. Biochem.* 350:1–23.
54. Royle, L., Campbell, M. P., Radcliffe, C. M., White, D. M., Harvey, D. J., Abrahams, J. L., Kim, Y. G., Henry, G. W., Shadick, N. A., and Weinblatt, M. E. 2008. HPLC-based analysis of serum N-glycans on a 96-well plate platform with dedicated database software. *Anal. Biochem.* 376:1–12.
55. Shen, Z. and Warren, C. D. 2000. Newburg DS. High-performance capillary electrophoresis of sialylated oligosaccharides of human milk. *Anal. Biochem.* 279:37–45.
56. Wu, S., Tao, N., German, J. B., Grimm, R., and Lebrilla, C. B. 2010. Development of an annotated library of neutral human milk oligosaccharides. *J. Proteome Res.* 9:4138–4151.
57. Caesar Jr., J. P., Sheeley, D. M., and Reinhold, V. N. 1990. Femtomole oligosaccharide detection using a reducing-end derivative and chemical ionization mass spectrometry. *Anal. Biochem.* 191:247–252.

58. Ciucanu, I. K. 1984. A simple and rapid method for the permethylation of carbohydrate. *Carbohydr. Res.* 131:209–217.
59. Thanawiroon, C., Rice, K. G., Toida, T., and Linhardt, R. J. 2004. Liquid chromatography/mass spectrometry sequencing approach for highly sulfated heparin-derived oligosaccharides. *J. Biol. Chem.* 279:2608–2615.
60. Eldridge, G. R., Vervoort, H. C., Lee, C. M., Cremin, P. A., Williams, C. T., Hart, S. M., Goering, M. G., O'Neil-Johnson, M., and Zeng, L. 2002. High-throughput method for the production and analysis of large natural product libraries for drug discovery. *Anal. Chem.* 74:3963–3971.
61. Wu, S., Grimm, R., German, J. B., and Lebrilla, C. B. 2011. Annotation and structural analysis of sialylated human milk oligosaccharides. *J. Proteome. Res.* 10:856–68.
62. Albrecht, S., Schols, H. A., vanden Heuvel, E. G., Voragen, A. G., and Gruppen, H. 2010. CE-LIF-MS n profiling of oligosaccharides in human milk and feces of breast-fed babies. *Electrophoresis.* 31:1264–73.
63. Fenn, J. B., Mann, M., Meng, C. K., Wong, S. F., and Whitehouse, C. M. 1989. Electrospray ionization for mass spectrometry of large biomolecules. *Science* 246:64–71.
64. Rashed, M. S., Bucknall, M. P., and Little, D. 1997. Screening blood spots for inborn errors of metabolism by electrospray tandem mass spectrometry with a microplate batch process and a computer algorithm for automated flagging of abnormal profiles. *Clin. Chem.* 43:1129–41.
65. Kwak, H. S. and Jeon, I. J. 1988. Comparison of high performance liquid chromatography and enzymatic method for the measurement of lactose in milk: A research note. *J. Food Sci.* 53:975–976.
66. Casadio, Y. S., Williams, T. M., Lai, C. T., Olsson, S. E., Hepworth, A. R., and Hartmann, P. E. 2010. Evaluation of a mid-infrared analyzer for the determination of the macronutrient composition of human milk. *J. Hum. Lact.* 26:376–383.
67. Torres, D. P. M., Goncalves, M. D. F., Teixeira, J. A., and Rodrigues, L. R. 2010. Galacto-oligosaccharides: Production, properties, applications, and significance as prebiotics. *Compr. Rev. Food Sci. F.* 9:438–454.
68. Kunz, C., Rudloff, S., Baier, W., Klein, N., and Strobel, S. 2000. Oligosaccharides in human milk: Structural, functional, and metabolic aspects. *Annu. Rev. Nutr.* 20:699–722.
69. Boehm, G., Stahl, B., Jelinek, J., Knol, J., Miniello, V., and Moro, G.E. 2005. Prebiotic carbohydrates in human milk and formulas. *Acta Paediatr. Suppl.* 94:18–21.

16 Fast and Reliable Analysis of Phenolic Compounds in Fruits and Vegetables by UHPLC–MS

M. I. Alarcón-Flores, R. Romero-González, J. L. Martínez Vidal, and A. Garrido Frenich

CONTENTS

16.1 INTRODUCTION

Functional foods represent an emerging market of growing economic importance [1,2], but they are usually related to processed food. However, vegetables and fruits are considered functional foods, as they provide beneficial health effects, also known as nutritional value, [3,4] due to phytochemicals such as vitamins, glucosinolates and phenolic compounds that can be found in their tissues [5]. Among the different groups of naturally occurring substances from plants, phenolic compounds could be

TABLE 16.1
Classification of Families of Phenolic Compounds

Class	Basic Skeleton	Basic Structure	Sources
Simple phenols	C_6	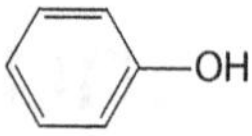 —OH	
Benzoquinones	C_6	 O= =O	
Phenolic acids	C_6–C_1	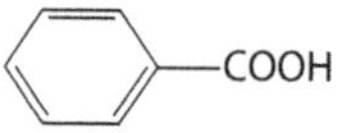 —COOH	Cranberry
Acetophenones	C_6–C_2	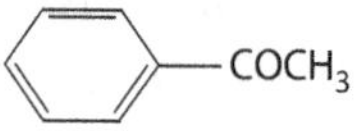 —$COCH_3$	Apple, apricot, banana, cauliflower
Phenylacetic acids	C_6–C_2	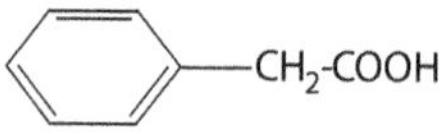 —CH_2-COOH	
Hydroxycinnamic acids	C_6–C_3	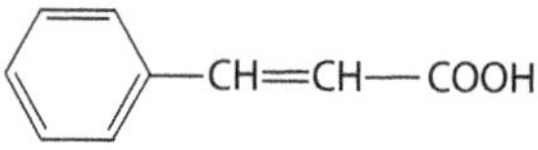 —CH=CH—COOH	Carrot, citrus, tomato, peaches, pears, eggplant
Coumarins, isocoumarins	C_6–C_3	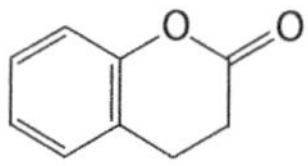 O, O	Carrot, celery, citrus, parsley
Chromones	C_6–C_3	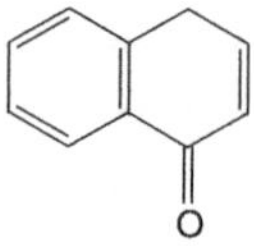 O	
Naftoquinones	C_6–C_4	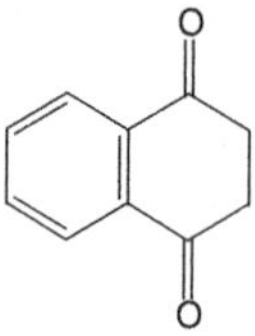 O, O	Spinach, cauliflower, orange
Xanthones	C_6–C_1–C_6	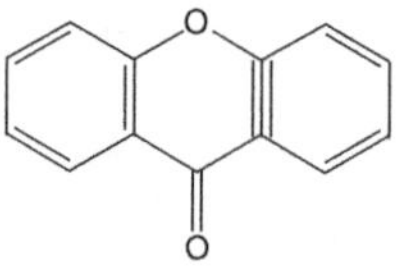 O, O	Mango
Stilbenes	C_6–C_2–C_6	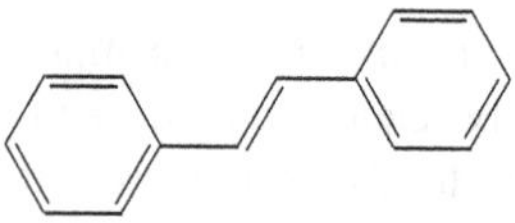	Grape

TABLE 16.1 (continued)
Classification of Families of Phenolic Compounds

Class	Basic Skeleton	Basic Structure	Sources
Flavonoids	C_6–C_3–C_6	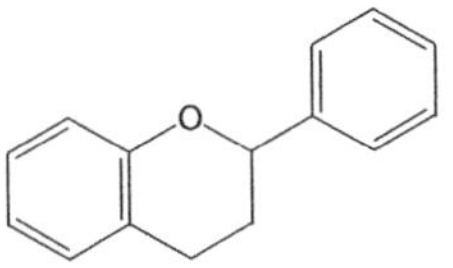	Widely distributed
Lignans, neolignans	$(C_6$–$C_3)_2$	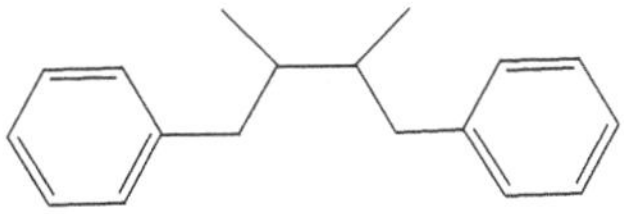	Pomegranate, raspberry, broccoli, cabbage
Lignins	$(C_6$–$C_3)_n$	Highly crosslinked aromatic polymer	

considered the most important [6]. Table 16.1 shows the chemical structures of typical phenolic compounds.

These compounds constitute one of the most numerous and widely distributed groups of natural products in the plant kingdom. More than 8000 phenolic structures are currently known, and among them, more than 4000 flavonoids have been identified [4,7,8]. In plants, most phenolic compounds can be found as glycosides, and several types and numbers of sugars can be combined at different positions of the aglycones, forming numerous structures. The most prevalent glycosylation is glucose, although rhamnose, galactose, xylose, and arabinose are also present [9], and the number of sugar ring substituents on the aglycon usually ranges from one to four. Disaccharides and occasionally tri- and even tetrasaccharides are often found combined with flavonoids [10]. In this sense, it has been demonstrated that the bioavailability of these compounds depends on the sugar moiety attached to the phenolic structure [11].

Phenolic compounds encompass analytes wherein chemical structures possess an aromatic ring with one or more hydroxyl functional groups. These are loosely classified into flavonoids and non-flavonoids. Flavonoids, within which structures are based on a C6-C3-C6 skeleton, are the most abundant group of phenolic compounds, and they are sub-divided into different classes, depending on the oxidation state of the central heterocyclic ring [6]. These comprise chalcones, flavonols, flavones, isoflavones, flavanones, anthocyanidins, and flavanols (catechins and tannins), among others. Tannins are classified into hydrolyzable and nonhydrolyzable or condensed tannins (proanthocyanidins). Hydrolyzable tannins are esters of phenolic acids and sugars, or their derivatives, that yield a sugar and a phenolic acid moiety upon hydrolysis. Proanthocyanidins are oligomers and polymers of flavan-3-ol monomeric units, which form colored anthocyanidins when they are subjected to acidic hydrolysis [6,12]. The classification of flavonoids can be observed in Table 16.2.

TABLE 16.2
Classification of Food Flavonoids

Flavonoid	Basic Structure
Chalcones	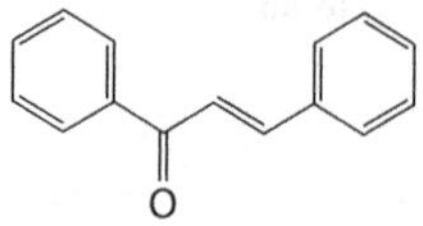
Dihydrochalcones	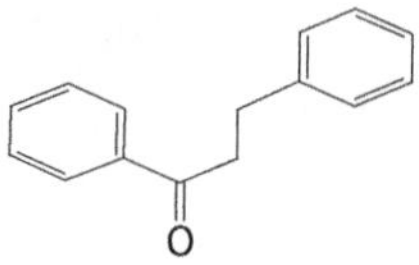
Flavones	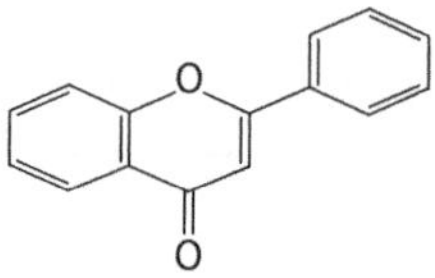
Flavonols	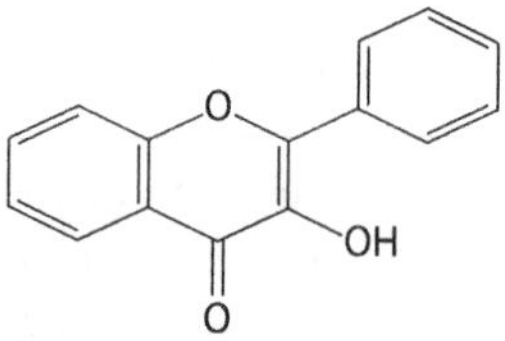
Dihydroflavonol	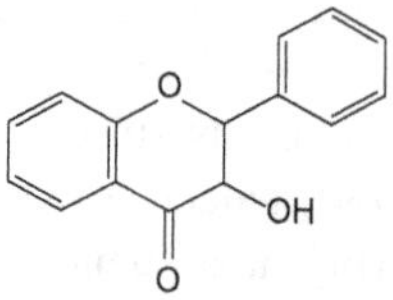
Flavanones	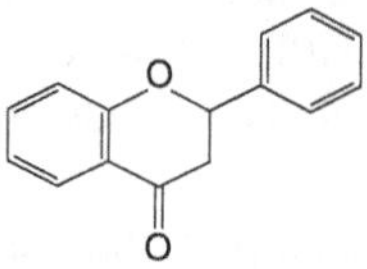
Flavanol	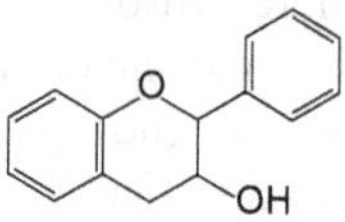
Flavandiol	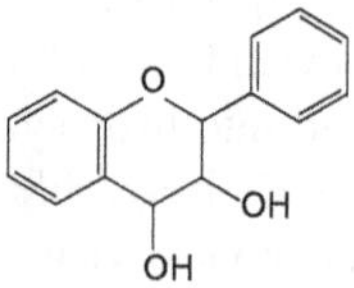

TABLE 16.2 (continued)
Classification of Food Flavonoids

Flavonoid	Basic Structure
Anthocyanidin	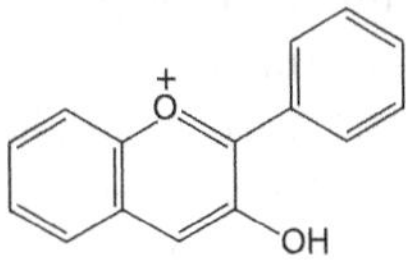
Isoflavonoids	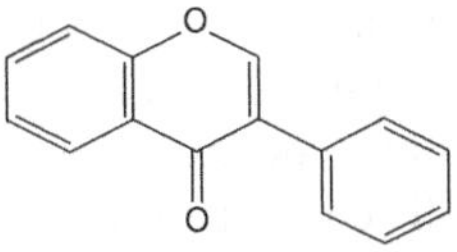
Proanthocyanidins or condensed tanins	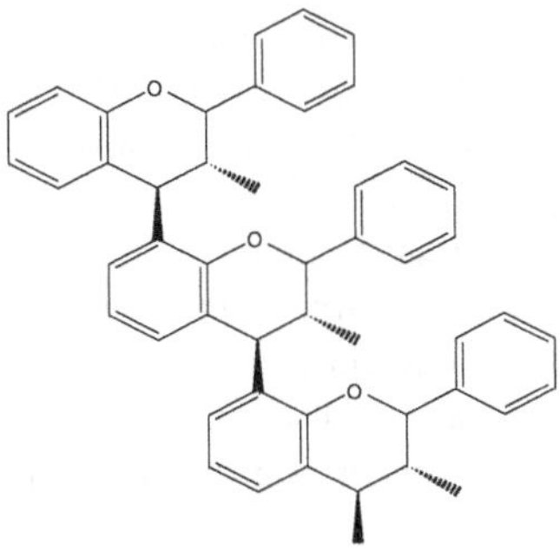

Nonflavonoid compounds comprise simple phenols, phenolic acids, coumarins, xanthones, stilbenes, lignins, and lignans. Phenolic acids are further divided into benzoic acid derivatives, based on a C6-C1 skeleton and cinnamic acid derivatives, which are based on a C6-C3 skeleton [6]. The coumarins are phenolic acid derivatives composed of a benzene ring fused with an oxygen heterocycle. Xanthones consist of a C6-C1-C6 basic structure, and stilbenes are composed of a C6-C2-C6 skeleton with various hydroxylation patterns [13]. Lignins are polymers of C6-C3 units, whereas lignans are made up of two phenylpropane units [13]. The structures of each of these classes are shown in Table 16.1.

These phenolic compounds seem to play a role against the development of different types of cancer and cardiovascular diseases, because these compounds provide, at least in part, antioxidant capacity (AOC) [14], anti-inflammation [15,16], lipid profile modification [17,18] and antitumor effects [19–22]. Besides these beneficial properties of phenolic compounds in human health, these compounds are responsible of color, flavor, and smell in fruits and vegetables [23], and their contents are influenced by variety, crop type, environmental conditions, location, germination, maturity, processing, and storage [24–26]. Therefore, several procedures have been developed for the analysis of this type of compound in the last few years.

In this sense, works regarding the extraction of the phenolic compounds from vegetables and fruits have attracted a special interest. Common extraction procedures, such as sonication-assisted solid–liquid extraction (SLE) [27,28], have been employed for the extraction of phenolic compounds from vegetables or fruits, applying solvents as methanol or mixtures of methanol:water at different ratios [28,29].

Moreover, there is an increasing demand for sensitive and selective analytical methods for the determination of phenolic compounds. The most commonly used technique is high-performance liquid chromatography (HPLC) coupled to ultraviolet (UV) detection [30,31], photodiode-array detection (DAD) [32], mass spectrometry (MS) [33], DAD-MS [34–36], UV-MS [37], or tandem MS (MS/MS) [38–41]. Among all of them, MS/MS has emerged as one of the preferred analytical techniques for quantification purposes, offering sufficient sensitivity, as well as capability of unambiguous evidence for phenolic compounds identification and quantification at a broad concentration range from a single injection. However, HPLC analysis of phenolic compounds requires a running time of 20 min or longer. To reduce analysis time, ultra-high-performance liquid chromatography (UHPLC) can be used instead. The advantages of this technique are better resolution, shorter running time, and higher sensitivity. Furthermore, UHPLC can be coupled to MS/MS for routine analysis, because it allows a rapid detection of more compounds in shorter running time [42].

Bearing in mind that the determination of phenolic compounds in fruits and vegetables has been of increasing interest [43], the objective of this chapter is to provide an overview of the current analytical methods, mainly based on UHPLC to MS, for the determination of these analytes in matrices from vegetal origin, achieving the analytical goals concerning the separation, identification, and quantification of phenolic compounds.

16.2 SAMPLE PREPARATION

Most of the extraction techniques of phenolic compounds from vegetables are based on ultrasound-assisted extraction (UAE) [27,44,45]. In addition, other techniques have been successfully applied to the pretreatment of phenolic compounds in fruits and vegetables, including pressurized liquid extraction (PLE) [46], solid-phase extraction (SPE) [47], supercritical fluid extraction (SFE) [48], microwave-assisted extraction (MAE) [49], rotary shaker-assisted extraction (RAE), [50] and QuEChERS (acronym of quick, easy, cheap, effective, rugged and safe) [51], as can be observed in Tables 16.3 and 16.4. In some cases, an acid treatment [52] was applied to hydrolyze the glycosides in order to determine the content of free and conjugated flavonoids as aglycons.

16.2.1 Ultrasound-Assisted Extraction

UAE is a cheap, fast, and effective method for extracting phenolic compounds from fruits and vegetables. Table 16.3 shows some characteristics of the developed methods based on this technique. This is an efficient way of extracting compounds from different matrices, bearing in mind that the extraction time is shorter than with other techniques [53]. Therefore, the higher yields achieved in these UAE processes have provoked an increasing interest from an industrial standpoint [54].

TABLE 16.3
Ultrasound-Assisted Procedures Used for the Extraction of Phenolic Compounds from Fruits and Vegetables

Family of Compounds	N° of Studied Compounds	Matrix	Solvent	Extraction Conditions	Recoveries	Reference
Dihydrochalcones Flavones Flavonols Isoflavones Phenolic acids Procyanidins	25	Apples Pear Beans	Methanol:water (80:20, v/v)	120 min Room temperature	95%–103%	[57]
Flavonols Flavonones Isoflavones Phenolic acids	11	Tomato	Methanol	60 min	81%–89%	[59]
Chalcones Cinnamic acids Flavonols Flavonones Isoflavones Phenolic acids	10	Tomato Pepper Eggplant	Methanol:water (80:20, v/v)	20 min Room temperature	NS[a]	[28]
Flavones Flavonols	6	Tomato, mango Onion, apple Garlic, orange Potato, chilies strawberries	Methanol:water (80:20, v/v)	60 min Room temperature	NS	[45]
Flavonols	14	Seed tomato	Methanol:water (50:50, v/v)	120 min	NS	[44]

continued

TABLE 16.3 (continued)
Ultrasound-Assisted Procedures Used for the Extraction of Phenolic Compounds from Fruits and Vegetables

Family of Compounds	N° of Studied Compounds	Matrix	Solvent	Extraction Conditions	Recoveries	Reference
Flavones Flavonols Flavonones Phenolic acids	56	Tomato	Methanol	180 min	NS	[27]
Flavonols Flavonones Phenolic acids	15	Tomato	Methanol:water (80:20, v/v)	60 min Room temperature	NS	[56]
Flavonones Phenolic acids	10	Citrus fruit	Methanol:dimethylsulfoxide (50:50, v/v)	20 min, 37°C	92–98%	[35]
Flavonols Phenolic acids	7	Tomato	Methanol:water (50:50, v/v)	60 min Room temperature	NS	[58]
Phenolic acids	6	Carrot	Acetone:water:acetic acid (70:29.5:0.5, v/v/v)	10 min, 37°C	NS	[60]

Note: NS, not specified.

TABLE 16.4
Extraction Techniques Most Widely Used for the Extraction of Phenolic Acids from Fruits and Vegetables

Family Compounds	N° Studied Compounds	Matrix	Extraction Type	Solvent	Extraction Time (min)	Recoveries (%)	Reference
Dihydrochalcones Flavanols Flavonols Phenolic acids	13	Apple	PLE	Methanol	15	92–100	[46]
Phenolic acids	10	Parsley	PLE	Ethanol:water (50:50, v/v) and/or Acetone:water (50:50, v/v)	15	NS	[64]
Flavonols Phenolic acids	9	Onion, potato, carrot	PLE + clean up	Methanol:water (65:35, v/v)	16	98–99	[62]
Phenolic acids	4	Potato peels	PLE	Methanol:water (90:10, v/v)	15	NS	[63]
Flavones Flavonols	10	Apple juice	SPE	Methanol	NE	96–98	[65]
Anthocyanidins	14	Tomato	SPE	Methanol:formic acid (95:5)	NE	92–105	[47]
Flavanols Phenolic acids	3	Grape seeds	SFE	CO_2 + Methanol	15	NS	[67]
Anthocyanidins	4	Grape	SFE	CO_2 + ethylenalcohol (25–30%)	NE	80–85	[68]
Flavanols Phenolic acids	8	Grape pomace	SFE	CO_2 + ethanol	60	NS	[48]

continued

TABLE 16.4 (continued)
Extraction Techniques Most Widely Used for the Extraction of Phenolic Acids from Fruits and Vegetables

Family Compounds	N° Studied Compounds	Matrix	Extraction Type	Solvent	Extraction Time (min)	Recoveries (%)	Reference
Chalcones Cinnamic acids Flavonols Flavonones Isoflavones Phenolic acids	10	Tomato, pepper, eggplant	SFE	CO_2 + methanol	20	NS	[28]
Coumarins Flavanols Flavonols Phenolic acids Stilbens	22	Grape	MAE	Methanol	20	94–108	[49]
Flavonols Flavonones Phenolic acids	15	Tomato	MAE	Ethanol:water (66.2:33.8, v/v)	2	94–105	[72]
Anthocyanidins Flavanols Flavonols	23	Beans	RAE	Methanol (100%)	Overnight, 20°C	NS	[74]
Flavonols glycosides Phenolic acids	18	Eggplant	RAE	Methanol:water (80:20, v/v)	Overnight Room temperature	NS	[50]
Phenolic acids (Glucosinolates)	32	Brassica vegetables	RAE	Methanol:water (70:30, v/v) 70 C	30	NS	[29]
Anthocyanidins Phenolic acids	18	Tomato	RAE	Methanol:water:HCl (80:20:0.1, v/v/v)	Overnight Room temperature	NS	[72]

Flavones Flavonols Isoflavones Phenolic acids (Glucosinolates)	30	Tomato, broccoli grape, carrot, eggplant	RAE	Methanol:water (80:20, v/v)	30	60–110	[75]
Flavanols Flavonols Cinnamic acid Phenolic acids Stilbenoid	15	Carrot, tomato, broccoli, onion, garlic, pepper, beetroot	QuECHERS + clean up	Acetonitrile:ethylacetate (50:50)	15 min + clean up	78–99	[51]
Flavonols Flavones	6	Tomato, onions, lettuce	Hydrolysis	1.2 M HCl in 50% aqueous methanol	2 h, 90°C	NS	[79]
Flavonols Flavones	6	Celery	Hydrolysis	2 M HCl in 50% aqueous methanol	4 h, 90°C	NS	[79]
Flavonols	6	Tomato	Hydrolysis	1.2 M HCl in 50% aqueous methanol	2 h, 90°C	NS	[52]
Flavones Flavonols Flavonones	7	Tomato	Hydrolysis	1.2 M HCl in 60% aqueous methanol	2 h, 90°C	NS	[82]
Flavonols	3	Tomato, pepper	Hydrolysis	2 M HCl in 50% aqueous methanol	4 h, 100°C	92–93	[80]
Flavonols	3	Peas, cabbages, spinach, cauliflower, turnip, onion, garlic, ginger, apple, plum, apricot, strawberry, mulberry, carrot,	Hydrolysis	1.2 M HCl	2 h, 90°C	NS	[84]
Flavonols Flavones	5	100 vegetables and fruits	Hydrolysis	Ethanol:water:HCl (80:10:10)	3 h, 90°C	93–98	[81]

Note: Abbreviations: MAE: microwave-assisted extraction; NS: not specified; PLE: pressurized liquid extraction; RAE: rotatory shaker-assisted extraction; SFE: supercritical fluid extraction; SPE: solid-phase extraction.

The extraction yield of phenolic compounds by ultrasound is attributed to the phenomenon of cavitation produced in the solvent by the ultrasonic wave. Cavitation bubbles are produced, which are filled with solvent vapour. During the compression cycle, the bubbles and the gas inside them are also compressed, resulting in a significant increase in temperature and pressure. This finally results in the collapse of the bubble with a resultant "shock wave" passing through the solvent and enhanced mixing occurs. Ultrasound also exerts a mechanical effect, allowing greater penetration of solvent into the plant tissues [55]. Therefore, the extraction efficiency of phenolic compounds from fruits and vegetables is improved by ultrasound, but it also depends on the type of solvent used during the extraction procedure. The most common solvents are those using mixtures of methanol:water at several ratios, as (80:20) [28,45,56,57] or (50:50) [44,58] or pure methanol [27,59]. However, Vilkhu et al. indicated that the use of methanol during UAE produced the lowest recovery, and these results were not statistically different from maceration with 70% ethanol. Other solvents as a mixture of water:acetone:acetic acid (70/29.5/0.5, v/v/v) [60] or methanol:dimethylsulfoxide (50:50, v/v) [35] were less frequently used, as it can be observed in Table 16.3.

Although tomato was the matrix most studied with this technique, other matrices such as pepper, eggplant, apple, pear, beans, mango, orange, citrus fruit, onion, garlic, potato, chillies, and strawberries were also studied by UAE. The extracted phenolic compounds included phenolic acids, isoflavones, flavonols, flavonones, flavones, flavonones, cinnamic acids, dihydrochalcones, chalcones, flavanols, and procyanidins (Table 16.3).

In general, when this technique is applied, recoveries ranged from 81% to 103% [35,57,59], although in most of the cases, recovery values were not provided [27,28,44,45,56,58].

One of the disadvantages of the UAE is the coextraction of others substances, such as proteins, sugars, and organic acids, which might interfere with the quantification of the phenolic compound [53]. To remove these interferences, the use of C18 cartridges is widely recommended [43] as a clean-up step.

16.2.2 Pressurized Liquid Extraction

A relatively new automated extraction method is PLE, also called accelerated solvent extraction (ASE), which is based on an extraction under elevated temperature (50–200°C) and pressure (3–205 bar) during a short period of time (5–15 min). This technique has been used for the extraction of phenolic compounds from foods such as vegetables and fruits. In PLE, a solid sample is packed into the extraction cell and analytes are extracted from the matrix with conventional low-boiling solvents or solvent mixtures at elevated temperatures up to 200°C and pressure (30–200 bar) to maintain the solvent in the liquid state [61].

A very interesting feature of this technique is the possibility of full automation, and many samples can be extracted sequentially. However, the extraction efficiency of PLE is dramatically influenced by extraction pressure and temperature conditions, and a sample matrix can also affect the extraction efficiency [61]. Therefore, the optimization of the extraction conditions should be carefully evaluated in order to get suitable results.

In this sense, some authors [49] evaluated different parameters, such as percentage of methanol, temperature, pressure, and static extraction time for the extraction of phenolic compounds from apple samples, obtaining the best results when the following conditions were used: pure methanol as solvent, extraction temperature of 40°C, pressure 69 bars, and 5 min static extraction time.

On the other hand, Søltoft et al. [62] evaluated different extraction procedures, such as UAE, water bath, MAE and PLE. This last technique was selected for the extraction of five quercetin glycosides and two isorhamnetin glycosides from onion, chlorogenic and caffeic acid from potato, and chlorogenic acid from carrot. The method can be automated and only small amounts of solvents were used, obtaining clean extracts. This allowed the extraction of light and oxygen-sensitive flavonoids in an inert atmosphere protected from light. Søltoft et al. also studied different parameters, such as extraction temperature, sample weight, flush volume, and solvent type. A clean-up step by in-cell addition of C18-material to the extraction cells was also integrated in the PLE procedure, which slightly improved the recovery (98–99%) and reproducibility (3.1–11%) of the method.

According to Luthria [63], PLE is effective for the extraction of phenolic acids from potato peels, and a variety of parameters were optimized. The results indicated that the optimum recoveries were obtained when phenolic acids were extracted using methanol:water (90:10, v/v) at 160°C. It has been shown that static time variations (5–15 min), pressure (35–138 bars), and flush volume (10–100%) variations did not significantly improve the extraction. Luthria [64] also optimized the extraction conditions of phenolic acids in parsley, with similar results as in potato peels, but in this case, all extractions were carried out with either one or two solvent mixtures, ethanol:water (50:50, v/v) and/or acetone:water (50:50, v/v), as it has been indicated in Table 16.4.

16.2.3 Solid-Phase Extraction

This technique is used for the extraction of semi-volatile or nonvolatile analytes from liquid samples or from solid samples, which have been previously extracted. This is based on the adsorption of the analytes into a solid sorbent and then eluted with a suitable solvent. The type of solvent chosen for the elution of the target compounds depends on the kind of sorbent and the polarity of each analyte. SPE technique can be performed in the off- and online modes, although the online approach is preferred due to advantages such as higher sensitivity and less sample handling.

Promising results were obtained by Ali et al. [65], who used C18 SPE cartridges, which achieved good recoveries (>95%) for the extraction of flavonoids from apple juice. On the other hand, anthocyanidins can be extracted from tomato using polymeric cartridges, utilizing a mixture of methanol:formic acid (95:5) [47] as elution solvent.

16.2.4 Supercritical Fluid Extraction

Carbon dioxide has been used as a supercritical fluid for the extraction of a variety of phenolic compounds from plant samples. The main advantages of this technique are the use of nontoxic, nonflammable, and inexpensive fluid as CO_2, the automation due to the available instruments and the ability of being coupled to chromatographic systems. Moreover, selective and fast extraction procedures can be performed by modifying the density of the supercritical fluid. Thus, increasing the density of the fluid, which can be modified by changing the temperature and pressure, the extraction of high molecular weight compounds can be improved. However, the main drawback of this technique is that it is limited to compounds of low or medium polarity. To increase the polarity of this technique, organic modifiers, such as ethanol [66], ethyl acetate [66], and methanol [28,67] can be used. Therefore, this technique can be used for the extraction of phenolic compounds, making necessary the optimization of the percentage of the organic modifier.

SFE is widely used for the extraction of phenolic compounds from grapes or derivatives [48,67,68] (Table 16.4). Furthermore, this technique has been applied in other matrices such as pepper, tomato, and eggplant by Helmja et al., who compared SFE and UAE. This study indicated that SFE provided the poorest results in comparison with the data obtained by UAE [28]. SFE was also compared with traditional techniques such as Soxhlet extraction using ethyl acetate and ethanol as solvents in guava. The best results, in terms of extraction yield (total and fraction) and product quality (antioxidant activity and total phenolic content), were obtained when SFE was applied using ethanol as an organic modifier [66]. The effect of pressure and temperature on the SFE was also evaluated, observing that the most appropriate conditions for the extraction of phenolic compounds was an extraction temperature of 60°C and pressure of 102 bar [66].

SFE has also been used in combination with other sample treatment processes, such as nanofiltration approach, for the effective extraction of polyphenols from cocoa seeds [69].

16.2.5 Microwave-Assisted Extraction

The success of this technique is related to the development of specialized microwave instruments that provide temperature-controlled and closed-system operation, permitting the processing of many samples simultaneously.

The main advantage is that the simultaneous extraction of four to six samples could be performed quicker than in Soxhlet extraction, and similar recoveries to those obtained with SFE were obtained [70]. Another advantage is the possibility to perform extractions in the absence of light, which is important because the presence of light may cause some phenolic compounds to be degraded or transformed to the inactive form [71]. However, special care must be taken with flammable solvents or with samples that contain constituents that can strongly be coupled with microwave radiation, provoking a rapid increase of the temperature, thus leading to potentially hazardous situations [70]. Another important factor to be considered is the extraction temperature. According to Liazid et al., MAE can be used in up to 100°C for 20 min without degradation of phenolic compounds. Moreover, it has been observed that the fewer the substituents present in the aromatic ring of the compound, the higher the stability of phenolic compounds during MAE. Li et al. [72], proposed a similar working temperature but with shorter extraction time (2 min). However, Sutivisedsak et al. [73], also observed good results when 150°C was used as the extraction temperature during 15 min for the determination of the total phenolic content in eight bean types.

16.2.6 Rotary Shaker-Assisted Extraction

RAE is considered one of the simplest extraction techniques because it is easy to perform with common laboratory equipment (i.e., a rotary shaker). When RAE is used, the sample is mixed with the appropriate solvent and placed into the rotary shaker. Solvents such as methanol [74] or mixtures of methanol and water [29,50,75] at different proportions have been used for the extraction of phenolic compounds from several matrices, although for some compounds as anthocyanins, acidified organic solvents can be used [72] (see Table 16.4).

This technique had been used to determine several families of phenolic compounds in different vegetables, such as tomato [72,75] broccoli [29,75], grape [75], carrot [75], eggplant [50,75], and beans [74]. Several procedures can be found for the extraction of phenolic compounds by RAE, and the extraction time usually ranged from 30 min [29,50] to overnight [29,47,74]. This technique is used at room temperature to avoid the degradation of phenolic compounds and the main advantage of this technique is that it is possible to perform the extraction of a large number of samples simultaneously.

16.2.7 QuEChERS

In the last few years, the QuEChERS method has been used for the extraction of a wide variety of compounds [76–78] in different matrices, since this approach allows quantitative extraction of compounds with different physic-chemical properties in a short period of time, avoiding the use of toxic organic solvents. This procedure involves initial single-phase extraction with acetonitrile, followed by salting-out extraction/partitioning by addition of $MgSO_4$ plus NaCl, and finally dispersive solid-phase extraction (dSPE) is used for clean-up. Silva et al. reported QuEChERS-based extraction procedure for multifamily analysis of phenolic compounds in vegetables for the first time, and the process involves two simple steps. First, the homogenized samples were extracted and partitioned using a mixture of acetonitrile:ethyl acetate (50:50, v/v) as organic solvent and salt solution (trisodium citrate dihydrate, disodium hydrogencitrate sesquihydrate, sodium chloride, and anhydrous $MgSO_4$). Then, the supernatant is further extracted and cleaned using $MgSO_4$, primary-secondary amine (PSA), and C18. The use of C18, PSA and $MgSO_4$ to remove lipids is important to maximize the sensitivity of phenolic compounds and to minimize the presence of interfering compounds in the extract. Despite the advantages of this methodology, as far as we know only one paper has described the use of this procedure for the extraction of phenolic compounds from tomato, onion, carrot, broccoli, green and red peppers, and beetroot [51].

16.2.8 Hydrolysis

This treatment was used to determine the phenolic compounds (free and conjugated) as aglycones, by hydrolysis of glycosides. The most conventional hydrolysis procedure involves the addition of an inorganic acid (HCl) at reflux or above reflux temperatures in aqueous or alcoholic solvents (being methanol the most common used). For hydrolysis, the extraction conditions are usually 1.2 M HCl in 50% aqueous methanol, using a reaction time of 2 h at 90°C [52,79]. However there are some exceptions (see Table 16.4), as proven by A. Chassy et al. [80], who used 4 h at 100°C for the determination of quercetin, kaempferol, and luteolin in tomato and pepper; Crozier et al., who used 2 M HCl in 50% aqueous and methanol during 4 h at 90°C for the extraction of phenolic compounds from celery [79]; or Cao et al. [81], who used a mixture of ethanol:water:HCl (80:10:10) at reflux during 3 h at 90°C, for the extraction of flavonoids in more than 100 vegetables and fruits, obtaining recoveries ranging from 93% to 98%.

In order to avoid the oxidation of the compounds, several antioxidants can be added as diethyldithiocarbamate [52,79,82], tert-butyl-hydroquinone (TBHQ), [80,83,84] or butylated-hydroxy-toluene (BHT) [81]. Furthermore, internal standards are usually used, such as morin [52,80], luteolin [80], isorhamnetin, [79,82] or kaempferol [79] depending on the phenolic compounds analyzed to monitor losses during sample preparation.

16.3 UHPLC–MS/MS DETERMINATION OF PHENOLIC COMPOUNDS

16.3.1 LC versus UHPLC

In the last 30 years, HPLC is the analytical technique that has been widely used for the separation and characterization of phenolic compounds [84]. However, the development of faster chromatographic methods has attracted the interest of many researchers, providing new tools with higher sensitivity and peak resolution, increased efficiency, and shorter analysis time than those obtained with conventional LC. Thus, alternative strategies have been developed to obtain increased efficiency, together with short analysis time. Among these alternatives, UHPLC has emerged as a valuable tool, and the number of applications based on the use of this methodology has been growing in recent years [85].

As it can be observed in Tables 16.5 and 16.6, the column chosen for the determination of phenolic compounds both in HPLC and UHPLC are almost exclusively composed of a C18 stationary phase. In HPLC, internal diameter ranged from 2.1 mm [38,86] to 4.6 mm, although the most common is 4.6 mm [35,40,41,87–89]; column length ranges from 50 mm [38] to 250 mm [35,40,41,87,89], and the particle size is 3 μm [38,86] or 5 μm [40,41,87–89]. On the other hand, UHPLC uses sub-2-μm particles for the stationary phase, although other aspects of the column geometry can be kept. This allowed faster separation and/or increased peak capacity (i.e., the number of peaks that can be separated in a given time window) [90]. In HPLC, run time ranged from 25 min [86] to 95 min [88], whereas if UHPLC has been used, it is usually shorter than 30 min. Regarding flow rates, they usually ranged from 0.3 to 1 mL/min when HPLC is used, with 1 mL/min the most commonly used [35,40,41,87,88,91]. Longer analysis time, together with an elevated flow rate of HPLC, generated a higher amount of waste than UHPLC. Therefore, UHPLC is more environmentally friendly and has lower analyses costs. Injection volume usually ranged from 10 μL [35,87,88] to 50 μL [86] for HPLC, whereas UHPLC

TABLE 16.5
HPLC Conditions for Separation of Phenolic Compounds

Compounds	Stationary Phase	Mobile Phase A	Mobile Phase B	Temperature (°C)	Flow Rate (mL/min)	Run Time (min)	Injection Volume (μL)	Reference
Anthocyanidins	Zorbax SB C18 250 × 4.6 mm, 5 μm	Water (5% formic acid, v/v)	Methanol	NS	1	90	NS	[40]
Flavonoles Phenolic acids	Star C18 250 × 4.6 mm, 5 μm	Water (0.1% formic acid, v/v)	Acetonitrile:water (80:20)	NS	1	55	10	[87]
Isoflavones	Zorbax SB C18 100 × 2.1 mm, 3 μm	Water (0.05% acetic acid, v/v)	Methanol	36	0.3	25	50	[86]
Flavones Flavonones	Eclipse XDB C18 150 × 4.6 mm, 5 μm	Water (0.5% acetic acid, v/v)	Acetonitrile	25	1	95	10	[88]
Flavonoles Phenolic acids	Luna C18 250 × 4.6 mm, 5 μm	Water (1% formic acid, v/v)	Methanol	25	0.5	90	20	[89]
Phenolic acids	C18 250 × 4 mm, 5 μm	Water (2% acetic acid, v/v)	Methanol	NS	1	35	20	[91]
Flavonones Phenolic acids	Hypersil RP-C18 250 × 4.6 mm, 5 μm	Water (0.1% formic acid, v/v)	Acetonitrile	NS	1	35	10	[35]
Flavones Flavonoles Isoflavones Lignans	Eclipse XDB C18 50 × 2.1 mm, 3 μm	Water (Ammonium formate 5 mM, and formic acid 0.01%,w/v)	Acetonitrile	35	NS	NS	NS	[38]

Note: NS, not specified.

TABLE 16.6
UHPLC Conditions for the Separation of Phenolic Compounds

Compounds	Stationary Phase	Mobile Phase		Temperature (°C)	Flow Rate (mL/min)	Run Time (min)	Injection Volume (μL)	Reference
		A	B					
Flavanols Flavonones Phenolic acids	HSS T3 100 × 2.1 mm, 1.8 μm	Water (0.2% acetic acid, v/v)	Acetonitrile	30	0.4	20	2.5	[103]
Flavonols Phenolic acids	HSS T3 100 × 2.1 mm, 1.8 μm	Water:methanol: formic acid (94.9:5:0.1, v/v/v)	Methanol: water:formic acid (60:39.9:0.1, v/v/v)	35	0.5	30	10	[94]
Flavanols Flavones Flavonones Phenolic acids	RSLC 120 C18, 100 × 2.1 mm, 1.8 μm	Water (0.1% formic acid, v/v)	Acetonitrile (0.1% formic acid)	50	0.6	45.5	1	[92]
Anthocyanidins Flavonols Phenolic acids	XB-C18 100 × 2.1 mm, 1.7 μm	Water:methanol: formic acid (95:2:3, v/v/v)	Methanol: water:formic acid (95:2:3, v/v/v)	30	0.3	15	NS	[47]
Anthocyanidins Flavonols Hydroxycinnamics acids	Hypersil Gold RP C18 200 × 2.1 mm, 1.9 μm	Water (0.1% formic acid, v/v)	Acetonitrile (0.1% formic acid)	NS	0.3	65	NS	[100]
Flavonols Flavonones	BEH C18 100 × 2.1 mm, 1.7 μm	Water (0.05% formic acid, v/v)	Acetonitrile (0.1% formic acid)	40	0.6	5.5	6	[99]

Flavanols Flavonols Phenolic acids	BEH RP C18 50 × 2.1 mm, 1.7 μm	Water (0.1% formic acid, v/v)	Methanol	40	0.3	10	NS	[101]
Flavanols	Zorbax SB C18 50 × 2.1 mm, 1.8 μm	Water (0.1% formic acid, v/v)	Acetonitrile (0.1% formic acid)	30	0.4	27	NS	[104]
Flavonols Phenolic acids	C18 100 × 3 mm, 2,6 μm	Water (0.1% formic acid, v/v)	Acetonitrile:water: formic acid (89.9:10:0.1, v/v/v)	21	0.75	23.5	5	[95]
Isoflavones	Hypersil Gold C18 100 × 2.1 mm, 1.9 μm	Water (0.1% acetic acid, v/v)	Methanol (0.1% acetic acid)	40	0.3	13	1	[93]
Coumarins Isoflavones Flavonols Flavonones	C8 150 × 2.1 mm, 1.7 μm	Water (10 mM formic acid w/v)	Methanol	40	0.2	17	2	[102]
Flavones Flavonols Isoflavones Phenolic acids (Glucosinolates)	Zorbax Eclipse Plus C18 100 × 2.1 mm, 1.8 μm	Water (ammonium acetate 5 mM, ajusted to pH 5 with formic acid, w/v)	Methanol	30	0.2	13	5	[75]

Note: Abbreviations: BEH: ethylene bridged hybrid; HSS: high-strength silica; NS: not specified; RP: reversed phase; RSLC: rapid separation liquid chromatography; SB: stable bond.

uses the lowest injection volumes, 1 μL [92,93] to 10 μL [94]. In relation to column temperature, reported values range from 20°C [95] to 50°C [92].

Spáčil et al. [96] compared HPLC and UHPLC for the analysis of 34 phenolic compounds belonging to several families, such as phenolic acids, flavonoids, catechins, and coumarins. Retention time, peak area, asymmetry factor, resolution, and peak capacity were calculated for all components with both techniques. In the case of phenolic acids, all studied parameters were significantly better for UHPLC. Figure 16.1 shows that UHPLC analyses were performed 4.6 times faster than those by HPLC. Furthermore, it can be observed that UHPLC peak shape is better than HPLC peaks, and UHPLC provides better resolution in shorter running time than HPLC.

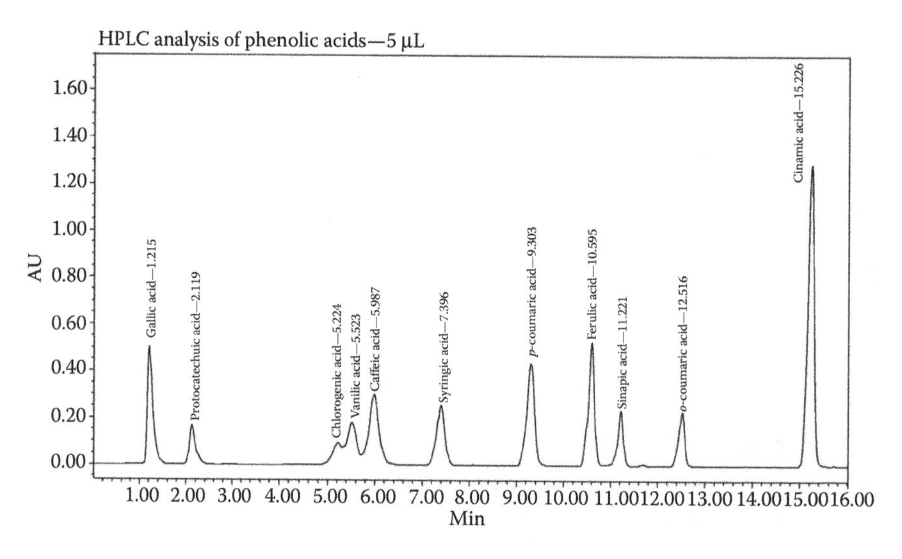

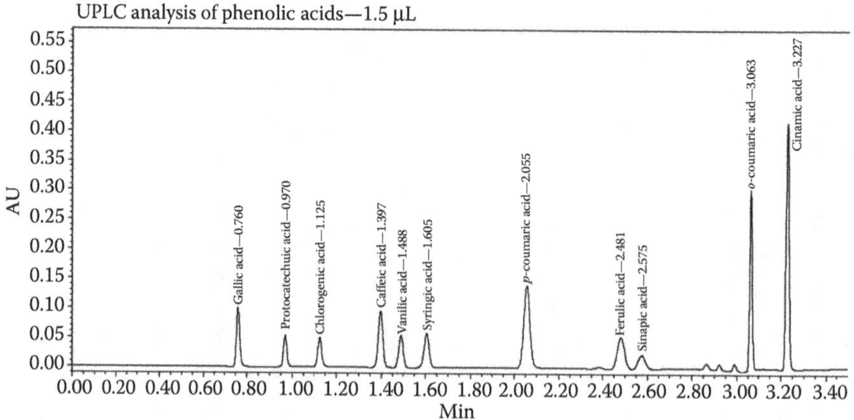

FIGURE 16.1 Comparison between UHPLC and HPLC chromatograms in analysis of phenolic acids: HPLC, 0.1% formic acid-methanol, from 85:15 to 50:50 (v/v), 1.0 mL/min; UHPLC, 0.1% formic acid-methanol, from 88.5:11.5 to 30:70 (v/v), 0.450 mL/min. (Adapted from Spáčil, Z., Nováková, L. and Solich, P. 2008. *Talanta* 76:189–199. With permission.)

Moreover, the HPLC method presented problems with the resolution of chlorogenic, vanillic, and caffeic acid. This poor resolution was not observed when UHPLC was used. In relation to catechins, sensitivity was 1.5 times higher for UHPLC than HPLC, but peak retention time and the area repeatability parameters for UHPLC analysis resembled the data gained for the HPLC method [96]. On the other hand, all peaks were separated by UHPLC method with satisfactory resolution, even the signals of epigallocatechin and catechin, whereas for HPLC, a suitable separation was not achieved. For coumarins, Spáčil et al. indicated that the UHPLC analyses showed an excellent repeatability for retention time, but slightly worse repeatability for peak area. Other studied parameters provided better values when UHPLC was used. Spáčil et al. published that the flavonoids also showed very good sensitivity when UHPLC was used (1.7 times higher), highlighting that the UHPLC system had significantly higher peak capacity than the HPLC system [96].

Moreover, other studies indicated that UHPLC was also characterized by better efficiency and sensitivity. For example, procyanidin oligomers up to hexamers could be detected by HPLC, whereas the UHPLC method allowed the detection of oligomers up to nonamers [97], as can be observed in Figure 16.2.

Finally, the excellent retention time and peak area repeatability of the UHPLC method should be highlighted, which are important in routine analyses.

16.3.2 UHPLC

UHPLC shows a variety of advantages in the analysis of phenolic compounds and valuable characteristics for routine laboratories, such as reduction of the required time during the chromatographic separation higher peak efficiency and reduced solvent consumption [98]. Moreover, UHPLC allowed low dead volume and high pressure (1000 bar), providing new strategies to improve resolution [42]. In this sense, Figure 16.3 demonstrates the growing interest in UHPLC in the field of analysis of flavonoids since 2006.

In the context of chromatographic separation, several methods have been developed for the determination of phenolic compounds by UHPLC. For instance, chromatographic separation was carried out by Alarcón-Flores et al., using a 100 mm, 1.8 μm C18 column, and an aqueous solution of ammonium acetate (0.03 mol/L) adjusted to pH 5 with formic acid and methanol as mobile phase at 30°C for the analysis of phenolic acids, flavonols, flavones, and isoflavones in the extracts of different vegetables [75]. The separation of more than 30 compounds was achieved in less than 13 min. On the other hand, Medina-Remon et al. [99] used a 100 mm C18 stationary phase with 1.7 μm particle size, an aqueous solution acidified with 0.05% formic acid, and acetonitrile acidified with 0.1% formic acid to resolve 11 compounds in 5.5 min. A total of 200 compounds were tentatively identified by Hurtado-Fernandez et al., utilizing a 1.8 μm C18 column and aqueous and organic mobile (acetonitrile) phase acidified with 0.1% formic acid at 50°C for the simultaneous separation of phenolic acids, flavonones, flavones, and flavanols [92]. The same mobile phase was used by Lin et al. to tentatively identify a total of 129 phenolic compounds in different *brassica* vegetables extracts, using a 65 min gradient analysis on a 200 mm, 1.8 μm column C18 [100]. Engels et al. separated 21 phenolic compounds using a

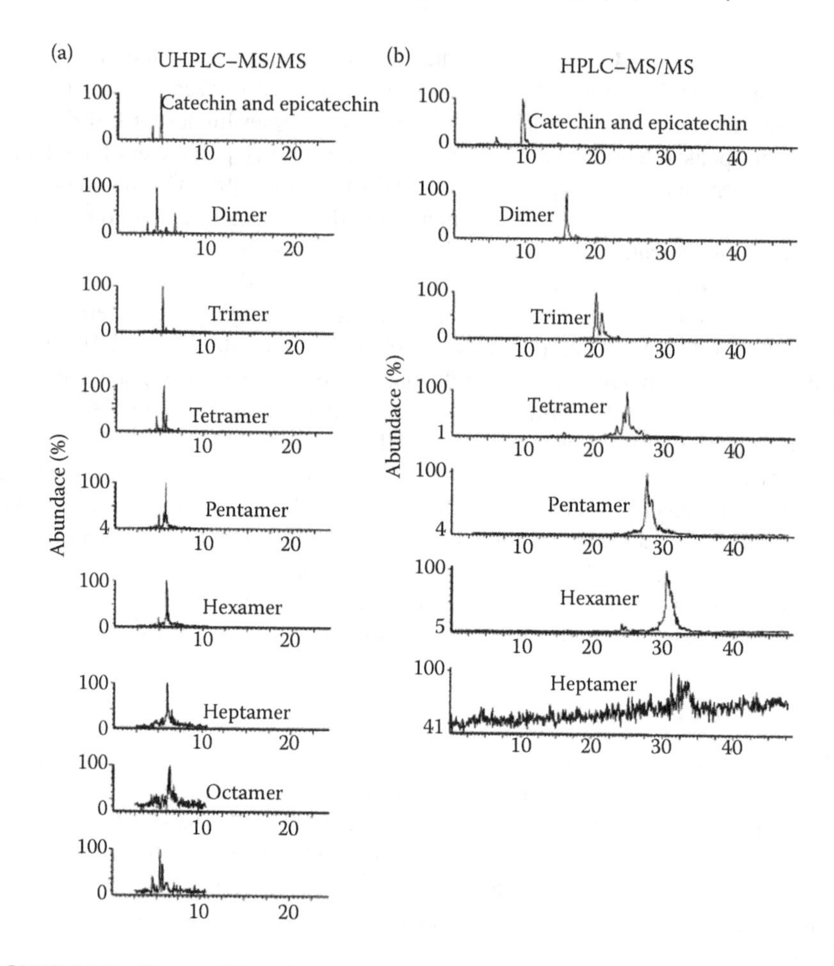

FIGURE 16.2 Extracted ion chromatogram of flavanols for the analysis of cocoa sample by: (a) UHPLC–MS/MS and (b) HPLC–MS/MS. (Adapted from Motilva, M.J., Serra, A., and Macià, A. 2013. *J. Chromatogr. A* 1292:66–82. With permission.)

100 × 3.0 mm, 2.6 μm C18 column set at 21°C. The compounds were eluted with 0.1% (v/v) formic acid in water and 0.1% formic acid in an acetonitrile:water mixture (90:10, v/v) at a flow rate of 0.75 mL/min. Ceymann at al. employed a short (5 mm) 1.7 μm C18 reversed column and a column temperature of 40°C for the analysis of flavanols in apple extracts, achieving a suitable separation of 12 compounds in 10 min [101]. Prokudina et al. developed a UHPLC method for analysis of 26 phenolic compounds, using 150 × 2.1 mm 1.7 μm C_8 column set at 40°C [102].

It is well known that in reversed phase the retention time of phenolic compounds is higher for substances that are less polar (myricetin, quercetin, kaempferol), while polar molecules as gallic acid, protocatechuic acid, epigallocatechin are eluted faster [43]. Alternative high-strength silica (HSS) T3 columns have also been investigated.

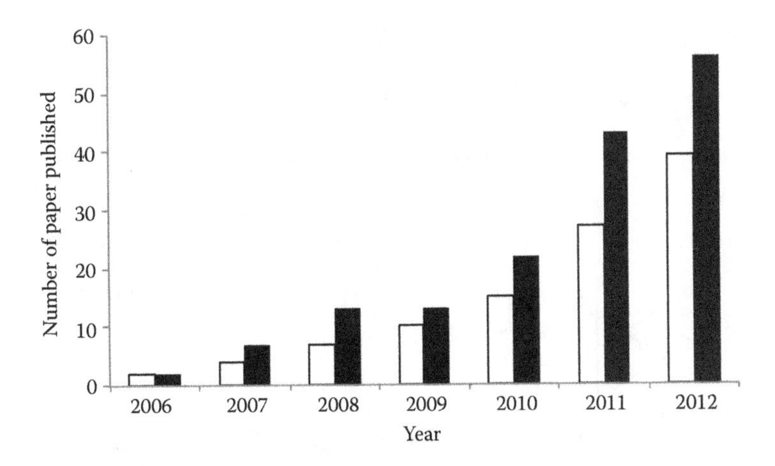

FIGURE 16.3 Number of papers published each year in the field of flavonoids by UHPLC and UHPLC–MS, since 2006. Black bars were obtained with keywords "UPLC" or "UHPLC" and "flavonoids," while white bars were obtained with key words "UPLC–MS" or "UHPLC–MS" and "flavonoids."

For instance, Ribas-Agustí et al. used a 100 mm, 1.8 μm HSS T3 column and mixture of methanol–water–formic acid as mobile phase for the determination of phenolic acids and flavonols in lettuce, achieving a separation of 11 compounds in 30 min [94]. Ortega et al. used the same column in combination with acidified water with 0.2% acetic acid, and acetonitrile as mobile phase for the determination of 28 compounds (including phenolic acids, flavanols, and flavonones) in 20 min [103]. These parameters as well as other important factors of the chromatographic separation have been indicated in Table 16.6.

In general, most UHPLC methods analyzed several families of phenolic compounds [75,92,94,95,99–103], except the work performed by Lojza et al. [93], which only analyzed isoflavones, such as genistein, daidzein, glycitein, and their respective acetyl, malonyl, and glycoside forms. Delcambre et al. [104] only analyzed flavanols, and they partially identified 14 new flavanol monoglycosides based on the exact mass of the molecular ions and their specific retro-Diels–Alder fragmentation, heterocyclic ring fragmentation, benzofuran forming fragmentation, and glycoside fragmentations. Figure 16.4 shows an example of retro-Diels–Alder fragmentation. This family (flavanols) presents one the most severe challenges in terms of chromatographic separation, bearing in mind that when the molecular weight of the compound increases, the number of potential isomers exponentially increases [97], making difficult the chromatographic resolution of these isomers.

Moreover, due to their chemical complexity and similarity of phenolic compounds in fruits and vegetables, they are usually eluted using a gradient elution instead of the isocratic mode, where the mobile phase is generally a binary system. The most suitable solvents for UHPLC are water, methanol, and acetonitrile. Consequently, the mobile phases employed for the separation of a single family

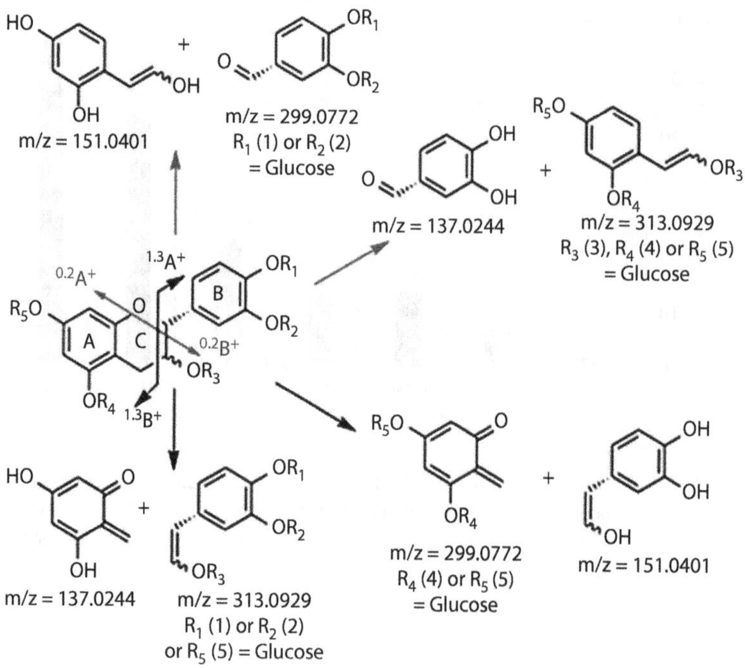

FIGURE 16.4 Example of retro-Diels–Alder fission (RDA) and OC9/C2C3 cleavage. (Adapted from Delcambre, A., and Saucier, C. 2012. *J. Mass Spectrom.* 47:727–736. With permission.)

or multiple families of phenolic compounds have been mixtures of water–methanol or water–acetonitrile. In addition, some additives or modifiers of the mobile phase have been employed to enhance ion abundance, diminishing the formation of sodium adducts and improving chromatographic peak shape, such as formic acid [72,92,94,95,99–102,104], acetic acid [93,103] and ammonium acetate [75]. Mobile phases containing nonvolatile compounds such as phosphate buffers should be avoided because they can clog the interface and produce build-up of deposits in the ion source if MS analyzers were used.

In general, the mobile phase is usually selected as a compromise between optimal chromatographic separation, adequate ionization efficiency, and overall MS performance. Some UHPLC procedures for determining some classes of phenolic compounds in vegetables or fruits are presented in Table 16.6.

Finally, it should be noted that UHPLC coupled to MS systems can increase the possibilities of this technology and the development of fast analytical methods.

16.3.3 MS(/MS) Determination

The combination of UHPLC with MS detection seems to be an appropriate approach to carry out key requirements in terms of selectivity, sensitivity, and peak-assignment certainty for the rapid determination of analytes in complex matrices [85].

Although the limiting factors in the use of MS as analytical tools are the high cost of equipment, complex laboratory requirements, and limitations in the type of solvents used in extraction and separation, in recent years this technique has become very popular, because it allows a reduction in sample treatment, and it is a universal, selective, and sensitive detection mode. Furthermore, the reliability of the obtained results is usually increased.

MS has a very important role for research and its analytical power is relevant for structural elucidation of phenolic compounds. The MS principle consists of ionizing chemical compounds to generate charged molecules or molecule fragments and measure their mass-to-charge ratios. MS detector is critical for the identification of phenolic compounds, but selectivity and sensitivity can be increased using tandem MS (MS/MS) or MS^n. MS/MS and MS^n produce more fragmentations of the precursor and product ions and, therefore, they provide additional structural information for the identification of these compounds.

Basic characteristics of the quality of mass analyzer are resolving power (RP), which has been used to specify the ability of mass analyzers performing high-resolution mass analysis and it is defined and/or calculated by using the mass (M) and the full-width-at-half-maximum (FWHM) of a mass spectral peak, and mass accuracy (MA) is the measured error in m/q divided by the accurate m/q [105].

Currently, definitions of high-RP and also high-MA are not sufficient to differentiate between high-resolution and ultra-high-resolution mass analyzers. Therefore, Holčapek et al. [106] suggest updated definitions for low, high, and ultra-high RP and MA. They suggest three basic categories of RP: low RP (<10,000), high RP (10,000–100,000), and ultra-high RP (>100,000). In fact, it means that simple quadrupole (Q), triple quadrupole (QqQ), and ion-trap (IT) mass analyzers belong to the low-RP category, time-of-flight (TOF)-based analyzers to the high RP, and the ultra-high RP contains mass analyzers as Orbitrap [106].

For the determination of phenolic compounds in fruits and vegetables, mass analyzer as QqQ has been widely used [75,99,101–103,107], although others such as IT [47], QTrap [95], TOF [92], QTOF [104], and Orbitrap [93,100] are also applied.

Atmospheric pressure ionization (API) has been mainly used for the ionization of phenolic compounds, applying either electrospray (ESI) or atmospheric pressure chemical ionization (APCI), as it can be observed in Table 16.7. However, ESI is more often used to ionize the different families of phenolic compounds. Moreover, APCI and ESI can be operated under both negative and positive ion modes. Although positive ionization mode [47,93,102,103] was used for detection of various phenolic compounds, it was found that negative ionization mode [94,95,101,104,107] was excellent for phenolic compounds analysis. In this sense, the combination of both polarities in the same method provided good results [75,92,99,100] for the simultaneous determination of several families of compounds. Phenolic acids [75,94,95,101,103,107], flavonols [75,94,95,99,101,103,107], flavonones [99,103] flavanols [92,101,103,104], and flavones [75,103] were often detected in negative ion mode, although some of these families were also detected in positive mode [75,99,100,102]. Anthocyanidins [47,100], coumarins [102], and isoflavones [75,93,102] were detected in positive mode.

In general, deprotonated ions, $[M-H]^-$, were observed for the compounds ionized in negative mode, whereas for the compounds detected in positive mode, protonated

TABLE 16.7
Detection Conditions for Identification of Phenolic Compounds Using UHPLC Separation

Family Compounds	Matrix	Detector	Ionization Source	Ionization Mode	Limit of Detection (mg/kg)	Reference
Flavanols Flavonones Flavonols Phenolic acids	Cocoa	QqQ	ESI	Positive	NS	[103]
Flavonols Phenolic acids	Lettuce	QqQ-PAD	ESI	Negative	0.1–0.8	[94]
Flavonols (Glucosinolates)	Broccoli	QqQ-DAD	ESI	Negative	0.03	[107]
Flavanols Flavonones Phenolic acids	Avocado	TOF	ESI	Negative and positive	0.002–0.7	[92]
Anthocyanidins Flavonols Phenolic acids	Tomato	IT	ESI	Positive	0.25	[47]
Anthocyanidins Flavonols	Brassica vegetables	Orbitrap-PDA	ESI	Positive and negative	NS	[100]

Flavonols	Citrus fruit	QqQ	API turbo ion source	Positve and negative	0.02–0.23	[99]
Flavanones						
Flavanols	Apple	QqQ	ESI	Negative	2	[101]
Flavonols						
Phenolic acids						
Flavanols	Grape	QTOF-MS	ESI	Negative	NS	[104]
Flavonols	Spanish plum	QTRAP-DAD	ESI	Negative	NS	[95]
Phenolic acids						
Isoflavones	Soybeans	Orbitrap	Dart source (APCI)	Positive	5	[93]
Coumarins	Mung bean	QqQ	ESI	Positive	0.001–78	[102]
Isoflavones						
>Flavonols						
Flavonones						
Flavones	Tomato	QqQ	ESI	Positve and negative	0.01–0.1	[75]
Flavonols	Grape					
Isoflavones	Broccoli					
Phenolic acids	Eggplant					
(Glucosinolates)	Carrot					

Note: Abreviations: DAD: diode array detector; IT: ion trap; MS: mass; NS: Not specified; PAD: photodiode array detector; Q: simple quadrupole; QqQ: triple quadrupole; TOF: time of flight.

ions, $[M + H]^+$, were usually observed. In addition other ion in-source fragments such as $[M\text{-}NH_3]^-$, $[M\text{-}H_2O]^-$, $[M\text{-}CO]^-$, $[M\text{-}C_2H_2O]^-$, $[M\text{-}CHNO]^-$, $[M\text{-}CO_2]^-$, $[M\text{-}C_5H_{10}O_5]^-$, $[M\text{-}C_6H_{12}O_6]^-$ and to sodium $[M + Na]^+$ and potassium $[M + K]^+$ or formic acid $[M - H + CHOOH]^-$ adducts, can be monitored [92]. The loss of pentoses or hexoses occurs because most of the flavonoids are linked to monosaccharides (flavonoid glycosides), and glucose is the most common sugar moiety in flavonoids, whereas galactose, rhamnose, xylose, and arabinose are less common, or mannose, fructose, glucuronic, and galacturonic acids are scarcely linked to flavonoids [10].

In relation to structure analysis, the main fragmentation paths of flavonoids are mostly independent of the type of analyzer applied (QqQ, IT, or QTOF) and the ionization mode (ESI, APCI, or matrix-assisted laser desorption/ionization [MALDI]). On the other hand, relative fragment abundances significantly varied when different instrumentation was used. For these reasons, methods based on detecting the presence or absence of specific or selective fragment ions are preferred over techniques relying on changes in relative intensity. So, despite the advantages of UHPLC–MS/MS, in some cases it is not enough to resolve some isomers of phenolic compounds [75]. However, when IT [108] or TOF [109] analyzers are used, O-, C-, and C,O-glycosides can be distinguished by their collision-induced dissociation (CID) fragments. In the case of C-glycosides, low fragmentation energy does not provide adequate fragmentation. On the other hand, application of higher fragmentation energy leads to intraglycosidic cleavages rendering difficult spectral interpretation. For O-glycosides, the application of low or medium fragmentation energy results in heterolytic cleavage of their hemi-acetal O–C bonds, yielding distinctive fragments [10,108,110]. Regarding fragmentation of phenolic acids, they yielded product ions more efficiently at lower collision energies [95]. Figure 16.5 shows the UHPLC–ESI–Orbitrap MS chromatograms of the exact masses corresponding to daidzein, genistein, and glycitein. These chromatograms contained signals corresponding to the protonated molecules of particular aglycones, and other signals at lower retention times as compared with those of aglycones. When the exact masses for particular acetyl, malonyl, and glucoside forms were extracted, it can be observed that the signals of their protonated molecules had the same retention times as those of the earlier eluting compounds.

Normally, phenolic compounds are identified by comparing their retention time, MS, MS/MS, and MS^n (when it is possible), fragmentation spectra with those obtained from pure standard solutions when they are commercially available. When peak data did not agree with available standards, bibliographic data was used for their tentative identification [94,107].

In relation to quantification, standard addition methodology is usually used due to the presence of matrix effect [75]. For that, spiking samples with different amounts of standard compounds were prepared, in order to obtain a concentration range in agreement with the levels observed in assayed samples. Tentatively identified compounds (with no standard available) were quantified as equivalents of the most similar compound, taking into account their molecular weight [94]. In the other cases, triphenyl phosphate (TPP) was used as an internal standard for quantification purposes

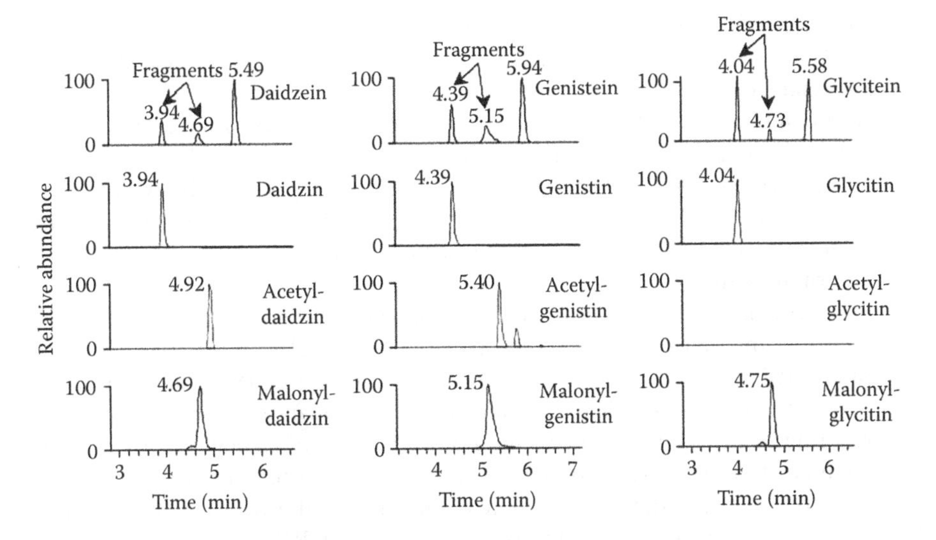

FIGURE 16.5 UHPLC–ESI–Orbitrap MS chromatogram documenting the presence of conjugate forms of isoflavones in soybean extract and their fragmentation yielding ions corresponding to aglycones. (Adapted from Lojza, J. et al. 2012. *J. Sep. Sci.* 35:476–481. With permission.)

[93]. Hurtado-Fernández et al. also used internal standard (taxifolin) to evaluate the reproducibility of the extraction system and the chromatographic runs [92].

16.4 CONCLUSION AND FUTURE TRENDS

Fruits and vegetables are excellent sources of phenolic compounds, which are widely recognized for their health benefits. Therefore, it is necessary to establish and consolidate phenolic compound databases in these matrices because these compounds are of importance to develop epidemiological studies, and their bioavailability should be deeply understood.

For analytical studies, sample preparation is necessary to determine the composition of phenolic compounds in these matrices. The most widely used extraction system is UAE, which is an inexpensive method since it involves the use of low organic solvent and requires short extraction times. Multi-analyte methods with minimal sample manipulation are demanded. However, vegetables and fruits are very complex matrices and hence sample extraction and clean-up treatments must be carefully developed to reduce manipulation and total analysis time. Therefore, QuEChERS could be a useful tool for this purpose in future applications.

Regarding UHPLC, due to the betterments performed in the preparation of sub-2 μm packing materials of the analytical columns, high-efficiency separations in shorter run time can be performed, improving the resolution. When UHPLC columns are used, less matrix effects can be observed, bearing in mind that the coelution of the compounds can be reduced and they can contribute to a reduction in ion suppression when MS is used as a detection system. The continuous progress in UHPLC allows for new possibilities in the analysis of phenolic compounds in vegetables and fruits.

The combination of UHPLC systems with MS analyzers is an additional and important approach, since the application of deconvolution tools does not require complete resolution of the phenolic compounds, improving the capability of the system: more compounds in less time. Moreover, MS/MS detection enhances specificity compared to single-stage MS, minimizing sample handling and allowing the reduction of running time.

The utilization of low-RP analyzers (QqQ, IT) is common in routine analysis and it is the first choice when MS is used. Nevertheless, high-RP (TOF) and ultrahigh-RP analyzers (Orbitrap) are becoming more frequently utilized, either as screening tools or for identification/confirmation/quantification purposes. The monitoring of full-scan spectra brings along a number of advantages, such as the possibility of performing retrospective analysis (monitoring of nontargeted compounds) or the establishment of analyte degradation pathways. However, the high price of this technology in comparison to typical low-RP instruments is still an important drawback.

However, it should be noted that the identification and quantification of phenolic compounds using low- and high-mass resolution is not normalized in food analysis, and therefore guidelines for the development and validation of analytical methodologies should be established for the analysis of phenolic compounds in vegetal matrices, as the SANCO guidelines do for pesticides in food and feed.

ACKNOWLEDGMENTS

The authors are grateful to the Andalusian Regional Government (Regional Ministry of Innovation, Science and Enterprise) and FEDER, as well as the Centre for Industrial Technological Development (CDTI) for financial support Project Ref. P11-AGR-7034 and IDI-20110017, respectively. MIAF acknowledges her grant (FPU, Ref: AP 2009–2074) from the Spanish Ministry of Education. RRG is also grateful for personal funding through Ramon y Cajal Program (Spanish Ministry of Economy and Competitiveness-European Social Fund).

REFERENCES

1. Siró, I., Kápolna, E., Kápolna, B., and Lugasi, A. 2008. Functional food. Product development, marketing and consumer acceptance—A review. *Appetite* 51: 456–467.
2. Verbeke, W. 2005. Consumer acceptance of functional foods: Socio-demographic, cognitive and attitudinal determinants. *Food Qual. Prefer.* 16:45–57.
3. Hasler, C.M. 2002. Functional foods: Benefits, concerns and challenges—A position paper from the American council on science and health. *J. Nutr.* 132:3772–3781.
4. Rodríguez, E.B., Flavier, M.E., Rodriguez-Amaya, D.B., and Amaya-Farfán, J. 2006. Phytochemicals and functional foods. Current situation and prospect for developing countries. *Segurança Alimentar e Nutricional* 13:1–22.
5. Mudgal, V., Madaan, N., Mudgal, A., and Mishra, S. 2010. Dietary polyphenols and human health. *Assian J. Biochem.* 5:154–62.

6. Tsao, R. 2010. Chemistry and biochemistry of dietary polyphenols. *Nutrients* 2:1231–246.
7. Aisling-Aherne, S. and O'Brien, N.M. 2002. Dietary flavonols: Chemistry, food content, and metabolism. *Nutrition* 18:75–81.
8. Dicko, M.H., Gruppen, H., Traoré, A.S., Voragen, A.G.J., and Van Berkel, W.J.H. 2006. Phenolic compounds and related enzymes as determinants of sorghum for food use. *Biotechnol. Mol. Biol. Rev.* 1:21–38.
9. Sakakibara, H., Honda, Y., Nakagawa, S., Ashida, H., and Kanazawa, K. 2003. Simultaneous determination of all polyphenols in vegetables, fruits, and teas. *J. Agric. Food Chem.* 51:571–581.
10. Stefano, V.D., Avellone, G., Bongiorno, D., Cunsolo, V., Muccilli, V., Sforza, S., Dossena, A., Drahos, L., and Vékey, K. 2012. Applications of liquid chromatography–mass spectrometry for food analysis. *J. Chromatogr. A* 1259: 74– 85.
11. Hollmanm, P.C.H. and Katan, M.B. 1999. Dietary flavonoids: Intake, health effects and bioavailability. *Food Chem. Toxicol.* 37:937–942.
12. Santos-Buelga, C. and Scalbert, A. 2000. Review proanthocyanidins and tannin-like compounds—Nature, occurrence, dietary intake and effects on nutrition and health. *J. Sci. Food Agric.* 80:1094–1017.
13. Bravo, L. 1998. Polyphenols: Chemistry, dietary sources, metabolism and nutritional significance. *Nutr. Rev.* 56:317–333.
14. Kim, D.O., Padilla-Zakour, O.I., and Griffiths, P.D. 2004. Flavonoids and antioxidant capacity of various cabbage genotypes at juvenile stage. *J. Food Sci.* 69:685–689.
15. González-Gallego, J., García-Mediavilla, M.V., Sánchez-Campos, S., and Tuñón, M.J. 2010. Fruit polyphenols, immunity and inflammation. *Br. J. Nutr.* 104:15–27.
16. Vincent, H.K., Bourguignon, C.M., and Taylor, A.G. 2010. Relationship of the dietary phytochemical index to weight gain, oxidative stress and inflammation in overweight young adults. *J. Hum. Nutr. Diet.* 23:20–29.
17. Perez-Vizcaino, F. and Duarte, J. 2010. Flavonols and cardiovascular disease. *Mol. Aspects Med.* 31:478–494.
18. Wang, S., Melnyk, J.P., Tsao, R., and Marcone, M.F. 2011. How natural dietary antioxidants in fruits, vegetables and legumes promote vascular health. *Food Res. Int.* 44:14–22.
19. Collins, A.R. 2005. Antioxidant intervention as a route to cancer prevention. *Eur. J. Cancer* 41:1923–1930.
20. Pietta, P., Minoggio, M., and Bramati, L. 2003. Plant polyphenols: structure, occurrence and bioactivity. *Stud. Natural Products Chem.* 56:257–312.
21. Prasain, J.K. and Barnes, S. 2007. Metabolism and bioavailability of flavonoids in chemoprevention: Current analytical strategies and future prospectus. *Mol. Pharmaceut.* 4:846–864.
22. Stan, S.D., Kar, S., Stoner, G.D., and Singh, S.V. 2008. Bioactive food components and cancer risk reduction. *J. Cell. Biochem.* 104:339–356.
23. Miglio, C., Chiavaro, E., Visconti, A., Flogliano, V., and Pellegrini, N. 2008. Effects of different cooking methods on nutritional and physicochemical characteristics of selected vegetables. *J. Agric. Food Chem.* 56:139–147.
24. Björkman, M., Klingen, I., Birch, A.N.E., Bones, A.M., Bruce, T.J.A., Johansen, T.J., Meadow, R. et al. 2011. Phytochemicals of Brassicaceae in plant protection and human health—Influences of climate, environment and agronomic practice. *Phytochemistry* 72:538–556.
25. Carbone, K., Giannini, B., Picchi, V., Lo Scalzo, R., and Cecchini, F. 2011. Phenolic composition and free radical scavenging activity of different apple varieties in relation to the cultivar, tissue type and storage. *Food Chem.* 127:493–500.
26. Ruiz-Rodriguez, A., Marín, F.R., Ocaña, A., and Soler-Rivas, C. 2008. Effect of domestic processing on bioactive compounds. *Phytochem. Rev.* 7:345–384.

27. Gómez-Romero, M., Segura-Carretero, A., and Fernández-Gutiérrez, A. 2010. Metabolite profiling and quantification of phenolic compounds in methanol extracts of tomato fruit. *Phytochemistry* 71:1848–1864.
28. Helmja, K., Vaher, M., Püssa, T., Raudsepp, P., and Kaljurand, M. 2008. Evaluation of antioxidative capability of the tomato (*Solanum lycopersicum*) skin constituents by capillary electrophoresis and high-performance liquid chromatography. *Electrophoresis* 29:3980–3988.
29. Velasco, P., Francisco, M., Moreno, D.A., Ferreres, F., García-Viguera, C., and Cartea, M.E. 2011. Phytochemical fingerprinting of vegetable *Brassica oleracea* and *Brassica napus* by simultaneous identification of glucosinolates and phenolics. *Phytochem. Anal.* 22:144–152.
30. Hubert, J., Berger, M., and Daydeä, J. 2005. Use of a dimplified HPLC-UV analysis for soya saponin B determination: Study of saponin and isoflavone variability in soybean cultivars and soy-based health food products. *J. Agric. Food Chem.* 53:3923–3930.
31. Rostagno, M.A., Palma, M., and Barroso, C.G. 2003. Ultrasound-assisted extraction of soy isoflavones. *J. Chromatogr. A* 1012:119–128.
32. Mattila, P. and Kumpulainen, J. 2002. Determination of free and total phenolic acids in plant-derived foods by HPLC with diode-array detection. *J. Agric. Food Chem.* 50:3660–3667.
33. Volpi, N. and Bergonzini, G. 2006. Analysis of flavonoids from propolis by on-line HPLC–electrospray mass spectrometry. *J. Pharmaceut. Biomed.* 42:354–361.
34. Corrales, M., Toepfl, S., Butz, P., Knorr, D., and Tauscher, B. 2008. Extraction of anthocyanins from grape by-products assisted by ultrasonics, high hydrostatic pressure or pulsed electric fields: A comparison. *Innov. Food Sci Emer.* 9:85–91.
35. He, D., Shan, Y., Wu, Y., Kiu, G., Chen, B., and Yao, S. 2011. Simultaneous determination of flavanones, hydroxycinnamic acids and alkaloids in citrus fruits by HPLC-DAD–ESI/MS. *Food Chem.* 127:880–885.
36. González-Gómez, D., Lozano, M., Fernández-León, M.F., Bernalte, M.J., Ayuso, M.C., and Rodríguez, A.B. 2010. Sweet cherry phytochemicals: Identification and characterization by HPLC-DAD/ESI-MS in six sweet-cherry cultivars grown in Valle del Jerte (Spain). *J. Food Compos. Anal.* 23:533–539.
37. El-Hela, A.A., Al-Amier, H.A., and Ibrahim, T.A. 2010. Comparative study of the flavonoids of some Verbena species cultivated in Egypt by using high-performance liquid chromatography coupled with ultraviolet spectroscopy and atmospheric pressure chemical ionization mass spectrometry. *J. Chromatogr. A* 1217:6388–6393.
38. Konar, N., Poyrazoğlu, E.S., Demir, K., and Artik, N. 2012. Effect of different sample preparation methods on isoflavone, lignan, coumestan and flavonoid contents of various vegetables determined by triple quadrupole LC–MS/MS. *J. Food Compos. Anal.* 26: 26–35.
39. Kuhnle, G.G.C., Dell'Aquila, C., Aspinall, S.M., Runswick, S.A., Mulligan, A.A., and Bigham, S.A. 2008. Phytoestrogen content of beverages, nuts, seeds, and oils. *J. Agric. Food Chem.* 56:7311–7315.
40. Wu, X. and Prior, R. L. 2005. Identification and characterization of anthocyanins by high-performance liquid chromatography-electrospray ionization-tandem mass spectrometry in common foods in the United States: Vegetables, nuts, and grains. *J. Agric. Food Chem.* 53:3101–3113.
41. Wu, X. and Prior, R.L. 2005. Systematic identification and characterization of anthocyanins by HPLC-ESI-MS/MS in common foods in the United States: Fruits and berries. *J. Agric. Food Chem.* 53:2589–2599.
42. Ferrer, I. and Thurman, E.M. 2009. *Liquid Chromatography Time-of-Flight Mass Spectrometry. Principles, Tools and Applications for Accurate Mass Analysis*. John Wiley & Sons, New Jersey.
43. Haminiuk, C.W.I., Maciel, G.M., Plata-Oviedo, M.S.V., and Peralta, R.M. 2012. Phenolic compounds in fruits—An overview. *Int. J. Food Sci. Tech.* 47:2023–2044.

44. Ferreres, F., Taveira, M., Pereira, D.M., Valentão, P., and Andrade, P.B. 2010. Tomato *Lycopersicon esculentum*) seeds: New flavonols and cytotoxic effect. *J. Agric. Food Chem.* 58:2854–2861.
45. Taskeen, A., Naeem, I., Bakhtawar, S., and Mehmood, T. 2010. A comparative study of flavonoids in fruits and vegetables with their products using reverse phase high performance liquid chromatography (RP-HPLC). *EJEAFChem.* 9: 1372–1377.
46. Alonso-Salces, R.M., Korta, E., Barranco, A., Berrueta, L.A., Gallo, B., and Vicente, F. 2001. Pressurized liquid extraction for the determination of polyphenols in apple. *J. Chromatogr. A* 933:37–43.
47. Li, H., Deng, Z., Liu, R., Young, J.C., Zhu, H., Loewen, S., and Tsao, R. 2011. Characterization of phytochemicals and antioxidant activities of a purple tomato (*Solanum lycopersicum* L.). *J. Agric. Food Chem.* 59:11803–11181.
48. Pinelo, M., Ruiz-Rodríguez, A., Sineiro, J., Señoráns, F.J., Reglero, G., and Núñez, M.J. 2007. Supercritical fluid and solid–liquid extraction of phenolic antioxidants from grape pomace: A comparative study. *Eur. Food Res. Tech.* 226:199–205.
49. Liazid, A., Palma, M., Brigui, J., and Barroso, C.G. 2007. Investigation on phenolic compounds stability during microwave-assisted extraction. *J. Chromatogr. A* 1140:29–34.
50. Singh, A.P., Luthria, D., Wilson, T., Vorsa, N., Singh, V., Banuelos, G.S., and Pasakdee, S. 2009. Polyphenols content and antioxidant capacity of eggplant pulp. *Food Chem.*114:955–961.
51. Silva, C.L., Haesenb, N., and Câmara, J.S. 2012. A new and improved strategy combining a dispersive-solid phase extraction-based multiclass method with ultra high pressure liquid chromatography for analysis of low molecular weight polyphenols in vegetables. *J. Chromatogr. A* 1260:154–163.
52. Stewart, A.J., Bozonnet, S., Mullen, W., Jenkins, G.I., Lean, M.E.J., and Crozier, A. 2000. Occurrence of flavonols in tomatoes and tomato-based products. *J. Agric. Food Chem.* 48:2663–2669.
53. Ignat, I., Volf, I., and Popa, V.I. 2011. A critical review of methods for characterisation of polyphenolic compounds in fruit and vegetables. *Food Chem.* 126:1821–1835.
54. Vilkhu, K., Mawson, R., Simons, L., and Bates, D. 2008. Applications and opportunities for ultrasound assisted extraction in the food industry—A review. *Innov. Food Sci. Emerg. Technol.* 9:161–169.
55. Paniwnyk, L., Beaufoy, E., Lorimer, J.P., and Mason, T.J. 2001. The extraction of rutin from floer buds of *Sophora japonica. Ultrason. Sonochem.* 8:299–301.
56. Choi, S.H., Kim, H.R., Kim, H.J., Lee, I.S., Kozukue, N., Levin, C.E., and Friedman, N. 2011. Free amino acid and phenolic contents and antioxidative and cancer cell-inhibiting activities of extracts of 11 greenhouse-grown tomato varieties and 13 tomato-based foods. *J. Agric. Food Chem.* 59:12801–12814.
57. Escarpa, A. and González, M.C. 2000. Optimization strategy and validation of one chromatographic method as approach to determine the phenolic compounds from different sources. *J. Chromatogr. A* 897:161–170.
58. Sánchez-Rodríguez, E., Moreno, D.A., Ferreres, F., Rubio-Wilhelmi, M.M., and Ruiz, J.M. 2011. Differential responses of five cherry tomato varieties to water stress: Changes on phenolic metabolites and related enzymes. *Phytochemistry* 72:723–729.
59. Schindler, M., Solar, S., and Sontag, G. 2005. Phenolic compounds in tomatoes. Natural variations and effect of gamma-irradiation. *Eur. Food Res. Technol.* 221:439–445.
60. Simões, A.D.N., Allende, A., Tudela, J.A., Puschmann, R., and Gil, M.I. 2011. Optimum controlled atmospheres minimise respiration rate and quality losses while increase phenolic compounds of baby carrots. *LWT—Food Sci. Technol.* 44:277–283.
61. Mustafa, A. and Turner, C. 2011. Pressurized liquid extraction as a green approach in food and herbal plants extraction: A review. *Anal. Chim. Acta* 703:8–18.

62. Søltoft, M., Christensen, J.H., Nielsen, J., and Knuthsen, P. 2009. Pressurised liquid extraction of flavonoids in onions. Method development and validation. *Talanta* 80:269–278.
63. Luthria, D.L. 2012. Optimization of extraction of phenolic acids from a vegetable waste product using a pressurized liquid extractor. *J. Funct. Foods* 4:842–850.
64. Luthria, D.L. 2008. Influence of experimental conditions on the extraction of phenolic compounds from parsley (*Petroselinum crispum*) flakes using a pressurized liquid extractor. *Food Chem.* 107:745–752.
65. Ali, I., Al-kindy, S.M.Z, Suliman, F.O., and Alam, S.D. 2011. Fast analysis of flavonoids in apple juice on new generation halo column by SPE-HPLC. *Anal. Methods* 3:2836–2841.
66. Castro-Vargas, H.I., Rodríguez-Varela, L.I., Ferreira, S.R.S., and Parada-Alfonso, F. 2010. Extraction of phenolic fraction from guava seeds (*Psidium guajava* L.) using supercritical carbon dioxide and co-solvents. *J. Supercrit. Fluid* 51:19–24.
67. Palma, M. and Taylor, L.T. 1999. Fractional extraction of compounds from grape seeds by supercritical fluid extraction and analysis for antimicrobial and agrochemical activities. *J. Agric. Food Chem.* 47:5044–5048.
68. Bleve, M., Ciurlia, L., Erroi, E., Lionetto, G., Longo, L., Rescio, L., Schettino, T., and Vasopollo, G. 2008. An innovative method for the purification of anthocyanins from grape skin extracts by using liquid and sub-critical carbon dioxide. *Sep. Purif. Technol.* 64:192–197.
69. Sarmento, L.A.V., Machado, R.A.F., Petrus, J.C.C., Tamanini, T.R., and Bolzan, A. 2008. Extraction of polyphenols from cocoa seeds and concentration through polymeric membranes. *J. Supercrit. Fluid* 45:64–69.
70. Jáuregui, O. and Galceran, M.T. 2001. Phenols (Chapter 6). In: *Environmental Analysis Handbook of Analytical Separations.* Ed. W. Kleiböhmer, Elsevier, Amsterdam, pp. 175–236.
71. Wang, Y., Catana, F., Yang, Y., Roderick, R., and Van Breemen, R.B. 2002. An LC-MS method for analyzing total resveratrol in grape juice, cranberry juice, and in wine. *J. Agric. Food Chem.* 50:431–435.
72. Li, H., Deng, Z., Wu, T., Liu, R., Loewen, S., and Tsao, R. 2012. Microwave-assisted extraction of phenolics with maximal antioxidant activities in tomatoes. *Food Chem.* 130:928–936.
73. Sutivisedsak, N., Cheng, H.N., Willett, J.L., Lesch, W.C., Tangsrud, R.R., and Biswas, A. 2010. Microwave-assisted extraction of phenolics from bean (*Phaseolus vulgaris* L.). *Food Res. Int.* 43: 516–519.
74. Aparicio-Fernández, X., Yousef, G.G., Loarca-Pina, G., De Mejia, E., and Lila, M.A. 2005. Characterization of polyphenolics in the seed coat of black jamapa bean (*Phaseolus vulgaris* L.). *J. Agric. Food Chem.* 53:4615–4622.
75. Alarcón-Flores, M.I., Romero-González, R., Garrido Frenich, A., and Martinez Vidal, J.L. 2013. Rapid determination of phytochemicals in vegetables and fruits by ultra high performance liquid chromatography coupled to tandem mass spectrometry. *Food Chem.* 141:1120–1129.
76. Alarcón-Flores, M.I., Romero-González, R., Garrido Frenich, A., and Martínez Vidal, J.L. 2011. QuEChERS-based extraction procedure for multifamily analysis of phytohormones in vegetables by UHPLC-MS/MS. *J. Sep. Sci.* 34:1517–1524.
77. Garrido Frenich, A., Aguilera-Luiz, M.M., Martínez Vidal, J.L., and Romero-González, R. 2010. Comparison of several extraction techniques for multiclass analysis of veterinary drugs in eggs using ultra-high pressure liquid chromatography-tandem mass spectrometry. *Anal. Chim. Acta* 66:150–160.
78. Padilla-Sánchez, J. A., Plaza-Bolaños, P., Romero-González, R., Garrido-Frenich, A., and Martínez Vidal, J.L. 2010. Application of a quick, easy, cheap, effective, rugged and

safe-based method for the simultaneous extraction of chlorophenols, alkylphenols, nitrophenols and cresols in agricultural soils, analyzed by using gas chromatography-triple quadrupole-mass spectrometry/mass spectrometry. *J. Chromatogr. A* 1217:5724–5731.

79. Crozier, A., Lean, M.E.J., McDonald, M.S., and Black, C. 1997. Quantitative analysis of the flavonoid content of commercial tomatoes, onions, lettuce, and celery. *J. Agric. Food Chem.* 45:590–595.
80. Chassy, A.W., Bui, L., Renaud, E.N.C., Van Horn, M., and Mitchell, A.E. 2006. Three-year comparison of the content of antioxidant microconstituents and several quality characteristics in organic and conventionally managed tomatoes and bell peppers. *J. Agric. Food Chem.* 54:8244–8252.
81. Cao, J., Chen, W., Zhang, Y., and Zhao, X. 2010. Content of selected flavonoids in 100 edible vegetables and fruits. *Food Sci. Technol. Res.* 16:395–402.
82. Martínez-Valverde, I., Periago, M.J., Provan, G., and Chesson, A. 2002. Phenolic compounds, lycopene and antioxidant activity in commercial varieties of tomato (*Lycopersicum esculentum*). *J. Sci. Food Agric.* 82:323–330.
83. Nama Medoua, G. and Oldewage-Theron, W.H. 2011. Bioactive compounds and antioxidant properties of selected fruits and vegetables available in the Vaal region, South Africa. *J. Food Biochem.* 35:1424–1433.
84. Sultana, B. and Anwar, F. 2008. Flavonols (kaempeferol, quercetin, myricetin) contents of selected fruits, vegetables and medicinal plants. *Food Chem.* 108:879–884.
85. Guillarme, D., Schappler, J., Rudaz, S., and Veuthey, J.L. 2010. Coupling ultra-high-pressure liquid chromatography with mass spectrometry. *Trends Anal. Chem.* 29:15–27.
86. Koblovská,R., Macková, Z., Vítková, M., Kokoska, L., Klejdus, B., and Lapcik, O. 2008. Isoflavones in the Rutaceae family: Twenty selected representatives of the genera citrus, fortunella, poncirus, ruta and severinia. *Phytochem. Anal.* 19:64–70.
87. Fang, Z., Zhang, M., and Wang, L. 2007. HPLC-DAD-ESIMS analysis of phenolic compounds in bayberries (*Myrica rubra* Sieb. et Zucc.). *Food Chem.* 100:845–852.
88. Ren, D.M., Qu, Z., Wang, X.N., Shi, J., and Lou, H.X. 2008. Simultaneous determination of nine major active compounds in *Dracocephalum rupestre* by HPLC. *J. Pharm. Biomed. Anal.* 48:1441–1445.
89. Simirgiotis, M.J., Caligari, P.D.S., and Schmeda-Hirschmann, G. 2009. Identification of phenolic compounds from the fruits of the mountain papaya *Vasconcellea pubescens* A. DC. grown in Chile by liquid chromatography–UV detection–mass spectrometry. *Food Chem.*115:775–784.
90. Nováková, L. and Vlĉková, H. 2009. A review of current trends and advances in modern bio-analytical methods: Chromatography and sample preparation. *Anal. Chim. Acta* 656:8–35.
91. Pfundstein, B., El Desouky, S.K., Hull, W.E., Haubner, R., Erben, G., and Owen, R.W. 2010. Polyphenolic compounds in the fruits of Egyptian medicinal plants (*Terminalia bellerica, Terminalia chebula* and *Terminalia horrida*): Characterization, quantitation and determination of antioxidant capacities. *Phytochemistry* 71:1132–1148.
92. Hurtado-Fernández, E., Pacchiarotta, T., Gómez-Romero, M., Schoenmaker, B., Derks, R., Deelder, A.M., Mayboroda, O.A., Carrasco-Pancorbo, A., and Fernández-Gutiérrez, A. 2011. Ultra high performance liquid chromatography-time of flight mass spectrometry for analysis of avocado fruit metabolites: Method evaluation and applicability to the analysis of ripening degrees. *J. Chromatogr. A* 1218:7723– 7738.
93. Lojza, J., Cajka, T., Schulzova, V., Riddellova, K., and Hajslova, J. 2012. Analysis of isoflavones in soybeans employing direct analysis in real-time ionization–high-resolution mass spectrometry. *J. Sep. Sci.* 35:476–481.
94. Ribas-Agustí, A., Gratacós-Cubarsí, M., Sárraga, C., García-Regueiro, J.A., and Castellari, M. 2011. Analysis of eleven phenolic compounds including novel *p*-coumaroyl derivatives in lettuce (*Lactuca sativa* L.) by ultra-high-performance liquid

chromatography with photodiode array and mass spectrometry detection. *Phytochem. Anal.* 22:555–563.

95. Engels, C., Gräter, D., Esquivel, P., Jiménez, V.M., Gänzle, M.G., and Schieber, A. 2012. Characterization of phenolic compounds in jocote (*Spondias purpurea* L.) peels by ultra high-performance liquid chromatography/electrospray ionization mass spectrometry. *Food Res. Int.* 46:557–562.
96. Spáčil, Z., Nováková, L., and Solich, P. 2008. Analysis of phenolic compounds by high performance liquid chromatography and ultra performance liquid chromatography. *Talanta* 76:189–199.
97. Kalili, K.M. and De Villiers, A. 2011. Recent developments in the HPLC separation of phenolic compounds. *J. Sep. Sci.* 34:854–876.
98. Motilva, M.J., Serra, A., and Macià, A. 2013. Analysis of food polyphenols by ultra high-performance chromatography coupled to mass spectrometry: An overview. *J. Chromatogr. A* 1292:66–82.
99. Medina-Remón, A., Tulipani, S. Rotchés-Ribalta, M., Mata-Bilbao, M.D.L., Andres-Lacueva, C., and Lamuela-Raventos, R.M. 2011. A fast method coupling ultrahigh performance liquid chromatography with diode array detection for flavonoid quantification in citrus fruit extracts. *J. Agric. Food Chem.* 59:6353–6559.
100. Lin, L.Z., Sun, J., Chen, P., and Harnly, J. 2011. UHPLC-PDA-ESI/HRMS/MS^n analysis of anthocyanins, flavonol glycosides, and hydroxycinnamic acid derivatives in red mustard greens (*Brassica juncea* Coss Variety). *J. Agric. Food Chem.* 59:12059–12072.
101. Ceymann, M., Arrigoni, E., Schärer, H., Bozzi Nising, A., and Hurrell, R.F. 2012. Identification of apples rich in health-promoting flavan-3-ols and phenolic acids by measuring the polyphenol profile. *J. Food Compos. Anal.* 26:128–135.
102. Prokudina, E.A., Havlíček, L., Al-Maharik, N., Lapcik, O., Strnad, M., and Gruz, J. 2012. Rapid UPLC-ESI-MS/MS method for the analysis of isoflavonoids and other phenylpropanoids. *J. Food Compos. Anal.* 26: 36–42.
103. Ortega, N., Romero, M.P., Macià, A., Reguant, J., Anglés, N., Morelló, J.R., and Motilva, M.J. 2008. Obtention and characterization of phenolic extracts from different cocoa sources. *J. Agric. Food Chem.* 56:9621–9627.
104. Delcambre, A. and Saucier, C. 2012. Identification of new flavan-3-ol monoglycosides by UHPLC-ESI-TOF in grapes and wine. *J. Mass Spectrom.* 47:727–736.
105. Dass, C. 2007. Fundamentals of contemporary mass spectrometry. 2007. John Wiley & Sons, New Jersey, pp. 68–69.
106. Holčapek, M., Jirasko, R., and Lisa, M. 2012. Recent developments in liquid chromatography–mass spectrometry and related techniques. *J. Chromatogr. A* 1259:3–15.
107. Gratacos-Cubarsi, M., Ribas-Agusti, A., Garcia-Regueiro, J.A., and Castellari, M. 2010. Simultaneous evaluation of intact glucosinolates and phenolic compounds by UHPLC-DAD-MS/MS in *Brassica oleracea* L. var. *botrytis*. *Food Chem.* 121:257–263.
108. Cuyckens, F., Rozenberg, R., De Hoffmann, E., and Claeys, M. 2001. Structure characterization of flavonoid *O*-diglycosides by positive and negative nano-electrospray ionization ion trap mass spectrometry. *J. Mass Spectrom.* 36: 1203–1210.
109. Pereira, C.A.M., Yariwake, J.H., and McCullagh, M. 2005. Distinction of the glycosylflavone isomer pairs orientin/isoorientin and vitexin/isovitexin using HPLC-MS exact mass measurement and in-source CID. *Phytochem. Anal.* 16:295–301.
110. March, R.E., Li, H., Belgacem., O., and Papanastasiou, D. 2007. High-energy and low-energy collision-induced dissociation of protonated flavonoids generated by MALDI and by electrospray ionization. *Int. J. Mass Spectrom.* 262:51–66.

Index

D

E

F

I

J

K

L

M

N

O

Q

R

S

V

W

X